Handbook of Tight Junctions

Handbook of Tight Junctions

Editor: Gwen Johnston

www.callistoreference.com

Callisto Reference,
118-35 Queens Blvd., Suite 400,
Forest Hills, NY 11375, USA

Visit us on the World Wide Web at:
www.callistoreference.com

ISBN: 978-1-64116-763-5 (Hardback)

Cataloging-in-Publication Data

Handbook of tight junctions / edited by Gwen Johnston.
p. cm.
Includes bibliographical references and index.
ISBN 978-1-64116-763-5
1. Tight junctions (Cell biology). 2. Cell junctions.
3. Junctional complexes (Epithelium). 4. Cytology.
I. Johnston, Gwen.
QH603.C4 H36 2023
571.6--dc23

Table of Contents

Preface

Tight junctions (TJ) close the gap between endothelial and epithelial cells against undesirable passage of water and solutes. The major protein families of the TJs include junctional adhesion molecules (JAM), claudins, angulins, and TJ-associated MARVEL proteins such as tricellulin and occludin. The majority of these proteins are related to the cytoskeleton through adapters such as zonula occludens (ZO) proteins. TJ proteins not only act as barriers but some of them also create water channels or paracellular ion channels. They are engaged in a variety of processes aside from barrier and channel functions. They may act as pathogen receptors and mediate immunological responses. TJ proteins are involved in a variety of inflammatory disorders as well as bacterial infections. They act as targets in the diagnostics and treatment of cancer tumor and they can also facilitate epithelial-mesenchymal transition, enabling metastasis and tumorigenesis. This book provides significant information to help develop a good understanding of tight junctions and its functions. It will serve as a reference to a broad spectrum of readers.

This book is the end result of constructive efforts and intensive research done by experts in this field. The aim of this book is to enlighten the readers with recent information in this area of research. The information provided in this profound book would serve as a valuable reference to students and researchers in this field.

At the end, I would like to thank all the authors for devoting their precious time and providing their valuable contribution to this book. I would also like to express my gratitude to my fellow colleagues who encouraged me throughout the process.

Editor

1

Temporal Effects of Quercetin on Tight Junction Barrier Properties and Claudin Expression and Localization in MDCK II Cells

Enrique Gamero-Estevez [1,2], Sero Andonian [3], Bertrand Jean-Claude [2,4], Indra Gupta [1,2,5] and Aimee K. Ryan [1,2,5,*]

1 Department of Human Genetics, McGill University, Montreal, QC H4A 3J1, Canada; egameroestevez@gmail.com (E.G.-E.); guptalab@gmail.com (I.G.)
2 Research Institute of the McGill University Health Centre, Glen Site, Montreal, QC H4A 3J1, Canada; bertrandj.jean-claude@mcgill.ca
3 Division of Urology, McGill University, Montreal, QC H4A 3J1, Canada; sero.andonian@mcgill.ca
4 Department of Medicine, McGill University, Montreal, QC H4A 3J1, Canada
5 Departments of Pediatrics, McGill University, Montreal, QC H4A 3J1, Canada
* Correspondence: aimee.ryan@mcgill.ca

Abstract: Kidney stones affect 10% of the population. Yet, there is relatively little known about how they form or how to prevent and treat them. The claudin family of tight junction proteins has been linked to the formation of kidney stones. The flavonoid quercetin has been shown to prevent kidney stone formation and to modify claudin expression in different models. Here we investigate the effect of quercetin on claudin expression and localization in MDCK II cells, a cation-selective cell line, derived from the proximal tubule. For this study, we focused our analyses on claudin family members that confer different tight junction properties: barrier-sealing (Cldn1, -3, and -7), cation-selective (Cldn2) or anion-selective (Cldn4). Our data revealed that quercetin's effects on the expression and localization of different claudins over time corresponded with changes in transepithelial resistance, which was measured continuously throughout the treatment. In addition, these effects appear to be independent of PI3K/AKT signaling, one of the pathways that is known to act downstream of quercetin. In conclusion, our data suggest that quercetin's effects on claudins result in a tighter epithelial barrier, which may reduce the reabsorption of sodium, calcium and water, thereby preventing the formation of a kidney stone.

Keywords: tight junctions; claudins; kidney stones; paracellular transport; ion reabsorption; quercetin

---

1. Introduction

Kidney stones affect 10% of the population and recur in 50% of adults. They are associated with renal failure in children and adults. They cause extreme pain and have a significant financial burden to society [1–3]. Given these morbidities, surprisingly little is known about why stones recur and how they can be prevented. Most stones are calcium-based with calcium oxalate more frequently observed than calcium phosphate stones [1]. Kidney stones result from the accumulation of salts along the kidney nephron, where an important factor in salt reabsorption is the epithelial tight junction barrier. Tight junctions are the most apical junction between apposing cells, where they compartmentalize the apical and basolateral intercellular space and regulate the passive paracellular movement of water and solutes [4]. The ion specificity of the tight junction barrier is determined by the claudin family of tetraspanin proteins, which contains close to 30 members [5–7]. Claudins can bind hetero- and homotypically with claudins in the same cell through their transmembrane domains or to claudins in the apposing cell through their extracellular loops [6]. The combination of claudins expressed within an epithelium determines the tightness and selectivity of the tight junction barriers [8].

The claudin composition of tight junctions varies along the different segments of the nephron and, consequently, ions and salts are differentially reabsorbed in the different nephron segments [8]. Reabsorption of calcium from the urine filtrate predominantly occurs in two segments: the proximal tubule reabsorbs 70% and the thick ascending limb reabsorbs 25% of the calcium. Several claudin family members are implicated in calcium reabsorption. A common sequence variant in CLDN14 and rare mutations in CLDN16 and -19, all of which are expressed in the thick ascending limb, are risk factors for the formation of calcium-based kidney stones in humans [9–11]. Both Cldn2 and -10 null mice develop nephrocalcinosis [12,13], while Cldn16 and -19 knockdown mice develop hypercalciuria [14,15]. Cldn7 is also essential for salt reabsorption: Cldn7 null mice die shortly after birth due to salt loss and dehydration [16]. Although not known to participate in ion exchange, Cldn3 has been shown to form a complex with Cldn16 and -19 in the thick ascending limb, which is essential for calcium and magnesium reabsorption in this segment [17]. Cldn4 has a critical role in chloride reabsorption in the kidney collecting duct where it forms a pore with Cldn8 [18]. The role of Cldn1 in ion reabsorption is less clear; however, it is present in different kidney segments and may be important in diabetic nephropathy [19]. Because claudins are essential for sodium, calcium, and water reabsorption in the nephron, targeting this claudin-based epithelial barrier may be a successful approach to decrease salt reabsorption and prevent kidney stone formation.

The flavonoid quercetin prevents kidney stone formation in a rat model [20]. Quercetin alters claudin expression in the Caco-2 intestine-derived cell line and in LLC-PK1 cells, an anion-selective cell line derived from the renal proximal tubule [21–24]. To date, no one has studied the effect of quercetin in a cation-selective cell line that models an epithelial barrier which is permeable to calcium and sodium. We hypothesize that quercetin may prevent kidney stones through its effects on claudins at tight junctions in the epithelial barrier of the nephron. For our study, we used Madin-Darby Canine Kidney cells (MDCK II), which are cation-selective and derived from the proximal tubule where the majority of sodium and calcium reabsorption occurs. MDCK II cells express Cldn1, -2, -3, -4, and -7 [25,26], which allows us to investigate the effect of quercetin on claudins with different properties: barrier-sealing (Cldn1, -3 and -7), cation-selective (Cldn2), or anion-selective (Cldn4) within a cation-selective cell line [27–30]. The MDCK II cell line has been widely studied and its TER and cation permeability properties are well characterized. Cldn2 confers the leaky barrier properties of MDCK II cells. If Cldn2 is depleted, TER increases and sodium and chloride potentials are significantly reduced [31].

We found that quercetin differentially affected both the expression and the localization of some claudins over time. Quercetin also significantly increased transepithelial resistance. Our data suggest that after treatment with quercetin, the barrier becomes tighter, which could lead to a decrease in cation and water reabsorption. This may result in a more favorable urinary filtrate that is less prone to crystal supersaturation and the formation of a stone.

2. Results

2.1. *Quercetin Increased Transepithelial Resistance of MDCK II Cells*

Transepithelial resistance (TER) is a measure of the electrical resistance of a cell monolayer and is modulated by cell confluence, barrier permeability, and tight junction composition. For these studies, we used the cellZscope (NanoAnalytics) to continuously measure TER in MDCK II cells from the time of plating until several days after confluence when stable, mature tight junctions are formed. In cation-selective cell lines, TER increases as cells become closer or when paracellular movement of cations is reduced [30]. As predicted, there was an increase in TER immediately after seeding, when cells are dividing and confluence is increasing (Figure S1A). Under control conditions, MDCK II cells reached a peak resistance of 130 $\Omega\cdot cm^2$ ~40 h after seeding. Once maximum confluence is achieved, contact inhibition and tight junction remodeling takes place, which leads to a decrease in the TER that eventually stabilizes and becomes constant as the cells form stable mature tight junctions [32,33]. In MDCK II cells, this translated into a decrease in TER that then remained at 50–80 $\Omega \cdot cm^2$, which is characteristic for mature tight junctions in this model [33]. Therefore, this was the time point selected for treatment with quercetin since it best represents the mature renal proximal tubule epithelium.

We confirmed that 400 μM quercetin, the concentration used in previous studies [21,22], was also the most optimal for our experiments through a dose response curve (Figure S1). We monitored the effects of quercetin on TER for ~96 h after treatment. Figure 1A shows the TER profile for one of the three biological replicates, which were each done in triplicate. Analysis of all data showed a small expected increase in TER immediately after treatment, due to the removal of the cells from the incubator and change of media. In control cells, the TER stabilized within 4–5 h and then remained at a resistance of ~50 $\Omega \cdot cm^2$ (Figure 1A). Two-way ANOVA analysis showed that TER changed significantly over time and in a treatment-dependent manner ($P_{int} = 0.0405$; $P_{time} < 0.0001$; $P_{treat} < 0.0001$). Cells treated with 400 μM quercetin exhibited a progressive increase in TER, reaching significance at 3 h when the TER was ~15 $\Omega \cdot cm^2$ higher than the control cells ($p = 0.049$; quercetin-treated = 90.04 ± 4.01 $\Omega \cdot cm^2$ versus control cells = 70.7 ± 1.62 $\Omega \cdot cm^2$). The TER remained significantly increased until 5 h post-treatment ($p = 0.046$; quercetin-treated = 86.33 ± 2.94 $\Omega \cdot cm^2$ versus control cells = 66.86 ± 3.59 $\Omega \cdot cm^2$) and then progressively decreased to ~5 $\Omega \cdot cm^2$ below control levels 15 h after treatment, which was not significant (15 h: $p > 0.99$; control 60.08 ± 3.61, quercetin 62.21 ± 2.37). Following the decrease, TER increased again, reaching a steady state level of 10 $\Omega \cdot cm^2$ above control, 36 h after treatment, which was statistically significant and remained significantly increased for the duration of the experiment (36 h: $p = 0.0071$; control 54.7 ± 2.31, quercetin 78.05 ± 5.19) (48 h: $p < 0.0001$; control 53.35 ± 1.8, quercetin 85.68 ± 2.55) (Figure 1).

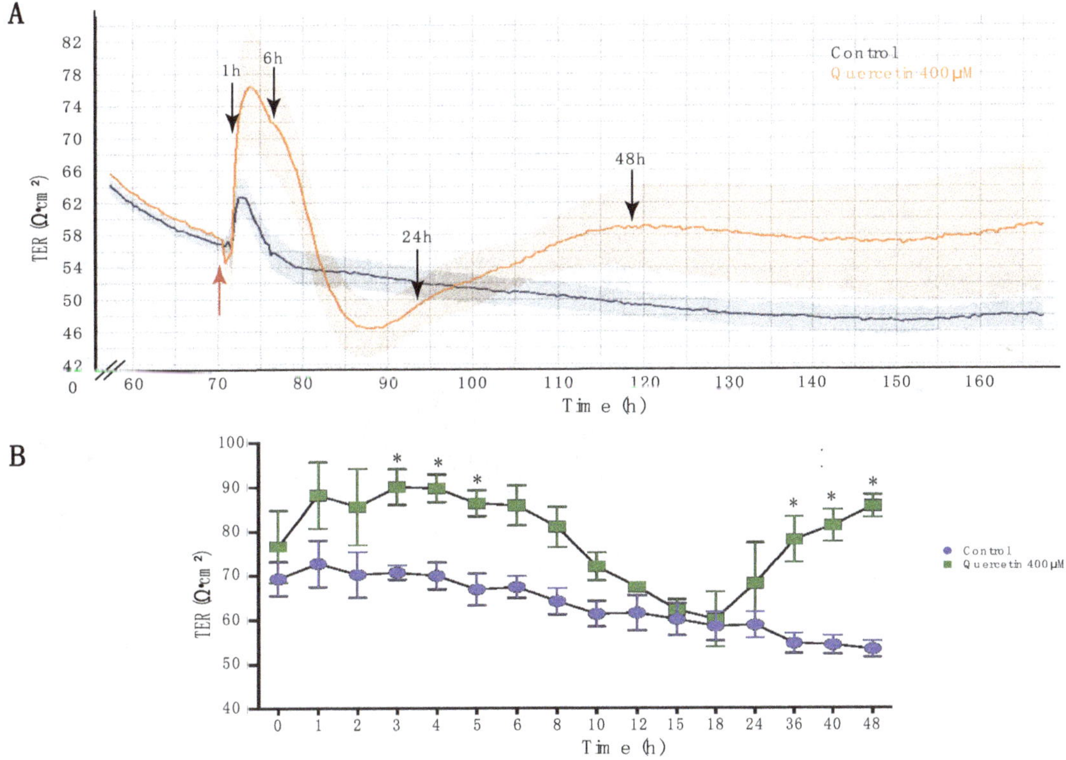

Figure 1. Quercetin caused oscillations in transepithelial resistance (TER) of MDCK II cells. (**A**) Representative plot of TER in control cells (black) and cells treated with 400 μM quercetin (orange) from one biological replicate performed in triplicate. Red arrow indicates when quercetin was added to the culture medium. Black arrows indicate the time points taken for western blot and immunofluorescence analysis. (**B**) TER of control and quercetin-treated cells at different time points after treatment from three independent experiments performed in triplicate. Two-way ANOVA was performed ($P_{int} = 0.04$; $P_{time} < 0.0001$; $P_{treat} < 0.0001$). Mean and SEM are plotted. * Denotes significance, $p < 0.05$.

2.2. Quercetin Treatment Caused Claudin-Specific Changes in Expression and Membrane Localization

To determine if the changes in TER caused by quercetin treatment corresponded with different claudin profiles, cells were collected 1, 6, 24, and 48 h after treatment. Claudin expression was assessed by western blot analysis and localization to the tight junction barrier was assessed by immunofluorescence. Immunofluorescence also provided a qualitative assessment of claudin expression. Five claudins expressed in MDCK II cells were studied: Cldn1, -3, and -7 that have barrier-sealing functions, Cldn2 that is involved in cation pore formation, and Cldn4 that is involved in anion pore formation. For all experiments, cells were cultured for 72 h before treatment with 400 μM quercetin to ensure that the cells had established mature tight junctions.

2.2.1. Cldn1

Western blot analysis revealed a significant decrease in Cldn1 expression over time in both controls and quercetin-treated cells ($P_{time} = 0.021$). Quercetin treatment significantly lowered Cldn1 levels at 48 h compared to controls ($p = 0.038$; control 1.47 ± 0.55; quercetin 0.44 ± 0.18). A change in the relative abundance in the two migratory bands was observed at 24 h, although the total amount of Cldn1 was not affected (Figure 2A,B). Immunofluorescence analysis revealed decreased levels of Cldn1 at 1, 6, and 48 h (Figure 2C). A reduction in Cldn1 co-localization with ZO1 can be seen at 1 and 48 h, although it was not significant ($p = 0.3$ and $p = 0.2$, respectively) (Figure 2D). These data suggest than even though general levels of Cldn1 were decreased, the remaining Cldn1 still co-localized with ZO1.

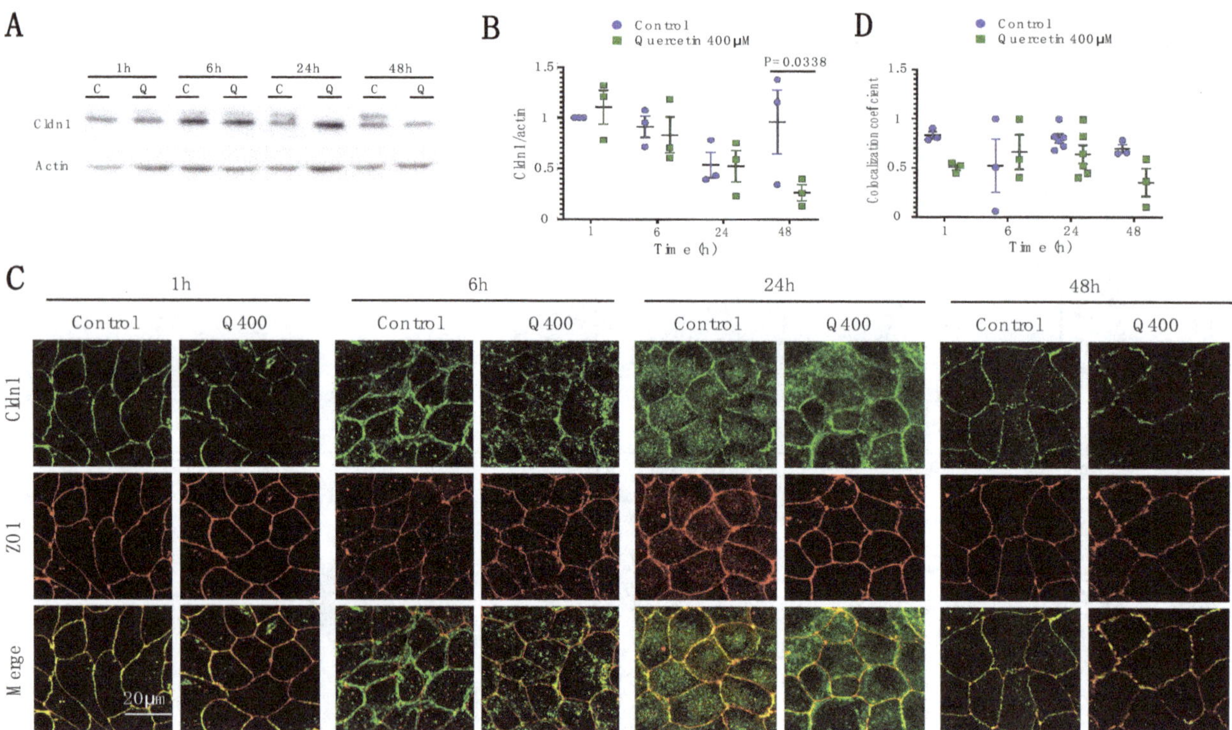

Figure 2. Analysis of Cldn1 expression and localization in MDCK II cells following quercetin treatment. (**A**) Western blot analysis of Cldn1 expression in cell lysates from control and 400 μM quercetin-treated MDCK II cells. Actin was used as a loading control. (**B**) Densitometry measurements of Cldn1 and actin intensity on western blot were normalized to expression at 1h. Normalized Cldn1 expression relative to normalized actin expression is plotted ($P_{int} = 0.11$; $P_{time} = 0.022$; $P_{treat} = 0.16$). (**C**) Immunofluorescence of control and 400 μM of quercetin-treated MDCK II cells at 1, 6, 24, and 48 h. Cldn1 is shown in green and ZO1 is shown in red. (**D**) Localization of claudin at the tight junction was assessed by determining the amount of ZO1 that was co-localized with claudin expression ($P_{int} = 0.24$; $P_{time} = 0.35$; $P_{treat} = 0.055$). For all graphs, each point corresponds to an independent experiment; mean and SEM are shown. C = Control and Q = Quercetin. Scale bar, 20 μm.

2.2.2. Cldn2

Two-way ANOVA analysis on Cldn2 expression showed that Cldn2 decreased significantly with quercetin treatment ($P_{treat} = 0.0021$), while changes in time and interaction only suggested significance ($P_{int} = 0.0525$; $P_{time} = 0.096$). Sidak's multiple comparison test showed that the decrease observed at 48 h after treatment with quercetin was significant ($p = 0.0011$; control 2.0 ± 0.79; quercetin 0.06 ± 0.057). At 24 h, no Cldn2 was observed by western blot analysis in quercetin-treated cells (Figure 3A,B). As previously reported, Cldn2 detection by immunofluorescence was patchy within the epithelial monolayer of control MDCK II cells [34], with clusters of cells showing high levels of Cldn2 expression and adjacent groups of cells showing virtually no expression (Figure 3C). Localization of Cldn2 was significantly changed after treatment with quercetin ($P_{treat} = 0.0055$). At 1 h and 6 h after quercetin treatment, no changes in localization of Cldn2 were observed. However, at 24 h and 48 h after treatment, quercetin-treated cells showed a significant reduction both in the amount of Cldn2 and in the portion of Cldn2 that co-localized with ZO1 (24h: $p = 0.036$; control 0.54 ± 0.11, quercetin 0.14 ± 0.08; 48 h: $p = 0.034$; control 0.52 ± 0.12, quercetin 0.12 ± 0.05) (Figure 3C,D).

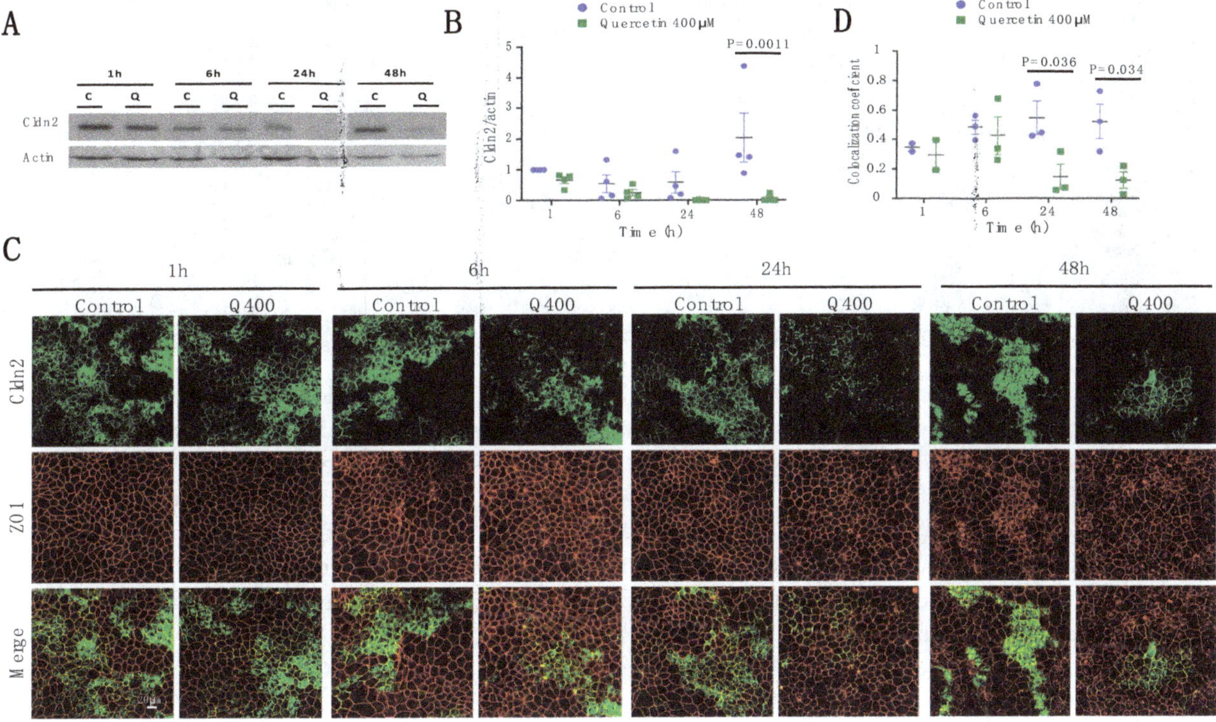

Figure 3. Analysis of Cldn2 expression and localization in MDCK II cells following quercetin treatment. (**A**) Western blot analysis of Cldn2 expression in cell lysates from control and 400 μM quercetin-treated MDCK II cells. Actin was used as a loading control. (**B**) Densitometry measurements of Cldn2 and actin intensity on western blot were normalized to control at 1 h. Normalized Cldn2 expression relative to normalized actin is plotted ($P_{int} = 0.052$; $P_{time} = 0.096$; $P_{treat} = 0.0021$). (**C**) Immunofluorescence of control and 400 μM of quercetin-treated MDCK II cells at 1, 6, 24, and 48 h. Cldn2 is shown in green and ZO1 is shown in red. (**D**) Localization of claudin at the tight junction was assessed by determining the amount of ZO1 that co-localized with claudin expression ($P_{int} = 0.15$; $P_{time} = 0.46$; $P_{treat} = 0.0055$). For all graphs, each point corresponds to an independent experiment; mean and SEM are shown. C = Control and Q = Quercetin. Scale bar, 20 μm.

2.2.3. Cldn3

Western blot analysis showed that quercetin caused a significant treatment-dependent increase in Cldn3 expression ($P_{treat} = 0.0395$). In marked contrast to the effects on Cldn1 and -2, Cldn3 expression was increased at 24 h and 48 h after treatment, although it was only statistically significant at 48 h

($p = 0.04$; control 0.29 ± 0.1; quercetin 2.4 ± 0.98) (Figure 4A,B). Although there appeared to be a general decrease in Cldn3 detection by immunofluorescence after 1 h and 6 h of treatment (Figure 4C), co-localization analysis showed that there was no effect on Cldn3 co-localization with ZO1 at 1 h and 6 h ($p = 0.79$ and $p = 0.99$, respectively) (Figure 4D). Cldn3 also co-localized with ZO1 at later time points.

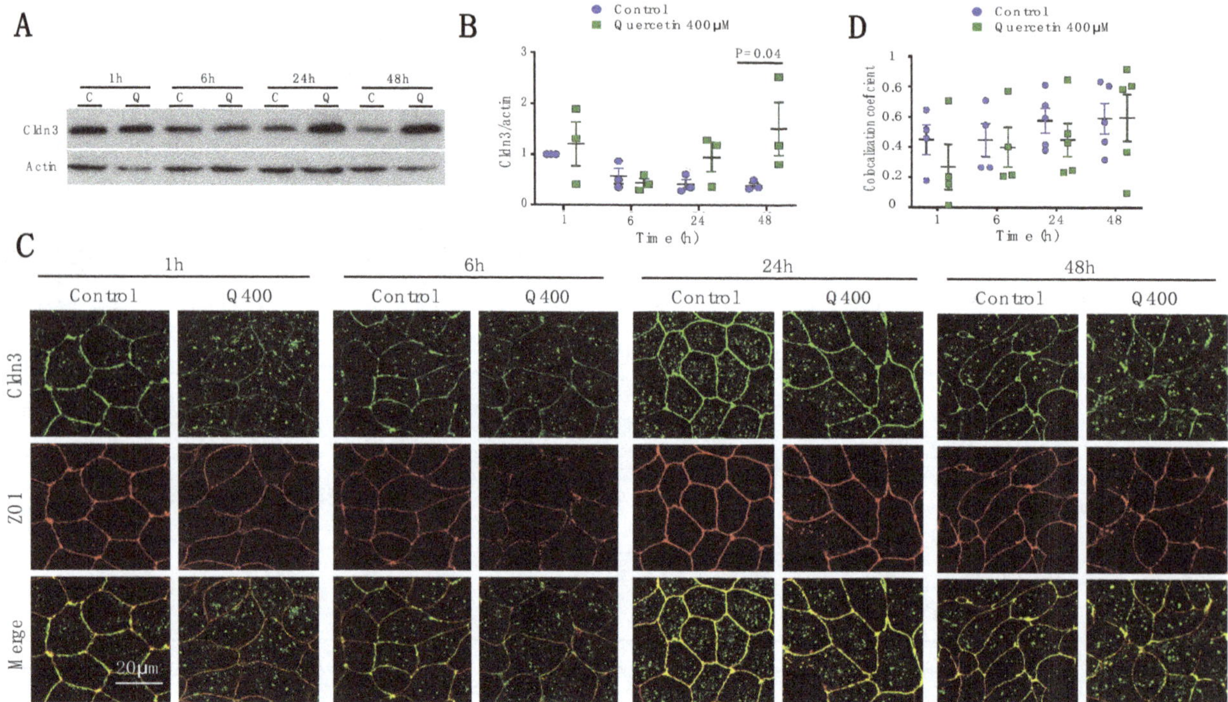

Figure 4. Analysis of Cldn3 expression and localization in MDCK II cells following quercetin treatment. (**A**) Western blot analysis of Cldn3 expression in cell lysates from control and 400 μM quercetin-treated MDCK II cells. Actin was used as a loading control. (**B**) Densitometry measurements of Cldn3 and actin were normalized to control at 1 h. Normalized Cldn3 expression relative to normalized actin is plotted ($P_{int} = 0.16$; $P_{time} = 0.16$; $P_{treat} = 0.039$). (**C**) Immunofluorescence of control and 400 μM of quercetin-treated MDCK II cells at 1, 6, 24, and 48 h. Cldn3 is shown in green and ZO1 is shown in red. (**D**) Localization of claudin at the tight junction was assessed by determining the amount of ZO1 that co-localized with claudin expression ($P_{int} = 0.87$; $P_{time} = 0.24$; $P_{treat} = 0.31$). For all graphs each point corresponds to an independent experiment; mean and SEM are shown. C = Control and Q = Quercetin. Scale bar, 20 μm.

2.2.4. Cldn4

Cldn4 expression changed over time following quercetin treatment ($P_{int} < 0.0001$; $P_{time} = 0.0017$; $P_{treat} < 0.0001$). At 24 h and 48 h, Cldn4 expression was significantly increased by western blot analysis (control = 0.67 ± 0.37; quercetin = 2.74 ± 0.36, $p < 0.0001$ and control = 0.79 ± 0.36; quercetin = 2.93 ± 0.37, $p < 0.0001$, respectively) (Figure 5A,B). In contrast, by immunofluorescence, qualitatively, Cldn4 was downregulated at 48 h (Figure 5C), but co-localized with ZO1 (Figure 5D). This could be a consequence of possible claudin modifications that may avoid correct recognition of the epitope by the antibody. At this point, we cannot explain the discrepancy in Cldn4 expression as assessed by western blot and immunofluorescence.

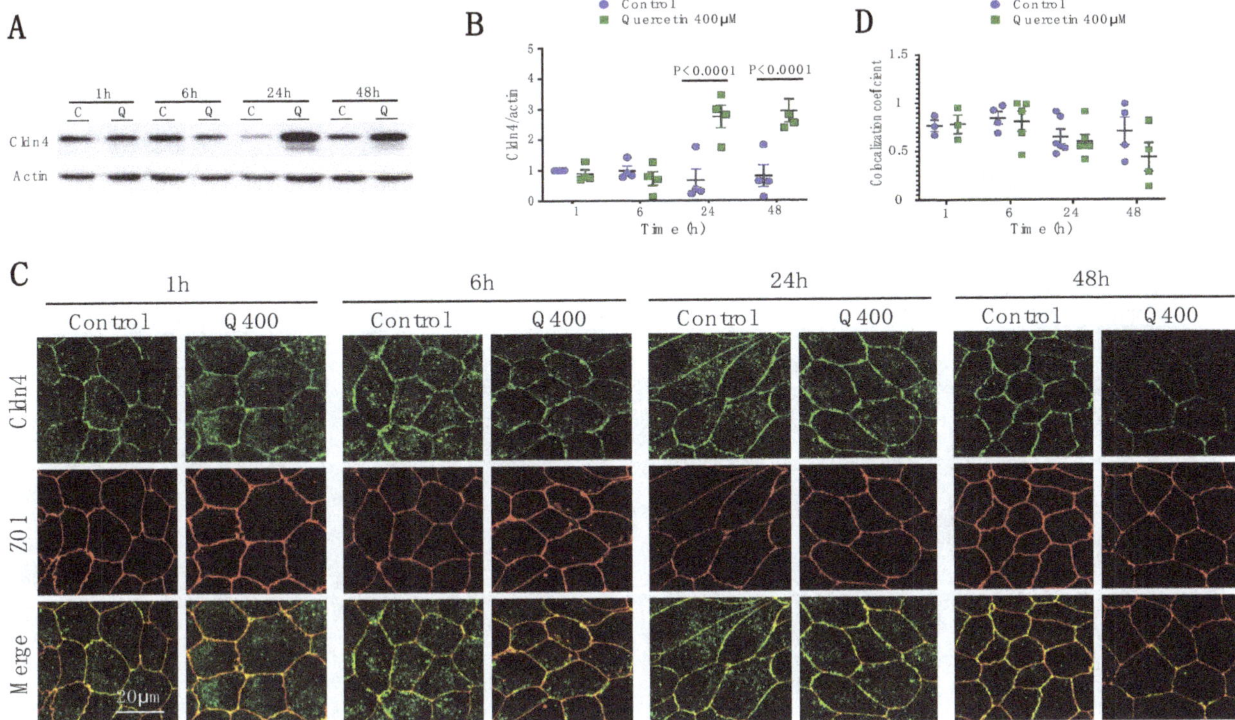

Figure 5. Analysis of Cldn4 expression and localization in MDCK II cells following quercetin treatment. (**A**) Western blot analysis of Cldn4 expression in cell lysates from control and 400 µM quercetin-treated MDCK II cells. Actin was used as a loading control. (**B**) Densitometry measurements of Cldn4 and actin intensity on western blot were normalized to control levels at 1 h. Normalized Cldn4 expression relative to normalized actin is plotted ($P_{int} < 0.0001$; $P_{time} = 0.0017$; $P_{treat} < 0.0001$). (**C**) Immunofluorescence of control and 400 µM of quercetin-treated MDCK II cells at 1, 6, 24, and 48 h. Cldn4 is shown in green and ZO1 is shown in red. (**D**) Localization of claudin at the tight junction was assessed by determining the amount of ZO1 that co-localized with claudin expression ($P_{int} = 0.54$; $P_{time} = 0.051$; $P_{treat} = 0.24$). For all graphs, each point corresponds to an independent experiment; mean and SEM are shown. C = Control and Q = Quercetin. Scale bar represent 20 µm.

2.2.5. Cldn7

In contrast to the other claudin family members examined, Cldn7 expression and localization were not affected by quercetin treatment (Figure 6). At 24h, there was an increased level of Cldn7 immunofluorescence in the cytoplasm (Figure 6C) but this was not associated with increased co-localization with ZO1 at the tight junction (Figure 6D). This may suggest that Cldn7 incorporation into the tight junction complex is saturated at this time point.

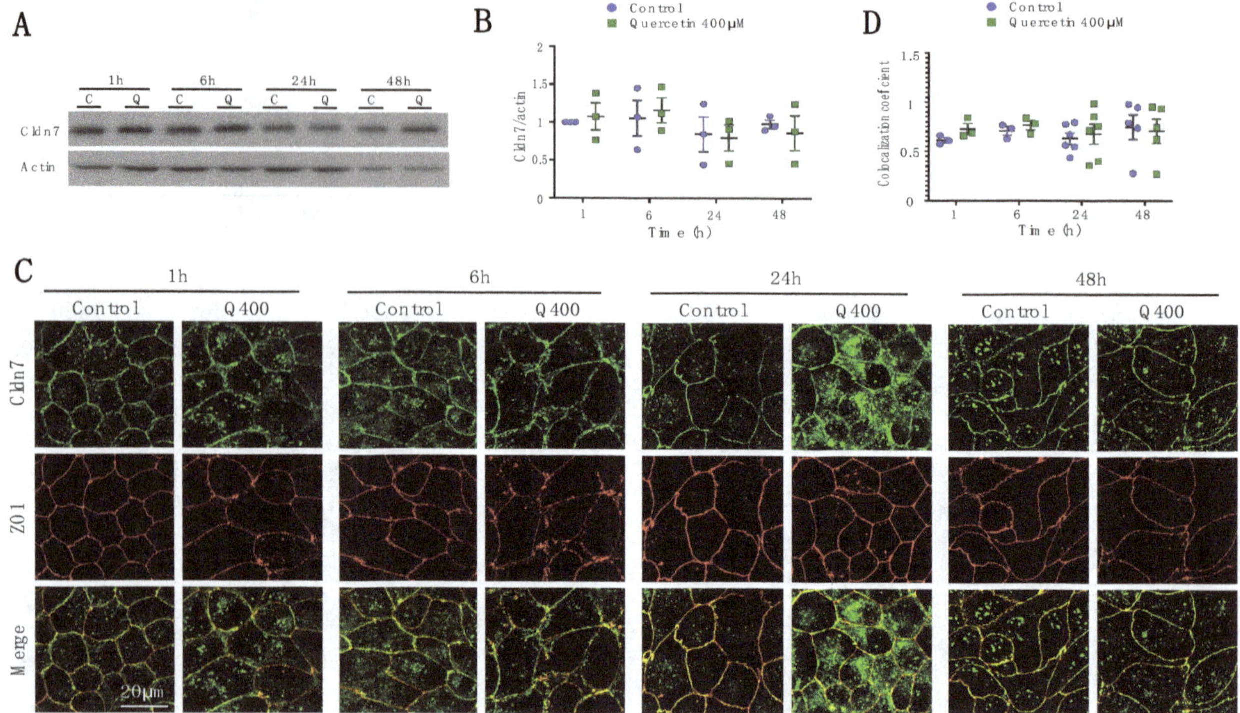

Figure 6. Analysis of Cldn7 expression and localization in MDCK II cells following quercetin treatment. (**A**) Western blot analysis of Cldn7 expression in cell lysates from control and 400 μM quercetin-treated MDCK II cells. Actin was used as a loading control. (**B**) Densitometry measurements of Cldn7 and actin intensity on western blot were normalized to control at 1h. Normalized Cldn7 expression relative to normalized actin is plotted ($P_{int} = 0.916$; $P_{time} = 0.41$; $P_{treat} = 0.98$). (**C**) Immunofluorescence of control and 400 μM of quercetin-treated MDCK II cells at 1, 6, 24, and 48 h. Cldn7 is shown in green and ZO1 is shown in red. (**D**) Localization of claudin at the tight junction was assessed by determining the amount of ZO1 that co-localized with claudin expression ($P_{int} = 0.89$; $P_{time} = 0.8$; $P_{treat} = 0.55$). For all graphs each point corresponds to an independent experiment; mean and SEM are shown. C = Control and Q = Quercetin. Scale bar, 20 μm.

2.3. Quercetin's Effects on Claudin Expression and Localization is Independent of the PI3K/AKT Pathway

Quercetin has been identified as a strong inhibitor of the PI3K/AKT pathway [35,36]. Given that receptor tyrosine kinases and pAKT phosphorylate and regulate claudins both directly and indirectly [37,38], we hypothesized that quercetin affects claudin expression through the PI3K/AKT pathway. Western blot analysis of pPI3K, AKT, and pAKT was performed following quercetin treatment. The levels of pPI3K and AKT in MDCK II cells did not change significantly after quercetin treatment. pAKT was reduced in quercetin-treated cells at 1 h and 6 h after treatment. pAKT levels could not be detected at 1 h and 6 h after quercetin treatment limiting the statistical analysis at these time points. At 24 h and 48 h, pAKT was decreased, but this was not statistically significant ($p = 0.99$ and $p = 0.7$, respectively). These data suggest that quercetin inhibits the AKT/pAKT pathway in MDCK type II cells (Figure 7A,B).

To query if quercetin's effects on claudin expression were due to impairment of the PI3K/AKT pathway, we inhibited or activated the PI3K/AKT pathways using wortmannin or SC79, respectively, alone or in combination with quercetin. As predicted, treatment with wortmannin caused a transient decrease in pAKT in MDCK II cells at 1 h and 6 h and SC79 increased pAKT at 1 h and 6 h (Figure 7C–F). In both cases, effects were greatly attenuated by 24 h. However, neither activation nor inhibition of the pAKT pathway recapitulated the effects on claudins observed following quercetin treatment (Figure 8), and neither wortmannin nor SC79 rescued the effects of quercetin. Therefore, our data suggest that quercetin is not acting through the PI3K/AKT pathway to effect changes in claudin protein expression.

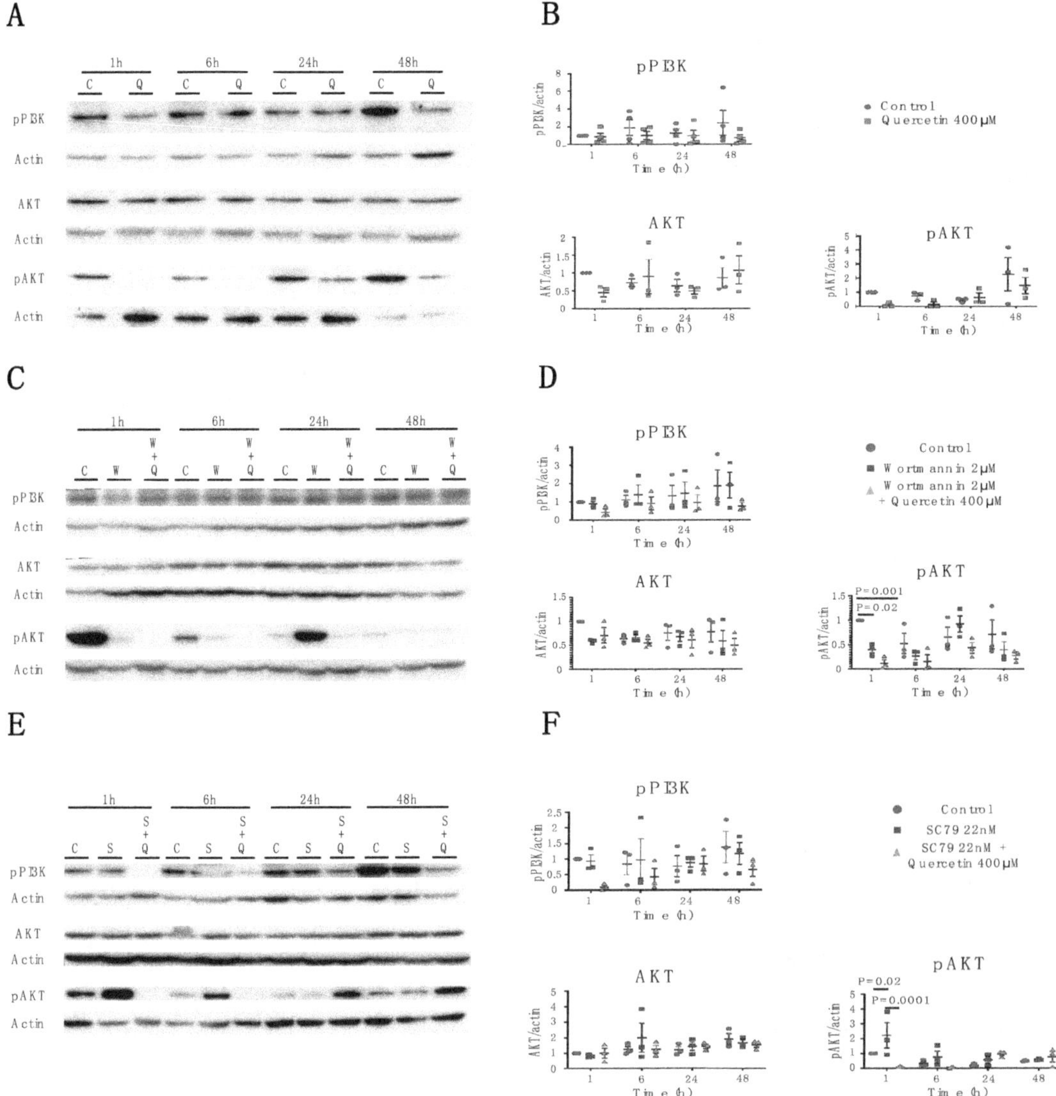

Figure 7. Analysis of pPI3K, pAKT, and AKT expression in MDCK II cells following quercetin treatment. (**A**) Western blot analysis of claudin expression in cell lysates from control and 400 μM quercetin-treated MDCK II cells. Actin was used as a loading control. (**B**) Densitometry measurements of band intensity on western blot were normalized to control at 1 h and protein expression was plotted relative to normalized actin expression. pPI3K (P_{int} = 0.67; P_{time} = 0.79; P_{treat} = 0.14), AKT (P_{int} = 0.44; P_{time} = 0.49; P_{treat} = 0.68), and pAKT (P_{int} = 0.68; P_{time} = 0.027; P_{treat} = 0.142) measurements were plotted. (**C**) Western blot analysis of pPI3K, pAKT, and AKT expression in cell lysates from control, 2 μM wortmannin-treated and 2 μM wortmannin plus 400 μM quercetin-treated MDCK II cells. Actin was used as a loading control. (**D**) Densitometry measurements of band intensity on western blot were normalized to control at 1 h and protein expression was plotted relative to normalized actin expression. pPI3K (P_{int} = 0.97; P_{time} = 0.03; P_{treat} = 0.13), AKT (P_{int} = 0.79; P_{time} = 0.53; P_{treat} = 0.099), and pAKT (P_{int} = 0.15; P_{time} = 0.064; P_{treat} = 0.0008) measurements were plotted. (**E**) Western blot analysis of pPI3K, pAKT, and AKT expression in cell lysates from control 22 nM SC79-treated and 22 nM SC79 plus 400 μM quercetin-treated MDCK II cells. Actin was used as a loading control. (**F**) Densitometry measurements of band intensity on western blot were normalized to control at 1 h and protein expression was plotted relative to normalized actin expression. pPI3K (P_{int} = 0.81; P_{time} = 0.51; P_{treat} = 0.078), AKT (P_{int} = 0.79; P_{time} = 0.52; P_{treat} = 0.099), and pAKT (P_{int} = 0.16; P_{time} = 0.063; P_{treat} = 0.0008) measurements were plotted. Each point corresponds to an independent experiment; mean and SEM are shown. W = Wortmannin; S = SC79; C = Control; and Q = Quercetin.

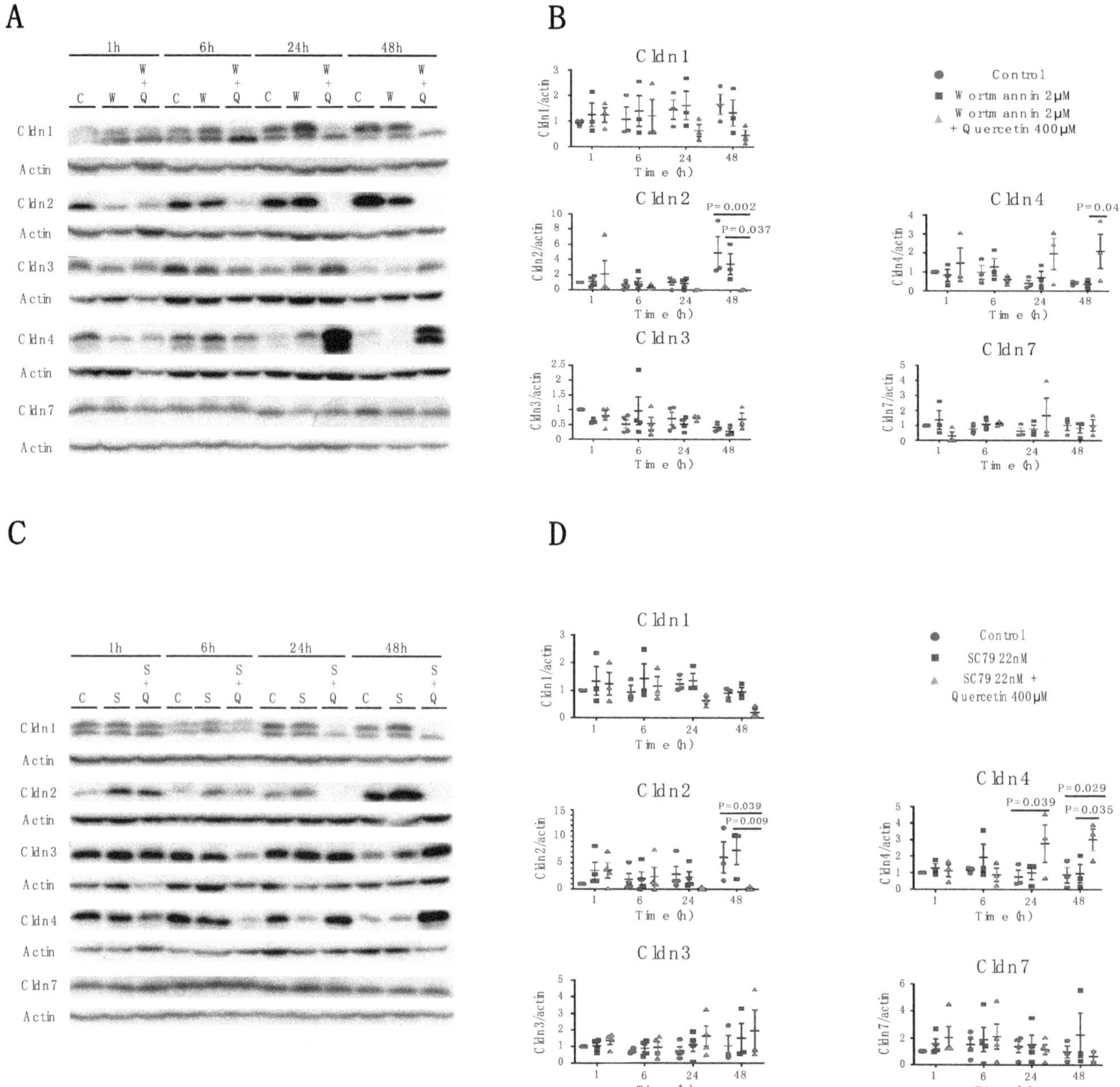

Figure 8. Analysis of Cldn1, -2, -3, -4, and -7 expression in MDCK II cells following wortmannin or SC79 treatment. (**A**) Western blot analysis of claudin expression in cell lysates from control, 2 µM wortmannin-treated and 2 µM wortmannin plus 400 µM quercetin-treated MDCK II cells. Actin was used as a loading control. (**B**) Densitometry measurements of band intensity on western blot were normalized to control at 1 h and protein expression was plotted relative to normalized actin expression. Cldn1 (P_{int} = 0.6; P_{time} = 0.98; P_{treat} = 0.24), Cldn2 (P_{int} = 0.059; P_{time} = 0.016; P_{treat} = 0.097), Cldn3 (P_{int} = 0.29; P_{time} = 0.26; P_{treat} = 0.78), Cldn4 (P_{int} = 0.21; P_{time} = 0.97; P_{treat} = 0.041), and Cldn7 (P_{int} = 0.41; P_{time} = 0.98; P_{treat} = 0.83) measurements were plotted. (**C**) Western blot analysis of claudin expression in cell lysates from control 22 nM SC79-treated and 22 nM SC79 plus 400 µM quercetin-treated MDCK II cells. Actin was used as a loading control. (**D**) Densitometry measurements of band intensity on western blot were normalized to control at 1 h and protein expression was plotted relative to normalized actin expression. Cldn1 (P_{int} = 0.6; P_{time} = 0.15; P_{treat} = 0.11), Cldn2 (P_{int} = 0.09; P_{time} = 0.14; P_{treat} = 0.11), Cldn3 (P_{int} = 0.99; P_{time} = 0.46; P_{treat} = 0.21), Cldn4 (P_{int} = 0.082; P_{time} = 0.73; P_{treat} = 0.047), and Cldn7 (P_{int} = 0.87; P_{time} = 0.78; P_{treat} = 0.5) measurements were plotted. Each point corresponds to an independent experiment; mean and SEM are shown. W = Wortmannin; S = SC79; C = Control; and Q = Quercetin.

3. Discussion

The purpose of this study was to determine the effect of quercetin on the cation-selective tight junction barrier in MDCK II cells, which models the proximal tubule of the kidney. We showed that the effects of quercetin on the MDCK barrier were stabilized 48 h after treatment, resulting in an increased TER of at least 20% over the TER observed in untreated cells. The quercetin-dependent increase in TER correlates with decreased expression of Cldn1 and decreased expression and localization of Cldn2 to the tight junction. Although a significant increase in Cldn3 and -4 was observed by western blot analysis, tight junction localization of Cldn3, -4, and -7 were not affected at this timepoint.

In the mature kidney, transepithelial resistance in the proximal tubule is <10 $\Omega\cdot cm^2$, indicating that this nephron segment has a leaky barrier, which correlates with its important role in salt and water reabsorption [30]. Therefore, the increase of 10–20 $\Omega\cdot cm^2$ observed following quercetin treatment of MDCK II cells would be very relevant in the context of the proximal tubule, where similar oscillations could completely change its barrier properties.

The TER oscillation observed during the first 48 h after quercetin treatment reflects dynamic tight junction remodeling and correlates with the changes in claudin expression that we observed. The decrease in TER observed at 6h coincided with a decrease in the immunofluorescent levels of barrier-sealing claudins, Cldn1 and -3. This is predicted to translate into a 'leakier' barrier, and consequently, a reduced TER. The increased TER at 24 h and 48 h coincided primarily with a decrease in the expression and tight junction localization of Cldn2, a cation-selective pore-forming claudin. Decreased Cldn2 at the tight junction is predicted to reduce paracellular movement of cations [31]—including sodium and calcium, and water—and may contribute to the increase in TER, as previously described for Cldn4 in MDCK [30].

Our data and those obtained by other groups, suggest that the effects of quercetin are cell-line dependent [21–24,31,39]. When Tokuda and Furuse depleted Cldn2 from MDCK II cells, they observed increased TER and decreased sodium and chloride potential [31]. The increase in TER was significantly higher than what we observed (>1000 Ω cm^2), perhaps due to compensation by other claudins in response to removal of Cldn2. In their experiment, depletion of Cldn2 led to increased tight junction localization of Cldn1, -3, -4, and -7, which would effectively tighten the barrier and increase TER. In our quercetin-treated cells, the 90% reduction of Cldn2, was not accompanied by an increased localization of other claudins to the tight junction: Cldn1 expression was decreased, Cldn3 and -4 expression was increased but localization to the tight junction was unchanged, and Cldn7 was unchanged. Thus, quercetin causes a different net effect on claudin expression/localization compared to depletion of Cldn2 alone and, as a result, there is a more modest increase in TER.

Quercetin also increased TER of the anion-selective barrier in LLC-PK1 cells, here, expression of Cldn2 and -3 were downregulated while Cldn4, -5, and -7 were upregulated [21]. However, claudin localization was not assessed. An oscillation and final increase in TER following quercetin treatment, has been observed in Caco-2 cells [23,24], where TER oscillated similarly to what we observed in MDCK II cells during the first 48 h after treatment. In this case Cldn1 was displaced to the cytoskeletal fraction, Cldn4 expression was increased, while Cldn3 was unperturbed. In contrast, Valenzano et al., did not report any effect on TER after 17 h of quercetin treatment on Caco-2 cells, although Cldn2, -4, and -5 were increased [22]. These data together with our findings indicate that quercetin is able to tighten tight junction barriers, but does so by differentially affecting different claudin family members, in a cell-type specific manner.

We observed two migratory bands for Cldn1 and -4 in our western blot analyses. Previous studies have linked these different migratory bands to either claudin degradation [40] or claudin phosphorylation [41,42]. Claudins are known to undergo post-translational modifications, including phosphorylation, glycosylation, palmitoylation, and ubiquitination [43]. These modifications are thought to rapidly change claudin stability at the tight junction. For instance, dephosphorylation of Cldn1 and -2 is a signal for degradation and reduces their presence at tight junctions [44,45]. In the case of Cldn4, studies have shown that the effects depend on the site of phosphorylation. In some cases, phosphorylation leads to the disruption of the tight junction [46,47], while in others, phosphorylation increases the stability of Cldn4 at the tight junction [41]. Further experiments are required to elucidate if the two migratory bands are a consequence of different posttranslational modifications, or if they correspond to degradation products [40].

Other studies that have looked at quercetin's effects on tight junction barrier properties and claudin expression have been based only on western blot analyses. The lack of congruence between our western blot and immunofluorescence expression data for some of the claudins may reflect the availability of claudin epitopes for detection using these two methods. It could also reflect disparities in how posttranslational modifications or protein degradation can be assessed by western blot versus immunofluorescence. For instance, in the case of Cldn1, at 24 h and 48 h, the western blot data shows a loss of the slower migrating band and decreased Cldn1 expression in the quercetin-treated cells. In contrast, the levels by immunofluorescence are equivalent to the control cells. This could suggest that the immunofluorescence signal is primarily due to recognition of the faster migrating band.

Cldn2 immunofluorescence exhibited quite a different pattern compared to other claudin family members. In contrast to most claudins, Cldn2 was highly expressed in some regions and almost absent in other regions within confluent cell layer. This has been previously described and high Cldn2 levels appear to correlate with increased cell confluence [34]. The increased Cldn2 during tight junction maturation, can be clearly seen in the control cells by both western blot analysis and by colocalization with ZO1 at tight junctions.

The mechanism(s) by which quercetin impacts the tight junction barrier through its effects on claudin expression and localization remains unresolved. Quercetin is known to bind to several membrane-bound nutraceutical receptors that signal through different pathways to reduce oxidative stress. Quercetin also interacts directly with several kinases [48–52]. Some studies show that it is a potent inhibitor of PI3K/AKT pathway [36,53]; although there is some controversy [49,54,55], while others show that it inhibits the ERK and NF-$\kappa\beta$ pathways [56], and activates the AMPK pathway [49,57]. However, these studies have been done on different cell lines, using different conditions and measuring different outcomes. Therefore, it remains unclear if quercetin acts upstream of all of these pathways, or if its effects are cell-type specific. We showed that quercetin inhibits PI3K/AKT pathway in MCDK II cells. However, inhibition the PI3K/AKT pathway was not sufficient to recapitulate quercetin's effects on claudin expression.

So, how does this translate to the potential therapeutic effects of quercetin to prevent kidney stone formation in the context of the proximal tubule? The site of kidney stone formation is not well understood. Some studies suggest that kidney stones are formed in the interstitium, while others suggest that they form in the lumen of the nephron [2,58,59]. Here, we showed that quercetin led to a relative increase in barrier claudins at the tight junction relative to the pore-forming claudins in MDCK II cells. Although direct measurements of cation potential and water permeability were not performed, our data suggest that quercetin may be beneficial by preventing sodium, calcium, and water within the urine from crossing the epithelial barrier to enter the interstitium. We believe that the effects seen for Cldn2 are essential to explain what may be happening in the nephron. Cldn2 is a cation-selective pore forming claudin that promotes the paracellular movement of sodium, calcium, and water [29]. A decrease in Cldn2 is predicted to lead to a reduction in reabsorption of these substances. This would then prevent the transport of sodium and calcium in the interstitium which can drive stone formation. Alternatively, quercetin may effectively tighten the proximal tubular epithelium to maintain the water content within the urinary lumen of this segment to keep calcium and sodium solubilized [60]. This mechanism could correlate with the beneficial effects that high-water diets or diuretic drugs have in the prevention of kidney stones. In vivo experiments are necessary to discern which specific scenario is taking place and whether this effect could be beneficial for the prevention of kidney stones.

4. Material and Methods

4.1. Cell Culture

Madin–Darby canine kidney cells II (MDCK II) cells were obtained from ATCC. They were incubated at 37 °C and 5% CO2 in Dulbecco's modified Eagle's medium (DMEM) supplemented with 10% FBS, 51 IU penicillin, 50 μg/mL streptomycin and 16 μg/mL of gentamicin (Wisent BioProducts, Quebec, QC, Canada).

4.2. Chemicals and Antibodies

Quercetin (Sigma-Aldrich, Darmstadt, Germany) was added to the culture media, and heated at 37 °C with continuous stirring for at least 30 min in order to ensure that it was dissolved. It was used at the working concentration of 400 μM. PI3K/AKT pathway modulators Wortmannin (Abcam, 120148, Cambridge, UK) and SC79 (Abcam, 146428, Cambridge, UK) were used at working concentrations of 2 μM and 22 nM respectively.

Primary antibodies used for immunofluorescence and western blotting were: Cldn1 (Invitrogen, 374900, Carlsbad, CA, USA), Cldn2 (Invitrogen, 516100, Carlsbad, CA, USA), Cldn3 (Abcam, 15102, Cambridge, UK), Cldn4 (Invitrogen, 364800, Carlsbad, CA, USA), Cldn7 (Spring Bioscience, E10594, Pleasanton, CA, USA), ZO1 (Invitrogen, 339100, Carlsbad, CA, USA), pPI3K (Cell signaling, 4228, Danvers, MA, USA), pAKT (Cell signaling, 9271, Danvers, MA, USA), AKT (Cell signaling, 9272, Danvers, MA, USA), and pan-actin (Cell signaling, 4968, Danvers, MA, USA,). In addition, secondary goat anti-rabbit (Alexa Fluor 595 and 488, Invitrogen, Carlsbad, CA, USA), goat anti mouse (Alexa Fluor 595 and 488, Invitrogen, Carlsbad, CA, USA), goat anti-rabbit-HRT conjugated (Cell Signaling, 70748, Danvers, MA, USA), and goat anti mouse peroxidase conjugated (Jackson ImmunoResearch, 115-035-146, West Grove, PA, USA) were used.

4.3. Transepithelial Resistance (TER)

Cells were seeded on sterile 0.4 μm pore size, 12-well transparent polyethylene terephthalate inserts (Corning, NY, USA) at a cell density of 0.15×10^6 cells/insert and placed in the automated cell monitoring CellZscope® system (NanoAnalytics, Münster, Germany). The CellZscope® system provides noninvasive, continuous monitoring of cell monolayers impedance, capacitance, and resistance.

The tissue culture inserts were placed into a 12-well cell module, and the system was incubated at 37 °C and 5% CO2. TER measurements, expressed in ohm square centimeters, were recorded in real time every 20 min. Insert background resistance was automatically subtracted by the CellZscope® system. DMEM in both the apical and basal compartments was replaced with fresh DMEM (control) or DMEM plus 400 μM quercetin at 72 h, at which point mature tight junctions were established and stabilized (Resistance $\cong 70\ \Omega{\cdot}\mathrm{cm}^2$). Measurements were recorded in real time every 20 min for at least 90 h following treatment. Each time the machine was removed from the incubator or the media was changed, a small increase in TER is expected which then stabilizes.

4.4. Immunofluorescence Staining

Cells were seeded on coverslips in 12-well plates at a cell density of 0.1×10^6 cells/well. Once confluence and TJ maturity were achieved ($\approx$ 72 h), cells were treated with DMEM (control) or DMEM plus 400 μM quercetin and collected at 1, 6, 24, or 48 h. Cell layers were rinsed with phosphate buffer saline (PBS) and fixed in 10% trichloroacetic acid for 15 min at 4 °C. Cells were then blocked with 10% normal goat serum in phosphate buffer saline and 0.3% Triton-100 (PBST) and incubated overnight at 4 °C with 5% normal goat serum in PBST and primary antibodies: Cldn1, -2, -3, and -4 (1:100), Cldn7 (1:50), and ZO1 (1:150). Cells were washed with PBS. Alexa Fluor-conjugated secondary antibodies (1:500) in PBST were added for 1 h at RT. Coverslips were washed with PBS and then placed on slides with Slowfade Gold with DAPI (Invitrogen, Carlsbad, CA, USA). Z-stacks were imaged using a Zeiss LSM780 laser scanning confocal microscope. To determine the amount of claudin localized in the membrane, maximum intensity projections were obtained for each stack of images and then ZEN software colocalization analysis was performed. Only M2 results are presented; M2 makes reference to how many of the red pixels (ZO1) coincide with green pixels (claudin). A high M2 indicates high claudin at the membrane while a low M2 indicates high ZO1 alone, which relates to a low claudin at the membrane. M2 is not a reading of abundance or intensity levels and therefore it only provides information about localization.

4.5. Western Blot Analysis

Cells were seeded in 10 cm plates at a cell density of 2×10^6 cells/plate. Once confluence and TJ maturity was reached (≈ 72 h), cells were treated with DMEM (control) or DMEM plus 400 μM quercetin for 1, 6, 24, or 48 h, and then collected by physical scraping into lysis buffer (25 mM Tris-HCl Ph7.4; 10 mM sodium pyrophosphate; 25 mM NaCl; 10 mM sodium fluoride; 2 mM EGTA; 2 mM EDTA; 1% NP40; 0.1% SDS; 1.45 nM Pepstatin A; 2.1 nM Leupeptin; 0.15 nM Aprotinin, and 0.57 μM PMSF). Protein concentration was determined by Bradford assay and 50 μg per sample were separated by 12.5% SDS-PAGE and transferred onto PVDF membranes. Membranes were blocked for 1 h with 5% non-fat milk in PBST or with 5% BSA in Tris-NaCl buffer and 0.3% of Tween-20 (TBST). Membranes were incubated overnight at 4 °C with primary antibodies in PBST or TBST: Cldn1, -2, -3, -4, -7, pPI3K, pAKT, and AKT (1:1000) or pan-actin (1:2000). Membranes were then washed with PBST or TBST and then incubated with goat-anti-rabbit-HRT conjugated or goat-anti-mouse-peroxidase conjugated (1:5000) secondary antibodies for 1h at RT. Membranes were revealed using Clarity Western ECL substrate (Biorad laboratories, 1705061, Hercules, CA, USA), and imaged using Amersham imager 600 (GE Healthcare, Little Chalfont, United Kingdom). Band densities were quantified by densitometry using ImageJ, normalized against the 1 h control sample on each blot and then to normalized actin. Because some variability was observed in actin, we performed a statistical analysis of actin expression from all experiments and determined that there was no significant change over time or in response to quercetin treatment.

4.6. Statistics

All plotted results are mean ± SEM. Statistical analyses were performed using Graph-Pad PRISM (version 8.02, GraphPad Software, Inc., California, USA). Two-way ANOVA analysis was used for effects of treatment, time, and interaction. Sidak's multiple comparison test was used to compare the effect of treatment at the different time points. For TER Figure S1B, unpaired T-test were performed at the specific time points. Significance is considered if $p < 0.05$.

Author Contributions: Conceptualization, E.G.-E., S.A., B.J.-C., I.G. and A.K.R.; Formal analysis, E.G.-E.; Funding acquisition, S.A., I.G. and A.K.R.; Investigation, E.G.-E., I.G. and A.K.R.; Methodology, E.G.-E., I.G. and A.K.R.; Writing—original draft, E.G.-E.; Writing—review and editing, I.G. and A.K.R.

Acknowledgments: We thank L. Jerome-Majewska, S. Kumar, L. McCaffrey and Y. Yamanaka for helpful discussions, A. Naumova for help with the statistical analysis, and M. Fu and Shi-Bo Feng (RI-MUHC Imaging Platform) for technical assistance. EGE is the recipient of a doctoral studentship from Fonds de recherche du Québec – Santé (FRQS). EGE has also been supported by CRRD, RI-MUHC and McGill University Faculty of Medicine studentships.

References

1. Walker, V. Phosphaturia in kidney stone formers: Still an enigma. *Adv. Clin. Chem.* **2019**, *90*, 133–196. [CrossRef] [PubMed]
2. Khan, S.R.; Pearle, M.S.; Robertson, W.G.; Gambaro, G.; Canales, B.K.; Doizi, S.; Traxer, O.; Tiselius, H.G. Kidney stones. *Nat. Rev. Dis. Primers* **2016**, *2*, 16008. [CrossRef] [PubMed]
3. Evan, A.P. Physiopathology and etiology of stone formation in the kidney and the urinary tract. *Pediatr. Nephrol.* **2010**, *25*, 831–841. [CrossRef] [PubMed]
4. Anderson, J.M.; Van Itallie, C.M. Physiology and function of the tight junction. *Cold Spring Harb. Perspect. Biol.* **2009**, *1*, a002584. [CrossRef] [PubMed]

5. Van Itallie, C.M.; Anderson, J.M. Claudins and epithelial paracellular transport. *Annu. Rev. Physiol.* **2006**, *68*, 403–429. [CrossRef] [PubMed]
6. Gupta, I.R.; Ryan, A.K. Claudins: Unlocking the code to tight junction function during embryogenesis and in disease. *Clin. Genet.* **2010**, *77*, 314–325. [CrossRef] [PubMed]
7. Tsukita, S.; Tanaka, H.; Tamura, A. The Claudins: From Tight Junctions to Biological Systems. *Trends Biochem. Sci.* **2019**, *44*, 141–152. [CrossRef]
8. Yu, A.S. Claudins and the kidney. *J. Am. Soc. Nephrol.* **2015**, *26*, 11–19. [CrossRef] [PubMed]
9. Simon, D.B.; Lu, Y.; Choate, K.A.; Velazquez, H.; Al-Sabban, E.; Praga, M.; Casari, G.; Bettinelli, A.; Colussi, G.; Rodriguez-Soriano, J.; et al. Paracellin-1, a renal tight junction protein required for paracellular Mg^{2+} resorption. *Science* **1999**, *285*, 103–106. [CrossRef]
10. Konrad, M.; Schaller, A.; Seelow, D.; Pandey, A.V.; Waldegger, S.; Lesslauer, A.; Vitzthum, H.; Suzuki, Y.; Luk, J.M.; Becker, C.; et al. Mutations in the tight-junction gene claudin 19 (CLDN19) are associated with renal magnesium wasting, renal failure, and severe ocular involvement. *Am. J. Hum. Genet.* **2006**, *79*, 949–957. [CrossRef]
11. Thorleifsson, G.; Holm, H.; Edvardsson, V.; Walters, G.B.; Styrkarsdottir, U.; Gudbjartsson, D.F.; Sulem, P.; Halldorsson, B.V.; de Vegt, F.; d'Ancona, F.C.; et al. Sequence variants in the CLDN14 gene associate with kidney stones and bone mineral density. *Nat. Genet.* **2009**, *41*, 926–930. [CrossRef] [PubMed]
12. Breiderhoff, T.; Himmerkus, N.; Stuiver, M.; Mutig, K.; Will, C.; Meij, I.C.; Bachmann, S.; Bleich, M.; Willnow, T.E.; Muller, D. Deletion of claudin-10 (Cldn10) in the thick ascending limb impairs paracellular sodium permeability and leads to hypermagnesemia and nephrocalcinosis. *Proc. Natl. Acad. Sci. USA* **2012**, *109*, 14241–14246. [CrossRef] [PubMed]
13. Muto, S.; Hata, M.; Taniguchi, J.; Tsuruoka, S.; Moriwaki, K.; Saitou, M.; Furuse, K.; Sasaki, H.; Fujimura, A.; Imai, M.; et al. Claudin-2-deficient mice are defective in the leaky and cation-selective paracellular permeability properties of renal proximal tubules. *Proc. Natl. Acad. Sci. USA* **2010**, *107*, 8011–8016. [CrossRef] [PubMed]
14. Himmerkus, N.; Shan, Q.; Goerke, B.; Hou, J.; Goodenough, D.A.; Bleich, M. Salt and acid-base metabolism in claudin-16 knockdown mice: Impact for the pathophysiology of FHHNC patients. *Am. J. Physiol. Renal Physiol.* **2008**, *295*, F1641–F1647. [CrossRef] [PubMed]
15. Hou, J.; Renigunta, A.; Gomes, A.S.; Hou, M.; Paul, D.L.; Waldegger, S.; Goodenough, D.A. Claudin-16 and claudin-19 interaction is required for their assembly into tight junctions and for renal reabsorption of magnesium. *Proc. Natl. Acad. Sci. USA* **2009**, *106*, 15350–15355. [CrossRef] [PubMed]
16. Tatum, R.; Zhang, Y.; Salleng, K.; Lu, Z.; Lin, J.J.; Lu, Q.; Jeansonne, B.G.; Ding, L.; Chen, Y.H. Renal salt wasting and chronic dehydration in claudin-7-deficient mice. *Am. J. Physiol. Renal Physiol.* **2010**, *298*, F24–F34. [CrossRef]
17. Plain, A.; Alexander, R.T. Claudins and nephrolithiasis. *Curr. Opin. Nephrol. Hypertens.* **2018**, *27*, 268–276. [CrossRef]
18. Gong, Y.; Hou, J. Claudins in barrier and transport function-the kidney. *Pflugers Arch.* **2017**, *469*, 105–113. [CrossRef]
19. Wakino, S.; Hasegawa, K.; Itoh, H. Sirtuin and metabolic kidney disease. *Kidney Int.* **2015**, *88*, 691–698. [CrossRef]
20. Zhu, W.; Xu, Y.F.; Feng, Y.; Peng, B.; Che, J.P.; Liu, M.; Zheng, J.H. Prophylactic effects of quercetin and hyperoside in a calcium oxalate stone forming rat model. *Urolithiasis* **2014**, *42*, 519–526. [CrossRef]
21. Mercado, J.; Valenzano, M.C.; Jeffers, C.; Sedlak, J.; Cugliari, M.K.; Papanikolaou, E.; Clouse, J.; Miao, J.; Wertan, N.E.; Mullin, J.M. Enhancement of tight junctional barrier function by micronutrients: Compound-specific effects on permeability and claudin composition. *PLoS ONE* **2013**, *8*, e78775. [CrossRef] [PubMed]
22. Valenzano, M.C.; DiGuilio, K.; Mercado, J.; Teter, M.; To, J.; Ferraro, B.; Mixson, B.; Manley, I.; Baker, V.; Moore, B.A.; et al. Remodeling of Tight Junctions and Enhancement of Barrier Integrity of the CACO-2 Intestinal Epithelial Cell Layer by Micronutrients. *PLoS ONE* **2015**, *10*, e0133926. [CrossRef] [PubMed]
23. Suzuki, T.; Hara, H. Quercetin enhances intestinal barrier function through the assembly of zonula [corrected] occludens-2, occludin, and claudin-1 and the expression of claudin-4 in Caco-2 cells. *J. Nutr.* **2009**, *139*, 965–974. [CrossRef] [PubMed]

24. Suzuki, T.; Tanabe, S.; Hara, H. Kaempferol enhances intestinal barrier function through the cytoskeletal association and expression of tight junction proteins in Caco-2 cells. *J. Nutr.* **2011**, *141*, 87–94. [CrossRef]
25. Dukes, J.D.; Whitley, P.; Chalmers, A.D. The MDCK variety pack: Choosing the right strain. *BMC Cell Biol.* **2011**, *12*, 43. [CrossRef] [PubMed]
26. Hou, J.; Gomes, A.S.; Paul, D.L.; Goodenough, D.A. Study of claudin function by RNA interference. *J. Biol. Chem.* **2006**, *281*, 36117–36123. [CrossRef]
27. Gunzel, D.; Fromm, M. Claudins and other tight junction proteins. *Compr. Physiol.* **2012**, *2*, 1819–1852. [CrossRef] [PubMed]
28. Borovac, J.; Barker, R.S.; Rievaj, J.; Rasmussen, A.; Pan, W.; Wevrick, R.; Alexander, R.T. Claudin-4 forms a paracellular barrier, revealing the interdependence of claudin expression in the loose epithelial cell culture model opossum kidney cells. *Am. J. Physiol. Cell Physiol.* **2012**, *303*, C1278–C1291. [CrossRef]
29. Rosenthal, R.; Gunzel, D.; Krug, S.M.; Schulzke, J.D.; Fromm, M.; Yu, A.S. Claudin-2-mediated cation and water transport share a common pore. *Acta Physiol.* **2017**, *219*, 521–536. [CrossRef]
30. Van Itallie, C.M.; Fanning, A.S.; Anderson, J.M. Reversal of charge selectivity in cation or anion-selective epithelial lines by expression of different claudins. *Am. J. Physiol. Renal Physiol.* **2003**, *285*, F1078–F1084. [CrossRef]
31. Tokuda, S.; Furuse, M. Claudin-2 knockout by TALEN-mediated gene targeting in MDCK cells: Claudin-2 independently determines the leaky property of tight junctions in MDCK cells. *PLoS ONE* **2015**, *10*, e0119869. [CrossRef] [PubMed]
32. Srinivasan, B.; Kolli, A.R.; Esch, M.B.; Abaci, H.E.; Shuler, M.L.; Hickman, J.J. TEER measurement techniques for in vitro barrier model systems. *J. Lab. Autom.* **2015**, *20*, 107–126. [CrossRef] [PubMed]
33. Dembla, S.; Hasan, N.; Becker, A.; Beck, A.; Philipp, S.E. Transient receptor potential A1 channels regulate epithelial cell barriers formed by MDCK cells. *FEBS Lett.* **2016**, *590*, 1509–1520. [CrossRef] [PubMed]
34. Amoozadeh, Y.; Anwer, S.; Dan, Q.; Venugopal, S.; Shi, Y.; Branchard, E.; Liedtke, E.; Ailenberg, M.; Rotstein, O.D.; Kapus, A.; et al. Cell confluence regulates claudin-2 expression: Possible role for ZO-1 and Rac. *Am. J. Physiol. Cell Physiol.* **2018**, *314*, C366–C378. [CrossRef] [PubMed]
35. Welker, M.E.; Kulik, G. Recent syntheses of PI3K/Akt/mTOR signaling pathway inhibitors. *Bioorg. Med. Chem.* **2013**, *21*, 4063–4091. [CrossRef]
36. Jiang, W.; Luo, T.; Li, S.; Zhou, Y.; Shen, X.Y.; He, F.; Xu, J.; Wang, H.Q. Quercetin Protects against Okadaic Acid-Induced Injury via MAPK and PI3K/Akt/GSK3beta Signaling Pathways in HT22 Hippocampal Neurons. *PLoS ONE* **2016**, *11*, e0152371. [CrossRef]
37. Khan, N.; Binder, L.; Pantakani, D.V.K.; Asif, A.R. MPA Modulates Tight Junctions' Permeability via Midkine/PI3K Pathway in Caco-2 Cells: A Possible Mechanism of Leak-Flux Diarrhea in Organ Transplanted Patients. *Front. Physiol.* **2017**, *8*, 438. [CrossRef]
38. Lin, X.; Shang, X.; Manorek, G.; Howell, S.B. Regulation of the Epithelial-Mesenchymal Transition by Claudin-3 and Claudin-4. *PLoS ONE* **2013**, *8*, e67496. [CrossRef]
39. Piegholdt, S.; Pallauf, K.; Esatbeyoglu, T.; Speck, N.; Reiss, K.; Ruddigkeit, L.; Stocker, A.; Huebbe, P.; Rimbach, G. Biochanin A and prunetin improve epithelial barrier function in intestinal CaCo-2 cells via downregulation of ERK, NF-kappaB, and tyrosine phosphorylation. *Free Radic. Biol. Med.* **2014**, *70*, 255–264. [CrossRef]
40. Horng, S.; Therattil, A.; Moyon, S.; Gordon, A.; Kim, K.; Argaw, A.T.; Hara, Y.; Mariani, J.N.; Sawai, S.; Flodby, P.; et al. Astrocytic tight junctions control inflammatory CNS lesion pathogenesis. *J. Clin. Investig.* **2017**, *127*, 3136–3151. [CrossRef]
41. Aono, S.; Hirai, Y. Phosphorylation of claudin-4 is required for tight junction formation in a human keratinocyte cell line. *Exp. Cell Res.* **2008**, *314*, 3326–3339. [CrossRef] [PubMed]
42. Fujibe, M.; Chiba, H.; Kojima, T.; Soma, T.; Wada, T.; Yamashita, T.; Sawada, N. Thr203 of claudin-1, a putative phosphorylation site for MAP kinase, is required to promote the barrier function of tight junctions. *Exp. Cell Res.* **2004**, *295*, 36–47. [CrossRef] [PubMed]
43. Shigetomi, K.; Ikenouchi, J. Regulation of the epithelial barrier by post-translational modifications of tight junction membrane proteins. *J. Biochem.* **2018**, *163*, 265–272. [CrossRef] [PubMed]
44. Fujii, N.; Matsuo, Y.; Matsunaga, T.; Endo, S.; Sakai, H.; Yamaguchi, M.; Yamazaki, Y.; Sugatani, J.; Ikari, A. Hypotonic Stress-induced Down-regulation of Claudin-1 and -2 Mediated by Dephosphorylation and Clathrin-dependent Endocytosis in Renal Tubular Epithelial Cells. *J. Biol. Chem.* **2016**, *291*, 24787–24799. [CrossRef] [PubMed]

45. Van Itallie, C.M.; Tietgens, A.J.; LoGrande, K.; Aponte, A.; Gucek, M.; Anderson, J.M. Phosphorylation of claudin-2 on serine 208 promotes membrane retention and reduces trafficking to lysosomes. *J. Cell Sci.* **2012**, *125*, 4902–4912. [CrossRef] [PubMed]
46. Tanaka, M.; Kamata, R.; Sakai, R. EphA2 phosphorylates the cytoplasmic tail of Claudin-4 and mediates paracellular permeability. *J. Biol. Chem.* **2005**, *280*, 42375–42382. [CrossRef] [PubMed]
47. D'Souza, T.; Indig, F.E.; Morin, P.J. Phosphorylation of claudin-4 by PKCepsilon regulates tight junction barrier function in ovarian cancer cells. *Exp. Cell Res.* **2007**, *313*, 3364–3375. [CrossRef] [PubMed]
48. Zhu, Y.; Teng, T.; Wang, H.; Guo, H.; Du, L.; Yang, B.; Yin, X.; Sun, Y. Quercetin inhibits renal cyst growth in vitro and via parenteral injection in a polycystic kidney disease mouse model. *Food Funct.* **2018**, *9*, 389–396. [CrossRef]
49. Jiang, H.; Yamashita, Y.; Nakamura, A.; Croft, K.; Ashida, H. Quercetin and its metabolite isorhamnetin promote glucose uptake through different signalling pathways in myotubes. *Sci. Rep.* **2019**, *9*, 2690. [CrossRef]
50. Gamero-Estevez, E.; Baumholtz, A.I.; Ryan, A.K. Developing a link between toxicants, claudins and neural tube defects. *Reprod. Toxicol.* **2018**, *81*, 155–167. [CrossRef]
51. Chuenkitiyanon, S.; Pengsuparp, T.; Jianmongkol, S. Protective effect of quercetin on hydrogen peroxide-induced tight junction disruption. *Int. J. Toxicol.* **2010**, *29*, 418–424. [CrossRef] [PubMed]
52. Tan, C.; Meng, F.; Reece, E.A.; Zhao, Z. Modulation of nuclear factor-kappaB signaling and reduction of neural tube defects by quercetin-3-glucoside in embryos of diabetic mice. *Am. J. Obstet. Gynecol.* **2018**, *219*, 197.e1–197.e8. [CrossRef] [PubMed]
53. Maurya, A.K.; Vinayak, M. PI-103 and Quercetin Attenuate PI3K-AKT Signaling Pathway in T- Cell Lymphoma Exposed to Hydrogen Peroxide. *PLoS ONE* **2016**, *11*, e0160686. [CrossRef] [PubMed]
54. Wang, X.Q.; Yao, R.Q.; Liu, X.; Huang, J.J.; Qi, D.S.; Yang, L.H. Quercetin protects oligodendrocyte precursor cells from oxygen/glucose deprivation injury in vitro via the activation of the PI3K/Akt signaling pathway. *Brain Res. Bull.* **2011**, *86*, 277–284. [CrossRef] [PubMed]
55. Du, G.; Zhao, Z.; Chen, Y.; Li, Z.; Tian, Y.; Liu, Z.; Liu, B.; Song, J. Quercetin attenuates neuronal autophagy and apoptosis in rat traumatic brain injury model via activation of PI3K/Akt signaling pathway. *Neurol. Res.* **2016**, *38*, 1012–1019. [CrossRef] [PubMed]
56. Wu, L.; Zhang, Q.; Dai, W.; Li, S.; Feng, J.; Li, J.; Liu, T.; Xu, S.; Wang, W.; Lu, X.; et al. Quercetin Pretreatment Attenuates Hepatic Ischemia Reperfusion-Induced Apoptosis and Autophagy by Inhibiting ERK/NF-kappaB Pathway. *Gastroenterol. Res. Pract.* **2017**, *2017*, 9724217. [CrossRef] [PubMed]
57. Qiu, L.; Luo, Y.; Chen, X. Quercetin attenuates mitochondrial dysfunction and biogenesis via upregulated AMPK/SIRT1 signaling pathway in OA rats. *Biomed. Pharmacother.* **2018**, *103*, 1585–1591. [CrossRef]
58. Bird, V.Y.; Khan, S.R. How do stones form? Is unification of theories on stone formation possible? *Arch. Esp. Urol.* **2017**, *70*, 12–27. [PubMed]
59. Coe, F.L.; Evan, A.P.; Worcester, E.M.; Lingeman, J.E. Three pathways for human kidney stone formation. *Urol. Res.* **2010**, *38*, 147–160. [CrossRef] [PubMed]
60. Tokuda, S.; Hirai, T.; Furuse, M. Effects of Osmolality on Paracellular Transport in MDCK II Cells. *PLoS ONE* **2016**, *11*, e0166904. [CrossRef]

2

A Novel Claudinopathy based on Claudin-10 Mutations

Susanne Milatz *

Institute of Physiology, Kiel University, Christian-Albrechts-Platz 4, 24118 Kiel, Germany

Abstract: Claudins are key components of the tight junction, sealing the paracellular cleft or composing size-, charge- and water-selective paracellular channels. Claudin-10 occurs in two major isoforms, claudin-10a and claudin-10b, which constitute paracellular anion or cation channels, respectively. For several years after the discovery of claudin-10, its functional relevance in men has remained elusive. Within the past two years, several studies appeared, describing patients with different pathogenic variants of the *CLDN10* gene. Patients presented with dysfunction of kidney, exocrine glands and skin. This review summarizes and compares the recently published studies reporting on a novel autosomal-recessive disorder based on claudin-10 mutations.

Keywords: tight junction; paracellular permeability; paracellular sodium transport; thick ascending limb; nephropathy; HELIX syndrome; hypokalemia; hypermagnesemia; anhidrosis; gland dysfunction

---

1. Introduction

1.1. Claudin-10

The protein family of claudins is a key component of the tight junction (TJ). Claudins comprise four transmembrane segments (TM1-4), two extracellular segments (ECS1 and 2) and intracellular N- and C-termini. Embedded in the plasma membranes of adjacent cells, they interact with each other within the same plasma membrane but also across the paracellular cleft, with claudins of the neighboring cell (cis- or trans-interaction, respectively). By this means, they form a complex strand meshwork and determine tightness and selectivity of the bicellular TJ. Whereas most claudins exhibit a mainly sealing function, some claudins form size-, charge- and water-selective channels through the TJ [1–18].

The claudin family encompasses at least 24 members in mammals. The human gene encoding claudin-10 (*CLDN10*) contains six exons and gives rise to two major isoforms: claudin-10a and -10b. According to their usage of either exon 1a or 1b, they differ in their TM1 and ECS1 (Figure 1). As ECS1 acts as main determinant of charge selectivity, claudin-10a and -10b strands exhibit contrarian permeability properties.

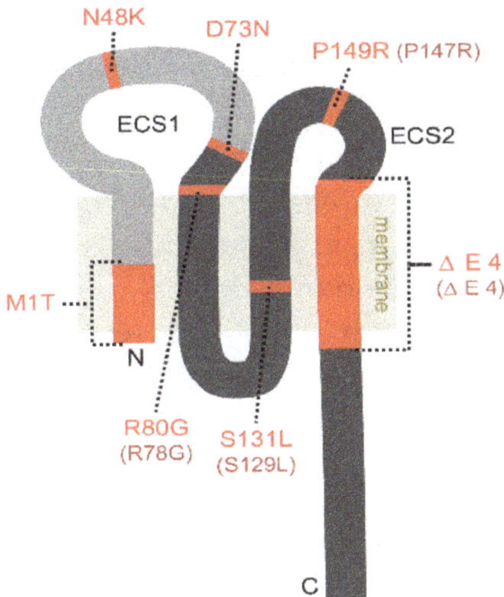

Figure 1. Predicted topology of claudin-10. The two major isoforms claudin-10a and claudin-10b differ in their first transmembrane segment and most of the first extracellular segment (ECS1), both shown in light grey. The remaining protein sequence is identical (shown in dark grey). The mutations discovered to date (red) comprise single amino acid substitutions or large deletions (M1T, ΔE4) and affect either both claudin-10a and claudin-10b or only claudin-10b. The existing claudin-10a variants with respective residue numbering are depicted in parentheses.

Due to claudin-10a's equipment with seven positive and only one negative amino acid in ECS1, it is predestined to form a paracellular anion channel [10,12]. Expression of human claudin-10a in the poorly ion permeable cell line MDCK C7 resulted in a decrease in transepithelial resistance (TER) without alteration in preference for Na^+ or Cl^- [12]. Moreover, claudin-10a expression increased the relative NO_3– permeability but decreased the permeability to the anion pyruvate, suggesting that claudin-10a modifies the anion preference of the paracellular pathway [12].

ECS1 of claudin-10b comprises four positive and five negative amino acids. In most cell culture models, heterologous expression of human or murine claudin-10b led to a marked decrease in TER that was based on an increase in Na^+ permeability over Cl^- permeability ($P_{Na}{}^+/P_{Cl}{}^-$). Further studies revealed a relative strong permeability to all monovalent cations with preference for Na^+, a lesser permeability to divalent cations and impermeability to larger molecules (4 kDa dextran) or water of the claudin-10b-based paracellular channel [10,12,13,19].

Claudin-10a and -10b do not only differ significantly in their function but also with respect to their expression in the body. Whereas claudin-10a appears to be restricted to the kidney, claudin-10b has been detected in many tissues, including kidney, skin, salivary glands, sweat glands, brain, lung and pancreas [10,12,20]. Along the kidney tubule, claudin-10a is expressed in the proximal tubule to the S3 segment whereas the main expression site of claudin-10b is the thick ascending limb of Henle's loop (TAL). Claudin-10b is found along the whole medullary–cortical axis of TAL from inner stripe of outer medulla (ISOM) to outer stripe of outer medulla (OSOM) to the renal cortex. In ISOM TAL, to current knowledge, solely claudin-10b constitutes the bicellular TJ, where it facilitates Na^+ reabsorption [21,22]. Towards OSOM and cortex, a TAL mosaic claudin expression is found. Claudin-10b equips part of the TJs, whereas the remaining TJs contain a complex of claudin-3, -16, -19 and to a smaller extent claudin-14 [21–23]. This complex is involved in the reabsorption of divalent cations such as Ca^{2+} and Mg^{2+} in the TAL. The thin limb of Henle also incorporates claudin-10, as yet the identity of the present isoform is unknown [24].

An important insight into the physiological role of claudin-10b was provided by the mouse model generated by Breiderhoff et al., lacking claudin-10 in the entire loop of Henle. These mice featured a strongly reduced paracellular Na^+ selectivity in the TAL that led to a urinary concentration defect and was accompanied by hypermagnesemia, polyuria, polydipsia, elevated plasma urea levels and compensatorily increased K^+ and H^+ secretion [25,26]. This misbalanced TAL electrolyte handling was accompanied by a severe medullary nephrocalcinosis in these animals.

1.2. Claudinopathies

So far, a number of human hereditary diseases based on defects in claudin-1, -14, -16 and -19 have been reported [27–32]. However, for 18 years after the discovery of claudin-10, its functional relevance in men has remained unclear. On the one hand, defects in the *CLDN10* gene are rare and clinical manifestations occur mainly in patients with biallelic defects (autosomal recessive disorder). On the other hand, some patients presented with symptoms many years ago but were originally misdiagnosed with Bartter syndrome or Gitelman syndrome. Both diseases are characterized by a salt-losing nephropathy with an imbalance in Ca^{2+} and Mg^{2+} homeostasis. In Bartter syndrome, the transcellular NaCl reabsorption in the TAL is disrupted, due to mutations in the Na-K-Cl cotransporter 2 (NKCC2), the renal outer medullary potassium channel (ROMK1) or the Cl^- channel Kb (ClC-Kb) [33–36]. Gitelman syndrome is caused by mutations affecting the Na^+-Cl^- cotransporter (NCC) in the distal convoluted tubule [37]. Nowadays, whole-exome sequencing is available at lower cost and increasingly used to identify the cause of rare mendelian disorders. In 2017, three studies reported on patients with different pathogenic variants of *CLDN10* [38–40]. Hadj-Rabia et al. coined a novel disease syndrome, summarizing the clinical manifestations of their patients (HELIX for hypohidrosis, electrolyte imbalance, lacrimal gland dysfunction, ichthyosis, xerostomia) [39]. This review aims to summarize and compare the data of Bongers et al., Hadj-Rabia et al., Klar et al. and a case report by Meyers et al., all describing a novel claudinopathy based on *CLDN10* mutations [38–41].

2. Clinical Manifestations

To date, a total of 22 patients carrying homozygous or compound heterozygous *CLDN10* mutations have been reported, their ages ranging from 4 to 53 years. Patients were mostly born in consanguineous families and first symptoms often occurred in early childhood, sometimes directly after birth. Patients presented with first symptoms as anhidrosis, xerostomia, alacrima, muscle cramps, falls, or chest pain. Table 1 provides a summary of all patient groups and their clinical manifestations.

Table 1. Summary of all patient groups, their *CLDN10* variants and clinical manifestations.

Publication	Bongers et al. [38]	Bongers et al. [38]	Klar et al. [40]	Hadj-Rabia et al. [39]	Hadj-Rabia et al. [39]	Meyers et al. [41]
Number of patients	1 (patient 1)	1 (patient 2)	13	4 (family A)	2 (family B)	1
Claudin-10a variant	P147R, ΔE4	wildtype, P147R	wildtype	S129L	wildtype	R78G
Claudin-10b variant	P149R, ΔE4	D73N, P149R	N48K	S131L	M1T	R80G
Loss of claudin-10b function	partial/complete	partial	partial	complete	complete *	not analyzed
			Extrarenal manifestations			
Xerostomia	yes	yes	yes	yes	yes	yes
Alacrima	n.r.	yes	yes	yes	yes	yes
Hypohidrosis	yes	yes	yes	yes	yes	yes
Dermatological manifestations in addition to hypohidrosis	yes	yes	no	yes	yes	no

Table 1. *Cont.*

Publication	Bongers et al. [38]	Bongers et al. [38]	Klar et al. [40]	Hadj-Rabia et al. [39]	Hadj-Rabia et al. [39]	Meyers et al. [41]
			Renal function			
Hypermagnesemia	yes	no	yes	yes	yes **	yes
Urinary magnesium	rather high ***	normal ***	low	low or normal	rather high	n.r.
Plasma/serum calcium	upper normal range	upper normal range	normal	1 increased, 3 normal	normal	n.r.
Hypocalciuria	yes	yes	yes	yes	yes	yes
Hypokalemia	yes	yes	no	yes **	yes **	yes
Metabolic alkalosis	yes	yes	n.r.	n.r.	n.r.	n.r.
Plasma aldosterone	n.r.	n.r.	n.r.	normal or high	normal or high	high
Polyuria	yes	no	no	inconsistent	inconsistent	yes
Polydipsia	n.r.	n.r.	n.r.	yes	yes	yes
Estimated Estimated glomerular filtration rate	decreased	normal	lower normal range	normal or decreased	normal	decreased
Kidney form/size	small right kidney	normal	normal	n.r.	normal	normal
Nephrocalcinosis	no	no	n.r.	no	no	no
Blood pressure	lower normal range	low to normal	n.r.	low	low	normal

* with the exception of normal deposition in eccrine sweat glands. ** considering the patient's age. *** compared to heterozygous family members. n.r. not reported

2.1. *Hypohidrosis, Xerostomia and Alacrima*

Hypohidrosis with intolerance to heat was frequently reported as one of the first symptoms observed in early childhood. Apparently, all known patients suffer from hypohidrosis, including the two patients described by Bongers et al., who did not complain about reduced sweating at the outset but confirmed hypohidrosis subsequently [38], (personal communication with Tom Nijenhuis, Radboud University Medical Center, Nijmegen). Klar et al. evaluated heat intolerance in two patients by exposure to heat for 20 min, which resulted in a rapidly increased body temperature from 37 °C to 39.6 °C and an increase in heart rate [40]. Generalized hypohidrosis was verified using starch-iod test applied on different body parts, corroborating a severe dysfunction of eccrine sweat glands.

Likewise, xerostomia due to reduced saliva production is apparently a typical symptom of *CLDN10* defects as it has been documented in all known patients including the patients examined by Bongers et al. [39–41]; (personal communication with Tom Nijenhuis). Hadj-Rabia et al. analyzed xerostomia by saliva secretion rate measurements in three adult patients [39]. As a result, the flow of fluid was reduced by 98% in patients compared to healthy controls. Moreover, the fluid/mucus ratio of saliva was dramatically decreased in patients with *CLDN10* variants. Hadj-Rabia also documented a poor dental condition with severe enamel wear and generalized gingival inflammation of their patients [39]. This is attributed to aptyalism but might also be a consequence of disturbed enamel mineralization (amelogenesis imperfecta) as claudin-10 expression was found in ameloblasts of mice [42].

Alacrima (the inability to produce tears) was described in the majority of patients and was confirmed using Schirmer's test by Hadj-Rabia et al. [39–41].

2.2. *Dermatological Manifestations in Addition to Hypohidrosis*

The occurrence of ichthyosis and other dermatological manifestations among patients was rather inconsistent. Hadj-Rabia described mild forms of ichthyosis with a thickened stratum corneum, palmar hyperlinearity and plantar keratoderma in two unrelated families with different *CLDN10* variants. Histological analysis of skin biopsies revealed slight epidermal hyperplasia and an abnormally high number of dilated eccrine sweat glands. The patients examined by Bongers et al. reported dry skin in retrospect, and dermatological consultation showed palmar hyperlinearity and plantar hyperkeratosis (personal communication with Tom Nijenhuis). In contrast, patients examined by Klar et al. and Meyers et al. showed no dermatological manifestations apart from hypohidrosis [40,41]. Morphology and number of eccrine sweat glands appeared normal in patients examined by Klar et al. [40].

2.3. *Kidney Dysfunction*

A number of features revealing a renal dysfunction in patients has been reported. All patients showed a disturbance in electrolyte balance, becoming manifest in hypermagnesemia, hypokalemia and hypocalciuria. Hypermagnesemia was present in the majority of patients, most pronounced in childhood and decreasing with age (Figure 2A). In contrast, hypokalemia was most severe in adults with the exception of the patient group examined by Klar et al. [40],(Figure 2B). These patients had serum K^+ levels in the upper normal range or higher in adulthood. The reason for that discrepancy remains unclear. Admittedly, more longitudinal intraindividual data over a longer time period are required for confirmation of the apparent age dependencies depicted in Figure 2. Bongers et al. reported that hypokalemia was accompanied by metabolic alkalosis in their patients [38].

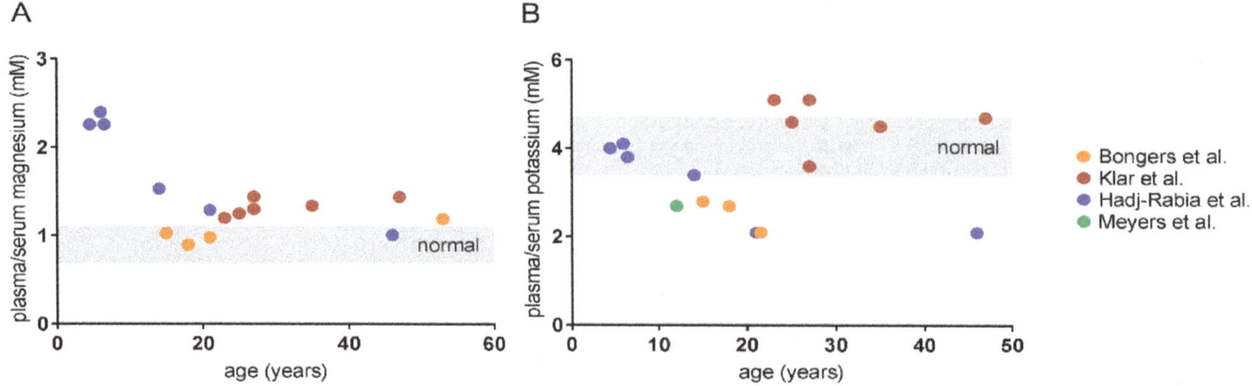

Figure 2. Available plasma/serum values of patients plotted against their age. (**A**) Hypermagnesemia is most pronounced in children and decreases with age. (**B**) Plasma/serum potassium levels decline with age, revealing a marked hypokalemia in adolescence. The only exception is the patient group examined by Klar et al [40]. Values were partly obtained by conversion into mmol/L. The publication by Bongers et al. provided data points of the same patients at different ages [38].

Of note, hypocalciuria appears to be a common manifestation as it was distinct in all tested patients and at all ages with plasma/serum Ca^{2+} in the normal or upper normal range. Despite the strong incidence of hypermagnesemia and hypocalciuria, most patients showed no decrease in urinary Mg^{2+} or an elevation of plasma/serum Ca^{2+}. The reason for this is not clear. However, it appears that renal Ca^{2+} and Mg^{2+} reabsorption and handling are tightly but differentially regulated.

In some patients, plasma aldosterone levels were determined and revealed hyperaldosteronism [39, 41]. The underlying cause of the electrolyte disorder and hyperaldosteronism is a NaCl wasting in the TAL, the main expression site of claudin-10b. Plasma/serum Na^+ appeared mostly normal if tested, presumably due to compensation in the more distal parts of the nephron. However, slight hypochloremia was present in few patients [39–41].

Bongers et al. tested the urine concentrating ability of one patient [38]. As a result, the patient showed a reduced response to thirsting and the application of the synthetic vasopressin analogue ddAVP. Verification of an adequate aquaporin-2 response in the collecting duct pointed to a dysfunction of NaCl reabsorption in the TAL as cause for an insufficient build-up of interstitial tonicity. In line with the reduced urine concentrating ability, this patient suffered from polyuria. Overall, the occurrence of polyuria and polydipsia among patients was rather inconsistent and might also be partially attributed to xerostomia.

The estimated glomerular filtration rate (eGFR) as indicator for kidney function was determined in most patients. It ranged from low to normal and decreased with time in some patients, indicating a progressive renal insufficiency. With one exception, kidneys analyzed by computer tomography scans were normal in form and size and, in contrast to the mouse model, nephrocalcinosis was not detected in any patient [38–41]. However, several patients of Klar et al. suffered from recurrent kidney pain due to nephrolithiasis [40]. If determined, the patient's blood pressure ranged from low to normal. Some patients complained of atypical chest pain, heart palpitations, collapse, falls, or muscle cramps. Klar et al. additionally analyzed lung and pancreas function without abnormal findings [40].

3. Protein Variants and Their Functional Analyses

Figure 1 displays all naturally occurring claudin-10 variants reported so far. Four mutations apply to both claudin-10a and claudin-10b proteins, whereas three defects affect only claudin-10b. Markedly, there are no obvious differences between the six patients with both defective isoforms and the 16 patients with defective claudin-10b and unaffected or only heterozygously affected claudin-10a, at least none that could be easily attributed to that special factor (Table 1). Of note, to date no patients with exclusive claudin-10a variants (without affection of claudin-10b) were described, also pointing to a minor role of claudin-10a defects in men.

Because of the predominant importance of claudin-10b defects in patients, most in vitro studies were carried out using the mutants of this isoform. Several analyses addressed localization, strand formation and channel function of mutated claudin-10b vs. wildtype protein. In the following, the denomination of claudin-10 variants refers to the alteration in claudin-10b protein, i.e., the amino acid exchange or deletion.

3.1. *Membrane Localization*

Correct trafficking to the membrane of a claudin variant is a fundamental prerequisite for a physiological function within the TJ. Similar to the wild type protein, the claudin-10b N48K mutant analyzed by Klar et al. resided in the cell membrane when heterologously expressed in the epithelial cell line MDCK C7. However, the subcellular distribution of N48K suggested an increased intracellular accumulation [40]. A very similar distribution in MDCK C7 cells was observed for the variants D73N and P149R investigated by Bongers et al. [38]. In sharp contrast, their third variant ΔE4 was very weakly expressed and did not localize to the cell membrane. This variant lacks TM4 which inevitably leads to exclusion from the plasma membrane. Likewise, the variant S131L with a missense mutation in TM3 analyzed by Hadj-Rabia et al. showed no detectable presence at the plasma membrane when expressed in a mouse TAL cell culture [39]. Contradictorily, the claudin-10 variant M1T probably lacking part of TM1 revealed normal deposition in eccrine sweat glands of a skin biopsy but was not localized in the TAL of a kidney biopsy [39]. Klar et al. also performed immunostainings of sweat glands and observed a normal localization of N48K in membranes facing the lumen but also an abnormally strong intracellular distribution without accumulation in canaliculi, pointing to a reduced delivery to the TJs [40].

Overall, the analyzed claudin-10b mutations resulted either in complete absence from the plasma membrane (especially those affecting one of the transmembrane segments) or, on the other hand, can basically insert into the plasma membrane, although an increased distribution outside the TJ suggests a reduced function.

3.2. TJ Strand Formation

Strand formation is the consequence of cis- and trans-interaction between claudins. In case of claudin-10b and its mutants, interaction with itself (homophilic interaction) is of particular importance as claudin-10b is not capable of interaction with any other claudin in the TAL [22]. Trans-interaction (with claudins in the opposing plasma membrane) can be detected by heterologous expression of the appropriate claudin in TJ-free HEK 293 cells and subsequent microscopic analysis of so-called contact enrichment. If the claudin is capable of autonomous strand formation, it enriches at cell–cell-contacts of two transfected cells compared to the remaining cell membrane. Bongers et al. reported a basal capability of trans-interaction and TJ formation for variants D73N and P149R but not for the truncated variant ΔE4 [38]. Klar et al. did not detect a significant contact enrichment of their mutant N48K, tantamount to a perturbed homophilic trans-interaction [40]. Moreover, homophilic cis-interaction was analyzed by Förster resonance energy transfer (FRET) and revealed an exaggerated oligomerization of N48K proteins. Ultrastructural analysis using freeze fracture electron microscopy showed that N48K formed few particle-typed TJ strands with less compact meshworks, compared to wildtype claudin-10b [40].

Overall, the few claudin-10b variants analyzed with respect to interaction properties mostly showed a fundamental capability of homophilic interaction and strand formation. Nonetheless, quality and/or quantity of interaction and strand assembly were impaired. As anticipated, the lack of a transmembrane segment resulted in a complete loss of function.

3.3. Channel Function

Analyses of claudin-10b mutant channel properties by means of electrophysiological measurements are available only for the N48K variant, as yet. Klar et al. stably transfected MDCK C7 cells and compared claudin-10b N48K-with wildtype-expressing cells [40]. The cation selectivity (P_{Na}/P_{Cl}) of N48K-expressing cells was first similar to that of claudin-10b wildtype-transfected cells but progressively decreased with passaging, more and more resembling that for clones with a weak expression of claudin-10b wildtype. Moreover, N48K-expressing cells showed a higher TER and altered relative permeabilities to other monovalent cations, compared to claudin-10 wildtype-expressing cells. Together, the results suggest that the channels formed by N48K have a subnormal Na^+ permeability and that the overall number of channels is markedly reduced [40].

In an attempt to mimic sweat secretion, Klar et al. used a 3D cell culture model by growing MDCK C7 cells in Matrigel. Epithelial cells formed three-dimensional cysts with the apical side towards the lumen. Cysts formed by wildtype-claudin-10b-expressing cells increased their lumen by transcellular Cl^- secretion, followed by claudin-10b-mediated Na^+ permeation and subsequent water transport. N48K-expressing cysts, in contrast, showed considerably less lumen expansion, indicating a reduced overall Na^+ conductance. In line with electrophysiological studies, this is probably caused by a combination of a reduction in single channel Na^+ conductance and a reduction in the overall number of channels [40].

3.4. Protein Structure

In order to determine the cause of the impairments brought about by single amino acid substitutions in claudin-10b, Klar et al. and Hadj-Rabia et al. provided 3D homology models of N48K or S131L, respectively, based on the crystal structure of murine claudin-15 [39,40,43]. The N48K mutation, localized in ECS1, appears to disrupt an intramolecular bridging between different backbones in ECS1 and membrane–ECS1-transition. Moreover, a potential electrostatic interaction could be disturbed by the replacement of an uncharged residue by a positively charged one. These alterations are presumed to perturb protein interaction and function indirectly [40]. The S131L mutation analyzed by Hadj-Rabia et al. affects TM3 and is suggested to clash sterically with several of the surrounding residues, especially of TM1 and -2 and by this perturbing the compactness of the helical bundle [39].

Furthermore, an intrahelical stabilizing hydrogen bond is lost by the amino acid substitution. As a consequence of the helical bundle destabilization, the S131L variant fails to insert into the cell membrane and is retained in the cytosol.

The impact of the amino acid substitutions D73N, R80G and P149R on protein folding or pore formation remain temporarily unsolved, also because functional data on the particular variants are scarce or absent. In general, substitution of a charged residue by an uncharged one in ECS1 (D73N, R80G) is considered critical for ion pore formation. On the other hand, neither D73 nor R80 have been shown to be important for charge selectivity of pore-forming claudins (for review see [44]). However, R80 is localized at the predicted transition between ECS1 and TM2 and might be crucial for the formation of the helical bundle. P149 in ECS2 is conserved in the majority of claudins. It is suggested to stabilize ECS2 conformation and to play a role in correct TJ strand arrangement [45].

3.5. *Short Summary of Functional Analyses*

Taken together, claudin-10b variants with truncations in one of the transmembrane segments necessarily show a complete loss of function as they are not inserted into the plasma membrane whereas variants with point mutations in one of the loops often keep a certain residual function. Nevertheless, localization, interaction and channel function can be impaired to a certain extent. Point mutations in transmembrane segments still have a high probability to severely affect membrane localization, probably depending on the substituting residue. The comparatively poor viability of cells expressing claudin-10b mutant proteins compared to wildtype-expressing cells and the increased degradation of mutant proteins due to suboptimal folding and localization as observed for several variants appear to be a common incidence and presumably contribute pivotally to the harmfulness of mutations [38–40].

Of course, the question arises to which extent site and type of claudin-10b mutation determine the clinical outcome. The most prominent differences between patient groups with particular *CLDN10* variants concern plasma/serum potassium values and the dermatological phenotype. However, with the actual number of patients and the information available it is difficult to assess a possible correlation and future studies will probably help to clarify that issue.

4. Mechanisms of Disease

The naturally occurring claudin-10 mutations discovered during the last years result in a complete or partial loss of function of the claudin-10b isoform and subsequently in a reduced or absent paracellular Na^+ permeation in the TAL as well as in sweat, salivary and lacrimal glands. Figures 3 and 4 illustrate the mechanistic principles, particularly the role of claudn-10b in epithelial transport, underlying a correct organ function.

4.1. *Kidney*

As mentioned above, the exact role of claudin-10a in the proximal tubule is not understood, as yet. Based on cell culture experiments, a function as paracellular anion channel is assumed (Figure 3A). However, the data presented in the reviewed studies indicate a rather minor role in the pathogenesis of claudin-10 defects. It appears likely that possible impairments affecting the proximal tubule can be distally compensated to a certain degree as it was shown for the loss of claudin-2 in a mouse model [46]. After all, future studies will have to clarify the physiological relevance of claudin-10a.

In the TAL, half of the Na^+ is reabsorbed paracellularly via the claudin-10b paracellular channel as a consequence of transcellular net uptake of Cl^- involving NKCC2 (Figure 3B). In TAL of the ISOM, claudin-10b dominates the TJ, although claudin-3, -16 and -19 are expressed intracellularly. Towards OSOM and cortex, an additional epithelial cell type occurs, expressing claudin-3, -16 and -19 but no claudin-10b. These cells form TJ complexes that are spatially separated from claudin-10b TJs and are involved in the reabsorption of Mg^{2+} and Ca^{2+}. The residual TJs assembled exclusively by claudin-10b enable a backflux of Na^+ into the lumen, due to its concentration gradient, thus adding to the lumen-positive potential and supporting the paracellular reabsorption of Mg^{2+} and Ca^{2+} [21,22,47].

An impaired claudin-10b function would reduce the Na^+ reabsorption in the TAL by the paracellular portion and lead to a compensatorily increased electrogenic Na^+ reabsorption via the epithelial sodium channel (ENaC) in the more distal nephron segments. This, in turn, would promote K^+ and H^+ loss. Whereas hyperaldosteronism, hypokalemia and metabolic alkalosis in patients with pathogenic claudin-10b variants are the consequence of compensatory mechanisms in the distal nephron, hypermagnesemia and hypocalciuria can be attributed to an exaggerated paracellular reabsorption of Mg^{2+} and Ca^{2+} in the TAL. In the mouse model described by Breiderhoff et al., the lack of claudin-10b in the TAL results in an increased distribution of the TJ complex containing claudin-16 and -19 over all TJs, including those in the ISOM, where normally only claudin-10b is constituting the TJ [22,25,26,48]. If a similar process takes place in men with *CLDN10* variants that lead to a reduced claudin-10b insertion into TAL TJs, it would explain the hypermagnesemia, in line with the mouse model. Apparently, in patients, the reabsorption of Mg^{2+} and Ca^{2+} is shifted from cortex/OSOM TAL to ISOM TAL and the lack of claudin-10b-based Na^+ backflux as additional driving force in cortex/OSOM TAL is thereby overcompensated.

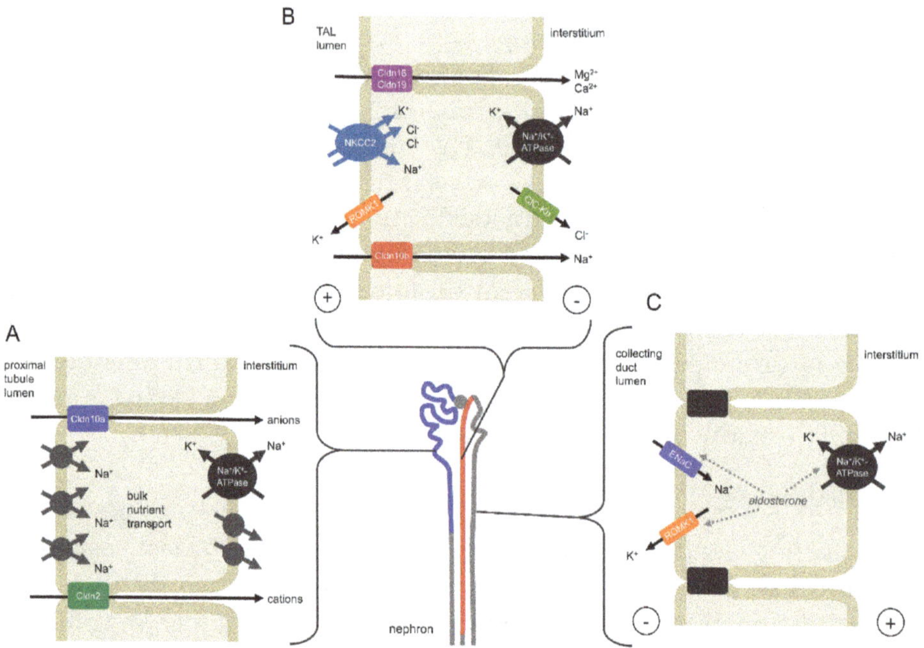

Figure 3. Mechanisms of claudin-10 function in the kidney. (**A**) Claudin-10a is expressed in the renal proximal tubule to the S3 segment (shown in blue) where it forms a paracellular anion pathway. Patients with pathogenic variants of both claudin-10a and claudin-10b appear to have no clinical manifestations different from patients with only claudin-10b variants, suggesting a minor role of an impaired claudin-10a function and/or compensation in more distal nephron segments. (**B**) Claudin-10b is mainly localized in the thick ascending limb of Henle's loop (TAL, shown in red) where it facilitates the paracellular reabsorption of Na^+. Na^+, 2 Cl^- and K^+ enter the epithelial cell by the secondary active NKCC2 transporter. Whereas Na^+ and Cl^- leave the cell basolaterally, K^+ is recirculated via the ROMK1 channel. This results in a net reabsorption of Cl^- and causes a lumen-positive transepithelial potential, which in turn drives paracellular Na^+ reabsorption via claudin-10b. Towards the distal TAL segment the expression of claudin-16 and -19 is involved in the paracellular reabsorption of Mg^{2+} and Ca^{2+}. Here, Na^+ can backflux into the lumen along its concentration gradient, thus adding to the lumen-positive potential as a driving force for Mg^{2+}- and Ca^{2+} reabsorption. An impairment of claudin-10b function would result in a NaCl wasting in the TAL and an expansion of claudin-16 and -19 over all TJs, causing hypermagnesemia and hypocalciuria. (**C**) Na^+ wasting in the TAL is compensated in the more distal nephron involving electrogenic transport via ENaC, promoting secretion of K^+ and causing hypokalemia.

However, the reasons for the magnitude of hypokalemia and hypermagnesemia depending on the patients' age remain unsolved.

4.2. Glands

In glands, a concerted secretion of ions and water across the glandular epithelium is required in order to produce sweat, saliva or lacrimal fluid, respectively (Figure 4). Molecular mechanisms in the secretory coil of sweat glands as well as in salivary and lacrimal acinar cells are partially understood and share certain mechanistic similarities [49–56]. Stimulation of G-protein-coupled muscarinergic receptors by acetylcholine results in an increased intracellular concentration of free Ca^{2+}, which activates apical Cl^- channels and basolateral K^+ channels. Subsequent activation of NKCC1 in the basolateral membrane leads to an influx of Na^+, K^+ and Cl^- into secretory cells. Na^+ and K^+ circulate across the basolateral membrane via NKCC1, the Na^+/K^+-ATPase and/or K^+ channels, respectively. The transcellular net flux of Cl^- via apical channels and basolateral NKCC1 generates a lumen-negative transepithelial potential and drives the paracellular secretion of Na^+ via claudin-10b channels. This is followed by transcellular water secretion involving aquaporin-5 water channels. Although an additional paracellular water secretion is frequently suggested, the strict water impermeability of claudin-10b-based TJs [13] and the implausibility of claudin-10b to form heterogeneous TJs with other claudins [22] rather argue against it.

An impairment of claudin-10b distribution or function as reported by Klar et al. would result in a complete abrogation of Na^+ and subsequent water secretion in sweat, salivary and lacrimal glands and become manifest in hypohidrosis, aptyalism and alacrima as observed in patients [40].

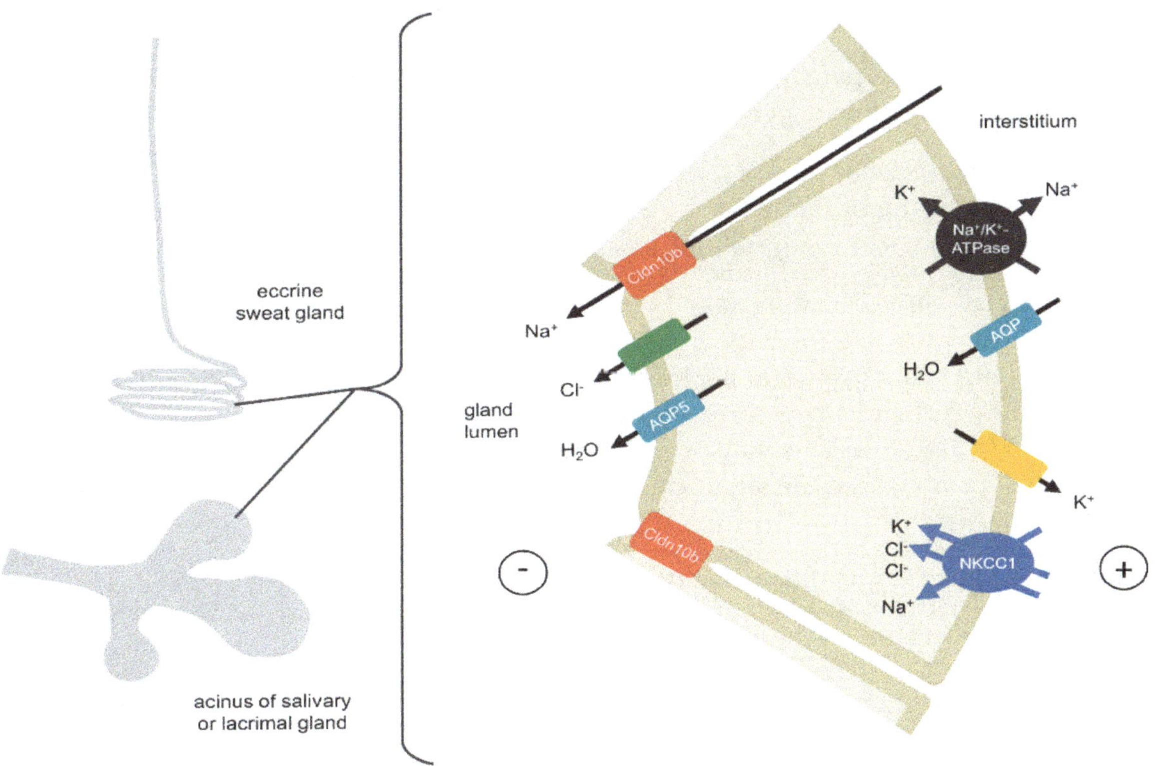

Figure 4. Mechanisms of claudin-10b function in the secretory portion of sweat glands and the acinar cells of salivary and lacrimal glands. During secretion, apical Cl^- channels and basolateral K^+ channels are activated. Na^+, 2 Cl^- and K^+ enter the cell via the basolaterally expressed NKCC1 transporter. Cl^- is secreted into the lumen whereas Na^+ and K^+ leave the cell basolaterally. The secretion of Cl^- drives Na^+ transport via the paracellular claudin-10b channel and transcellular water transport via aquaporins. An abrogation of the claudin-10b-mediated Na^+ transport would result in a complete dysfunction of sweat, saliva and tear secretion.

4.3. Skin

At present it remains unsolved whether dermatological manifestations such as ichthyosis and dry skin are a mere consequence of sweat gland dysfunction or may also be augmented by defective claudin-10 proteins in the epidermis. The mechanical barrier function of the skin is maintained on the one hand by the stratum corneum containing dead, cornified cells (corneocytes) and by the TJs on the other hand (for review see [57]). Functional TJs are localized in the stratum granulosum where they form a liquid–liquid interface barrier. The role of claudins in epidermal barrier function is illustrated by the phenotype of patients with pathogenic variants of the *CLDN1* gene. Patients lacking functional claudin-1 suffer from NISCH syndrome (neonatal ichthyosis–sclerosing cholangitis). The thick, scaly skin of ichthyosis is assumed to result from the skin compensating for barrier dysfunction [58]. On the other hand, claudins are expressed not only in the stratum granulosum but in all living layers of the epidermis and are probably involved in the differentiation of stratum granulosum cells to corneocytes.

Claudin-10 mRNA was detected in human skin biopsies and was found to be downregulated in patients with psoriasis vulgaris [59–61]. Immunohistochemical staining of rodent skin revealed an intracellular localization in the stratum basale [20] or in the stratum corneum [62]. The finding that claudin-10 is expressed beyond the stratum granulosum points to a possible function aside from its role as an ion pathway in the epidermis. However, its particular function in the skin has yet to be unraveled.

5. Summary

The four studies of Bongers et al., Klar et al., Hadj-Rabia et al. and Meyers et al. describe a novel claudinopathy that is based on mutations in the *CLDN10* gene and is characterized by an impaired function of mainly claudin-10b. Patients show a salt-losing nephropathy without hypercalciuria (as in Bartter syndrome) or hypomagnesemia (as in Gitelman syndrome). Main symptoms probably occurring in the majority of present and future patients are:

- age-dependent hypermagnesemia, age-dependent hypokalemia, hypocalciuria;
- hypohidrosis, xerostomia, alacrima;
- possibly dermatological abnormalities.

The severity of clinical manifestations may depend on site and type of mutation. Patients expressing a claudin-10b variant with a defect in one of the transmembrane regions, especially a truncation, are at high risk to develop the full severity of the phenotype. In this respect, genotyping of future patients may further help to correlate mutation type and course of disease. Further functional analysis of found *CLDN10* variants may provide a more detailed insight into the protein function. The reviewed studies demonstrate the significance of proper claudin-10b function for kidney, skin and gland physiology.

Acknowledgments: I thank Nina Himmerkus (Institute of Physiology, Kiel University, Kiel, Germany) and Tom Nijenhuis (Radboud University Medical Center, Nijmegen, The Netherlands) for reading the manuscript draft and for helpful discussions and suggestions.

References

1. Van Itallie, C.; Rahner, C.; Anderson, J.M. Regulated expression of claudin-4 decreases paracellular conductance through a selective decrease in sodium permeability. *J. Clin. Invest.* **2001**, *107*, 1319–1327. [CrossRef] [PubMed]
2. Furuse, M.; Furuse, K.; Sasaki, H.; Tsukita, S. Conversion of zonulae occludentes from tight to leaky strand type by introducing claudin-2 into Madin-Darby canine kidney I cells. *J. Cell Biol.* **2001**, *153*, 263–272. [CrossRef] [PubMed]
3. Furuse, M.; Hata, M.; Furuse, K.; Yoshida, Y.; Haratake, A.; Sugitani, Y.; Noda, T.; Kubo, A.; Tsukita, S. Claudin-based tight junctions are crucial for the mammalian epidermal barrier: a lesson from claudin-1-deficient mice. *J. Cell Biol.* **2002**, *156*, 1099–1111. [CrossRef] [PubMed]

4. Amasheh, S.; Meiri, N.; Gitter, A.H.; Schöneberg, T.; Mankertz, J.; Schulzke, J.D.; Fromm, M. Claudin-2 expression induces cation-selective channels in tight junctions of epithelial cells. *J. Cell. Sci.* **2002**, *115*, 4969–4976. [CrossRef] [PubMed]
5. Colegio, O.R.; Van Itallie, C.M.; McCrea, H.J.; Rahner, C.; Anderson, J.M. Claudins create charge-selective channels in the paracellular pathway between epithelial cells. *Am. J. Physiol., Cell Physiol.* **2002**, *283*, C142–147. [CrossRef]
6. Nitta, T.; Hata, M.; Gotoh, S.; Seo, Y.; Sasaki, H.; Hashimoto, N.; Furuse, M.; Tsukita, S. Size-selective loosening of the blood-brain barrier in claudin-5-deficient mice. *J. Cell Biol.* **2003**, *161*, 653–660. [CrossRef]
7. Van Itallie, C.M.; Fanning, A.S.; Anderson, J.M. Reversal of charge selectivity in cation or anion-selective epithelial lines by expression of different claudins. *Am. J. Physiol. Ren. Physiol.* **2003**, *285*, F1078–1084. [CrossRef]
8. Yu, A.S.L.; Enck, A.H.; Lencer, W.I.; Schneeberger, E.E. Claudin-8 expression in Madin-Darby canine kidney cells augments the paracellular barrier to cation permeation. *J. Biol. Chem.* **2003**, *278*, 17350–17359. [CrossRef]
9. Amasheh, S.; Schmidt, T.; Mahn, M.; Florian, P.; Mankertz, J.; Tavalali, S.; Gitter, A.H.; Schulzke, J.-D.; Fromm, M. Contribution of claudin-5 to barrier properties in tight junctions of epithelial cells. *Cell Tissue Res.* **2005**, *321*, 89–96. [CrossRef]
10. Van Itallie, C.M.; Rogan, S.; Yu, A.; Vidal, L.S.; Holmes, J.; Anderson, J.M. Two splice variants of claudin-10 in the kidney create paracellular pores with different ion selectivities. *Am. J. Physiol. Ren. Physiol.* **2006**, *291*, F1288–F1299. [CrossRef]
11. Angelow, S.; Schneeberger, E.E.; Yu, A.S.L. Claudin-8 expression in renal epithelial cells augments the paracellular barrier by replacing endogenous claudin-2. *J. Membr. Biol.* **2007**, *215*, 147–159. [CrossRef] [PubMed]
12. Günzel, D.; Stuiver, M.; Kausalya, P.J.; Haisch, L.; Krug, S.M.; Rosenthal, R.; Meij, I.C.; Hunziker, W.; Fromm, M.; Muller, D. Claudin-10 exists in six alternatively spliced isoforms that exhibit distinct localization and function. *J. Cell. Sci.* **2009**, *122*, 1507–1517. [CrossRef] [PubMed]
13. Rosenthal, R.; Milatz, S.; Krug, S.M.; Oelrich, B.; Schulzke, J.-D.; Amasheh, S.; Günzel, D.; Fromm, M. Claudin-2, a component of the tight junction, forms a paracellular water channel. *J. Cell. Sci.* **2010**, *123*, 1913–1921. [CrossRef] [PubMed]
14. Milatz, S.; Krug, S.M.; Rosenthal, R.; Günzel, D.; Müller, D.; Schulzke, J.-D.; Amasheh, S.; Fromm, M. Claudin-3 acts as a sealing component of the tight junction for ions of either charge and uncharged solutes. *Biochim. Biophys. Acta* **2010**, *1798*, 2048–2057. [CrossRef] [PubMed]
15. Tamura, A.; Hayashi, H.; Imasato, M.; Yamazaki, Y.; Hagiwara, A.; Wada, M.; Noda, T.; Watanabe, M.; Suzuki, Y.; Tsukita, S. Loss of claudin-15, but not claudin-2, causes Na+ deficiency and glucose malabsorption in mouse small intestine. *Gastroenterology* **2011**, *140*, 913–923. [CrossRef] [PubMed]
16. Krug, S.M.; Günzel, D.; Conrad, M.P.; Rosenthal, R.; Fromm, A.; Amasheh, S.; Schulzke, J.D.; Fromm, M. Claudin-17 forms tight junction channels with distinct anion selectivity. *Cell. Mol. Life Sci.* **2012**, *69*, 2765–2778. [CrossRef]
17. Tanaka, H.; Yamamoto, Y.; Kashihara, H.; Yamazaki, Y.; Tani, K.; Fujiyoshi, Y.; Mineta, K.; Takeuchi, K.; Tamura, A.; Tsukita, S. Claudin-21 Has a Paracellular Channel Role at Tight Junctions. *Mol. Cell. Biol.* **2016**, *36*, 954–964. [CrossRef]
18. Rosenthal, R.; Günzel, D.; Piontek, J.; Krug, S.M.; Ayala-Torres, C.; Hempel, C.; Theune, D.; Fromm, M. Claudin-15 forms a water channel through the tight junction with distinct function compared to claudin-2. *Acta Physiol (Oxf)* **2019**, e13334. [CrossRef]
19. Inai, T.; Kamimura, T.; Hirose, E.; Iida, H.; Shibata, Y. The protoplasmic or exoplasmic face association of tight junction particles cannot predict paracellular permeability or heterotypic claudin compatibility. *Eur. J. Cell Biol.* **2010**, *89*, 547–556. [CrossRef]
20. Inai, T.; Sengoku, A.; Guan, X.; Hirose, E.; Iida, H.; Shibata, Y. Heterogeneity in expression and subcellular localization of tight junction proteins, claudin-10 and -15, examined by RT-PCR and immunofluorescence microscopy. *Arch. Histol. Cytol.* **2005**, *68*, 349–360. [CrossRef]
21. Plain, A.; Wulfmeyer, V.C.; Milatz, S.; Klietz, A.; Hou, J.; Bleich, M.; Himmerkus, N. Corticomedullary difference in the effects of dietary Ca^{2+} on tight junction properties in thick ascending limbs of Henle's loop. *Pflugers Arch.* **2016**, *468*, 293–303. [CrossRef] [PubMed]

22. Milatz, S.; Himmerkus, N.; Wulfmeyer, V.C.; Drewell, H.; Mutig, K.; Hou, J.; Breiderhoff, T.; Müller, D.; Fromm, M.; Bleich, M.; et al. Mosaic expression of claudins in thick ascending limbs of Henle results in spatial separation of paracellular Na^+ and Mg^{2+} transport. *Proc. Natl. Acad. Sci. USA* **2017**, *114*, E219–E227. [CrossRef] [PubMed]
23. Gong, Y.; Renigunta, V.; Himmerkus, N.; Zhang, J.; Renigunta, A.; Bleich, M.; Hou, J. Claudin-14 regulates renal Ca^{++} transport in response to CaSR signalling via a novel microRNA pathway. *EMBO J.* **2012**, *31*, 1999–2012. [CrossRef]
24. Lee, J.W.; Chou, C.-L.; Knepper, M.A. Deep Sequencing in Microdissected Renal Tubules Identifies Nephron Segment-Specific Transcriptomes. *J. Am. Soc. Nephrol.* **2015**, *26*, 2669–2677. [CrossRef] [PubMed]
25. Breiderhoff, T.; Himmerkus, N.; Stuiver, M.; Mutig, K.; Will, C.; Meij, I.C.; Bachmann, S.; Bleich, M.; Willnow, T.E.; Müller, D. Deletion of claudin-10 (Cldn10) in the thick ascending limb impairs paracellular sodium permeability and leads to hypermagnesemia and nephrocalcinosis. *Proc. Natl. Acad. Sci. U.S.A.* **2012**, *109*, 14241–14246. [CrossRef]
26. Seker, M.; Fernandez-Rodriguez, C.; Martinez-Cruz, L.A.; Müller, D. Mouse models of human claudin-associated disorders: benefits and limitations. *Int. J. Mol. Sci.* **2019**. submitted for publication.
27. Weber, S.; Hoffmann, K.; Jeck, N.; Saar, K.; Boeswald, M.; Kuwertz-Broeking, E.; Meij, I.I.; Knoers, N.V.; Cochat, P.; Suláková, T.; et al. Familial hypomagnesaemia with hypercalciuria and nephrocalcinosis maps to chromosome 3q27 and is associated with mutations in the PCLN-1 gene. *Eur. J. Hum. Genet.* **2000**, *8*, 414–422. [CrossRef]
28. Wilcox, E.R.; Burton, Q.L.; Naz, S.; Riazuddin, S.; Smith, T.N.; Ploplis, B.; Belyantseva, I.; Ben-Yosef, T.; Liburd, N.A.; Morell, R.J.; et al. Mutations in the gene encoding tight junction claudin-14 cause autosomal recessive deafness DFNB29. *Cell* **2001**, *104*, 165–172. [CrossRef]
29. Hadj-Rabia, S.; Baala, L.; Vabres, P.; Hamel-Teillac, D.; Jacquemin, E.; Fabre, M.; Lyonnet, S.; De Prost, Y.; Munnich, A.; Hadchouel, M.; et al. Claudin-1 gene mutations in neonatal sclerosing cholangitis associated with ichthyosis: a tight junction disease. *Gastroenterology* **2004**, *127*, 1386–1390. [CrossRef]
30. Thorleifsson, G.; Holm, H.; Edvardsson, V.; Walters, G.B.; Styrkarsdottir, U.; Gudbjartsson, D.F.; Sulem, P.; Halldorsson, B.V.; de Vegt, F.; d'Ancona, F.C.H.; et al. Sequence variants in the CLDN14 gene associate with kidney stones and bone mineral density. *Nat. Genet.* **2009**, *41*, 926–930. [CrossRef]
31. Konrad, M.; Schaller, A.; Seelow, D.; Pandey, A.V.; Waldegger, S.; Lesslauer, A.; Vitzthum, H.; Suzuki, Y.; Luk, J.M.; Becker, C.; et al. Mutations in the tight-junction gene claudin 19 (CLDN19) are associated with renal magnesium wasting, renal failure, and severe ocular involvement. *Am. J. Hum. Genet.* **2006**, *79*, 949–957. [CrossRef] [PubMed]
32. Kausalya, P.J.; Amasheh, S.; Günzel, D.; Wurps, H.; Müller, D.; Fromm, M.; Hunziker, W. Disease-associated mutations affect intracellular traffic and paracellular Mg2+ transport function of Claudin-16. *J. Clin. Invest.* **2006**, *116*, 878–891. [CrossRef] [PubMed]
33. Simon, D.B.; Karet, F.E.; Hamdan, J.M.; DiPietro, A.; Sanjad, S.A.; Lifton, R.P. Bartter's syndrome, hypokalaemic alkalosis with hypercalciuria, is caused by mutations in the Na-K-2Cl cotransporter NKCC2. *Nat. Genet.* **1996**, *13*, 183–188. [CrossRef] [PubMed]
34. Simon, D.B.; Karet, F.E.; Rodriguez-Soriano, J.; Hamdan, J.H.; DiPietro, A.; Trachtman, H.; Sanjad, S.A.; Lifton, R.P. Genetic heterogeneity of Bartter's syndrome revealed by mutations in the K+ channel, ROMK. *Nat. Genet.* **1996**, *14*, 152–156. [CrossRef] [PubMed]
35. Watanabe, S.; Fukumoto, S.; Chang, H.; Takeuchi, Y.; Hasegawa, Y.; Okazaki, R.; Chikatsu, N.; Fujita, T. Association between activating mutations of calcium-sensing receptor and Bartter's syndrome. *Lancet* **2002**, *360*, 692–694. [CrossRef]
36. Vargas-Poussou, R.; Huang, C.; Hulin, P.; Houillier, P.; Jeunemaître, X.; Paillard, M.; Planelles, G.; Déchaux, M.; Miller, R.T.; Antignac, C. Functional characterization of a calcium-sensing receptor mutation in severe autosomal dominant hypocalcemia with a Bartter-like syndrome. *J. Am. Soc. Nephrol.* **2002**, *13*, 2259–2266. [CrossRef]
37. Simon, D.B.; Nelson-Williams, C.; Bia, M.J.; Ellison, D.; Karet, F.E.; Molina, A.M.; Vaara, I.; Iwata, F.; Cushner, H.M.; Koolen, M.; et al. Gitelman's variant of Bartter's syndrome, inherited hypokalaemic alkalosis, is caused by mutations in the thiazide-sensitive Na-Cl cotransporter. *Nat. Genet.* **1996**, *12*, 24–30. [CrossRef]

38. Bongers, E.M.H.F.; Shelton, L.M.; Milatz, S.; Verkaart, S.; Bech, A.P.; Schoots, J.; Cornelissen, E.A.M.; Bleich, M.; Hoenderop, J.G.J.; Wetzels, J.F.M.; et al. A Novel Hypokalemic-Alkalotic Salt-Losing Tubulopathy in Patients with *CLDN10* Mutations. *JASN* **2017**, *28*, 3118–3128. [CrossRef]
39. Hadj-Rabia, S.; Brideau, G.; Al-Sarraj, Y.; Maroun, R.C.; Figueres, M.-L.; Leclerc-Mercier, S.; Olinger, E.; Baron, S.; Chaussain, C.; Nochy, D.; et al. Multiplex epithelium dysfunction due to CLDN10 mutation: the HELIX syndrome. *Genet Med* **2018**, *20*, 190–201. [CrossRef]
40. Klar, J.; Piontek, J.; Milatz, S.; Tariq, M.; Jameel, M.; Breiderhoff, T.; Schuster, J.; Fatima, A.; Asif, M.; Sher, M.; et al. Altered paracellular cation permeability due to a rare CLDN10B variant causes anhidrosis and kidney damage. *PLoS Genet* **2017**, *13*, e1006897. [CrossRef]
41. Meyers, N.; Nelson-Williams, C.; Malaga-Dieguez, L.; Kaufmann, H.; Loring, E.; Knight, J.; Lifton, R.P.; Trachtman, H. Hypokalemia Associated With a Claudin 10 Mutation: A Case Report. *Am. J. Kidney Dis.* **2019**, *73*, 425–428. [CrossRef] [PubMed]
42. Hata, M.; Kawamoto, T.; Kawai, M.; Yamamoto, T. Differential expression patterns of the tight junction-associated proteins occludin and claudins in secretory and mature ameloblasts in mouse incisor. *Med. Mol. Morphol.* **2010**, *43*, 102–106. [CrossRef] [PubMed]
43. Suzuki, H.; Nishizawa, T.; Tani, K.; Yamazaki, Y.; Tamura, A.; Ishitani, R.; Dohmae, N.; Tsukita, S.; Nureki, O.; Fujiyoshi, Y. Crystal structure of a claudin provides insight into the architecture of tight junctions. *Science* **2014**, *344*, 304–307. [CrossRef] [PubMed]
44. Milatz, S.; Breiderhoff, T. One gene, two paracellular ion channels-claudin-10 in the kidney. *Pflugers Arch.* **2017**, *469*, 115–121. [CrossRef] [PubMed]
45. Piontek, J.; Winkler, L.; Wolburg, H.; Müller, S.L.; Zuleger, N.; Piehl, C.; Wiesner, B.; Krause, G.; Blasig, I.E. Formation of tight junction: determinants of homophilic interaction between classic claudins. *FASEB J.* **2008**, *22*, 146–158. [CrossRef]
46. Pei, L.; Solis, G.; Nguyen, M.T.X.; Kamat, N.; Magenheimer, L.; Zhuo, M.; Li, J.; Curry, J.; McDonough, A.A.; Fields, T.A.; et al. Paracellular epithelial sodium transport maximizes energy efficiency in the kidney. *J. Clin. Invest.* **2016**, *126*, 2509–2518. [CrossRef]
47. Shan, Q.; Himmerkus, N.; Hou, J.; Goodenough, D.A.; Bleich, M. Insights into driving forces and paracellular permeability from claudin-16 knockdown mouse. *Ann. N. Y. Acad. Sci.* **2009**, *1165*, 148–151. [CrossRef]
48. Breiderhoff, T.; Himmerkus, N.; Drewell, H.; Plain, A.; Günzel, D.; Mutig, K.; Willnow, T.E.; Müller, D.; Bleich, M. Deletion of claudin-10 rescues claudin-16-deficient mice from hypomagnesemia and hypercalciuria. *Kidney Int.* **2018**, *93*, 580–588. [CrossRef]
49. Murota, H.; Matsui, S.; Ono, E.; Kijima, A.; Kikuta, J.; Ishii, M.; Katayama, I. Sweat, the driving force behind normal skin: an emerging perspective on functional biology and regulatory mechanisms. *J. Dermatol. Sci.* **2015**, *77*, 3–10. [CrossRef]
50. Cui, C.-Y.; Schlessinger, D. Eccrine sweat gland development and sweat secretion. *Exp. Dermatol.* **2015**, *24*, 644–650. [CrossRef]
51. Roussa, E. Channels and transporters in salivary glands. *Cell Tissue Res.* **2011**, *343*, 263–287. [CrossRef] [PubMed]
52. Baker, O.J. Current trends in salivary gland tight junctions. *Tissue Barriers* **2016**, *4*, e1162348. [CrossRef] [PubMed]
53. Proctor, G.B. The physiology of salivary secretion. *Periodontol. 2000* **2016**, *70*, 11–25. [CrossRef] [PubMed]
54. Tsubota, K.; Hirai, S.; King, L.S.; Agre, P.; Ishida, N. Defective cellular trafficking of lacrimal gland aquaporin-5 in Sjögren's syndrome. *Lancet* **2001**, *357*, 688–689. [CrossRef]
55. Walcott, B.; Birzgalis, A.; Moore, L.C.; Brink, P.R. Fluid secretion and the Na+-K+-2Cl- cotransporter in mouse exorbital lacrimal gland. *Am. J. Physiol. Cell Physiol.* **2005**, *289*, C860–C867. [CrossRef]
56. Rocha, E.M.; Alves, M.; Rios, J.D.; Dartt, D.A. The aging lacrimal gland: changes in structure and function. *Ocul Surf* **2008**, *6*, 162–174. [CrossRef]
57. Yokouchi, M.; Kubo, A. Maintenance of tight junction barrier integrity in cell turnover and skin diseases. *Exp. Dermatol.* **2018**, *27*, 876–883. [CrossRef]
58. Segre, J.A. Epidermal barrier formation and recovery in skin disorders. *J. Clin. Invest.* **2006**, *116*, 1150–1158. [CrossRef]

59. Brandner, J.M.; Kief, S.; Grund, C.; Rendl, M.; Houdek, P.; Kuhn, C.; Tschachler, E.; Franke, W.W.; Moll, I. Organization and formation of the tight junction system in human epidermis and cultured keratinocytes. *Eur. J. Cell Biol.* **2002**, *81*, 253–263. [CrossRef]
60. Suárez-Fariñas, M.; Fuentes-Duculan, J.; Lowes, M.A.; Krueger, J.G. Resolved psoriasis lesions retain expression of a subset of disease-related genes. *J. Invest. Dermatol.* **2011**, *131*, 391–400. [CrossRef]
61. Kast, J.I.; Wanke, K.; Soyka, M.B.; Wawrzyniak, P.; Akdis, D.; Kingo, K.; Rebane, A.; Akdis, C.A. The broad spectrum of interepithelial junctions in skin and lung. *J. Allergy Clin. Immunol.* **2012**, *130*, 544–547.e4. [CrossRef] [PubMed]
62. Troy, T.-C.; Rahbar, R.; Arabzadeh, A.; Cheung, R.M.-K.; Turksen, K. Delayed epidermal permeability barrier formation and hair follicle aberrations in Inv-Cldn6 mice. *Mech. Dev.* **2005**, *122*, 805–819. [CrossRef] [PubMed]

3

Structural Features of Tight Junction Proteins

Udo Heinemann [1,*] and Anja Schuetz [2,*]

1 Macromolecular Structure and Interaction Laboratory, Max Delbrück Center for Molecular Medicine, 13125 Berlin, Germany
2 Protein Production & Characterization Platform, Max Delbrück Center for Molecular Medicine, 13125 Berlin, Germany
* Correspondence: heinemann@mdc-berlin.de (U.H.); anja.schuetz@mdc-berlin.de (A.S.);

Abstract: Tight junctions are complex supramolecular entities composed of integral membrane proteins, membrane-associated and soluble cytoplasmic proteins engaging in an intricate and dynamic system of protein–protein interactions. Three-dimensional structures of several tight-junction proteins or their isolated domains have been determined by X-ray crystallography, nuclear magnetic resonance spectroscopy, and cryo-electron microscopy. These structures provide direct insight into molecular interactions that contribute to the formation, integrity, or function of tight junctions. In addition, the known experimental structures have allowed the modeling of ligand-binding events involving tight-junction proteins. Here, we review the published structures of tight-junction proteins. We show that these proteins are composed of a limited set of structural motifs and highlight common types of interactions between tight-junction proteins and their ligands involving these motifs.

Keywords: tight junction; protein structure; protein domain; claudins; occludin; tricellulin; junctional adhesion molecule; *zonula occludens*; MAGUK proteins; PDZ domain

---

1. Introduction

A classical paper published more than half a century ago [1] clearly demonstrated that the epithelia of several glands and cavity-forming internal organs of the rat and guinea pig all share characteristic tripartite junctional complexes between adjacent cells. These junctional complexes were found in the epithelia of the stomach, intestine, gall bladder, uterus, oviduct, liver, pancreas, parotid, thyroid, salivary ducts, and kidney. Progressing from the apical to the basal side of the endothelial cell layer, the elements of the junctional complex were characterized as tight junctions (*zonulae occludens*), adherens junctions, and desmosomes. As most apical elements of the junctional complex, tight junctions (TJs) were distinct by the apparent fusion of adjacent cell membranes over variable distances and appeared as a diffuse band of dense cytoplasmic material in the electron microscope. TJs formed a continuous belt-like structure, whereas desmosomes displayed discontinuous button-like structures, and adherens junctions (AJs) were intermediate in appearance. The molecular composition of TJs was revealed in subsequent work by many laboratories, e.g., [2–5], and shown to include at least 40 different proteins.

In the pioneering work of Farquhar and Palade, TJs were proposed to function as effective diffusion barriers or seals [1]. The sealing function of TJs contributes to the formation and physiological function of the blood-brain barrier (BBB), which consists of endothelial cells sealed by apical junctional complexes including TJs. Functions of transmembrane TJ proteins at the BBB are well documented [6–8]. BBB dysfunction is linked to a number of diseases including multiple sclerosis, stroke, brain tumors, epilepsy, and Alzheimer's disease [9,10].

Here, we review the current literature regarding three-dimensional structures of TJ proteins, their domains and intermolecular interactions. We do not primarily aim at presenting each and every structure in detail, but attempt to distill common structural principles that underlie the architecture of the TJ. We apologize to those authors whose work may not have been covered in this paper for reasons of space and readability.

2. Structural Insight into Tight-Junction Proteins, Their Domains, and Interactions

In the most general, birds-eye description, the TJ consists of a set of transmembrane (TM) proteins and the cytoplasmic plaque, a complex network of scaffolding and effector proteins that connects the TM proteins to the actomyosin cytoskeleton of the cell (Figure 1). The TM proteins interact with their extracellular domains in the paracellular space, and the connection to the cytoskeleton inside the cell is structurally as yet uncharacterized [2–5,11,12]. The *zonula occludens* proteins ZO-1, ZO-2, and ZO-3 and the two mammalian polarity complexes PAR-3/PAR-6/aPKC and Crumbs/ PALS1/PATJ are central players of the cytoplasmic plaque and are described in more detail in this review together with the transmembrane TJ proteins.

2.1. *Tight-Junction Transmembrane Proteins*

TJ transmembrane proteins contain either one, three, or four TM segments. The Crumbs proteins (CRBs), the junctional adhesion molecules (JAMs), the angulin proteins, and the coxsackievirus–adenovirus receptor (CAR) are representatives of single-span TJ membrane proteins. BVES (blood-vessel epicardial substance, also known as POPDC1 for Popeye domain-containing protein-1) is a TJ-associated protein with three TM regions [13,14]. The claudins and the TAMPs (tight junction-associated MARVEL-domain proteins) occludin (MARVELD1), tricellulin (MARVELD2), and MARVELD3 are tetra-span TM proteins. MARVEL is used as a common acronym for MAL (myelin and lymphocyte) and related proteins for vesicle trafficking and membrane link [15]. Where crystal structures are available, for example the claudin family (Section 2.1.2, [16]), the TM segments were shown to be α-helical.

2.1.1. Junctional Adhesion Molecules and Other Ig-Like TJ Proteins

The JAMs are a family of adhesion molecules with immunoglobulin (Ig)-like ectodomains, localized in epithelial and endothelial cells, leukocytes, and myocardial cells [17]. The 2.5-Å crystal structure of the soluble extracellular part of mouse JAM-A provided the first structural insight into a TJ transmembrane protein [18]. In this structure, two Ig domains are connected by a short linker peptide, and a U-shaped dimer is formed by symmetrical interaction of the N-terminal Ig domains (Figure 2). This structure provided the basis for a model of homophilic interactions between the N-domains to explain the adhesive function of JAMs in the TJ. The crystal structure of the extracellular Ig domains of coxsackievirus–adenovirus receptor CAR, another component of the epithelial apical junction complex that is essential for TJ integrity [19], suggests a very similar mode of CAR homodimer formation through symmetrical interaction of its N-terminal Ig domain [20].

The extracellular portions of JAMs serve as viral attachment sites. Reoviruses attach to human cells by binding to cell–surface carbohydrates and the junctional adhesion molecule JAM-A. The crystal structure of reovirus attachment protein σ1 bound to the soluble form of JAM-A shows that σ1 disrupts the native JAM-A dimer to form a heterodimer via the same interface as used in JAM-A homodimers, but with a 1000-fold lower dissociation constant of the σ1/JAM-A heterodimer as compared to the JAM-A homodimer [21]. In cat, infection with calicivirus is initiated by binding of the minor capsid protein VP2 to feline junctional adhesion molecule A (JAM-A). High-resolution cryo-EM structures of VP2 and soluble JAM-A-decorated VP2 show formation of a large portal-like assembly, which is hypothesized to serve as a channel for the transfer of the viral genome [22].

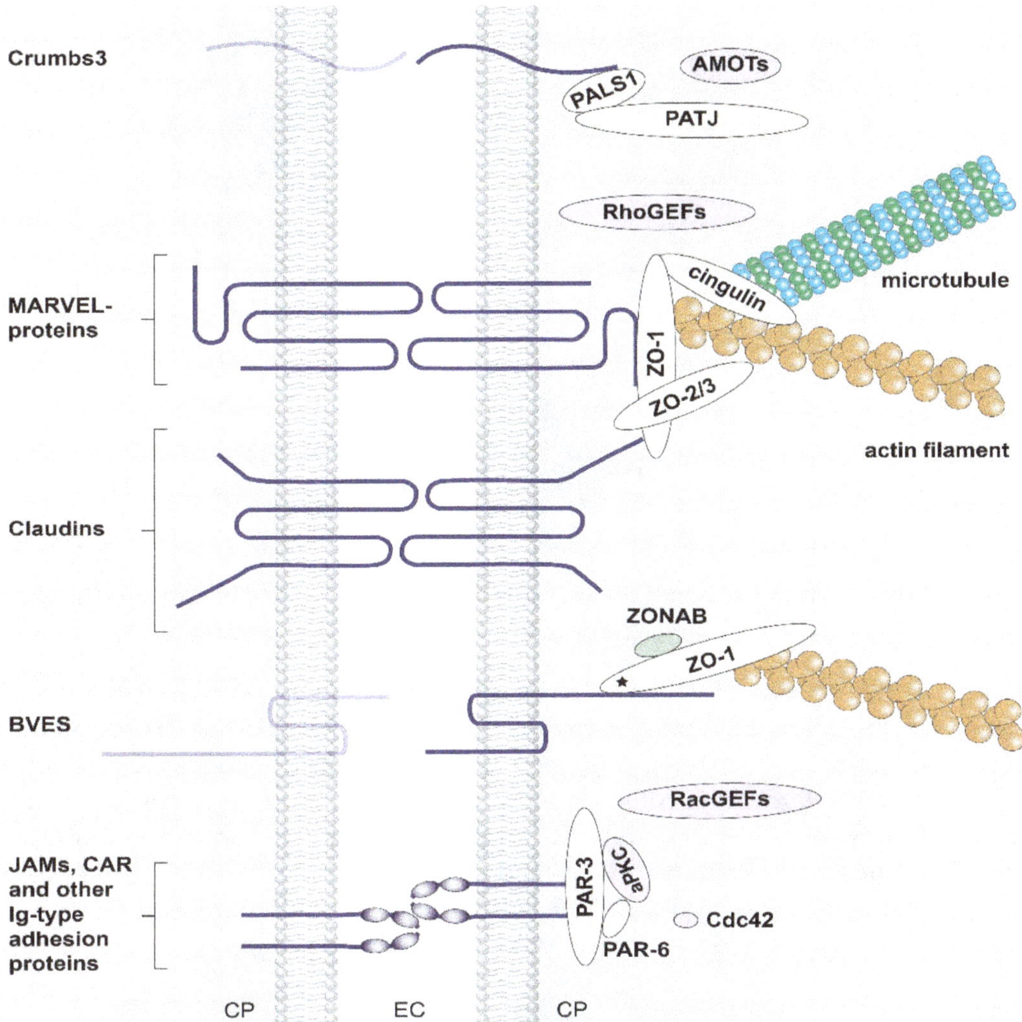

Figure 1. The tight-junction core structure. TM proteins of the TJ (dark blue) interact with a complex cytoplasmic protein network, the cytoplasmic plaque (shown on the right), providing a physical link to the cytoskeleton (microtubules, actin filaments). Cross-membrane interactions between TM proteins are indicated schematically. For some TM proteins (shown in pale blue) there is no direct evidence for a *trans* pairing interaction between TM proteins of opposing cells. The cytoplasmic plaque is composed of scaffolding proteins (yellow ovals) that associate with signaling proteins (purple ovals) and (post)transcriptional regulators (green oval), forming the zonular signalosome [23]. The three major protein complexes located in the cytoplasmic plaque are depicted. Within the ZO complex, ZO proteins are present as homodimers or ZO-1/ZO-2 and ZO-1/ZO-3 heterodimers [24] that directly associate with integral TJ membrane proteins through multiple interactions. The polarity complexes PAR-3/PAR-6/aPKC and Crumbs/PALS1/PATJ are responsible for the development of the apico-basal axis of epithelial cells and act as apical components of TJs. TM proteins: Crumbs homolog 3 (CRB3); MARVEL-domain containing proteins occludin, tricellulin, and MARVEL domain-containing protein 3 (MARVELD3); the claudins; the protein blood vessel epicardial substance (BVES); immunoglobulin (Ig) superfamily members such as junctional adhesion molecules (JAMs) and the coxsackievirus–adenovirus receptor (CAR). Cytoplasmic scaffolding proteins: *Zonula occludens* (ZO) proteins ZO-1, ZO-2, and ZO-3; partitioning defective 3/6 homologs (PAR-3, PAR-6); protein associated with Lin-7 1 (PALS1); PALS1-associated tight junction (PATJ) protein; cytoskeletal linker cingulin. Signaling proteins: Atypical protein kinase C (aPKC); proteins of the angiomotin family (AMOTs) [25]; the small Rho-GTPase Cdc42, and guanine nucleotide exchange factors for the Rho-GTPases RhoA (RhoGEFs, e.g., ARHGEF11 [26]) and Rac1 (RacGEFs, e.g., Tiam-1 [27]), respectively. Transcriptional regulator: ZO-1–associated nucleic acid-binding protein (ZONAB, YBX3 in human) [28]. Figure modified and updated after Zihni et al. [12].

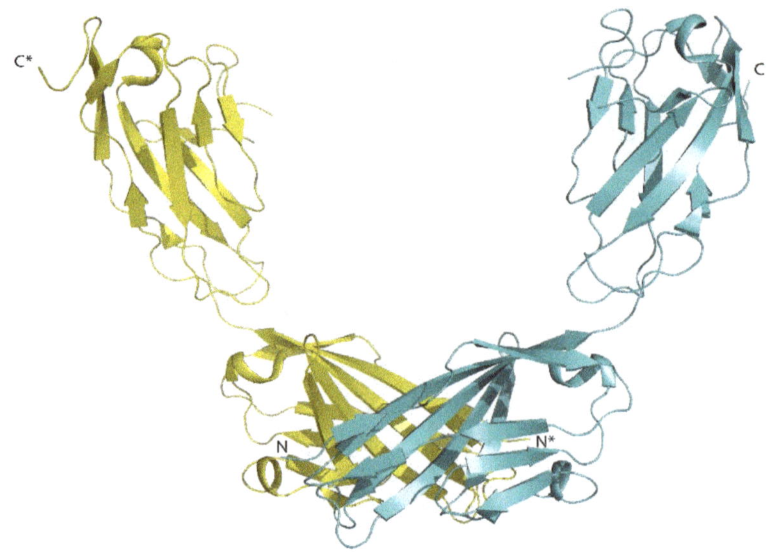

Figure 2. JAM-A dimerization via extracellular Ig domains. Crystal structure of murine JAM-A (PDB entry 1F97) [18]. The dimer is generated by crystallographic two-fold symmetry.

In addition to JAMs and CAR, other single-span Ig-like adhesion molecules such as the endothelial cell-selective adhesion molecule (ESAM), the coxsackievirus and adenovirus receptor-like membrane protein (CLMP), the brain- and testis-specific immunoglobulin superfamily protein (BT-IgSF or IgSF11) [19], and the angulin family of proteins are present at TJs. The latter comprise the proteins LSR (lipolysis-stimulated lipoprotein receptor), ILDR1, and ILDR2 (Ig-like domain-containing receptors) that complement each other at tricellular TJs and co-operate with tricellulin to mediate full barrier function in epithelial sheets [29,30]. Loss of LSR is linked to cell invasion and migration in human cancer cells [31]. To date, however, no structural data are available for any of these proteins.

2.1.2. Claudins

Members of the claudin family are the most abundant TM proteins of the TJ [32]. Claudin genes are expressed across all epithelial tissues, and in all epithelia various different claudins are expressed at the same time [12,33–36]. Tissue-specific expression of claudin genes has been documented, for example, for claudins in the kidney, inner ear, and eye [33].

At TJs, claudins are arranged to form extended strands by homophilic or heterophilic *cis* pairing within the same membrane or *trans* pairing across membranes [37,38]. In humans, 23 claudins and two claudin-like proteins are currently known (Figure 3). Although clearly homologous, the claudins share only a small number of strictly conserved residues and differ in the lengths of their N- and C-termini and the loops connecting their four TM helices. Highest sequence conservation is observed within the first extracellular loop (ECL1) where a tryptophan and two cysteine residues are strictly conserved in all sequences, suggesting formation of a disulfide bond in this region, which was experimentally verified by crystal structure analysis [16]. Various schemes for grouping claudins have been proposed. Based on sequence conservations, a grouping into classical claudins (1–10, 14, 15, 17, 19) and non-classical claudins (11–13, 16, 18, 20–24) was suggested [39], but the alignment shown in Figure 3 does not clearly support a separation into these groups. Based on function within the TJ, claudins may be grouped according to their barrier or channel forming properties with respect to different solutes [34,35,40]. Claudins 1, 3, 5, 11, 14, and 19, for example, have been characterized as predominantly sealing, whereas claudins 2, 10a, 10b, 15, and 17 were described as predominantly channel forming. For claudins 4, 7, 8, and 16 a sealing or channel-forming function has not been unequivocally determined [34]. Moreover, assignment of these functions to TM proteins of the TJ may be difficult when individual TJ proteins are functionally replaced by paralogs in certain epithelia or in the presence of post-translational modification.

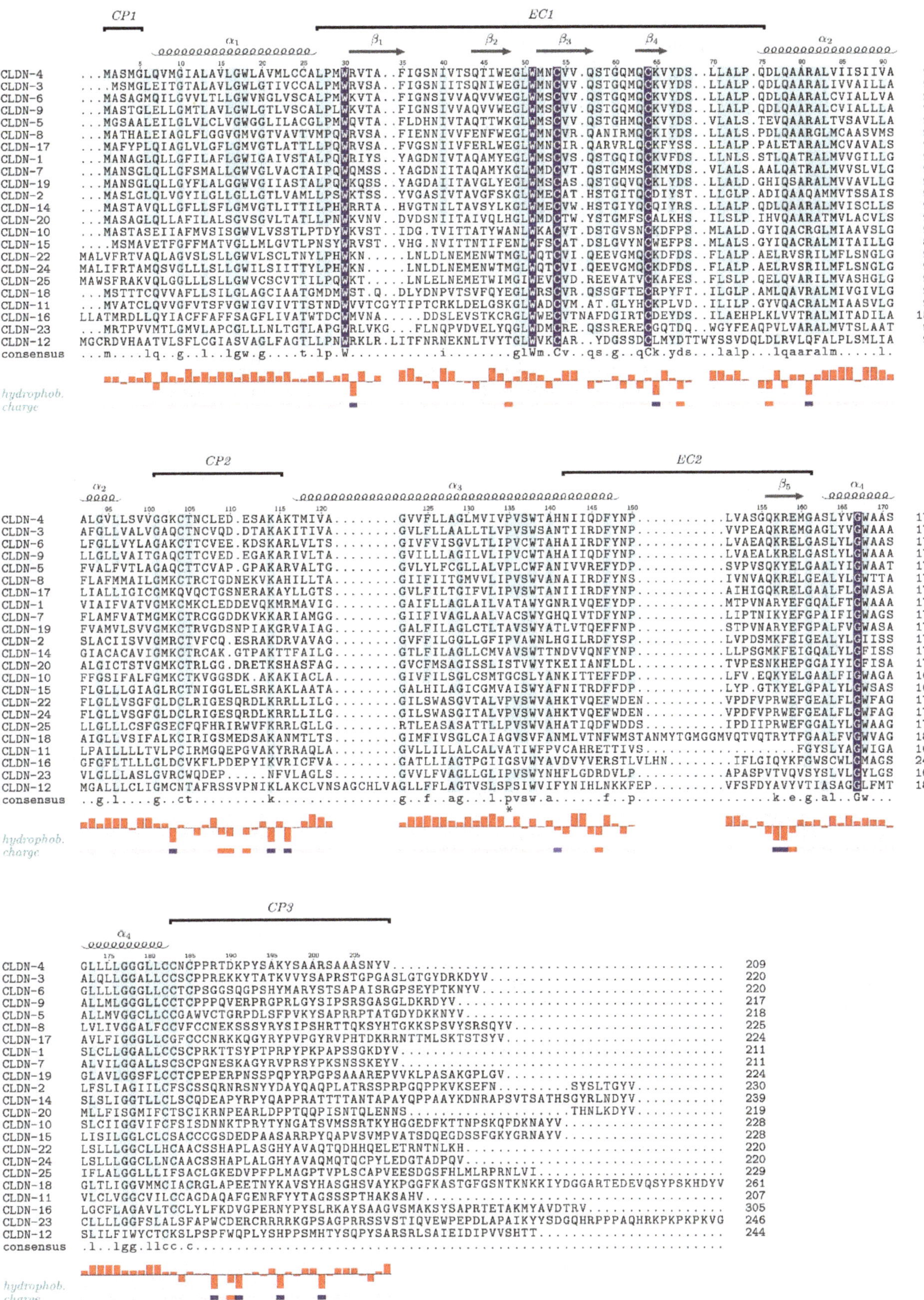

Figure 3. Sequence alignment of human claudins. The Uniprot database [41] lists 23 human claudin sequences belonging to CLDN-1 to CLDN-12 and CLDN-14 to CLDN-25; CLDN-21 is sometimes referred to as CLDN-24. In addition, there are two claudin domain-containing proteins (CLDND1 und

CLDND2), which are not included in the alignment. 66 N-terminal residues of CLDN-16 and 46 C-terminal residues of CLDN-23 are omitted from the alignment, because they have no match in any other human claudin sequence. Domain and secondary-structure annotation follows CLDN-4 for which a crystal structure is known in the presence of a bound toxin and loop EC1 is fully ordered [16,42], and the claudin sequences are listed in order of their match with the CLDN-4 sequence. Residues conserved across all human claudins are highlighted on dark blue background and residues conserved in ≥ 50% of the sequences are shown on a light blue background. EC: Extracellular, CP: Cytoplasmic. Conservation of hydrophobicity and charge (blue, positive; red, negative) is indicated at the bottom of the alignment. The asterisk marks a proline residue within α3 of claudin-3, which induces a kink in this helix and probably most other claudins. The amino-acid sequences were aligned using the Clustal Omega server [43], and TEXshade [44] was used for illustration.

A major breakthrough in TJ research was made in 2014 when the 2.4-Å resolution crystal structure of full-length mouse claudin-15 was reported [45]. As predicted from the sequence, the polypeptide chain was organized into four antiparallel TM helices with the N- and C-termini on the cytoplasmic side. On the extracellular side, a five-stranded up-and-down antiparallel β-sheet was formed by the long ECL1 (strands β1-β4) and the short ECL2 (strand β5, pairing with β1). In this crystal structure, ECL1 is partially disordered, because the loop (v1) connecting strands β1 and β2 is not represented in electron density [45], but it is ordered in the presence of bound ligand (see below). A molecular dynamics study based on the claudin-15 crystal structure [45] suggested that the protein forms a tetrameric channel in which a cage of four aspartate-15 residues acts as a selectivity filter that favors cation flux over anion flux [46].

This and other claudin structures are hoped to provide a basis for the targeted disruption of epithelial barriers in the administration of drugs [47]. The subsequently published crystal structure of mouse claudin-3 showed that proline 134 in TM helix α3 induces a bend in this helix, which is alleviated by the corresponding alanine or glycine mutations. A proline residue at this position is present in the majority of human claudin sequences; a helix bend brought about by this residue is likely to modulate the morphology and adhesiveness of TJ strands [48]. Three-dimensional structures of claudins provide the basis for *in silico* modeling of claudin based TJ self-assembly, their barrier and/or channel forming potential [49–51].

Much as the extracellular Ig domains of the JAMs are attachment sites for viruses, the extracellular loops of claudins serve as landing sites for bacterial toxins such as the *Clostridium perfringens* enterotoxin (CpE). A crystal structure of full-length claudin-19 bound to the soluble, claudin-binding C-terminal fragment of CpE (C-CPE) was determined at 3.7 Å resolution [52]. This structure showed that ligand binding leads to a stabilization of loop v1, which is now ordered, and indicated how C-CPE binding to selected claudins may lead to the disintegration of TJs and increased permeability across epithelial layers. C-CPE appears to bind different claudins with a conserved geometry and to disrupt the lateral interactions of their extracellular parts in the same way [16,52] as suggested by the crystal structure of C-CPE-bound human claudin-4 (Figure 4) [42]. Human claudin-9 (hCLDN-9) is highly expressed in the inner ear, essential for hearing and a high-affinity receptor of CpE. Two recently published 3.2-Å crystal structures of hCLDN-9 bound to C-CPE reveal structural changes in claudin epitopes involved in claudin self-assembly and suggest a mechanism for the disruption of claudin and TJ dissociation by CpE [53].

2.1.3. Occludin

Occludin and the other TAMPs of the TJ, tricellulin, and MARVELD3, share with the claudins the general architecture as tetraspan TM proteins with cytoplasmic N- and C-termini. However, the TAMPs are not homologous with claudins and differ in the length and structure of their cytoplasmic domains and extracellular loops.

The occludin cytosolic C-terminus forms a coiled-coil structure, dimerizes, and associates with all three ZO-proteins from the TJ cytoplasmic plaque [24,54,55]. Disulfide formation within the coiled-coil domain was proposed as a mechanism to influence the oligomerization of occludin [56,57]. The 1.45-Å crystal structure of the cytosolic C-terminus of occludin comprises three helices that form two separate anti-parallel coiled-coils and a loop that packs tightly against one of the coiled-coils (Figure 5a). This structure revealed a large positively charged surface that binds ZO-1 [58]. The cytoplasmic C-terminal coiled-coil region of occludin associates with mainly the GUK region of ZO-1 as shown by SAXS, NMR, and in vitro binding studies [59], which also revealed that serine phosphorylation within the acidic binding motif of the occludin coiled-coil significantly increases the binding affinity. Notably, several occludin isoforms result from alternative splicing and alternate promoter use, but neither this structural polymorphism nor the multitude of known post-translational modifications from proteolysis and serine, threonine or tyrosine phosphorylation of occludin [60] have so far been studied by X-ray or NMR methods.

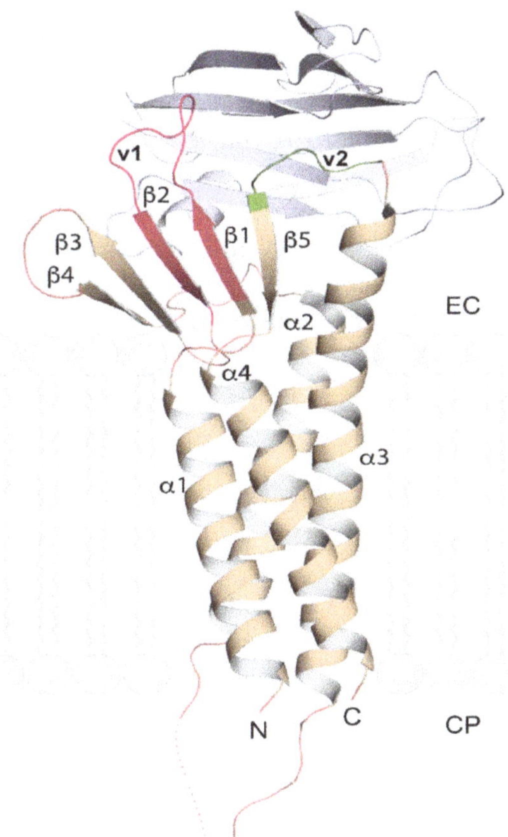

Figure 4. Crystal structure of human claudin-4. Cartoon model of the overall fold of human CLDN-4 (wheat color) in complex with the C-terminal fragment of *Clostridium perfringens* enterotoxin (C-CPE, light blue; PDB entry 5B2G) [16,42]. The extracellular variable regions of CLDN-4 that mediate hetero- and homotypic interactions are highlighted in magenta (v1, comprising β1 and β2) and green (v2, between TM-helix α3 and β5), respectively. The dotted line marks a segment of polypeptide chain not represented in electron density. The stylized lipid molecules indicate the cell membrane and are not part of the experimental structure. EC: Extracellular; CP: Cytoplasmic.

2.1.4. Tricellulin

The precise definition of TJ architecture through freeze–fracture microscopy of epithelial preparations from rat intestine revealed a modified structure at tricellular junctions [61]. Tricellular pores and bicellular strand opening contribute to allowing the passage of large molecules through the TJ in the "leak pathway" as suggested by computational structural dynamics studies [62]. TJs completely disappear during the epithelial–mesenchymal transition (EMT), where the transcriptional repressor Snail plays a central role. The protein tricellulin was identified in a screen using Snail-overexpressing epithelial cells as a protein concentrated at tricellular tight junctions (tTJs) and named for this property [63].

Tricellulin is downregulated during the EMT. The E3 ubiquitin ligase Itch binds the N-terminus of tricellulin via its WW domain (named after two signature tryptophan residues) to stimulate its ubiquitination, which is, however, not primarily involved in proteasomal breakdown of tricellulin [64]. During apoptosis, cells are extruded from epithelial cell layers. Loss of functional tricellulin contributes to dissociation of tTJs during apoptosis, when it is cleaved by caspase at aspartate residues 441 and 487 in the C-terminal coiled-coil [65]. Tricellulin is of key importance for hearing, as it was reported that mutations in the human *TRIC* gene are associated with deafness [66].

Tricellulin is localized to tTJs but also to bicellular TJs. When tricellulin is selectively overexpressed at tTJs, it decreases the permeability for large solutes up to 10 kDa, but not for ions. This seemingly paradoxical observation may be explained by the rare occurrence of tricellular junctions relative to bicellular junctions [67]. Tricellular TJs are regarded as potential weak points in the paracellular barrier. Tricellulin-dependent macromolecular passage is observed in both leaky and tight epithelia [68]. Tricellulin tightens tricellular junctions and regulates bicellular TJ proteins. The extracellular loops of tricellulin may be crucial for its sealing function, because it could be shown that a synthetic peptide (trictide) derived from the tricellulin ECL2 may increase the passage of solutes into human adenocarcinoma cells [69]. In MDCK cells, the tricellulin C-terminus is important for basolateral translocation, whereas the N-terminus directs tricellulin to tricellular contacts. There is evidence for the formation of heteromeric tricellulin–occludin contacts at elongating bicellular junctions and of homomeric tricellulin–tricellulin complexes at tricellular junctions [70].

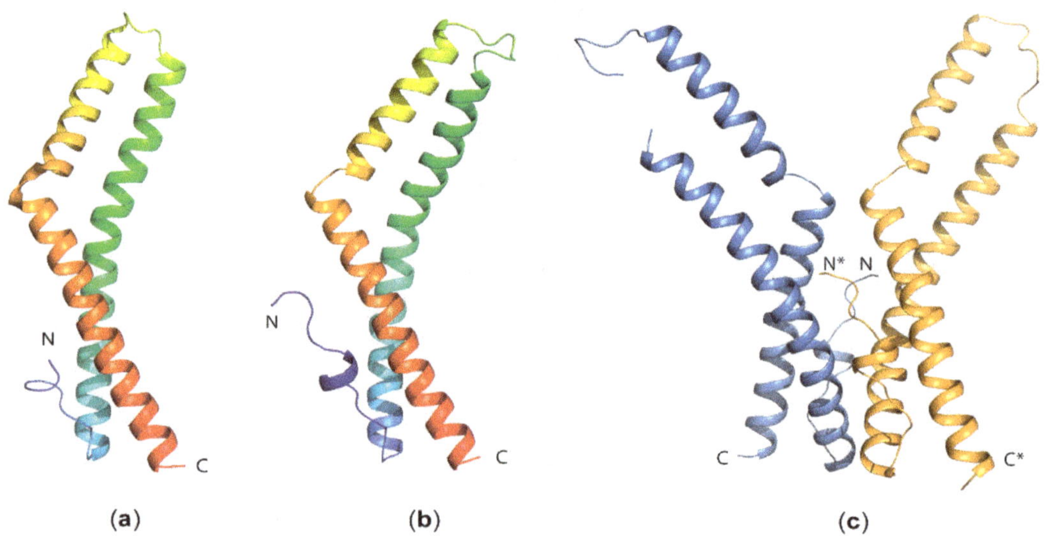

Figure 5. Structural insight into occludin and tricellulin function. Cartoon models of the overall fold of the coiled-coil domain of (**a**) human occludin (PDB entry 1XAW) [58] and (**b**) human tricellulin (PDB entry 5N7K) [71]. The molecules are colored in a gradient ranging from blue at the N-terminus (N) to red at the C-terminus. (**c**) Dimeric arrangement of the tricellulin C-terminal coiled-coil domain observed in the crystal structure [71]. The chain marked with an asterisk (*) corresponds to the second monomer within the dimer.

Tricellulin has an extended cytoplasmic N-terminus of 194 aa and a cytoplasmic C-terminal region of 195 aa, in marked contrast to occludin, where these regions include 66 aa and 256 aa, respectively. With the exception of the C-terminal coiled-coil domain, no cytoplasmic region carries a sequence signature suggesting a known domain structure in either protein. A crystal structure of the C-terminal coiled-coil domain of tricellulin was determined at 2.2-Å resolution (Figure 5b). This structure reveals a dimeric arrangement with an extended polar interface (Figure 5c), which may contribute to stabilizing tTJs [71].

2.1.5. Other Tight-Junction Transmembrane Proteins

With the exception of tricellulin, the extracellular loops and ectodomains of the abovementioned transmembrane TJ proteins are involved in *trans* pairing interactions of opposing cells (Figure 1) [5]. For the POPDC and Crumbs family of transmembrane TJ proteins, no such cross–membrane interactions are described. The POPDC family of tri-span TM proteins consists of BVES/POPDC1, POPDC2, and POPDC3. BVES protein dimers are mediated by the cytoplasmic Popeye domain, and BVES–BVES *cis* pairing interactions are necessary to maintain epithelial integrity and junctional stability. The cytoplasmic tail of BVES was shown to directly interact with ZO-1 [14], but structural information on the atomic level is still missing [14]. Crumbs was first described in *D. melanogaster* [72]; in mammals it has three homologs (CRB1, CRB2, CRB3) of which the latter is expressed in all epithelial tissues [73]. As the Crumbs protein family members are part of the cell polarity complex Crumbs/PALS1/PATJ, further information is included in Section 2.2.3.

2.2. Proteins of the Cytoplasmic Plaque

The proteins of the cytoplasmic plaque are characterized by recurrent protein–protein interaction (PPI) domains and frequently contain natively unfolded regions [74–76]. They are interconnected in a dynamic and multivalent PPI system, which has been partly mapped down to the domain level (Figure 6). In addition to the interactions displayed in the figure, there are multiple PPIs with regulatory and signaling proteins not covered in this review.

2.2.1. PDZ Domains

Many proteins of the cytoplasmic plaque contain one or multiple PDZ domains (Figure 6). We next discuss some key features of these ubiquitous PDZ domains. PDZ domains regulate multiple cellular processes by promoting protein–protein interactions and are abundant protein modules in TJ proteins, but also in many other proteins in all kingdoms of life. Frequently, PDZ domains are associated with WW, SH2, SH3 (Src homology 2 or 3), or PH (Pleckstrin homology) domains within one polypeptide chain [77]. The term PDZ is derived from the three founding members of the family, PSD-95 (postsynaptic density-95), the *Drosophila* tumor suppressor protein DLG-1 (discs large 1), and ZO-1. As early as 2010, > 900 PDZ domains were annotated in > 300 proteins encoded in the mouse genome, and > 200 X-ray or NMR structures of PDZ domains from various sources were known [78]. In August 2019, a PDB [79] search returned 533 entries with the keyword "PDZ domain" and 138 entries with the keyword "PDZ domain-like". Thus, extensive structural data are available for these domains. In general, PDZ domains are structured as a β-sandwich capped by two α-helices and bind ligand peptides in a shallow groove between helix α2 and strand β2 (Figure 7a). Their propensity to dimerize via domain swapping was first described for the second PDZ domain (PDZ2) of ZO-2 [80] and later also for PDZ2 of ZO-1 and ZO-3 (Figure 7b, see Section 2.2.2.).

Domain swapping is frequently observed in small β-sheet domains. Bacterial major cold-shock proteins [81,82], for example, were found to form domain-swapped dimers. A domain-swapped three-stranded segment of the *E. coli* cold-shock protein CspA is capable of recombining with a polypeptide region of ribosomal protein S1 to form a closed β-barrel recapitulating structural features of both parent proteins [73].

PDZ domains have been divided into three specificity classes according to the preferred amino acid residue at position –2 (P^{-2}) of the binding groove [72]. Typically, PDZ domains recognize sequence motifs at the extreme carboxy terminus of ligand proteins (Figure 7c), but binding of internal sequence motifs is also common (Figure 7d).

PDZ domains are regarded as promising drug targets for neurological and oncological disorders, as well as viral infections. Many structure-guided efforts are underway towards the development of small-molecule or peptidic modulators of PDZ domains [83,84], including the PDZ domains from Shank3, a central scaffolding protein of the post-synaptic density protein complex [85] and of the protein interacting with C kinase (PICK1), a regulator of AMPA receptor trafficking at neuronal synapses [86].

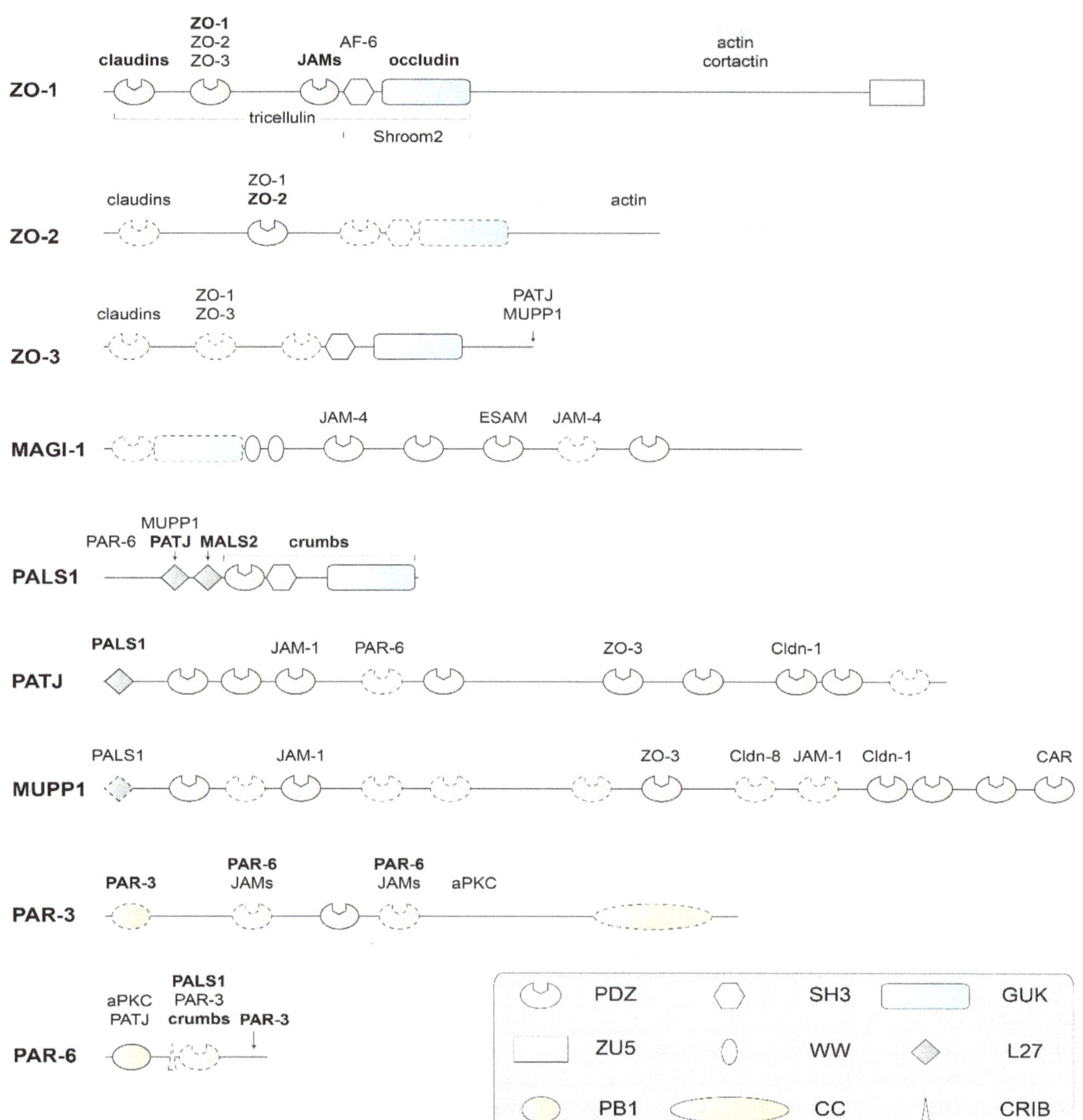

Figure 6. Domain structure of human PDZ domain-containing adapter proteins of the cytoplasmic plaque that interact physically with TM proteins of the tight junction. The proteins are scaled to the length of their amino-acid sequences. Experimental structures (usually by X-ray or NMR analysis) are available for protein domains drawn with solid contours, but not for domains drawn with dashed contours or for extended regions of polypeptide chains without domain annotation. Proteins binding to components of the human cytoplasmic plaque or their homologs are indicated above or below their interacting domains. With the exception of aPKC (as a subunit of the PAR-3/PAR-6/aPKC complex), only TM proteins or classical adapter proteins of the TJ are included as interacting proteins. Names of interacting proteins are written in bold letters, where the interaction is structurally characterized. Protein names and abbreviations are explained in the text or the legend of Figure 1. Domains are abbreviated as follows: PDZ: Initially identified in PSD-95 (postsynaptic density-95); DLG-1 (the *Drosophila* tumor suppressor protein discs large 1) and ZO-1; SH3: Src homology-3; PB1: Initially identified in PHOX and BEM1; ZU5: Present in ZO-1 and UNC5; L27: LIN-2/LIN-7; GUK: guanylate kinase homolog; WW: Named after two signature tryptophan residues; CC: coiled-coil; CRIB: CDC42/RAC interactive binding. Figure modified and updated after Guillemot et al. [3].

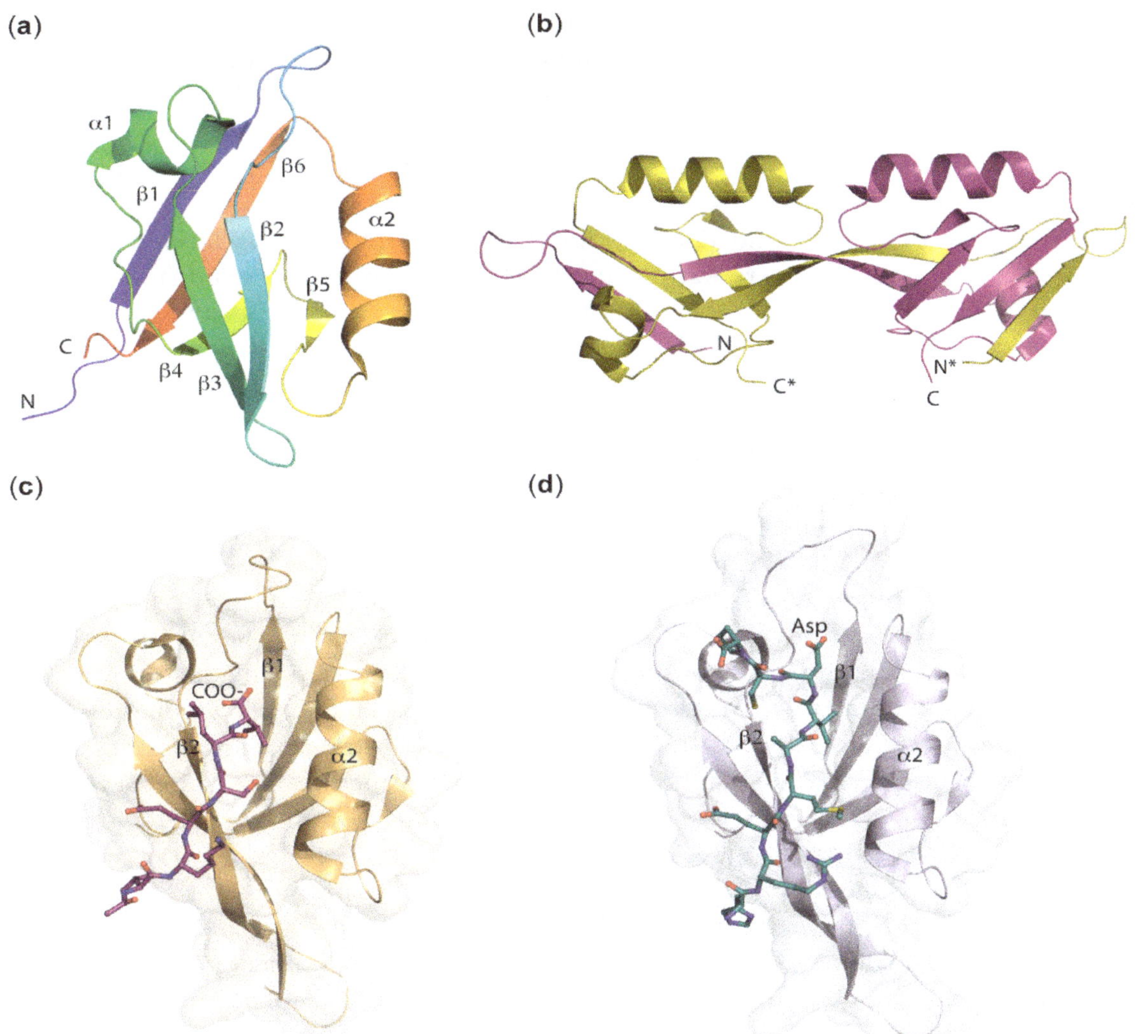

Figure 7. Structural features of PDZ domains. (**a**) Topology of a prototypical PDZ domain, here PALS 1 PDZ, PDB entry 4UU6 [87]. The polypeptide chain is drawn in rainbow colors changing from N- to C-terminus. A cleft for binding ligand peptides in an extended conformation is visible between strand β2 and helix α2. (**b**) Domain-swapped dimer formed by ZO-2 PDZ2, PDB entry 3E17 [88]. The two polypeptide chains are drawn in yellow and purple. The chain marked with an asterisk (*) corresponds to the second monomer within the dimer. Domain swapping moves the N-terminal β1 strand and half of β2 of one chain into the core structure of the other, leaving the ligand-binding geometry in both halves of the dimer intact. The PDZ2 domains in ZO-1, ZO-2, and ZO-3 are all found in the domain-swapped dimeric form [80,88–91]. (**c**) Canonical binding of a C-terminal ligand peptide to a PDZ domain, here PAR-6 PDZ bound to the hexapeptide VKRSLV, PDB entry 1RZX [92]. The terminal carboxy group is bound to the carboxylate-binding loop between strands β1 and β2 of PAR-6 PDZ and the extended ligand peptide aligns in antiparallel orientation with β2, extending the β-sheet. Note that this binding mode is energetically disfavored for PAR-6 PDZ and requires binding of CDC42 at a nearby CRIB domain (not shown). (**d**) Binding of an internal peptide to a PDZ domain, here PAR-6 PDZ bound to a dodecapeptide representing amino acids 29–40 of PALS1, PDB entry 1X8S [93]. The ligand peptide adopts an extended conformation with an aspartic acid side chain mimicking the carboxy group of the canonical C-terminal peptide ligand. Note the altered conformation of the carboxylate-binding loop.

2.2.2. MAGUK Proteins

Membrane-associated guanylate kinase homologs (MAGUKs) constitute a family of scaffolding molecules with a core MAGUK module consisting of a PDZ, SH3, and an enzymatically inactive guanylate kinase (GUK) domain [94–96]. The MAGUK protein family members ZO-1, ZO-2, and ZO-3 link the TJ membrane proteins to the cytoskeleton and provide the structural basis for the assembly of multiprotein complexes at the cytoplasmic side of TJs (Figure 1) [97]. ZO-1 is a cytoplasmic component of both TJs and AJs, and connects the TJ to the actin cytoskeleton via extended, presumably unstructured polypeptide regions near its C-terminus [98]. Direct actin binding was also reported for ZO-2 and ZO-3 [24,54,55].

ZO-1 and its paralogs ZO-2 and ZO-3 contain three N-terminal PDZ domains (Figure 6). The propensity of these PDZ domains to recognize specific C-terminal or other peptide motifs and assemble multicomponent TJ protein complexes will be highlighted below. Most of the claudins present at TJs have conserved C-terminal tails that bind to PDZ1 of ZO proteins. Compared to the single PDZ domain of the AJ protein Erbin, the ZO-1 PDZ1 domain has a broadened ligand specificity. Crystal structures of the Erbin PDZ and the ZO-1 PDZ1 revealed the structural basis for the different ligand specificities, where subtle conformational rearrangements are identified at multiple ligand-binding subsites, and support a model for ligand recognition by these domains [99].

The intracellular C-terminus of claudins binds to the N-terminal PDZ1 domain of ZO proteins with variable affinity. The affinity of claudin binding to ZO-1 PDZ1 depends on the absence or presence of a tyrosine residue at position -6 from the claudin C-terminus. Crystal structures of ZO-1 PDZ1 with empty ligand-binding groove, with a bound claudin-1 heptapeptide, which does not have a Tyr -6, or with a bound claudin-2 heptapeptide containing a Tyr -6 revealed significantly different binding geometries explaining the influence of the signature tyrosine residue on binding affinity [100]. In addition to claudin binding, the ZO-1 PDZ1 also mediates interactions with phosphoinositides. Mapping the inositol hexaphosphate binding site onto an NMR structure of ZO-1 PDZ1 revealed spatial overlap with the claudin binding surface and thus provided a structural rationale for the observed competition of both ligands for ZO-1 [101].

The second PDZ domain (PDZ2) of ZO-proteins is known to promote protein dimerization [24, 54,55]. A crystal structure shows that ZO-1—PDZ2 dimerization is stabilized by extensive domain swapping of β-strands. This structural rearrangement leaves the canonical peptide-binding groove intact in both subunits of PDZ2 dimer, which are composed of elements from both monomers [89]. Domain swapping of human ZO-1 PDZ2 was subsequently confirmed by solution NMR analysis. In this study, the importance of strand β2 for the domain exchange was demonstrated by insertion mutagenesis [91]. NMR analysis clearly demonstrated that PDZ2 of ZO-2 may also dimerize by domain swapping. A 1.75 Å resolution crystal structure of the ZO-2 PDZ2 confirms formation of a domain-swapped dimer with exchange of β-strands 1 and 2 (Figure 7b) [88], and there is evidence for the formation of PDZ2-promoted domain-swapped homodimers in all three ZO proteins. Based on this observation and the high sequence similarity between the ZO-1, -2, and -3 PDZ2 domains (66% sequence identity between ZO-1 and ZO-2, 50% between ZO-1 and ZO-3, 54% between ZO-2 and ZO-3), heterodimer formation between them was proposed as a potential mechanism of forming and stabilizing the cytoplasmic plaque [80]. Structural evidence for domain-swapped heterodimers of ZO proteins is, however, still lacking.

ZO-1 PDZ2 interacts with connexins, in particular the abundant connexin43, which functions in gap junction formation and regulation. X-ray and NMR analyses showed that domain swapping of ZO-1 PDZ2 preserves the carboxylate tail-binding pockets of the PDZ domains and creates a distinct interface for connexin43 binding [90].

The third PDZ domain (PDZ3) of ZO proteins is important for the interaction with the C-terminus of transmembrane JAMs. A crystal structure of ZO-1 PDZ3 was determined at 1.45 Å resolution. This study established that ZO-1 PDZ3 preferentially binds ligands of sequence type $X[D/E]X\Phi_{COOH}$ where X may be any amino acid and Φ is a hydrophobic residue [72].

Following the three N-terminal PDZ domains, ZO proteins contain a SH3-GUK module. Crystal structure analysis of the ZO-1 SH3-GUK tandem domain confirmed independent folding of the SH3 and GUK domains, and pulldown assays identified the downstream U6 loop as an intramolecular ligand of the SH3-GUK core with a potential role in regulating TJ assembly in vivo [102]. Crystal structures of the complete MAGUK core module of ZO-1 comprising the PDZ3-SH3-GUK region and its complex with the cytoplasmic tail of adhesion molecule JAM-A revealed that residues from the adjacent SH3 domain are involved in ligand binding to the ZO-1 PDZ3 [103].

ZO-1 is distinct from its paralogs ZO-2 and ZO-3 by the presence of an extended C-terminal region harboring a ZU5 domain [104,105] and described to mediate physical interaction with the CDC42 effector kinase MRCKβ. An NMR structure showed the ZO-1 ZU5 domain to adopt a β-barrel structure, which is incomplete in comparison with homologous proteins by lacking two β-strands. Attempts to analyze the structure of a ZO-1 ZU5/MRCKβ complex remained unsuccessful, but evidence could be provided that GRINL1A (glutamate receptor, ionotropic, N-methyl-D-aspartate-like 1A combined protein) binds ZO-1 ZU5 in a very similar way as MRCKβ. NMR analysis then showed that a 22-aa GRINL1A peptide hairpin associates with the ZO-1 ZU5 domain to form a complete canonical ZU5 domain [106].

In the MAGI proteins (MAGUKs with inverted domain structure), the characteristic arrangement of PPI domains present in common MAGUK proteins is inverted. Furthermore, the MAGI proteins contain two WW domains in place of the SH3 domain found in MAGUKs (Figure 6) [107]. The family member MAGI-1 is tethered to TJs through interactions of its PDZ domains with the C-terminus of the non-classical junctional adhesion molecule JAM-4 [108]. In addition to its function in the TJ, the first PDZ domain of MAGI-1 binds peptide ligands derived from the oncoprotein E6 of human papillomavirus and the ribosomal S6 kinase 1 (RSK1) [109,110]. NMR analysis suggests the involvement of peptide regions flanking the PDZ domain in ligand binding [109]. PALS1 (protein associated with Lin seven 1, also known as MPP5—membrane-associated palmitoylated protein 5) is another member of the MAGUK family (Figure 6) and described below as part of the Crumbs/ PALS1/PATJ complex.

2.2.3. The Crumbs/PALS1/PATJ Complex

The Crumbs/PALS1/PATJ complex is involved in establishing and maintaining cell polarity and located in the cytoplasmic plaque of TJs [2,3,111]. Crumbs is a single-span TM protein, whereas the other proteins present in this complex, PALS1 and PATJ (PALS-associated tight-junction protein), are cytoplasmic scaffolding proteins (Figure 1). PALS1 functions as an adaptor protein mediating indirect interactions between Crumbs and PATJ (Figure 6). Both PALS1 and PATJ share an N-terminal L27 domain. L27 domains organize scaffold proteins into supramolecular complexes by heteromeric L27 interactions. PATJ is recruited to TJs through interactions with the C-termini of claudin-1 and ZO-3 [112].

The PATJ–PALS1 interaction is mediated by the single L27 domain of PATJ and the N-terminal L27 domain (L27N) of PALS1 [113]. A crystal structure of the PALS1-L27N/PATJ-L27 heterodimer shows that each L27 domain is composed of three α-helices and that heterodimer formation is due to formation of a four-helix bundle by the first two α-helices of the L27 domains and coiled–coil interactions between the helices α3 [114]. NMR structure analysis revealed closely similar topologies for heterotetrameric mLin-2/mLin-7 and PATJ/PALS1 complexes, suggesting a general assembly mode for L27 domains [115]. A crystal structure of a heterotrimeric complex formed by the N-terminal L27 domain of PATJ, the N-terminal tandem L27 domains of PALS1, and the N-terminal L27 domain of MALS2 (mammalian homolog-2 of Lin-7) revealed an assembly of two cognate pairs of heterodimeric L27 domains. This structure is thought to reveal a novel mechanism for tandem L27 domain-mediated supramolecular complex assembly [116].

The intracellular functions of Crumbs3 (CRB3) are mediated by its conserved 37-aa cytoplasmic tail (Crb-CT) and its interaction with PALS1 and the actin-binding protein moesin. The crystal structure of a PALS1 PDZ-SH3-GUK/Crb-CT complex shows that all three domains of PALS1 contribute to Crb-CT binding [117]. A further crystal structure of human PALS1 PDZ bound to 17-aa C-terminal CRB1 peptide shows that only the very C-terminal tetrapeptide ERLI is involved in direct binding to PALS1 PDZ. Comparison with apo-PALS1 PDZ (Figure 7a) revealed that a key phenylalanine residue in the PALS1 PDZ controls access to the ligand-binding groove [87]. To reveal the nature of the Crumbs/moesin interaction, the FERM (protein 4.1/ezrin/radixin/moesin) domain of murine moesin was co-crystallized with the soluble C-terminus of *Drosophila* Crumbs. The 1.5-Å resolution crystal structure revealed that both the FERM-binding motif, as well as the PDZ-binding motif present in the Crumbs C-terminal peptide contribute to the interaction with moesin. Phosphorylation of the Crb-CT by atypical protein kinase C (aPKC) disrupts the Crumbs/moesin association but not the Crumbs/PALS1 interaction. Crumbs may therefore act as aPKC-mediated sensor in epithelial tissues [118].

2.2.4. The PAR-3/PAR-6/aPKC Complex

Similar as the Crumbs/PALS1/PATJ complex, the evolutionarily conserved PAR-3/PAR-6/aPKC complex is associated to TJs and crucial for establishing and maintaining cell polarity. The complex formed by the PAR (partition defective)-3 and PAR-6 proteins, as well as the atypical protein kinase C (aPKC) interacts with subunits from the Crumbs/PALS1/PATJ complex and is regulated by binding to the small GTPases CDC42 and RAC1. Composition and stoichiometry of the PAR-3/PAR-6/aPKC complex are linked to cell polarity and to the cell cycle [119].

Human PAR-6 contains a single PDZ domain, which mediates binding to the C-terminus of TM receptor CRB3. Binding of C-terminal ligands to the PAR-6 PDZ depends on binding of the Rho-GTPase CDC42 to a CRIB domain adjacent to the PAR-6 PDZ. In addition, the PAR-6 PDZ also binds internal peptides, e.g., from PALS1 and its *Drosophila* homolog Stardust. The regulation of ligand binding to PAR-6 PDZ by CDC42 has been structurally characterized in a number of studies. A 2.5-Å crystal structure of a PAR-6 PDZ-bound internal dodecapeptide derived from PALS1 revealed a characteristic deformation of the carboxylate-binding loop of PAR-6 PDZ relative to the structure with bound C-terminal ligand (Figure 7d) [93]. The structural adjustments associated with regulator and ligand binding to the PAR-6 PDZ were also highlighted in a 2.1-Å crystal structure and an NMR structure of the PAR-6 PDZ domain (Figure 7c), which revealed deviations from the canonical PDZ conformation that account for low-affinity binding of C-terminal ligands. CDC42 binding to the adjacent CRIB domain triggered a structural transition to the canonical PDZ conformation and was associated with a ~13-fold increase in affinity for C-terminal ligands [92]. NMR structures of the isolated PAR-6 PDZ domain and a disulfide-stabilized CRIB-PDZ fragment identified a conformational switch in the PAR-6 PDZ domain that is linked to the increase in ligand affinity induced by CDC42 binding to PAR-6 [120]. Finally, NMR analysis of a C-terminal Crumbs peptide binding to PAR-6 and the crystal structure of the PAR-6 PDZ/peptide complex indicated why the affinity of this interaction is 6-fold higher than in previously studied PAR-6/peptide binding studies [121].

PAR-3 acts as central organizer of the PAR-3/PAR-6/aPKC complex and is thus essential for establishment and maintenance of cell polarity. In *Caenorhabditis elegans*, PAR-3 mediates TJ binding through interaction with junctional adhesion molecule (JAM) [122,123]. In cultured endothelial cells, PAR-3 associates with JAM-2 and JAM-3, but neither with the related Ig-like TM proteins ESAM nor CAR [124]. PAR-3 contains an N-terminal oligomerization domain in addition to three PDZ domains. NMR analysis showed the monomeric PAR-3 N-terminal domain (NTD) to adopt a PB1-like fold and to oligomerize into helical filaments. This interaction was proposed to facilitate the assembly of higher-order PAR-3/PAR-6/aPKC complexes [125]. The ability of the PAR-3 NTD to self-associate and form filamentous structures was further studied by crystallographic analysis of the PAR-3 NTD and analysis of the filament structure by cryo-electron microscopy (cryo-EM). Here, it was revealed that both lateral and longitudinal packing within PAR-3 NTD filaments is primarily mediated by Coulomb interactions [126].

The second PDZ domain of PAR-3 binds phosphatidylinositol (PI) lipid membranes with high affinity as shown in a biochemical and NMR study of PAR-3 PDZ2. This study also showed that the lipid phosphatase PTEN (phosphatase and tensin homolog) binds PAR-3 PDZ3 and thus cooperates with PI in regulating cell polarity through PAR-3 [127]. A three-dimensional structure of the second PDZ domain of human PAR-3 was also determined as part of an NMR structure analysis automation study [128]. A previously unknown C-terminal PDZ-binding motif, identified in PAR-6 through crystal structures and NMR binding analyses, mediates interactions with PDZ1 and PDZ3, but not with PDZ2 of PAR-3. Evidently, PAR-3 has the ability to recruit two PAR-6 molecules simultaneously, possibly facilitating the assembly of polarity protein networks through these interactions [129].

2.2.5. Other Cytoplasmic Tight-Junction Proteins

In addition to the MAGUK proteins, the subunits and regulators of the Crumbs/PALS1/PATJ and PAR-3/PAR-6/aPKC complexes, various other cytoplasmic proteins are associated with TJs. The multiple PDZ domain protein 1 (MUPP1, also known as MPDZ) contains 13 PDZ domains. MUPP1 is a paralog of PATJ, shares several TJ-binding partners (Figure 6) and a similar subcellular localization, but displays a distinct selectivity in its interactions with claudins and is dispensable for TJ formation while PATJ is not [130,131]. A crystal structure of the twelfth PDZ domain of MUPP1 was determined in a structural genomics effort to crystallize PDZ domains with self-binding C-terminal extensions [132]. Several additional MUPP1-PDZ domains were analyzed within the research program of the Center for Eukaryotic Structural Genomics [133] and submitted to the Protein Data Bank (PDB) [79], but without functional annotation. Crystal structure analysis of the mouse MUPP1-PDZ4 domain revealed a canonical PDZ fold with six β-strands and three α-helices [60]. The angiomotin (AMOT) proteins [134] were reported to interact with MUPP1, PATJ [135], ZO-1 and MAGI-1b, and ascribed a role in the assembly of endothelial cell junctions [25]. The cytoskeletal linker cingulin, a predicted dimeric coiled-coil protein of unknown three-dimensional structure, was initially characterized as a peripheral TJ component [136]. Although its amino-terminal region was reported to interact with ZO-1 in cells [137], cingulin was later shown to be dispensable for TJ integrity and epithelial barrier function [138]. The ZO-1 associated nucleic acid-binding protein ZONAB (also referred to as YBX3 or CSDA1) is a transcription factor that shuttles between TJs and the nucleus and regulates epithelial cell proliferation [139,140]. Although a crystal or NMR structure of ZONAB is not available, the conformation and nucleic acid binding of its N-terminal cold shock domain may be inferred from structures of bacterial major cold shock proteins [141] or the homologous Y-box factor YB-1 [142].

The molecular composition of TJs varies significantly between different epithelia and determines their dual functions as effective barriers for solutes or channels for particular classes of solutes [143]. Therefore, the expression patterns of TJ proteins in various tissues are kept under tight control by various transcription factors in addition to ZONAB. A large number of TJ and AJ transmembrane (TM) proteins are under transcriptional control by the Grainyhead-like proteins GRHL1 and GRHL2 or by nuclear receptors [144]. These transcription factors therefore regulate a large subset of proteins, making up the apical junction complex. Frequently, one transcription factor controls the expression of multiple genes encoding TJ or AJ proteins. For example, GRHL2 acts as transcriptional activator of both AJ and TJ components including several claudins and thus functions as regulator of epithelial differentiation [145]. Equally frequently, one TJ protein is controlled by multiple transcription factors. The *Cldn4* gene, being under transcriptional control by GRHL2, GRHL3, the androgen receptor, retinoic acid/retinoid X receptors, and p63 in different tissues [144,146,147], serves as an impressive example here. Although structural information is available for several of these factors [148–151], these proteins will not be further discussed here, where the focus is placed on proteins of the TJ core structure. Equally, the large set of signaling and effector proteins acting on the TJ [2,11,12] will not be discussed further.

3. Conclusions

Here, we have reviewed three-dimensional structures of TJ proteins, focusing on the core TJ complex. It becomes clear that our knowledge of these structures is fairly incomplete, because most TJ proteins do not lend themselves easily to structure analysis due to their size and/or the presence of TM regions, natively unfolded polypeptide segments or heterogeneous post-translational modifications [74–76]. Our knowledge of protein–protein interactions within the TJ also does not go far beyond structural features of selected binary interactions, often involving small protein fragments or peptides. At a resolution that permits construction of atomic models, very little is known about the general architecture of the TJ. Herein lies a great challenge and opportunity for future research making use of new integrative methods in structural biology, including cryo-electron microscopy [152–155], cryo-electron tomography [156,157], cross-linking mass spectrometry [158,159], small-angle X-ray and neutron scattering [160,161], and others [162]. These rapidly emerging and developing methods can be informed by the available high-resolution structures of TJ proteins, protein domains and protein–protein interactions. We may expect to see exciting results along these lines in the near future, revealing the architecture of TJs at high resolution in defined functional states.

Author Contributions: U.H. wrote the first draft of manuscript and did the final editing. A.S. designed the figures and revised the manuscript text.

Acknowledgments: U.H. and A.S. express their gratitude to Michael Fromm, Dorothee Günzel, and Susanne Krug (Charité, Berlin) for long-standing collaboration and encouragement in our work on TJ proteins.

Abbreviations

aa	Amino acid
AJ	Adherens junction
BBB	Blood-brain barrier
CTD	C-terminal domain
ECL	Extracellular loop
EMT	Epithelial-mesenchymal transition
NTD	N-terminal domain
PDB	Protein Data Bank
PPI	Protein–protein interaction
TJ	Tight junction
tTJ	Tricellular tight junction
TM	Transmembrane

References

1. Farquhar, M.G.; Palade, G.E. Junctional complexes in various epithelia. *J. Cell Biol.* **1963**, *17*, 375–412. [CrossRef] [PubMed]
2. Balda, M.S.; Matter, K. Tight junctions at a glance. *J. Cell Sci.* **2008**, *121*, 3677–3682. [CrossRef]
3. Guillemot, L.; Paschoud, S.; Pulimeno, P.; Foglia, A.; Citi, S. The cytoplasmic plaque of tight junctions: A scaffolding and signalling center. *Biochim. Biophys. Acta* **2008**, *1778*, 601–613. [CrossRef] [PubMed]
4. Furuse, M. Molecular basis of the core structure of tight junctions. *Cold Spring Harb Perspect Biol.* **2010**, *2*, a002907. [CrossRef]
5. Van Itallie, C.M.; Anderson, J.M. Architecture of tight junctions and principles of molecular composition. *Semin. Cell Dev. Biol.* **2014**, *36*, 157–165. [CrossRef] [PubMed]
6. Haseloff, R.F.; Dithmer, S.; Winkler, L.; Wolburg, H.; Blasig, I.E. Transmembrane proteins of the tight junctions at the blood-brain barrier: Structural and functional aspects. *Semin. Cell Dev. Biol.* **2015**, *38*, 16–25. [CrossRef] [PubMed]
7. Irudayanathan, F.J.; Trasatti, J.P.; Karande, P.; Nangia, S. Molecular Architecture of the Blood Brain Barrier Tight Junction Proteins-A Synergistic Computational and In Vitro Approach. *J. Phys. Chem. B* **2016**, *120*, 77–88. [CrossRef] [PubMed]
8. Irudayanathan, F.J.; Wang, N.; Wang, X.Y.; Nangia, S. Architecture of the paracellular channels formed by claudins of the blood-brain barrier tight junctions. *Ann. New York Acad. Sci.* **2017**, *1405*, 131–146. [CrossRef] [PubMed]
9. Weiss, N.; Miller, F.; Cazaubon, S.; Couraud, P.O. The blood-brain barrier in brain homeostasis and neurological diseases. *Biochim. Biophys. Acta* **2009**, *1788*, 842–857. [CrossRef]

10. Luissint, A.C.; Artus, C.; Glacial, F.; Ganeshamoorthy, K.; Couraud, P.O. Tight junctions at the blood brain barrier: Physiological architecture and disease-associated dysregulation. *Fluids Barriers Cns.* **2012**, *9*, 23. [CrossRef]
11. Zihni, C.; Balda, M.S.; Matter, K. Signalling at tight junctions during epithelial differentiation and microbial pathogenesis. *J. Cell Sci.* **2014**, *127*, 3401–3413. [CrossRef] [PubMed]
12. Zihni, C.; Mills, C.; Matter, K.; Balda, M.S. Tight junctions: From simple barriers to multifunctional molecular gates. *Nat. Rev. Mol. Cell Biol.* **2016**, *17*, 564–580. [CrossRef] [PubMed]
13. Russ, P.K.; Pino, C.J.; Williams, C.S.; Bader, D.M.; Haselton, F.R.; Chang, M.S. Bves modulates tight junction associated signaling. *PLoS ONE* **2011**, *6*, e14563. [CrossRef] [PubMed]
14. Han, P.; Lei, Y.; Li, D.; Liu, J.; Yan, W.; Tian, D. Ten years of research on the role of BVES/ POPDC1 in human disease: A review. *Onco. Targets* **2019**, *12*, 1279–1291. [CrossRef]
15. Sanchez-Pulido, L.; Martin-Belmonte, F.; Valencia, A.; Alonso, M.A. MARVEL: A conserved domain involved in membrane apposition events. *Trends Biochem. Sci.* **2002**, *27*, 599–601. [CrossRef]
16. Suzuki, H.; Tani, K.; Fujiyoshi, Y. Crystal structures of claudins: Insights into their intermolecular interactions. *Ann. New York Acad. Sci.* **2017**, *1397*, 25–34. [CrossRef]
17. Garrido-Urbani, S.; Bradfield, P.F.; Imhof, B.A. Tight junction dynamics: The role of junctional adhesion molecules (JAMs). *Cell Tissue Res.* **2014**, *355*, 701–715. [CrossRef]
18. Kostrewa, D.; Brockhaus, M.; D'Arcy, A.; Dale, G.E.; Nelboeck, P.; Schmid, G.; Mueller, F.; Bazzoni, G.; Dejana, E.; Bartfai, T.; et al. X-ray structure of junctional adhesion molecule: Structural basis for homophilic adhesion via a novel dimerization motif. *Embo. J.* **2001**, *20*, 4391–4398. [CrossRef]
19. Matthaus, C.; Langhorst, H.; Schutz, L.; Juttner, R.; Rathjen, F.G. Cell-cell communication mediated by the CAR subgroup of immunoglobulin cell adhesion molecules in health and disease. *Mol. Cell Neurosci.* **2017**, *81*, 32–40. [CrossRef]
20. Patzke, C.; Max, K.E.; Behlke, J.; Schreiber, J.; Schmidt, H.; Dorner, A.A.; Kroger, S.; Henning, M.; Otto, A.; Heinemann, U.; et al. The coxsackievirus-adenovirus receptor reveals complex homophilic and heterophilic interactions on neural cells. *J. Neurosci.* **2010**, *30*, 2897–2910. [CrossRef]
21. Kirchner, E.; Guglielmi, K.M.; Strauss, H.M.; Dermody, T.S.; Stehle, T. Structure of reovirus sigma1 in complex with its receptor junctional adhesion molecule-A. *Plos Pathog.* **2008**, *4*, e1000235. [CrossRef] [PubMed]
22. Conley, M.J.; McElwee, M.; Azmi, L.; Gabrielsen, M.; Byron, O.; Goodfellow, I.G.; Bhella, D. Calicivirus VP2 forms a portal-like assembly following receptor engagement. *Nature* **2019**, *565*, 377–381. [CrossRef] [PubMed]
23. Citi, S.; Guerrera, D.; Spadaro, D.; Shah, J. Epithelial junctions and Rho family GTPases: The zonular signalosome. *Small Gtpases.* **2014**, *5*, 1–15. [CrossRef] [PubMed]
24. Wittchen, E.S.; Haskins, J.; Stevenson, B.R. Protein interactions at the tight junction. Actin has multiple binding partners, and ZO-1 forms independent complexes with ZO-2 and ZO-3. *J. Biol. Chem.* **1999**, *274*, 35179–35185. [CrossRef]
25. Bratt, A.; Birot, O.; Sinha, I.; Veitonmaki, N.; Aase, K.; Ernkvist, M.; Holmgren, L. Angiomotin regulates endothelial cell-cell junctions and cell motility. *J. Biol. Chem.* **2005**, *280*, 34859–34869. [CrossRef] [PubMed]
26. Itoh, M.; Tsukita, S.; Yamazaki, Y.; Sugimoto, H. Rho GTP exchange factor ARHGEF11 regulates the integrity of epithelial junctions by connecting ZO-1 and RhoA-myosin II signaling. *Proc. Natl. Acad. Sci. USA* **2012**, *109*, 9905–9910. [CrossRef] [PubMed]
27. Chen, X.; Macara, I.G. Par-3 controls tight junction assembly through the Rac exchange factor Tiam1. *Nat. Cell Biol.* **2005**, *7*, 262–269. [CrossRef]
28. Balda, M.S.; Matter, K. The tight junction protein ZO-1 and an interacting transcription factor regulate ErbB-2 expression. *Embo. J.* **2000**, *19*, 2024–2033. [CrossRef]
29. Higashi, T.; Tokuda, S.; Kitajiri, S.; Masuda, S.; Nakamura, H.; Oda, Y.; Furuse, M. Analysis of the 'angulin' proteins LSR, ILDR1 and ILDR2–tricellulin recruitment, epithelial barrier function and implication in deafness pathogenesis. *J. Cell Sci.* **2013**, *126*, 966–977. [CrossRef]
30. Shimada, H.; Satohisa, S.; Kohno, T.; Konno, T.; Takano, K.I.; Takahashi, S.; Hatakeyama, T.; Arimoto, C.; Saito, T.; Kojima, T. Downregulation of lipolysis-stimulated lipoprotein receptor promotes cell invasion via claudin-1-mediated matrix metalloproteinases in human endometrial cancer. *Oncol. Lett.* **2017**, *14*, 6776–6782. [CrossRef]

31. Shimada, H.; Abe, S.; Kohno, T.; Satohisa, S.; Konno, T.; Takahashi, S.; Hatakeyama, T.; Arimoto, C.; Kakuki, T.; Kaneko, Y.; et al. Loss of tricellular tight junction protein LSR promotes cell invasion and migration via upregulation of TEAD1/AREG in human endometrial cancer. *Sci. Rep.* **2017**, *7*, 37049. [CrossRef] [PubMed]
32. Tsukita, S.; Tanaka, H.; Tamura, A. The Claudins: From Tight Junctions to Biological Systems. *Trends Biochem. Sci.* **2019**, *44*, 141–152. [CrossRef] [PubMed]
33. Elkouby-Naor, L.; Ben-Yosef, T. Functions of claudin tight junction proteins and their complex interactions in various physiological systems. *Int. Rev. Cell Mol. Biol.* **2010**, *279*, 1–32. [PubMed]
34. Gunzel, D.; Fromm, M. Claudins and other tight junction proteins. *Compr. Physiol.* **2012**, *2*, 1819–1852. [PubMed]
35. Gunzel, D.; Yu, A.S. Claudins and the modulation of tight junction permeability. *Physiol. Rev.* **2013**, *93*, 525–569. [CrossRef] [PubMed]
36. Krause, G.; Protze, J.; Piontek, J. Assembly and function of claudins: Structure-function relationships based on homology models and crystal structures. *Semin. Cell Dev. Biol.* **2015**, *42*, 3–12. [CrossRef] [PubMed]
37. Milatz, S.; Piontek, J.; Schulzke, J.D.; Blasig, I.E.; Fromm, M.; Gunzel, D. Probing the cis-arrangement of prototype tight junction proteins claudin-1 and claudin-3. *Biochem. J.* **2015**, *468*, 449–458. [CrossRef]
38. Milatz, S.; Piontek, J.; Hempel, C.; Meoli, L.; Grohe, C.; Fromm, A.; Lee, I.F.M.; El-Athman, R.; Gunzel, D. Tight junction strand formation by claudin-10 isoforms and claudin-10a/-10b chimeras. *Ann. New York Acad. Sci.* **2017**, *1405*, 102–115. [CrossRef]
39. Krause, G.; Winkler, L.; Mueller, S.L.; Haseloff, R.F.; Piontek, J.; Blasig, I.E. Structure and function of claudins. *Biochim. Biophys. Acta* **2008**, *1778*, 631–645. [CrossRef]
40. Rosenthal, R.; Gunzel, D.; Theune, D.; Czichos, C.; Schulzke, J.D.; Fromm, M. Water channels and barriers formed by claudins. *Ann. New York Acad. Sci.* **2017**, *1397*, 100–109. [CrossRef]
41. The UniProt Consortium, UniProt: A worldwide hub of protein knowledge. *Nucleic Acids Res.* **2018**, *47*, D506–D515.
42. Shinoda, T.; Shinya, N.; Ito, K.; Ohsawa, N.; Terada, T.; Hirata, K.; Kawano, Y.; Yamamoto, M.; Kimura-Someya, T.; Yokoyama, S.; et al. Structural basis for disruption of claudin assembly in tight junctions by an enterotoxin. *Sci. Rep.* **2016**, 6. [CrossRef] [PubMed]
43. Sievers, F.; Wilm, A.; Dineen, D.; Gibson, T.J.; Karplus, K.; Li, W.; Lopez, R.; McWilliam, H.; Remmert, M.; Soding, J.; et al. Fast, scalable generation of high-quality protein multiple sequence alignments using Clustal Omega. *Mol. Syst. Biol.* **2011**, *7*, 539. [CrossRef] [PubMed]
44. Beitz, E. TEXshade: Shading and labeling of multiple sequence alignments using LATEX2 epsilon. *Bioinformatics* **2000**, *16*, 135–139. [CrossRef]
45. Suzuki, H.; Nishizawa, T.; Tani, K.; Yamazaki, Y.; Tamura, A.; Ishitani, R.; Dohmae, N.; Tsukita, S.; Nureki, O.; Fujiyoshi, Y. Crystal Structure of a Claudin Provides Insight into the Architecture of Tight Junctions. *Science* **2014**, *344*, 304–307. [CrossRef]
46. Samanta, P.; Wang, Y.T.; Fuladi, S.; Zou, J.J.; Li, Y.; Shen, L.; Weber, C.; Khalili-Araghi, F. Molecular determination of claudin-15 organization and channel selectivity. *J. Gen. Physiol.* **2018**, *150*, 949–968. [CrossRef]
47. Artursson, P.; Knight, S.D. Structural biology. Breaking the intestinal barrier to deliver drugs. *Science* **2015**, *347*, 716–717. [CrossRef]
48. Nakamura, S.; Irie, K.; Tanaka, H.; Nishikawa, K.; Suzuki, H.; Saitoh, Y.; Tamura, A.; Tsukita, S.; Fujiyoshi, Y. Morphologic determinant of tight junctions revealed by claudin-3 structures. *Nat. Commun.* **2019**, *10*. [CrossRef]
49. Weber, C.R.; Turner, J.R. Dynamic modeling of the tight junction pore pathway. *Ann. New York Acad. Sci.* **2017**, *1397*, 209–218. [CrossRef]
50. Alberini, G.; Benfenati, F.; Maragliano, L. Molecular Dynamics Simulations of Ion Selectivity in a Claudin-15 Paracellular Channel. *J. Phys. Chem. B* **2018**, *122*, 10783–10792. [CrossRef]
51. Irudayanathan, F.J.; Wang, X.Y.; Wang, N.; Willsey, S.R.; Seddon, I.A.; Nangia, S. Self-Assembly Simulations of Classic Claudins-Insights into the Pore Structure, Selectivity, and Higher Order Complexes. *J. Phys. Chem. B* **2018**, *122*, 7463–7474. [CrossRef] [PubMed]
52. Saitoh, Y.; Suzuki, H.; Tani, K.; Nishikawa, K.; Irie, K.; Ogura, Y.; Tamura, A.; Tsukita, S.; Fujiyoshi, Y. Structural insight into tight junction disassembly by Clostridium perfringens enterotoxin. *Science* **2015**, *347*, 775–778. [CrossRef] [PubMed]

53. Vecchio, A.J.; Stroud, R.M. Claudin-9 structures reveal mechanism for toxin-induced gut barrier breakdown. *Proc. Natl. Acad. Sci. USA* **2019**, 201908929. [CrossRef] [PubMed]
54. Fanning, A.S.; Jameson, B.J.; Jesaitis, L.A.; Anderson, J.M. The tight junction protein ZO-1 establishes a link between the transmembrane protein occludin and the actin cytoskeleton. *J. Biol. Chem.* **1998**, *273*, 29745–29753. [CrossRef]
55. Itoh, M.; Morita, K.; Tsukita, S. Characterization of ZO-2 as a MAGUK family member associated with tight as well as adherens junctions with a binding affinity to occludin and alpha catenin. *J. Biol. Chem.* **1999**, *274*, 5981–5986. [CrossRef]
56. Walter, J.K.; Castro, V.; Voss, M.; Gast, K.; Rueckert, C.; Piontek, J.; Blasig, I.E. Redox-sensitivity of the dimerization of occludin. *Cell Mol. Life Sci.* **2009**, *66*, 3655–3662. [CrossRef]
57. Walter, J.K.; Rueckert, C.; Voss, M.; Mueller, S.L.; Piontek, J.; Gast, K.; Blasig, I.E. The oligomerization of the coiled coil-domain of occludin is redox sensitive. *Ann. N. Y. Acad. Sci.* **2009**, *1165*, 19–27. [CrossRef]
58. Li, Y.; Fanning, A.S.; Anderson, J.M.; Lavie, A. Structure of the conserved cytoplasmic C-terminal domain of occludin: Identification of the ZO-1 binding surface. *J. Mol. Biol.* **2005**, *352*, 151–164. [CrossRef]
59. Tash, B.R.; Bewley, M.C.; Russo, M.; Keil, J.M.; Griffin, K.A.; Sundstrom, J.M.; Antonetti, D.A.; Tian, F.; Flanagan, J.M. The occludin and ZO-1 complex, defined by small angle X-ray scattering and NMR, has implications for modulating tight junction permeability. *P. Natl. Acad. Sci. USA* **2012**, *109*, 10855–10860. [CrossRef]
60. Zhu, H.; Liu, Z.; Huang, Y.; Zhang, C.; Li, G.; Liu, W. Biochemical and structural characterization of MUPP1-PDZ4 domain from Mus musculus. *Acta. Biochim. Biophys. Sin. (Shanghai)* **2015**, *47*, 199–206. [CrossRef]
61. Staehelin, L.A. Further observations on the fine structure of freeze-cleaved tight junctions. *J. Cell Sci.* **1973**, *13*, 763–786. [PubMed]
62. Tervonen, A.; Ihalainen, T.O.; Nymark, S.; Hyttinen, J. Structural dynamics of tight junctions modulate the properties of the epithelial barrier. *PLoS ONE* **2019**, *14*, e0214876. [CrossRef] [PubMed]
63. Ikenouchi, J.; Furuse, M.; Furuse, K.; Sasaki, H.; Tsukita, S.; Tsukita, S. Tricellulin constitutes a novel barrier at tricellular contacts of epithelial cells. *J. Cell Biol.* **2005**, *171*, 939–945. [CrossRef] [PubMed]
64. Jennek, S.; Mittag, S.; Reiche, J.; Westphal, J.K.; Seelk, S.; Dorfel, M.J.; Pfirrmann, T.; Friedrich, K.; Schutz, A.; Heinemann, U.; et al. Tricellulin is a target of the ubiquitin ligase Itch. *Ann. New York Acad. Sci.* **2017**, *1397*, 157–168. [CrossRef]
65. Janke, S.; Mittag, S.; Reiche, J.; Huber, O. Apoptotic Fragmentation of Tricellulin. *Int. J. Mol. Sci.* **2019**, *20*, 4882. [CrossRef]
66. Mariano, C.; Sasaki, H.; Brites, D.; Brito, M.A. A look at tricellulin and its role in tight junction formation and maintenance. *Eur. J. Cell Biol.* **2011**, *90*, 787–796. [CrossRef]
67. Krug, S.M.; Amasheh, S.; Richter, J.F.; Milatz, S.; Gunzel, D.; Westphal, J.K.; Huber, O.; Schulzke, J.D.; Fromm, M. Tricellulin forms a barrier to macromolecules in tricellular tight junctions without affecting ion permeability. *Mol. Biol. Cell.* **2009**, *20*, 3713–3724. [CrossRef]
68. Krug, S.M. Contribution of the tricellular tight junction to paracellular permeability in leaky and tight epithelia. *Ann. N. Y. Acad. Sci.* **2017**, *1397*, 219–230. [CrossRef]
69. Cording, J.; Arslan, B.; Staat, C.; Dithmer, S.; Krug, S.M.; Kruger, A.; Berndt, P.; Gunther, R.; Winkler, L.; Blasig, I.E.; et al. Trictide, a tricellulin-derived peptide to overcome cellular barriers. *Ann. N. Y. Acad. Sci.* **2017**, *1405*, 89–101. [CrossRef]
70. Westphal, J.K.; Dorfel, M.J.; Krug, S.M.; Cording, J.D.; Piontek, J.; Blasig, I.E.; Tauber, R.; Fromm, M.; Huber, O. Tricellulin forms homomeric and heteromeric tight junctional complexes. *Cell Mol. Life Sci.* **2010**, *67*, 2057–2068. [CrossRef]
71. Schuetz, A.; Radusheva, V.; Krug, S.M.; Heinemann, U. Crystal structure of the tricellulin C-terminal coiled-coil domain reveals a unique mode of dimerization. *Ann. N. Y. Acad. Sci.* **2017**, *1405*, 147–159. [CrossRef] [PubMed]
72. Ernst, A.; Appleton, B.A.; Ivarsson, Y.; Zhang, Y.; Gfeller, D.; Wiesmann, C.; Sidhu, S.S. A structural portrait of the PDZ domain family. *J. Mol. Biol.* **2014**, *426*, 3509–3519. [CrossRef] [PubMed]
73. de Bono, S.; Riechmann, L.; Girard, E.; Williams, R.L.; Winter, G. A segment of cold shock protein directs the folding of a combinatorial protein. *Proc. Natl. Acad. Sci. USA* **2005**, *102*, 1396–1401. [CrossRef] [PubMed]

74. Uversky, V.N.; Gillespie, J.R.; Fink, A.L. Why are "natively unfolded" proteins unstructured under physiologic conditions? *Proteins* **2000**, *41*, 415–427. [CrossRef]
75. Uversky, V.N. What does it mean to be natively unfolded? *Eur. J. Biochem.* **2002**, *269*, 2–12. [CrossRef]
76. Fink, A.L. Natively unfolded proteins. *Curr. Opin. Struct. Biol.* **2005**, *15*, 35–41. [CrossRef]
77. Ye, F.; Zhang, M. Structures and target recognition modes of PDZ domains: Recurring themes and emerging pictures. *Biochem. J.* **2013**, *455*, 1–14. [CrossRef]
78. Lee, H.J.; Zheng, J.J. PDZ domains and their binding partners: Structure, specificity, and modification. *Cell Commun. Signal.* **2010**, *8*, 8. [CrossRef]
79. Burley, S.K.; Berman, H.M.; Kleywegt, G.J.; Markley, J.L.; Nakamura, H.; Velankar, S. Protein Data Bank (PDB): The Single Global Macromolecular Structure Archive. *Methods Mol. Biol.* **2017**, *1607*, 627–641.
80. Wu, J.; Yang, Y.; Zhang, J.; Ji, P.; Du, W.; Jiang, P.; Xie, D.; Huang, H.; Wu, M.; Zhang, G.; et al. Domain-swapped dimerization of the second PDZ domain of ZO2 may provide a structural basis for the polymerization of claudins. *J. Biol. Chem.* **2007**, *282*, 35988–35999. [CrossRef]
81. Max, K.E.; Zeeb, M.; Bienert, R.; Balbach, J.; Heinemann, U. Common mode of DNA binding to cold shock domains. Crystal structure of hexathymidine bound to the domain-swapped form of a major cold shock protein from Bacillus caldolyticus. *Febs. J.* **2007**, *274*, 1265–1279. [CrossRef] [PubMed]
82. Mojib, N.; Andersen, D.T.; Bej, A.K. Structure and function of a cold shock domain fold protein, CspD, in Janthinobacterium sp. Ant5–2 from East Antarctica. *Fems. Microbiol. Lett.* **2011**, *319*, 106–114. [CrossRef] [PubMed]
83. Christensen, N.R.; Čalyševa, J.; Fernandes, E.F.A.; Lüchow, S.; Clemmensen, L.S.; Haugaard-Kedström, L.M.; Strømgaard, K. PDZ Domains as Drug Targets. *Adv. Ther.* **2019**, *2*, 1800143. [CrossRef]
84. Dev, K.K. Making protein interactions druggable: Targeting PDZ domains. *Nat. Rev. Drug Discov.* **2004**, *3*, 1047–1056. [CrossRef] [PubMed]
85. Saupe, J.; Roske, Y.; Schillinger, C.; Kamdem, N.; Radetzki, S.; Diehl, A.; Oschkinat, H.; Krause, G.; Heinemann, U.; Rademann, J. Discovery, structure-activity relationship studies, and crystal structure of nonpeptide inhibitors bound to the Shank3 PDZ domain. *Chem. Med. Chem.* **2011**, *6*, 1411–1422. [CrossRef] [PubMed]
86. Lin, E.Y.S.; Silvian, L.F.; Marcotte, D.J.; Banos, C.C.; Jow, F.; Chan, T.R.; Arduini, R.M.; Qian, F.; Baker, D.P.; Bergeron, C.; et al. Potent PDZ-Domain PICK1 Inhibitors that Modulate Amyloid Beta-Mediated Synaptic Dysfunction. *Sci. Rep.* **2018**, *8*, 13438. [CrossRef] [PubMed]
87. Ivanova, M.E.; Fletcher, G.C.; O'Reilly, N.; Purkiss, A.G.; Thompson, B.J.; McDonald, N.Q. Structures of the human Pals1 PDZ domain with and without ligand suggest gated access of Crb to the PDZ peptide-binding groove. *Acta. Cryst. D Biol. Cryst.* **2015**, *71*, 555–564. [CrossRef]
88. Chen, H.; Tong, S.; Li, X.; Wu, J.; Zhu, Z.; Niu, L.; Teng, M. Structure of the second PDZ domain from human zonula occludens 2. *Acta. Cryst. Sect. F Struct. Biol. Cryst. Commun.* **2009**, *65*, 327–330. [CrossRef]
89. Fanning, A.S.; Lye, M.F.; Anderson, J.M.; Lavie, A. Domain swapping within PDZ2 is responsible for dimerization of ZO proteins. *J. Biol. Chem.* **2007**, *282*, 37710–37716. [CrossRef]
90. Chen, J.; Pan, L.; Wei, Z.; Zhao, Y.; Zhang, M. Domain-swapped dimerization of ZO-1 PDZ2 generates specific and regulatory connexin43-binding sites. *Embo. J.* **2008**, *27*, 2113–2123. [CrossRef]
91. Ji, P.; Yang, G.; Zhang, J.; Wu, J.; Chen, Z.; Gong, Q.; Wu, J.; Shi, Y. Solution structure of the second PDZ domain of Zonula Occludens 1. *Proteins* **2011**, *79*, 1342–1346. [CrossRef] [PubMed]
92. Peterson, F.C.; Penkert, R.R.; Volkman, B.F.; Prehoda, K.E. Cdc42 regulates the Par-6 PDZ domain through an allosteric CRIB-PDZ transition. *Mol. Cell.* **2004**, *13*, 665–676. [CrossRef]
93. Penkert, R.R.; DiVittorio, H.M.; Prehoda, K.E. Internal recognition through PDZ domain plasticity in the Par-6-Pals1 complex. *Nat. Struct. Mol. Biol.* **2004**, *11*, 1122–1127. [CrossRef] [PubMed]
94. Woods, D.F.; Bryant, P.J. ZO-1, DlgA and PSD-95/SAP90: Homologous proteins in tight, septate and synaptic cell junctions. *Mech. Dev.* **1993**, *44*, 85–89. [CrossRef]
95. Anderson, J.M. Cell signalling: MAGUK magic. *Curr. Biol.* **1996**, *6*, 382–384. [CrossRef]
96. Gonzalez-Mariscal, L.; Betanzos, A.; Avila-Flores, A. MAGUK proteins: Structure and role in the tight junction. *Semin. Cell Dev. Biol.* **2000**, *11*, 315–324. [CrossRef]
97. Bauer, H.; Zweimueller-Mayer, J.; Steinbacher, P.; Lametschwandtner, A.; Bauer, H.C. The dual role of zonula occludens (ZO) proteins. *J. Biomed. Biotechnol.* **2010**, *2010*, 402593. [CrossRef]

98. Fanning, A.S.; Ma, T.Y.; Anderson, J.M. Isolation and functional characterization of the actin binding region in the tight junction protein ZO-1. *Faseb. J.* **2002**, *16*, 1835–1837. [CrossRef]
99. Appleton, B.A.; Zhang, Y.; Wu, P.; Yin, J.P.; Hunziker, W.; Skelton, N.J.; Sidhu, S.S.; Wiesmann, C. Comparative structural analysis of the Erbin PDZ domain and the first PDZ domain of ZO-1. Insights into determinants of PDZ domain specificity. *J. Biol. Chem.* **2006**, *281*, 22312–22320. [CrossRef]
100. Nomme, J.; Antanasijevic, A.; Caffrey, M.; Van Itallie, C.M.; Anderson, J.M.; Fanning, A.S.; Lavie, A. Structural Basis of a Key Factor Regulating the Affinity between the Zonula Occludens First PDZ Domain and Claudins. *J. Biol. Chem.* **2015**, *290*, 16595–16606. [CrossRef]
101. Hiroaki, H.; Satomura, K.; Goda, N.; Nakakura, Y.; Hiranuma, M.; Tenno, T.; Hamada, D.; Ikegami, T. Spatial Overlap of Claudin- and Phosphatidylinositol Phosphate-Binding Sites on the First PDZ Domain of Zonula Occludens 1 Studied by NMR. *Molecules* **2018**, *23*, 2465. [CrossRef] [PubMed]
102. Lye, M.F.; Fanning, A.S.; Su, Y.; Anderson, J.M.; Lavie, A. Insights into regulated ligand binding sites from the structure of ZO-1 Src homology 3-guanylate kinase module. *J. Biol. Chem.* **2010**, *285*, 13907–13917. [CrossRef] [PubMed]
103. Nomme, J.; Fanning, A.S.; Caffrey, M.; Lye, M.F.; Anderson, J.M.; Lavie, A. The Src homology 3 domain is required for junctional adhesion molecule binding to the third PDZ domain of the scaffolding protein ZO-1. *J. Biol. Chem.* **2011**, *286*, 43352–43360. [CrossRef] [PubMed]
104. Ackerman, S.L.; Kozak, L.P.; Przyborski, S.A.; Rund, L.A.; Boyer, B.B.; Knowles, B.B. The mouse rostral cerebellar malformation gene encodes an UNC-5-like protein. *Nature* **1997**, *386*, 838–842. [CrossRef]
105. Leonardo, E.D.; Hinck, L.; Masu, M.; Keino-Masu, K.; Ackerman, S.L.; Tessier-Lavigne, M. Vertebrate homologues of C. elegans UNC-5 are candidate netrin receptors. *Nature* **1997**, *386*, 833–838. [CrossRef]
106. Huo, L.; Wen, W.; Wang, R.; Kam, C.; Xia, J.; Feng, W.; Zhang, M. Cdc42-dependent formation of the ZO-1/MRCKbeta complex at the leading edge controls cell migration. *Embo. J.* **2011**, *30*, 665–678. [CrossRef]
107. Dobrosotskaya, I.; Guy, R.K.; James, G.L. MAGI-1, a membrane-associated guanylate kinase with a unique arrangement of protein-protein interaction domains. *J. Biol. Chem.* **1997**, *272*, 31589–31597. [CrossRef]
108. Hirabayashi, S.; Tajima, M.; Yao, I.; Nishimura, W.; Mori, H.; Hata, Y. JAM4, a junctional cell adhesion molecule interacting with a tight junction protein, MAGI-1. *Mol. Cell Biol.* **2003**, *23*, 4267–4282. [CrossRef]
109. Charbonnier, S.; Nomine, Y.; Ramirez, J.; Luck, K.; Chapelle, A.; Stote, R.H.; Trave, G.; Kieffer, B.; Atkinson, R.A. The structural and dynamic response of MAGI-1 PDZ1 with noncanonical domain boundaries to the binding of human papillomavirus E6. *J. Mol. Biol.* **2011**, *406*, 745–763. [CrossRef]
110. Gogl, G.; Biri-Kovacs, B.; Poti, A.L.; Vadaszi, H.; Szeder, B.; Bodor, A.; Schlosser, G.; Acs, A.; Turiak, L.; Buday, L.; et al. Dynamic control of RSK complexes by phosphoswitch-based regulation. *Febs. J.* **2018**, *285*, 46–71. [CrossRef]
111. Assemat, E.; Bazellieres, E.; Pallesi-Pocachard, E.; Le Bivic, A.; Massey-Harroche, D. Polarity complex proteins. *Biochim. Biophys. Acta* **2008**, *1778*, 614–630. [CrossRef] [PubMed]
112. Roh, M.H.; Liu, C.J.; Laurinec, S.; Margolis, B. The carboxyl terminus of zona occludens-3 binds and recruits a mammalian homologue of discs lost to tight junctions. *J. Biol. Chem.* **2002**, *277*, 27501–27509. [CrossRef] [PubMed]
113. Roh, M.H.; Makarova, O.; Liu, C.J.; Shin, K.; Lee, S.; Laurinec, S.; Goyal, M.; Wiggins, R.; Margolis, B. The Maguk protein, Pals1, functions as an adapter, linking mammalian homologues of Crumbs and Discs Lost. *J. Cell Biol.* **2002**, *157*, 161–172. [CrossRef] [PubMed]
114. Li, Y.; Karnak, D.; Demeler, B.; Margolis, B.; Lavie, A. Structural basis for L27 domain-mediated assembly of signaling and cell polarity complexes. *Embo. J.* **2004**, *23*, 2723–2733. [CrossRef] [PubMed]
115. Feng, W.; Long, J.F.; Zhang, M. A unified assembly mode revealed by the structures of tetrameric L27 domain complexes formed by mLin-2/mLin-7 and Patj/Pals1 scaffold proteins. *Proc. Natl. Acad. Sci. USA* **2005**, *102*, 6861–6866. [CrossRef] [PubMed]
116. Zhang, J.; Yang, X.; Wang, Z.; Zhou, H.; Xie, X.; Shen, Y.; Long, J. Structure of an L27 domain heterotrimer from cell polarity complex Patj/Pals1/Mals2 reveals mutually independent L27 domain assembly mode. *J. Biol. Chem.* **2012**, *287*, 11132–11140. [CrossRef] [PubMed]
117. Li, Y.; Wei, Z.; Yan, Y.; Wan, Q.; Du, Q.; Zhang, M. Structure of Crumbs tail in complex with the PALS1 PDZ-SH3-GK tandem reveals a highly specific assembly mechanism for the apical Crumbs complex. *Proc. Natl. Acad. Sci. USA* **2014**, *111*, 17444–17449. [CrossRef]

118. Wei, Z.; Li, Y.; Ye, F.; Zhang, M. Structural basis for the phosphorylation-regulated interaction between the cytoplasmic tail of cell polarity protein crumbs and the actin-binding protein moesin. *J. Biol. Chem.* **2015**, *290*, 11384–11392. [CrossRef]
119. Dickinson, D.J.; Schwager, F.; Pintard, L.; Gotta, M.; Goldstein, B. A Single-Cell Biochemistry Approach Reveals PAR Complex Dynamics during Cell Polarization. *Dev. Cell.* **2017**, *42*, 416–434 e11. [CrossRef]
120. Whitney, D.S.; Peterson, F.C.; Volkman, B.F. A conformational switch in the CRIB-PDZ module of Par-6. *Structure* **2011**, *19*, 1711–1722. [CrossRef]
121. Whitney, D.S.; Peterson, F.C.; Kittell, A.W.; Egner, J.M.; Prehoda, K.E.; Volkman, B.F. Binding of Crumbs to the Par-6 CRIB-PDZ Module Is Regulated by Cdc42. *Biochemistry* **2016**, *55*, 1455–1461. [CrossRef] [PubMed]
122. Ebnet, K.; Suzuki, A.; Horikoshi, Y.; Hirose, T.; Meyer Zu Brickwedde, M.K.; Ohno, S.; Vestweber, D. The cell polarity protein ASIP/PAR-3 directly associates with junctional adhesion molecule (JAM). *Embo. J.* **2001**, *20*, 3738–3748. [CrossRef] [PubMed]
123. Itoh, M.; Sasaki, H.; Furuse, M.; Ozaki, H.; Kita, T.; Tsukita, S. Junctional adhesion molecule (JAM) binds to PAR-3: A possible mechanism for the recruitment of PAR-3 to tight junctions. *J. Cell Biol.* **2001**, *154*, 491–497. [CrossRef] [PubMed]
124. Ebnet, K.; Aurrand-Lions, M.; Kuhn, A.; Kiefer, F.; Butz, S.; Zander, K.; Meyer zu Brickwedde, M.K.; Suzuki, A.; Imhof, B.A.; Vestweber, D. The junctional adhesion molecule (JAM) family members JAM-2 and JAM-3 associate with the cell polarity protein PAR-3: A possible role for JAMs in endothelial cell polarity. *J. Cell Sci.* **2003**, *116*, 3879–3891. [CrossRef] [PubMed]
125. Feng, W.; Wu, H.; Chan, L.N.; Zhang, M. The Par-3 NTD adopts a PB1-like structure required for Par-3 oligomerization and membrane localization. *Embo. J.* **2007**, *26*, 2786–2796. [CrossRef] [PubMed]
126. Zhang, Y.; Wang, W.; Chen, J.; Zhang, K.; Gao, F.; Gao, B.; Zhang, S.; Dong, M.; Besenbacher, F.; Gong, W.; et al. Structural insights into the intrinsic self-assembly of Par-3 N-terminal domain. *Structure* **2013**, *21*, 997–1006. [CrossRef]
127. Wu, H.; Feng, W.; Chen, J.; Chan, L.N.; Huang, S.; Zhang, M. PDZ domains of Par-3 as potential phosphoinositide signaling integrators. *Mol. Cell.* **2007**, *28*, 886–898. [CrossRef]
128. Jensen, D.R.; Woytovich, C.; Li, M.; Duvnjak, P.; Cassidy, M.S.; Frederick, R.O.; Bergeman, L.F.; Peterson, F.C.; Volkman, B.F. Rapid, robotic, small-scale protein production for NMR screening and structure determination. *Protein Sci.* **2010**, *19*, 570–578. [CrossRef]
129. Renschler, F.A.; Bruekner, S.R.; Salomon, P.L.; Mukherjee, A.; Kullmann, L.; Schutz-Stoffregen, M.C.; Henzler, C.; Pawson, T.; Krahn, M.P.; Wiesner, S. Structural basis for the interaction between the cell polarity proteins Par3 and Par6. *Sci. Signal.* **2018**, *11*. [CrossRef]
130. Poliak, S.; Matlis, S.; Ullmer, C.; Scherer, S.S.; Peles, E. Distinct claudins and associated PDZ proteins form different autotypic tight junctions in myelinating Schwann cells. *J. Cell Biol.* **2002**, *159*, 361–372. [CrossRef]
131. Adachi, M.; Hamazaki, Y.; Kobayashi, Y.; Itoh, M.; Tsukita, S.; Furuse, M.; Tsukita, S. Similar and distinct properties of MUPP1 and Patj, two homologous PDZ domain-containing tight-junction proteins. *Mol. Cell Biol.* **2009**, *29*, 2372–2389. [CrossRef] [PubMed]
132. Elkins, J.M.; Papagrigoriou, E.; Berridge, G.; Yang, X.; Phillips, C.; Gileadi, C.; Savitsky, P.; Doyle, D.A. Structure of PICK1 and other PDZ domains obtained with the help of self-binding C-terminal extensions. *Protein Sci.* **2007**, *16*, 683–694. [CrossRef] [PubMed]
133. Markley, J.L.; Aceti, D.J.; Bingman, C.A.; Fox, B.G.; Frederick, R.O.; Makino, S.; Nichols, K.W.; Phillips, G.N., Jr.; Primm, J.G.; Sahu, S.C.; et al. The Center for Eukaryotic Structural Genomics. *J. Struct. Funct. Genom.* **2009**, *10*, 165–179. [CrossRef] [PubMed]
134. Moleirinho, S.; Guerrant, W.; Kissil, J.L. The Angiomotins–from discovery to function. *Febs. Lett.* **2014**, *588*, 2693–2703. [CrossRef] [PubMed]
135. Sugihara-Mizuno, Y.; Adachi, M.; Kobayashi, Y.; Hamazaki, Y.; Nishimura, M.; Imai, T.; Furuse, M.; Tsukita, S. Molecular characterization of angiomotin/JEAP family proteins: Interaction with MUPP1/Patj and their endogenous properties. *Genes Cells* **2007**, *12*, 473–486. [CrossRef]
136. Citi, S.; Sabanay, H.; Jakes, R.; Geiger, B.; Kendrick-Jones, J. Cingulin, a new peripheral component of tight junctions. *Nature* **1988**, *333*, 272–276. [CrossRef]
137. D'Atri, F.; Nadalutti, F.; Citi, S. Evidence for a functional interaction between cingulin and ZO-1 in cultured cells. *J. Biol. Chem.* **2002**, *277*, 27757–27764. [CrossRef]

138. Guillemot, L.; Schneider, Y.; Brun, P.; Castagliuolo, I.; Pizzuti, D.; Martines, D.; Jond, L.; Bongiovanni, M.; Citi, S. Cingulin is dispensable for epithelial barrier function and tight junction structure, and plays a role in the control of claudin-2 expression and response to duodenal mucosa injury. *J. Cell Sci.* **2012**, *125*, 5005–5014. [CrossRef]
139. Balda, M.S.; Garrett, M.D.; Matter, K. The ZO-1-associated Y-box factor ZONAB regulates epithelial cell proliferation and cell density. *J. Cell Biol.* **2003**, *160*, 423–432. [CrossRef]
140. Lima, W.R.; Parreira, K.S.; Devuyst, O.; Caplanusi, A.; N'Kuli, F.; Marien, B.; Van Der Smissen, P.; Alves, P.M.; Verroust, P.; Christensen, E.I.; et al. ZONAB promotes proliferation and represses differentiation of proximal tubule epithelial cells. *J. Am. Soc. Nephrol.* **2010**, *21*, 478–488. [CrossRef]
141. Schindelin, H.; Marahiel, M.A.; Heinemann, U. Universal nucleic acid-binding domain revealed by crystal structure of the B. subtilis major cold-shock protein. *Nature* **1993**, *364*, 164–168. [CrossRef] [PubMed]
142. Yang, X.J.; Zhu, H.; Mu, S.R.; Wei, W.J.; Yuan, X.; Wang, M.; Liu, Y.; Hui, J.; Huang, Y. Crystal structure of a Y-box binding protein 1 (YB-1)-RNA complex reveals key features and residues interacting with RNA. *J. Biol. Chem.* **2019**, *294*, 10998–11010. [CrossRef] [PubMed]
143. Krug, S.M.; Schulzke, J.D.; Fromm, M. Tight junction, selective permeability, and related diseases. *Semin. Cell Dev. Biol.* **2014**, *36*, 166–176. [CrossRef] [PubMed]
144. Boivin, F.J.; Schmidt-Ott, K.M. Transcriptional mechanisms coordinating tight junction assembly during epithelial differentiation. *Ann. N. Y. Acad. Sci.* **2017**, *1397*, 80–99. [CrossRef]
145. Werth, M.; Walentin, K.; Aue, A.; Schonheit, J.; Wuebken, A.; Pode-Shakked, N.; Vilianovitch, L.; Erdmann, B.; Dekel, B.; Bader, M.; et al. The transcription factor grainyhead-like 2 regulates the molecular composition of the epithelial apical junctional complex. *Development* **2010**, *137*, 3835–3845. [CrossRef]
146. Kojima, T.; Kohno, T.; Kubo, T.; Kaneko, Y.; Kakuki, T.; Kakiuchi, A.; Kurose, M.; Takano, K.I.; Ogasawara, N.; Obata, K.; et al. Regulation of claudin-4 via p63 in human epithelial cells. *Ann. N. Y. Acad. Sci.* **2017**, *1405*, 25–31. [CrossRef]
147. Kaneko, Y.; Kohno, T.; Kakuki, T.; Takano, K.I.; Ogasawara, N.; Miyata, R.; Kikuchi, S.; Konno, T.; Ohkuni, T.; Yajima, R.; et al. The role of transcriptional factor p63 in regulation of epithelial barrier and ciliogenesis of human nasal epithelial cells. *Sci. Rep.* **2017**, *7*, 10935. [CrossRef]
148. Renaud, J.P.; Rochel, N.; Ruff, M.; Vivat, V.; Chambon, P.; Gronemeyer, H.; Moras, D. Crystal structure of the RAR-gamma ligand-binding domain bound to all-trans retinoic acid. *Nature* **1995**, *378*, 681–689. [CrossRef]
149. Matias, P.M.; Donner, P.; Coelho, R.; Thomaz, M.; Peixoto, C.; Macedo, S.; Otto, N.; Joschko, S.; Scholz, P.; Wegg, A.; et al. Structural evidence for ligand specificity in the binding domain of the human androgen receptor. Implications for pathogenic gene mutations. *J. Biol. Chem.* **2000**, *275*, 26164–26171. [CrossRef]
150. Chen, C.; Gorlatova, N.; Kelman, Z.; Herzberg, O. Structures of p63 DNA binding domain in complexes with half-site and with spacer-containing full response elements. *Proc. Natl. Acad. Sci. USA* **2011**, *108*, 6456–6461. [CrossRef]
151. Ming, Q.; Roske, Y.; Schuetz, A.; Walentin, K.; Ibraimi, I.; Schmidt-Ott, K.M.; Heinemann, U. Structural basis of gene regulation by the Grainyhead/CP2 transcription factor family. *Nucleic Acids Res.* **2018**, *46*, 2082–2095. [CrossRef] [PubMed]
152. Fernandez-Leiro, R.; Scheres, S.H. Unravelling biological macromolecules with cryo-electron microscopy. *Nature* **2016**, *537*, 339–346. [CrossRef] [PubMed]
153. Punjani, A.; Rubinstein, J.L.; Fleet, D.J.; Brubaker, M.A. cryoSPARC: Algorithms for rapid unsupervised cryo-EM structure determination. *Nat. Methods* **2017**, *14*, 290–296. [CrossRef] [PubMed]
154. Cheng, Y. Single-particle cryo-EM-How did it get here and where will it go. *Science* **2018**, *361*, 876–880. [CrossRef] [PubMed]
155. Terwilliger, T.C.; Adams, P.D.; Afonine, P.V.; Sobolev, O.V. A fully automatic method yielding initial models from high-resolution cryo-electron microscopy maps. *Nat. Methods* **2018**, *15*, 905–908. [CrossRef]
156. Beck, M.; Lucic, V.; Forster, F.; Baumeister, W.; Medalia, O. Snapshots of nuclear pore complexes in action captured by cryo-electron tomography. *Nature* **2007**, *449*, 611–615. [CrossRef]
157. Schaffer, M.; Pfeffer, S.; Mahamid, J.; Kleindiek, S.; Laugks, T.; Albert, S.; Engel, B.D.; Rummel, A.; Smith, A.J.; Baumeister, W.; et al. A cryo-FIB lift-out technique enables molecular-resolution cryo-ET within native Caenorhabditis elegans tissue. *Nat. Methods* **2019**, *16*, 757–762. [CrossRef]
158. O'Reilly, F.J.; Rappsilber, J. Cross-linking mass spectrometry: Methods and applications in structural, molecular and systems biology. *Nat. Struct. Mol. Biol.* **2018**, *25*, 1000–1008. [CrossRef]

159. Yu, C.; Huang, L. Cross-Linking Mass Spectrometry: An Emerging Technology for Interactomics and Structural Biology. *Anal. Chem.* **2018**, *90*, 144–165. [CrossRef]
160. Blanchet, C.E.; Svergun, D.I. Small-angle X-ray scattering on biological macromolecules and nanocomposites in solution. *Annu. Rev. Phys. Chem.* **2013**, *64*, 37–54. [CrossRef]
161. Tuukkanen, A.T.; Spilotros, A.; Svergun, D.I. Progress in small-angle scattering from biological solutions at high-brilliance synchrotrons. *IUCrJ* **2017**, *4*, 518–528. [CrossRef] [PubMed]
162. Nannenga, B.L.; Gonen, T. The cryo-EM method microcrystal electron diffraction (MicroED). *Nat. Methods* **2019**, *16*, 369–379. [CrossRef] [PubMed]

4

Tight Junction Proteins and the Biology of Hepatobiliary Disease

Natascha Roehlen [1,2], Armando Andres Roca Suarez [1,2], Houssein El Saghire [1,2], Antonio Saviano [1,2,3], Catherine Schuster [1,2], Joachim Lupberger [1,2] and Thomas F. Baumert [1,2,3,*]

1 Institut de Recherche sur les Maladies Virales et Hépatiques, Inserm UMR1110, F-67000 Strasbourg, France; natascha.roehlen@etu.unistra.fr (N.R.); andres.roca-suarez@etu.unistra.fr (A.A.R.S.); elsaghire@unistra.fr (H.E.S.); saviano@unistra.fr (A.S.); catherine.schuster@unistra.fr (C.S.); joachim.lupberger@unistra.fr (J.L.)

2 Université de Strasbourg, F-67000 Strasbourg, France

3 Pôle Hepato-digestif, Institut Hopitalo-universitaire, Hôpitaux Universitaires de Strasbourg, F-67000 Strasbourg, France

* Correspondence: thomas.baumert@unistra.fr.

Abstract: Tight junctions (TJ) are intercellular adhesion complexes on epithelial cells and composed of integral membrane proteins as well as cytosolic adaptor proteins. Tight junction proteins have been recognized to play a key role in health and disease. In the liver, TJ proteins have several functions: they contribute as gatekeepers for paracellular diffusion between adherent hepatocytes or cholangiocytes to shape the blood-biliary barrier (BBIB) and maintain tissue homeostasis. At non-junctional localizations, TJ proteins are involved in key regulatory cell functions such as differentiation, proliferation, and migration by recruiting signaling proteins in response to extracellular stimuli. Moreover, TJ proteins are hepatocyte entry factors for the hepatitis C virus (HCV)—a major cause of liver disease and cancer worldwide. Perturbation of TJ protein expression has been reported in chronic HCV infection, cholestatic liver diseases as well as hepatobiliary carcinoma. Here we review the physiological function of TJ proteins in the liver and their implications in hepatobiliary diseases.

Keywords: Claudin; occludin; blood-biliary barrier; chronic liver disease; hepatocellular carcinoma; cholangiocellular carcinoma; NISCH syndrome

---

1. Introduction

Tight junctions (TJ) are protein complexes on epithelial cells in all organs of the body and establish paracellular diffusion barriers between different compartments. The distinct cell polarity and selective paracellular diffusion hereby provides the molecular basis of tissue homeostasis [1]. Structurally, TJs consist of transmembrane proteins that function as the diffusion barriers and cytosolic proteins that interface the junctional complexes with the cytoskeleton [1]. While initially TJs were believed to serve as simple paracellular gates, in the past years, accumulating data have identified additional functions of TJs proteins. By maintaining cellular differentiation, intercellular communication as well as assembly of signaling proteins, TJ proteins have been shown to orchestrate inside-out and outside-in signaling, hereby affecting cell proliferation, migration, apoptosis, and inflammation [2–4]. On the other hand, several growth factors, cytokines, and signaling cascades induce and regulate localization and expression of TJ proteins, hereby affecting epithelial differentiation and barrier integrity [5,6].

In the healthy liver, TJ proteins are expressed on hepatocytes, cholangiocytes, and nonparenchymal cells such as endothelial cells [5,7,8]. While TJ proteins on hepatocytes build the blood-biliary barrier (BBIB) and are hijacked during hepatitis C virus (HCV) infection, TJ proteins on cholangiocytes line the intrahepatic bile ducts [7,9,10]. Besides their localization at the apical membrane, TJ proteins have also been described to be localized at the basolateral membrane and in the cytoplasm of hepatocytes. In these non-junctional localizations, TJ proteins regulate cell-matrix interactions, intracellular signaling and proliferation, migration, and invasion [11]. Perturbation of TJ structure, protein expression, and localization have frequently been described in chronic liver and biliary diseases, indicating their fundamental role in liver biology [12]. This review provides an overview of TJ proteins being expressed in the liver, their function in maintaining TJ structure and cell signaling outside of TJs, as well as their implication in hepatobiliary diseases.

2. Biology of Tight Junction Proteins

2.1. Structure and Composition of Tight Junctions

Tight junctions are shaped by intercellular protein-protein complexes connecting plasma membranes of neighboring cells. Thus, TJs often appear as "kissing points" by electron microscopy. Two models of TJ structure exist: the protein model and the protein-lipid hybrid model. The protein model postulates construction of the junctional diffusion barrier by transmembrane proteins on both sides, interacting in a homotypic or heterotypic way (shown in Figure 1a), whereas the hybrid model proposes membrane hemifusions built by inverted lipid micelles and stabilized by transmembrane proteins [1]. Yet no consensus on the ultrastructural appearance has been reached. However, in both cases, TJs build a regulatory semipermeable gate that enables selective paracellular diffusion depending on the size and charge of the corresponding molecule [1]. Moreover, TJs form an intramembrane barrier (also referred to as "fence function"), that restricts exchange between the cells' apical and basolateral surfaces [13]. However, whether the fence function of TJs is critical or not for the establishment of a polarized phenotype has been a matter of debate, taking into account that it has been observed how epithelial cells are able to polarize in the absence of cell-cell junctions [14,15].

The transmembrane domains of TJs on epithelial cells are mainly built by tetraspanin-associated proteins of the claudin (CLDN) family and the junctional proteins occludin (OCLN) and MarvelD3, which contain a MAL and related proteins for vesicle trafficking and membrane link (MARVEL) domain. Moreover, junctional adhesion molecules (JAMs) have been reported as integral membrane proteins in TJs [16,17]. Tricellular TJ proteins characterize cell adhesion between three neighboring cells and include tricellulin [18], lipolysis-stimulated lipoprotein receptor (LSR) [19], as well as immunoglobulin-like domain containing receptor (ILDR1 and ILDR2) [20]. Representatives of the cytosolic junctional plaque on the other hand are adapter proteins as Zonula occludens 1-3 (ZO1-3), membrane-associated guanylate kinase inverted (MAGI) proteins, and cingulin [1] (Figure 1a).

OCLN was the first identified transmembrane protein in TJs and belongs to the large protein family of Marvel-domain-containing proteins [21]. In contrast to the multiple and differentially expressed members of CLDN family, only one OCLN transcript has been described, which however occurs in differently spliced variants. With a size of 65 kDa, OCLN contains four transmembrane domains, one small intracellular loop, two extracellular loops, and intracellular localized C and N terminals (Figure 1a) [22].

The family of CLDN proteins comprises 27 members in mammals [23]. According to their physiological role in paracellular permeability, CLDNs can further be subgrouped into sealing CLDNs (CLDN1, 3, 5, 11, 14, and 19), cation-selective (CLDN2, 10b and 15) and anion-selective paracellular channel forming CLDNs (CLDN10a and 17), as well as water-permeable CLDNs (CLDN2 and 15). For the remaining CLDNs, their roles on epithelial barriers are not yet fully understood [24]. These 20–27 kDa proteins consist of four transmembrane domains, two extracellular loops, and a cytoplasmatic carboxyl tail (Figure 1a). As integral proteins of TJs, CLDNs are reported to regulate ion and water permeability of the paracellular barrier [1,25,26].

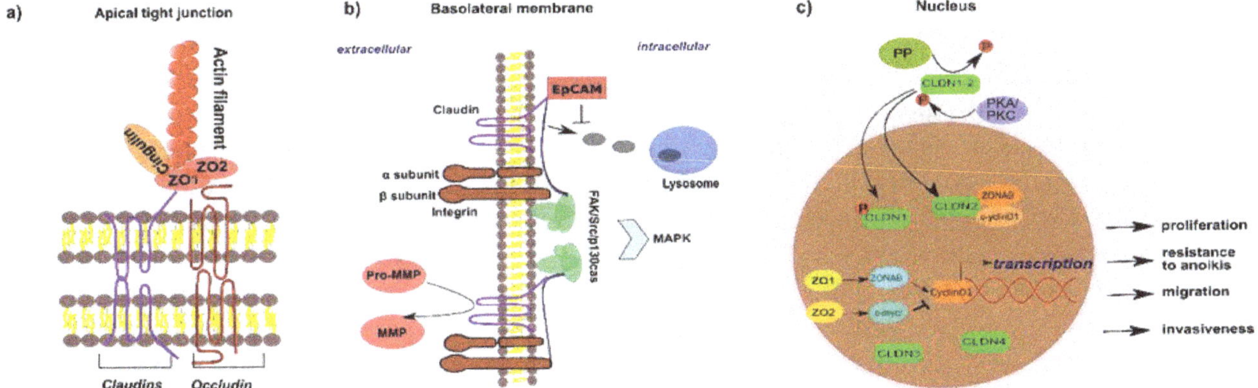

Figure 1. Functions of tight junction proteins at different subcellular localizations. Tight junction proteins are expressed at three different locations within epithelial cells with different functions including the apical membrane (**a**), the basolateral membrane (**b**), and in the nucleus (**c**). (**a**) At the apical membrane, tight junctions (TJs) are typically built by integral membrane proteins of the CLDN or Marvel-domain containing protein family (e.g., occludin—OCLN) that connect via C-terminus bound adapter proteins to intracellular actin filaments. (**b**) In the normal intestinal mucosa and in various cancer cell types, basolateral localized CLDNs have been found to regulate activation of pro-MMPs into MMPs and to interact with integrins at focal adhesion complexes, hereby affecting main intracellular signaling cascades such as the MAPK pathway. Investigations on colon cancer cell lines indicate EpCAM to specifically stabilize expression of CLDN1 and 7 at the basolateral membrane and to prevent their lysosomal degradation. (**c**) Nuclear localization has been reported for ZO1 and ZO2 as well as CLDN1-4 in various cancer cell types and is regulated by posttranslational modification. Within the nucleus, CLDN2 retains cyclinD1 and ZONAB hereby enhancing cell proliferation. Specific interaction of ZO1 with the transcription factor ZONAB regulates G1/S-phase progression by increasing cyclin D1, while ZO2 inhibits transcription of cyclin D1 by binding to c-myc. CLDN (Claudin); c-myc (MYC proto-oncogene); EpCAM (epithelial cell adhesion molecule); FAK (focal adhesion kinase); MAPK (Mitogen-activated protein kinase); MMP (Matrix-metalloproteinase); PKA (protein kinase A); PKC (protein kinase C); PP (protein phosphatase); Src (steroid receptor coactivator); ZO1 (Zonula occludens 1); ZO2 (Zonula occludens 2); ZONAB (ZO1-associated nucleic acid binding protein).

With four transmembrane domains, cytoplasmatic C- and N-terminals, and two extracellular loops, tricellulin shows strong structural similarity to CLDNs and OCLN [18,27]. While OCLN and CLDN represent the main transmembrane proteins of apical cell adhesions between two cells (bicellular tight junction, bTJ), tricellulin is mainly enriched at tricellular contact regions (tricellular tight junction, tTJ), although also been identified in bTJs [18]. LSR, ILDR1 and 2, which are commonly described as the angulin family, have been reported to recruit tricellulin to tTJ [20].

JAMs belong to the immunoglobulin superfamily (IgSF). Originally discovered on leucocytes as key players of leucocyte-endothelial cell interaction and trans-endothelial migration, JAM-A-C as well as the related IgSF members CAR, endothelial cell-selective adhesion molecule (ESAM), and JAM-4 were later described to be enriched in epithelial and endothelial TJs. Consisting of two IgSF domains, two Ig-like domains, one single transmembrane domain, and a PDZ-domain binding cytoplasmatic tail, these proteins contribute to barrier formation and TJ associated signaling [16,17].

Besides transmembrane proteins, TJs consist of junctional plaque components that connect the junctional membrane with the cytoskeleton. ZO proteins are the most important adapter proteins, that connect CLDN, OCLN, and tricellulin with the cytoskeleton, hereby enabling clustering of protein complexes to the intracellular domains of TJs (Figure 1a). Apart from TJs, ZO proteins have also been described in cadherin-based adherens junctions and gap junctions [28]. Three ZO proteins (ZO1-3) with high structural similarity have been discovered. ZO1, the best described member of the family of ZO proteins represents a 220 kDa scaffolding protein, that includes three types of functional domains, a Src homology 3 domain (SH3), three PDZ domains, a proline rich and a guanylate kinase domain [29,30].

ZO proteins directly interact with the intracellular actin filaments and the first PDZ domain has been shown to associate with the C-terminus of CLDN and OCLN proteins, hereby regulating TJ assembly (Figure 1a) [31,32]. Other representatives of the junctional plaque are cingulin and 7H6 [33,34]. For a detailed review regarding the general structure and composition of TJs see [35,36].

The TJ complex is known to be highly dynamic with continuous remodeling by clathrin-mediated endocytic recycling [37–40]. Recycled or newly produced TJ proteins are sorted in the Golgi-network and transported by specific trafficking proteins to the desired localizations [41,42]. On the other hand, several growth factors, cytokines, and signaling cascades induce and regulate localization and expression of TJ proteins, hereby affecting epithelial differentiation and barrier integrity [5,6].

Knockout (KO) studies in cultured epithelial cells indicate an increase of paracellular permeability by loss of single CLDN proteins [43,44]. In contrast, KO of OCLN does not alter baseline barrier function, but attenuates cytokine-induced increase in trans-epithelial resistance [45]. Knockdown of tricellulin using siRNA decreases trans-epithelial electrical resistance and increases the paracellular permeability in cultured epithelial cells [18]. JAM-A in vitro and in vivo KO studies revealed increased epithelial permeability potentially due to perturbed regulation of CLDN expression and induction of apoptosis [46,47]. Loss of ZO1 retards but not completely hampers TJ formation, probably due to compensatory upregulation of ZO2. Thus, assembly of CLDN and OCLN proteins to TJs takes longer in the absence of ZO1 but does not block eventual establishment of the polarized epithelial structure with functional TJs within hours in cell culture [15]. However, KO of ZO1 and knockdown of ZO2 by RNA interference results in diffuse distribution of integral TJ proteins in epithelial cells with severe perturbation of the paracellular barrier [48]. While to our knowledge KO of 7H6 in epithelial cells has not yet been analyzed, its localization would suggest a paracellular barrier function [49,50]. In mice invivo KO or knockdown of TJ proteins results in a wide variety of phenotypes, ranging from a normal phenotype without any disease to lethality [51–55]. Furthermore, there are differences in the phenotype of TJ protein loss of function in mice and humans: e.g., while CLDN1 KO in a mouse model has shown to be lethal [52], congenital CLDN1 KO loss-of function mutations in human patients can manifest in a highly variable phenotype ranging normal health without disease to neonatal sclerosing cholangitis and ichthyosis of variable severity (NISCH syndrome), potentially due to compensatory upregulation of other CLDN members [56]. This indicates differential functions of the TJ orthologs in mice and humans and suggests that a complete loss of TJ proteins can be functionally compensated as shown for CLDN1 in humans.

2.2. *Non-Junctional Localization of Tight Junction Proteins*

Several TJ proteins have been described to be also localized outside of TJs at the basolateral membrane, in the cytoplasm, and in the nucleus. Non-junctional TJ proteins exert key regulatory functions on cell proliferation, cell adhesion, as well as migration and invasion [11]. As an example, CLDN1, 2, and 7 regulate cell-matrix interaction by forming complexes with integrin proteins at focal adhesions on the basolateral membrane of human lung, melanoma, colon, as well as breast cancer cells (Figure 1b) [57–61]. These interactions have not only been shown to affect epithelial adhesion to the matrix and cell proliferation [59], but also to be associated with cancer progression and metastasis [61]. The epithelial cell adhesion molecule (EpCAM) specifically stabilizes this non-junctional CLDN expression and regulates its lysosomal degradation (Figure 1b) [62]. In line with the potential pro-oncogenic function of CLDN proteins at the basolateral membrane, interaction of EpCAM with CLDN7 was reported to promote tumor progression and cell dissemination [63].

Several studies link basolateral CLDN expression with expression and activity of matrix metalloproteinases (MMPs) [64–66]. At the basolateral membrane of epithelial cells, secreted MMPs are able to degrade extracellular matrix proteins [67]. Interestingly, CLDN proteins have been shown to recruit and activate pro-MMP, hereby promoting migration and invasion of the corresponding cancer cells (Figure 1b) [68].

Nuclear localization has been reported for ZO1/ZO2 [69,70] and CLDN1-4 [71–74] in several types of cancer cells. The conditions or inducers under which these TJ proteins localize in the nucleus are poorly understood. However, in the case of CLDN1, phosphorylation by protein kinase A and C (PKA and PKC) has been shown to promote nuclear import [75]. Nuclear import of CLDN2 on the other hand is induced by dephosphorylation [72]. Functional investigations in colon cancer cells indicate nuclear localization of CLDN proteins to be associated with resistance to anoikis as well as migration and invasiveness [71], while nuclear localization of ZO1/ZO2 affects cell cycle progression and cell proliferation by transcriptional regulation of cyclin D1 in tumorous and non-tumorous epithelial cells [76,77] (Figure 1c).

3. Tight Junction Proteins and Their Role in Signaling

In colon and liver cancer cells, TJ proteins functionally crosstalk with key cellular signaling pathways, including PI3K/AKT, Wnt/β-catenin, and EGFR/ERK signaling [78–80]. Proteomic analysis of OCLN and CLDNs revealed numerous binding partners, that are known to be involved in cell signaling and trafficking, such as kinases, phosphatases, signaling adaptors, and receptor proteins [81,82]. A strong body of evidence indicates functional crosstalk of CLDN proteins with the EGFR signaling pathway. Dhawan et al. reported CLDN2 overexpression to promote cell proliferation in an EGFR-dependent manner in colon tumor cells [79]. De Souza et al. found EGF to increase CLDN3 expression via ERK and PI3K signaling, hereby accelerating colorectal tumor cell migration in vitro [83]. Finally, EGFR signaling has been shown to mediate the formation of a CD81-CLDN1 complex, hereby enabling entry of HCV into hepatocytes [82,84] (Figure 2).

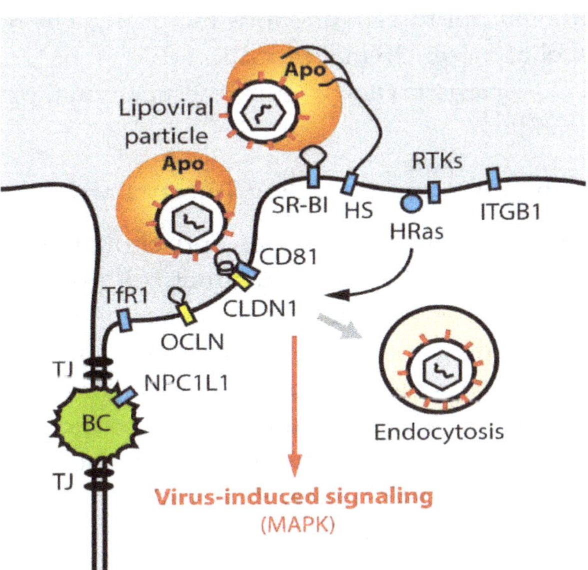

Figure 2. Hepatitis C virus (HCV) entry process and signaling. HCV lipoviral particle entry into hepatocytes requires a complex orchestration of entry factors that involves non-junctional TJ proteins CLDN1 and OCLN and virus-induced host signaling. Apo (Apolipoproteins), BC (Bile canaliculi), CD81 (Cluster of Differentiation 81), CLDN1 (Claudin-1), HRas (HRas Proto-Oncogene, GTPase), HS (Heparan sulfate), ITGB1 (Integrin Subunit Beta 1), MAPK (Mitogen-activated protein kinase), NPC1L1 (Niemann-Pick C1-like protein 1), OCLN (Occludin), RTK (Receptor tyrosine kinases), SR-BI (Scavenger Receptor Class B Member 1), TfR1 (Transferrin Receptor 1), TJ (Tight junction).

Several studies further associate CLDN proteins with proapoptotic signaling. Singh et al. indicated CLDN1 as a driver of resistance to anoikis in colon cancer cells, a form of self-programmed death in epithelial cells following detachment from the surrounding extracellular matrix. Mechanistically, CLDN1 was found to directly interact with steroid receptor coactivator (Src), a non-receptor tyrosine kinase that binds to extracellular matrix proteins and plays a pivotal role in cellular signal transduction, promoting survival, proliferation, and angiogenesis in its activated form. The authors postulated the presence of a multiprotein complex consisting of CLDN1, ZO1, and Src2 that regulates activation of Src downstream oncogenic signaling [85]. Another cellular self-defense mechanism, Fas-mediated apoptosis, has been shown to alter OCLN and ZO1 expression in lung epithelia [86].

Furthermore, several studies indicate TJ proteins to function as intracellular signaling platforms, involved in regulation of cell differentiation and growth. Indeed, Spadaro et al. reported conformational changes of ZO1 to induce recruitment of the transcription factor DbpA to TJs in epithelial (Eph4) cells, hereby affecting cell proliferation [87]. In lung cells, interaction between CLDN18 and the signaling molecule Yes-associated protein (YAP) has been shown to affect colony formation and progenitor cell proliferation [88].

Posttranslational modification of TJ transmembrane proteins by growth factor signaling pathways fine-tune the TJ barrier function. Mitogen-activated protein kinase (MAPK) [89] and PKA [90] have been shown to phosphorylate CLDN1 at TJs of cerebral and lung endothelial cells, hereby affecting TJ permeability. Phosphorylation of CLDN5, induced by cyclic-AMP potentiates the blood–brain barrier [90], while PKA mediated phosphorylation of CLDN16 affects Mg2+ transport in kidney cells [91]. Vascular endothelial growth factor (VEGF) signaling perturbs hepatocellular TJ integrity by targeting OCLN via the PKC pathway [92]. Moreover, several studies indicate that cytokines, which are upregulated during inflammation, affect TJ protein expression. For example, Ni and coworkers demonstrated that TNF-α-induced phosphorylation of OCLN in human cerebral endothelial cells via MAPK, modulates TJ permeability [93]. Moreover, OCLN phosphorylation regulates its interaction with ZO1 in kidney cancer cells [94]. Exposure of intestinal epithelial cells with TNF-α hampers TJ permeability via NF-κB-dependent downregulation of ZO1 expression and altered junctional localization [95]. Loss of epithelial cell-to-cell junctions including TJs, represents a typical and early event in the evolution of epithelial-mesenchymal transition (EMT). EMT describes a process by which epithelial cells lose epithelial characteristics and acquire mesenchymal properties including the ability of migration and invasion [96,97].

4. Tight Junction Proteins in the Liver and the Blood-Biliary Barrier

Epithelial cells in the liver, namely hepatocytes and cholangiocytes, form the parenchymal structure of the organ and are characterized by a distinct cell polarity. TJs between neighboring hepatocytes separate the hepatocyte cell membrane into basal (sinusoidal), basolateral, and apical (bile canalicular) domains. By sealing the paracellular space, TJs and other adhesion complexes build the physiological BBIB, that segregates blood-containing basal hepatic sinusoids from apical bile canaliculi [9]. The BBIB hereby enables simultaneous execution of two major functions of the liver: the production and secretion of bile and the continuous metabolic exchange with the portal and systemic circulation allowing detoxification and excretion of proteins and coagulation factors. In particular, the apical bile canalicular domain of hepatocytes is characterized by numerous bile transporters and microvilli, that are required for bile secretion and absorption, while the basolateral sinusoidal domain is specialized in metabolic exchange with the blood [98]. CLDNs 1-3 and OCLN are expressed in TJs of hepatocytes and cholangiocytes [53,99,100]. While transmembrane TJ proteins on hepatocytes build the BBIB and shape bile canaliculi, TJs on cholangiocytes line the intrahepatic bile ducts [7]. The gallbladder on the other hand, shows physiologically strong expression of CLDNs 2, 3, 7, and OCLN. The hepatic sinusoidal endothelium strongly expresses CLDN5 [8].

In the normal liver and in contrast to other TJ proteins, tricellulin expression in hepatocytes and biliary epithelial cells strongly variates between individuals but is accentuated at tricellular contacts in colocalization with CLDN1 and CLDN4 [101]. In contrast to their weak expression on hepatocytes, the junctional adaptor proteins 7H6 and ZO1 are enriched in bile canaliculi [33,102].

KO studies in mice suggest a crucial role of CLDN2 and 3 for the BBIB. Thus, KO of the channel-forming CLDN2 lead to cholesterol gallstone disease due to a decrease in paracellular water transport [53]. CLDN3 KO in mice on the other hand, increases the paracellular phosphate ion transport of hepatic tight junctions, resulting in calcium phosphate core formation. Cholesterol overdose causes the cholesterol gallstone disease in these mice [99].

5. Tight Junction Proteins in Chronic Hepatobiliary Diseases

Chronic liver diseases constitute a global health problem, associated with high mortality due to its complications of liver cirrhosis and cancer [103]. Major causes comprise chronic hepatitis B virus (HBV) and HCV infection, alcoholic and metabolic liver disease such as non-alcoholic steatohepatitis. Decompensated liver cirrhosis is the fourth most common cause of death in adults in central Europe [104,105]. Downregulated expression or impaired function of TJ proteins have frequently been associated with chronic liver diseases [12]. Loss of the BBIB, which is maintained by junctional adhesion complexes including TJs represents a common feature in mice models of chronic liver injury [106,107]. Takaki et al. observed loss of TJ protein expression, including CLDN3 and ZO1 following hepatectomy and reappearance several days after surgery. This suggests a functional role of TJ proteins in liver regeneration [108]. Moreover, alterations related to the expression of TJ proteins have been implicated in chronic HCV infection, biliary diseases, and liver cancer.

5.1. Tight Junction Proteins and HCV Infection

Chronic HCV infection represents a serious global health problem affecting more than 71 million people worldwide and potentially leads to liver fibrosis, cirrhosis, and hepatocellular carcinoma (HCC) [109–111]. Cell entry is a critical step in the HCV life cycle and involves a complex multi-step process consisting of viral attachment to the hepatocyte cell membrane and internalization [10,112]. HCV requires a complex orchestration of host dependency factors including among others CLDN1, OCLN, CD81, and SR-B1. Mechanistically, EGFR signaling promotes CLDN1-CD81 coreceptor association, which is a prerequisite for the internalization of the virus (Figure 2).

OCLN on the other hand, is believed to act downstream of the other cell entry factors CD81, CLDN1, and SRB1 during the HCV entry process [113,114]. OCLN interacts with HCV surface glycoprotein E2 via its extracellular loop 2 (ECL2) [115]. Of note, transgenic expression of human OCLN enables HCV infection of non-permissive species like mice [116–118]. However, the exact mechanism and localization of OCLN-HCV interaction is not fully understood. Considering its role for HCV cell entry, alterations in CLDN1 and OCLN expression levels and their functional consequences have been a focus of interest in the HCV field within the last years. Hepatic expression of CLDN1 and OCLN was found to be increased in liver biopsies of patients with chronic HCV infection [119]. In accordance, HCV liver graft infection is associated with OCLN and CLDN1 upregulation [120].

Anti-CLDN1 antibodies prevent and eliminate chronic HCV infection in cell-based and animal models without any detectable adverse effects and especially without disrupting TJ integrity or function [121–124]. The safety profile was further confirmed in human liver-chimeric mice and is most likely related to the molecular mechanism of action of CLDN1 monoclonal antibodies (mAbs) targeting the non-junctional expressed CLDN1 on hepatocytes without binding to CLDN1 localized in TJs [123–125]. Xiao et al. reported synergistic effects of anti-CLDN1 mAb with direct-acting antivirals as antiviral approaches for difficult-to-treat patients [126,127]. Confirming the functional role of OCLN in HCV entry, previous mechanistic monoclonal antibodies targeting ECL2 of OCLN were efficient in the prevention of infection both in cell culture and human liver chimeric mice without detectable side effects [114,128,129].

5.2. Tight Junction Proteins in Hepatocellular Carcinoma

Primary liver cancer is the sixth most frequent and second most deadly type of cancer in the world, with HCC being the most common histological subtype (75%–85%) [130]. Several members of the CLDN family have been reported to be perturbed during hepatocarcinogenesis. CLDN1, 4, 5, 7, and 10 are overexpressed in HCC [80,131–135]. Low levels of CLDN5 and high levels of CLDN7 were found to be independent prognostic factors [131]. Similarly, CLDN10 overexpression in HCC correlated with poor patients' outcome and tumor recurrence [133,136]. In contrast, CLDN14 downregulation in HCCs correlates with advanced tumor stage and poor overall survival [137] and CLDN3 expression is decreased in HCC [138]. Bouchagier and coworkers reported an overexpression of OCLN in HCC tumors compared to non-neoplastic liver tissues, which positively correlated with a favorable prognosis [131]. Orban et al. on the other hand, found decreased OCLN mRNA and protein levels in HCC [102]. These opposing findings may be due to different histological grading of the analyzed HCC samples and a potential dedifferentiation characterized by decreased OCLN levels. Decreased cell migration and proliferation following treatment of HCC cells with different compounds was accompanied by upregulation of OCLN expression, indicating mesenchymal-epithelial transition (MET) [139–141] and thus supporting the findings from Bouchagier et al. Expression of tricellulin is very heterogeneous in HCC tissues, but seems to be positively correlated with poor prognosis [101]. Downregulation of ZO1 on the other hand, associates with poor prognosis in HCC patients undergoing hepatectomy [142]. Collectively, these studies suggest a pathogenic role of TJ proteins in hepatocarcinogenesis.

Studies on TJ protein expression in chronic liver diseases together with clinical correlations are summarized in Table 1.

Table 1. Perturbation of TJ proteins in chronic liver diseases.

Liver Disease	Tight Junction Protein	Perturbation	Potential Clinical Impact	References
HCV infection	CLDN1	• Overexpression in chronically HCV-infected liver tissue • Upregulation upon HCV liver graft infection	• SNPs in *CLDN1* promoter confer susceptibility to HCV infection • Crucial HCV entry factor, antiviral target	[143,144] [119,121,124], [120,122,123]
	OCLN	• Overexpression in chronically HCV-infected liver tissue • Upregulation upon HCV liver graft infection	• Crucial HCV entry factor, antiviral target	[114,119,120,128,129,145]
HCC	CLDN1	• Upregulated in the large majority of HCCs	• Correlation of expression with patients' survival • Therapeutic target	[80,131,132,134,135]
	CLDN3, CLDN14	• Downregulated/low expression in HCC	• Unknown	[137,138]
	CLDN4, 5, 7 and 10	• Upregulated in HCC	• Unknown	[131] [133,136]
	OCLN	• Both downregulated and upregulated described in HCC	• Positive correlation of expression with good prognosis	[102,131]
	Tricellulin	• Heterogeneous	• Positive correlation with poor prognosis	[101]
	ZO1	-	• Low expression correlates with HCC recurrence after hepatic resection	[142]

6. Tight Junction Proteins in Biliary Diseases

Considering that TJ proteins on bile canaliculi are major contributors to the BBIB, TJ integrity has frequently been investigated in biliary diseases. Indeed, disruption of bile duct epithelial barrier plays a crucial role in the pathogenesis of chronic biliary diseases [7]. Studies in animal models of cholestatic disease hereby revealed secondary expressional and morphologic alterations of the tight junctional network upon cholestatic liver injury [146]. Perturbation of TJ proteins could further be found in human biliary liver diseases as primary sclerosing cholangitis (PSC) [147] and cholangiocellular carcinoma (CCA) [148]. Moreover, primary perturbation of TJ proteins caused by homozygous mutations have been identified to account for cholestatic syndromes, including progressive familial intrahepatic cholestasis (PFIC) type 4 [149,150] and the neonatal ichthyosis-sclerosing cholangitis (NISCH) syndrome [151].

6.1. Tight Junction Proteins in Primary Biliary Cirrhosis and Secondary Sclerosing Cholangitis

Primary biliary cirrhosis (PBC) and PSC represent etiologies of chronic liver disease that are characterized by cholestasis and an increased risk of developing liver cirrhosis and cancer. Mediated by immunological mechanisms of bile duct destruction, patients typically present with elevated serum levels of bile acids [152,153]. Ultrastructural studies of damaged bile ducts in PBC show electron-dense deposits in enlarged intercellular spaces, infiltrated by immune cells indicating perturbated barrier integrity [154]. TJ proteins are responsible for the main barrier formations maintaining the BBIB and preventing bile regurgitation from the biliary tract. In this context, downregulation of the TJ proteins 7H6 and ZO1 in bile ducts in PBC and in hepatocytes in PSC has been suggested to account for the increased paracellular permeability observed in chronic cholestatic liver diseases. Consequently, toxic bile acids can enter the periductal area and promote the infiltration of immune cells, eventually leading to inflammatory driven progression of bile injury. Interestingly, the expression of these TJ proteins is preserved in PBC patients treated with ursodeoxycholic acid [147].

6.2. Primary Perturbation of Tight Junction Proteins in Biliary Diseases: NISCH Syndrome and PFIC Type 4

NISCH syndrome represents an extremely rare autosomal-recessive ichthyosis syndrome caused by mutations in the *CLDN1* gene leading to its abolished expression in liver and skin (KO phenotype). First being described in 2002, only 12 cases have been reported [151,155–161]. The clinical manifestation is variable ranging from absent or regressive cholestasis to progressive liver disease with liver failure. The hepatic feature of this syndrome is characterized by neonatal sclerosing cholangitis with elevated serum bile acids and hepatomegaly. Additional non-hepatic manifestations can include dental anomalies, mild psychomotor delay, ichthyosis, and scalp hypotrichosis as well as scarring alopecia [56,151]. The human phenotype hereby strongly deviates from the one observed in CLDN1-KO mice that present severely wrinkled appearance of the skin and death within 24 h after birth [52], indicating differential function of CLDNs in mice and humans. Thus, increased paracellular permeability and secondary bile injury due to CLDN1 absence in patients with NISCH syndrome [44] may be compensated by overexpression of other TJ protein members in the liver, explaining the variable phenotype [56]. Alternatively, mutations in other genes may be responsible for part of the observed phenotype. In conclusion, these findings demonstrate that CLDN1 is not essential for life in humans and its absence has a variable clinical phenotype.

Loss of ZO2 on the other hand, is observed in PFIC type 4 [149,150]. Mechanistically, a mutation in the *ZO2* gene has been described to hamper proper localization of CLDN1 in TJs of cholangiocytes in the liver despite normal protein levels, hereby increasing paracellular permeability to bile acids [149].

Clinical signs of cholestasis appear within the first year of life in patients homozygous for this mutation and are typically contrasted by normal levels of γ-glutamyl transferase activity (GGT). Progressing into secondary biliary cirrhosis, affected patients present with severe liver disease at a young age, often requiring liver transplantation [149]. A missense mutation in the first PDZ domain of ZO2, that binds to CLDN1 in TJs has further been described in patients with familial hypercholanemia, characterized by pruritus and fat malabsorption but without progressive liver disease [162].

6.3. *Tight Junction Proteins in Cholangiocellular Carcinoma*

Cholangiocellular carcinoma (CCA) represents the second most common primary liver cancer type. With an overall incidence rate of 2/100000 it belongs to the rather rare cancer subtypes, though within the last few years, a dramatic increase in prevalence and mortality have been documented [163–165]. In contrast to the strong linkage of liver fibrosis/cirrhosis with HCC, most CCAs occur sporadically. However, known risk factors are PSC and HBV/HCV associated liver cirrhosis [166–169].

Several studies have reported evidence for potential functional implication of TJ proteins in CCA. CLDN3, 7, 8, and 10 expression were found to be decreased in intrahepatic CCAs compared to normal tissues. Significantly lower expression of CLDN1, 8, and 10 was also found in extrahepatic CCA, while CLDN1, 2, 3, 7, 8, and 10 are decreased in CCA of the gallbladder [148]. The most significant alteration of CLDN expression between CCA and adjacent liver tissue was found for CLDN10, as it was markedly decreased in all forms of bile duct cancers [148]. Moreover, in contrast to its restricted membrane localization in normal bile epithelia, intrahepatic CCA showed cytoplasmatic localization of CLDN10. Based on the negative staining in HCC and normal mature hepatocytes, CLDN4 and CLDN7 have been suggested as immunohistochemical markers of cholangiocellular differentiation in primary liver cancer [170,171]. In view of its preserved or even elevated expression in intra- and extrahepatic CCA, especially CLDN4 represents an attractive histological marker of CCA [148]. Interestingly, downregulation of CLDN4 by siRNA led to decreased migration and invasion of CCA cell lines [172]. CLDN18, that has been intensively studied in relation to gastric cancer is expressed in 40% of intrahepatic CCAs and is associated with lymph node metastasis and poor prognosis [173].

In intrahepatic CCA, tricellulin is decreased compared to adjacent tumor tissue, while patients with preserved tricellulin expression had significantly better clinical outcome and lower histological grading [101]. Downregulation of ZO1 and OCLN are associated with progression in biliary tract cancers [174].

All reported perturbations of TJ protein expressions in chronic hepatobiliary diseases are summarized in Table 2.

Table 2. Perturbation of TJ proteins in chronic biliary diseases.

Biliary Disease	TJ Protein	Perturbation	Potential Clinical Implication	References
Primary biliary cirrhosis (PBC)	ZO1	• Downregulation in bile ducts of patients with PBC	• Increased paracellular permeability • Preservation of ZO-1 expression in patients treated with ursodeoxycholic acid	[147]
Primary sclerosing cholangitis (PSC)	ZO1	• Downregulation on hepatocytes of patients with PSC	• Increased paracellular permeability	[147]
Progressive familial intrahepatic cholestasis (PFIC) type 4	ZO2	• Loss of expression	• Failed localization of CLDN1 to TJs on cholangiocytes despite normal CLDN1 protein levels • Increased paracellular permeability • Progressive chronic liver disease	[149,150]
Familial hypercholanemia	ZO2	• Missense mutation in the first PDZ domain of ZO2	• Perturbed localization of CLDN1 in TJs • Pruritus, fat malabsorption, elevated serum bile acid concentrations	[162]
NISCH syndrome	CLDN1	• Loss of CLDN1 expression due to homozygous *CLDN1* mutation (functional KO)	• Variable clinical outcome from mild to absent disease to neonatal sclerosing cholangitis and ichthyosis (with functional impact of additional mutations unknown) • Increased paracellular permeability	[56,151]
CCA	CLDN1–3, 7, 8, and 10	• Perturbed expression in intrahepatic, extrahepatic CCA, and/or CCA of the gallbladder	• CLDN7: suggested as histological marker to distinguish CCA from HCC	[148,170]
	CLDN4	• Perturbed expression in CCA	• Suggested as histological marker to distinguish HCC and CCA	[148,170–172]
	CLDN18	• Expressed in 40% of intrahepatic CCAs	• Expression is associated with lymph node metastasis and poor prognosis	[173]
	Tricellulin	• Downregulated in CCA	• Positive correlation of expression with clinical outcome and low staging	[101]
	OCLN	• Downregulated in CCA	• Correlation of downregulated expression with tumor progression	[148]
	ZO1	• Downregulated in CCA	• Correlation of downregulated expression with tumor progression	[148]

7. Summary

Tight junction proteins on hepatocytes and cholangiocytes play an important functional role as paracellular gatekeepers and represent the molecular basis of the BBIB, enabling exertion of two major function of the liver: production and secretion of bile as well as metabolic exchange and detoxification. Moreover, non-junctional TJ proteins at the basolateral membrane and in the nucleus exert key functions in cellular signaling, apoptosis, and migration. The TJ proteins CLDN1 and OCLN on the basolateral membrane of hepatocytes serve as entry factors for HCV—a major cause of liver disease and cancer worldwide. Highlighting its function as regulators of paracellular permeability enabling maintenance of the BBIB, secondary perturbation of TJ proteins has been described in biliary diseases, including PSC and PBC. In humans, the complete loss of distinct TJ proteins is not lethal, and the associated clinical phenotypes are highly variable as described for NISCH-syndrome or PFIC type 3. Finally, up- or downregulation of TJ protein expression in hepatobiliary cancer suggests a functional implication of TJ proteins in key cell regulatory signaling cascades potentially associated with carcinogenesis.

Author Contributions: N.R. and T.F.B. conceptualized the N.R. performed the literature review and wrote the manuscript. J.L., C.S. and T.F.B. revised the manuscript, J.L., A.S., H.E.S. and A.A.R.S. prepared original figures. All authors have read and agreed to the published version of the manuscript.

Abbreviations

Akt	AKT serine/threonine kinase
Apo	Apolipoprotein
BBIB	Blood-biliary barrier
bTJ	Bicellular tight junction
CCA	Cholangiocellular carcinoma
CD81	Cluster of differentiation 81
CLDN	Claudin
c-myc	MYC proto-oncogene
ECL2	Extracellular loop 2
EGFR	Epidermal growth factor receptor
EMT	Epithelial-mesenchymal transition
EpCAM	Epithelial cell adhesion molecule
ESAM	Endothelial cell-selective adhesion molecule
FAK	Focal adhesion kinase
GGT	γ-glutamyl transferase
HBV	Hepatitis B virus
HCC	Hepatocellular carcinoma
HCV	Hepatitis C virus
HRas	HRas proto-oncogene, GTPase
HS	Heparan sulfate
IgSF	Immunoglobulin superfamily
ILDR	Immunoglobulin-like domain containing receptor
ITGB1	Integrin subunit beta 1
JAM	Junctional adhesion molecules
KO	Knockout
LSR	Lipolysis-stimulated lipoprotein receptor
mAbs	Monoclonal antibodies

MAGI	Membrane-associated guanylate kinase inverted
MAPK	Mitogen-activated protein kinase
MARVEL	MAL and related proteins for vesicle trafficking and membrane link
MET	Mesenchymal-epithelial transition
MMP	Matrix metalloproteinase
NISCH	Neonatal ichthyosis-sclerosing cholangitis
NPC1L1	Niemann-Pick C1-like protein 1
OCLN	Occludin
PBC	Primary biliary cirrhosis
PFIC	Progressive familial intrahepatic cholestasis
PKA	Protein kinase A
PKC	Protein kinase C
PP	Protein phosphatase
PSC	Primary sclerosing cholangitis
RTK	Receptor tyrosine kinase
SH3	Src homology 3 domain
SNPs	Single nucleotide polymorphisms
SR-BI	Scavenger receptor class B member 1
Src	Steroid receptor coactivator
TfR1	Transferrin receptor 1
TJ	Tight junction
tTJ	Tricellular tight junction
TNF-α	Tumor necrosis factor alpha
VEGF	Vascular endothelial growth factor
YAP	Yes-associated protein
ZO	Zonula occludens
ZONAB	ZO1-associated nucleic acid binding protein

References

1. Zihni, C.; Mills, C.; Matter, K.; Balda, M.S. Tight junctions: From simple barriers to multifunctional molecular gates. *Nat. Rev. Mol. Cell Biol.* **2016**, *17*, 564–580. [CrossRef] [PubMed]
2. Severson, E.A.; Parkos, C.A. Mechanisms of outside-in signaling at the tight junction by junctional adhesion molecule A. *Ann. N. Y. Acad. Sci.* **2009**, *1165*, 10–18. [CrossRef] [PubMed]
3. Singh, A.B.; Uppada, S.B.; Dhawan, P. Claudin proteins, outside-in signaling, and carcinogenesis. *Pflug. Arch.* **2017**, *469*, 69–75. [CrossRef] [PubMed]
4. Farkas, A.E.; Capaldo, C.T.; Nusrat, A. Regulation of epithelial proliferation by tight junction proteins. *Ann. N. Y. Acad. Sci.* **2012**, *1258*, 115–124. [CrossRef]
5. Kojima, T.; Sawada, N. Expression and function of claudins in hepatocytes. *Methods Mol. Biol.* **2011**, *762*, 233–244. [CrossRef]
6. Gonzalez-Mariscal, L.; Tapia, R.; Chamorro, D. Crosstalk of tight junction components with signaling pathways. *Biochim. Biophys. Acta* **2008**, *1778*, 729–756. [CrossRef]
7. Rao, R.K.; Samak, G. Bile duct epithelial tight junctions and barrier function. *Tissue Barriers* **2013**, *1*, e25718. [CrossRef]
8. Sakaguchi, T.; Suzuki, S.; Higashi, H.; Inaba, K.; Nakamura, S.; Baba, S.; Kato, T.; Konno, H. Expression of tight junction protein claudin-5 in tumor vessels and sinusoidal endothelium in patients with hepatocellular carcinoma. *J. Surg. Res.* **2008**, *147*, 123–131. [CrossRef]

9. Kojima, T.; Yamamoto, T.; Murata, M.; Chiba, H.; Kokai, Y.; Sawada, N. Regulation of the blood-biliary barrier: Interaction between gap and tight junctions in hepatocytes. *Med. Electron. Microsc.* **2003**, *36*, 157–164. [CrossRef]
10. Miao, Z.; Xie, Z.; Miao, J.; Ran, J.; Feng, Y.; Xia, X. Regulated Entry of Hepatitis C Virus into Hepatocytes. *Viruses* **2017**, *9*, 100. [CrossRef]
11. Hagen, S.J. Non-canonical functions of claudin proteins: Beyond the regulation of cell-cell adhesions. *Tissue Barriers* **2017**, *5*, e1327839. [CrossRef]
12. Zeisel, M.B.; Dhawan, P.; Baumert, T.F. Tight junction proteins in gastrointestinal and liver disease. *Gut* **2018**. [CrossRef]
13. Markov, A.G.; Aschenbach, J.R.; Amasheh, S. The epithelial barrier and beyond: Claudins as amplifiers of physiological organ functions. *IUBMB Life* **2017**, *69*, 290–296. [CrossRef]
14. Baas, A.F.; Kuipers, J.; van der Wel, N.N.; Batlle, E.; Koerten, H.K.; Peters, P.J.; Clevers, H.C. Complete polarization of single intestinal epithelial cells upon activation of LKB1 by STRAD. *Cell* **2004**, *116*, 457–466. [CrossRef]
15. Umeda, K.; Matsui, T.; Nakayama, M.; Furuse, K.; Sasaki, H.; Furuse, M.; Tsukita, S. Establishment and characterization of cultured epithelial cells lacking expression of ZO-1. *J. Biol. Chem.* **2004**, *279*, 44785–44794. [CrossRef]
16. Ebnet, K.; Suzuki, A.; Ohno, S.; Vestweber, D. Junctional adhesion molecules (JAMs): More molecules with dual functions? *J. Cell Sci.* **2004**, *117*, 19–29. [CrossRef]
17. Ebnet, K. Junctional Adhesion Molecules (JAMs): Cell Adhesion Receptors with Pleiotropic Functions in Cell Physiology and Development. *Physiol. Rev.* **2017**, *97*, 1529–1554. [CrossRef]
18. Ikenouchi, J.; Furuse, M.; Furuse, K.; Sasaki, H.; Tsukita, S.; Tsukita, S. Tricellulin constitutes a novel barrier at tricellular contacts of epithelial cells. *J. Cell Biol.* **2005**, *171*, 939–945. [CrossRef]
19. Masuda, S.; Oda, Y.; Sasaki, H.; Ikenouchi, J.; Higashi, T.; Akashi, M.; Nishi, E.; Furuse, M. LSR defines cell corners for tricellular tight junction formation in epithelial cells. *J. Cell Sci.* **2011**, *124*, 548–555. [CrossRef]
20. Higashi, T.; Tokuda, S.; Kitajiri, S.; Masuda, S.; Nakamura, H.; Oda, Y.; Furuse, M. Analysis of the 'angulin' proteins LSR, ILDR1 and ILDR2–tricellulin recruitment, epithelial barrier function and implication in deafness pathogenesis. *J. Cell Sci.* **2013**, *126*, 966–977. [CrossRef]
21. Furuse, M.; Hirase, T.; Itoh, M.; Nagafuchi, A.; Yonemura, S.; Tsukita, S.; Tsukita, S. Occludin: A novel integral membrane protein localizing at tight junctions. *J. Cell Biol.* **1993**, *123*, 1777–1788. [CrossRef] [PubMed]
22. Cummins, P.M. Occludin: One protein, many forms. *Mol. Cell. Biol.* **2012**, *32*, 242–250. [CrossRef] [PubMed]
23. Mineta, K.; Yamamoto, Y.; Yamazaki, Y.; Tanaka, H.; Tada, Y.; Saito, K.; Tamura, A.; Igarashi, M.; Endo, T.; Takeuchi, K.; et al. Predicted expansion of the claudin multigene family. *FEBS Lett.* **2011**, *585*, 606–612. [CrossRef] [PubMed]
24. Gunzel, D.; Fromm, M. Claudins and other tight junction proteins. *Compr. Physiol.* **2012**, *2*, 1819–1852. [CrossRef]
25. Tamura, A.; Tsukita, S. Paracellular barrier and channel functions of TJ claudins in organizing biological systems: Advances in the field of barriology revealed in knockout mice. *Semin. Cell Dev. Biol.* **2014**, *36*, 177–185. [CrossRef]
26. Tanaka, H.; Tamura, A.; Suzuki, K.; Tsukita, S. Site-specific distribution of claudin-based paracellular channels with roles in biological fluid flow and metabolism. *Ann. N. Y. Acad. Sci.* **2017**, *1405*, 44–52. [CrossRef]
27. Chiba, H.; Osanai, M.; Murata, M.; Kojima, T.; Sawada, N. Transmembrane proteins of tight junctions. *Biochim. Biophys. Acta* **2008**, *1778*, 588–600. [CrossRef]
28. Bauer, H.; Zweimueller-Mayer, J.; Steinbacher, P.; Lametschwandtner, A.; Bauer, H.C. The dual role of zonula occludens (ZO) proteins. *J. Biomed. Biotechnol.* **2010**, *2010*, 402593. [CrossRef]
29. Tsukita, S.; Furuse, M.; Itoh, M. Molecular architecture of tight junctions: Occludin and ZO-1. *Soc. Gen. Physiol. Ser.* **1997**, *52*, 69–76.
30. Willott, E.; Balda, M.S.; Fanning, A.S.; Jameson, B.; Van Itallie, C.; Anderson, J.M. The tight junction protein ZO-1 is homologous to the Drosophila discs-large tumor suppressor protein of septate junctions. *Proc. Natl. Acad. Sci. USA* **1993**, *90*, 7834–7838. [CrossRef]
31. Guillemot, L.; Paschoud, S.; Pulimeno, P.; Foglia, A.; Citi, S. The cytoplasmic plaque of tight junctions: A scaffolding and signalling center. *Biochim. Biophys. Acta* **2008**, *1778*, 601–613. [CrossRef] [PubMed]

32. Li, Y.; Fanning, A.S.; Anderson, J.M.; Lavie, A. Structure of the conserved cytoplasmic C-terminal domain of occludin: Identification of the ZO-1 binding surface. *J. Mol. Biol.* **2005**, *352*, 151–164. [CrossRef] [PubMed]
33. Zhong, Y.; Saitoh, T.; Minase, T.; Sawada, N.; Enomoto, K.; Mori, M. Monoclonal antibody 7H6 reacts with a novel tight junction-associated protein distinct from ZO-1, cingulin and ZO-2. *J. Cell Biol.* **1993**, *120*, 477–483. [CrossRef] [PubMed]
34. Citi, S.; Sabanay, H.; Jakes, R.; Geiger, B.; Kendrick-Jones, J. Cingulin, a new peripheral component of tight junctions. *Nature* **1988**, *333*, 272–276. [CrossRef] [PubMed]
35. Tsukita, S.; Furuse, M.; Itoh, M. Multifunctional strands in tight junctions. *Nat. Rev. Mol. Cell Biol.* **2001**, *2*, 285–293. [CrossRef]
36. Tsukita, S.; Tanaka, H.; Tamura, A. The Claudins: From Tight Junctions to Biological Systems. *Trends Biochem. Sci.* **2019**, *44*, 141–152. [CrossRef] [PubMed]
37. Shen, L.; Weber, C.R.; Turner, J.R. The tight junction protein complex undergoes rapid and continuous molecular remodeling at steady state. *J. Cell Biol.* **2008**, *181*, 683–695. [CrossRef]
38. Chalmers, A.D.; Whitley, P. Continuous endocytic recycling of tight junction proteins: How and why? *Essays Biochem.* **2012**, *53*, 41–54. [CrossRef]
39. Ivanov, A.I.; Nusrat, A.; Parkos, C.A. Endocytosis of the apical junctional complex: Mechanisms and possible roles in regulation of epithelial barriers. *Bioessays* **2005**, *27*, 356–365. [CrossRef]
40. Ivanov, A.I.; Nusrat, A.; Parkos, C.A. Endocytosis of epithelial apical junctional proteins by a clathrin-mediated pathway into a unique storage compartment. *Mol. Biol. Cell* **2004**, *15*, 176–188. [CrossRef]
41. Lu, R.; Stewart, L.; Wilson, J.M. Scaffolding protein GOPC regulates tight junction structure. *Cell Tissue Res.* **2015**, *360*, 321–332. [CrossRef] [PubMed]
42. Lu, R.; Johnson, D.L.; Stewart, L.; Waite, K.; Elliott, D.; Wilson, J.M. Rab14 regulation of claudin-2 trafficking modulates epithelial permeability and lumen morphogenesis. *Mol. Biol. Cell* **2014**, *25*, 1744–1754. [CrossRef]
43. Gunzel, D.; Yu, A.S. Claudins and the modulation of tight junction permeability. *Physiol. Rev.* **2013**, *93*, 525–569. [CrossRef]
44. Grosse, B.; Cassio, D.; Yousef, N.; Bernardo, C.; Jacquemin, E.; Gonzales, E. Claudin-1 involved in neonatal ichthyosis sclerosing cholangitis syndrome regulates hepatic paracellular permeability. *Hepatology* **2012**, *55*, 1249–1259. [CrossRef] [PubMed]
45. Van Itallie, C.M.; Fanning, A.S.; Holmes, J.; Anderson, J.M. Occludin is required for cytokine-induced regulation of tight junction barriers. *J. Cell Sci.* **2010**, *123*, 2844–2852. [CrossRef] [PubMed]
46. Laukoetter, M.G.; Nava, P.; Lee, W.Y.; Severson, E.A.; Capaldo, C.T.; Babbin, B.A.; Williams, I.R.; Koval, M.; Peatman, E.; Campbell, J.A.; et al. JAM-A regulates permeability and inflammation in the intestine in vivo. *J. Exp. Med.* **2007**, *204*, 3067–3076. [CrossRef]
47. Vetrano, S.; Rescigno, M.; Cera, M.R.; Correale, C.; Rumio, C.; Doni, A.; Fantini, M.; Sturm, A.; Borroni, E.; Repici, A.; et al. Unique role of junctional adhesion molecule-a in maintaining mucosal homeostasis in inflammatory bowel disease. *Gastroenterology* **2008**, *135*, 173–184. [CrossRef] [PubMed]
48. Umeda, K.; Ikenouchi, J.; Katahira-Tayama, S.; Furuse, K.; Sasaki, H.; Nakayama, M.; Matsui, T.; Tsukita, S.; Furuse, M.; Tsukita, S. ZO-1 and ZO-2 independently determine where claudins are polymerized in tight-junction strand formation. *Cell* **2006**, *126*, 741–754. [CrossRef] [PubMed]
49. Satoh, H.; Zhong, Y.; Isomura, H.; Saitoh, M.; Enomoto, K.; Sawada, N.; Mori, M. Localization of 7H6 tight junction-associated antigen along the cell border of vascular endothelial cells correlates with paracellular barrier function against ions, large molecules, and cancer cells. *Exp. Cell Res.* **1996**, *222*, 269–274. [CrossRef]
50. Zhong, Y.; Enomoto, K.; Isomura, H.; Sawada, N.; Minase, T.; Oyamada, M.; Konishi, Y.; Mori, M. Localization of the 7H6 antigen at tight junctions correlates with the paracellular barrier function of MDCK cells. *Exp. Cell Res.* **1994**, *214*, 614–620. [CrossRef]
51. Kage, H.; Flodby, P.; Gao, D.; Kim, Y.H.; Marconett, C.N.; DeMaio, L.; Kim, K.J.; Crandall, E.D.; Borok, Z. Claudin 4 knockout mice: Normal physiological phenotype with increased susceptibility to lung injury. *Am. J. Physiol. Lung Cell. Mol. Physiol.* **2014**, *307*, L524–L536. [CrossRef] [PubMed]
52. Furuse, M.; Hata, M.; Furuse, K.; Yoshida, Y.; Haratake, A.; Sugitani, Y.; Noda, T.; Kubo, A.; Tsukita, S. Claudin-based tight junctions are crucial for the mammalian epidermal barrier: A lesson from claudin-1-deficient mice. *J. Cell Biol.* **2002**, *156*, 1099–1111. [CrossRef] [PubMed]

53. Matsumoto, K.; Imasato, M.; Yamazaki, Y.; Tanaka, H.; Watanabe, M.; Eguchi, H.; Nagano, H.; Hikita, H.; Tatsumi, T.; Takehara, T.; et al. Claudin 2 deficiency reduces bile flow and increases susceptibility to cholesterol gallstone disease in mice. *Gastroenterology* **2014**, *147*, 1134–1145. [CrossRef] [PubMed]
54. Katsuno, T.; Umeda, K.; Matsui, T.; Hata, M.; Tamura, A.; Itoh, M.; Takeuchi, K.; Fujimori, T.; Nabeshima, Y.; Noda, T.; et al. Deficiency of zonula occludens-1 causes embryonic lethal phenotype associated with defected yolk sac angiogenesis and apoptosis of embryonic cells. *Mol. Biol. Cell* **2008**, *19*, 2465–2475. [CrossRef]
55. Saitou, M.; Furuse, M.; Sasaki, H.; Schulzke, J.D.; Fromm, M.; Takano, H.; Noda, T.; Tsukita, S. Complex phenotype of mice lacking occludin, a component of tight junction strands. *Mol. Biol. Cell* **2000**, *11*, 4131–4142. [CrossRef]
56. Hadj-Rabia, S.; Baala, L.; Vabres, P.; Hamel-Teillac, D.; Jacquemin, E.; Fabre, M.; Lyonnet, S.; De Prost, Y.; Munnich, A.; Hadchouel, M.; et al. Claudin-1 gene mutations in neonatal sclerosing cholangitis associated with ichthyosis: A tight junction disease. *Gastroenterology* **2004**, *127*, 1386–1390. [CrossRef]
57. Izraely, S.; Sagi-Assif, O.; Klein, A.; Meshel, T.; Ben-Menachem, S.; Zaritsky, A.; Ehrlich, M.; Prieto, V.G.; Bar-Eli, M.; Pirker, C.; et al. The metastatic microenvironment: Claudin-1 suppresses the malignant phenotype of melanoma brain metastasis. *Int. J. Cancer* **2015**, *136*, 1296–1307. [CrossRef]
58. Ding, L.; Lu, Z.; Foreman, O.; Tatum, R.; Lu, Q.; Renegar, R.; Cao, J.; Chen, Y.H. Inflammation and disruption of the mucosal architecture in claudin-7-deficient mice. *Gastroenterology* **2012**, *142*, 305–315. [CrossRef]
59. Lu, Z.; Kim, D.H.; Fan, J.; Lu, Q.; Verbanac, K.; Ding, L.; Renegar, R.; Chen, Y.H. A non-tight junction function of claudin-7-Interaction with integrin signaling in suppressing lung cancer cell proliferation and detachment. *Mol. Cancer* **2015**, *14*, 120. [CrossRef]
60. Ding, L.; Wang, L.; Sui, L.; Zhao, H.; Xu, X.; Li, T.; Wang, X.; Li, W.; Zhou, P.; Kong, L. Claudin-7 indirectly regulates the integrin/FAK signaling pathway in human colon cancer tissue. *J. Hum. Genet.* **2016**, *61*, 711–720. [CrossRef]
61. Tabaries, S.; Dong, Z.; Annis, M.G.; Omeroglu, A.; Pepin, F.; Ouellet, V.; Russo, C.; Hassanain, M.; Metrakos, P.; Diaz, Z.; et al. Claudin-2 is selectively enriched in and promotes the formation of breast cancer liver metastases through engagement of integrin complexes. *Oncogene* **2011**, *30*, 1318–1328. [CrossRef] [PubMed]
62. Wu, C.J.; Mannan, P.; Lu, M.; Udey, M.C. Epithelial cell adhesion molecule (EpCAM) regulates claudin dynamics and tight junctions. *J. Biol. Chem.* **2013**, *288*, 12253–12268. [CrossRef] [PubMed]
63. Nubel, T.; Preobraschenski, J.; Tuncay, H.; Weiss, T.; Kuhn, S.; Ladwein, M.; Langbein, L.; Zoller, M. Claudin-7 regulates EpCAM-mediated functions in tumor progression. *Mol. Cancer Res* **2009**, *7*, 285–299. [CrossRef] [PubMed]
64. Agarwal, R.; D'Souza, T.; Morin, P.J. Claudin-3 and claudin-4 expression in ovarian epithelial cells enhances invasion and is associated with increased matrix metalloproteinase-2 activity. *Cancer Res.* **2005**, *65*, 7378–7385. [CrossRef]
65. Leotlela, P.D.; Wade, M.S.; Duray, P.H.; Rhode, M.J.; Brown, H.F.; Rosenthal, D.T.; Dissanayake, S.K.; Earley, R.; Indig, F.E.; Nickoloff, B.J.; et al. Claudin-1 overexpression in melanoma is regulated by PKC and contributes to melanoma cell motility. *Oncogene* **2007**, *26*, 3846–3856. [CrossRef]
66. Yoon, C.H.; Kim, M.J.; Park, M.J.; Park, I.C.; Hwang, S.G.; An, S.; Choi, Y.H.; Yoon, G.; Lee, S.J. Claudin-1 acts through c-Abl-protein kinase Cdelta (PKCdelta) signaling and has a causal role in the acquisition of invasive capacity in human liver cells. *J. Biol. Chem.* **2010**, *285*, 226–233. [CrossRef]
67. Conlon, G.A.; Murray, G.I. Recent advances in understanding the roles of matrix metalloproteinases in tumour invasion and metastasis. *J. Pathol.* **2019**, *247*, 629–640. [CrossRef]
68. Torres-Martinez, A.C.; Gallardo-Vera, J.F.; Lara-Holguin, A.N.; Montano, L.F.; Rendon-Huerta, E.P. Claudin-6 enhances cell invasiveness through claudin-1 in AGS human adenocarcinoma gastric cancer cells. *Exp. Cell Res.* **2017**, *350*, 226–235. [CrossRef]
69. Gottardi, C.J.; Arpin, M.; Fanning, A.S.; Louvard, D. The junction-associated protein, zonula occludens-1, localizes to the nucleus before the maturation and during the remodeling of cell-cell contacts. *Proc. Natl. Acad. Sci. USA* **1996**, *93*, 10779–10784. [CrossRef]
70. Islas, S.; Vega, J.; Ponce, L.; Gonzalez-Mariscal, L. Nuclear localization of the tight junction protein ZO-2 in epithelial cells. *Exp. Cell Res.* **2002**, *274*, 138–148. [CrossRef]
71. Dhawan, P.; Singh, A.B.; Deane, N.G.; No, Y.; Shiou, S.R.; Schmidt, C.; Neff, J.; Washington, M.K.; Beauchamp, R.D. Claudin-1 regulates cellular transformation and metastatic behavior in colon cancer. *J. Clin. Investig.* **2005**, *115*, 1765–1776. [CrossRef] [PubMed]

72. Ikari, A.; Watanabe, R.; Sato, T.; Taga, S.; Shimobaba, S.; Yamaguchi, M.; Yamazaki, Y.; Endo, S.; Matsunaga, T.; Sugatani, J. Nuclear distribution of claudin-2 increases cell proliferation in human lung adenocarcinoma cells. *Biochim. Biophys. Acta* **2014**, *1843*, 2079–2088. [CrossRef] [PubMed]
73. Todd, M.C.; Petty, H.M.; King, J.M.; Piana Marshall, B.N.; Sheller, R.A.; Cuevas, M.E. Overexpression and delocalization of claudin-3 protein in MCF-7 and MDA-MB-415 breast cancer cell lines. *Oncol. Lett.* **2015**, *10*, 156–162. [CrossRef] [PubMed]
74. Cuevas, M.E.; Gaska, J.M.; Gist, A.C.; King, J.M.; Sheller, R.A.; Todd, M.C. Estrogen-dependent expression and subcellular localization of the tight junction protein claudin-4 in HEC-1A endometrial cancer cells. *Int. J. Oncol.* **2015**, *47*, 650–656. [CrossRef] [PubMed]
75. French, A.D.; Fiori, J.L.; Camilli, T.C.; Leotlela, P.D.; O'Connell, M.P.; Frank, B.P.; Subaran, S.; Indig, F.E.; Taub, D.D.; Weeraratna, A.T. PKC and PKA phosphorylation affect the subcellular localization of claudin-1 in melanoma cells. *Int. J. Med. Sci.* **2009**, *6*, 93–101. [CrossRef] [PubMed]
76. Sourisseau, T.; Georgiadis, A.; Tsapara, A.; Ali, R.R.; Pestell, R.; Matter, K.; Balda, M.S. Regulation of PCNA and cyclin D1 expression and epithelial morphogenesis by the ZO-1-regulated transcription factor ZONAB/DbpA. *Mol. Cell. Biol.* **2006**, *26*, 2387–2398. [CrossRef]
77. Huerta, M.; Munoz, R.; Tapia, R.; Soto-Reyes, E.; Ramirez, L.; Recillas-Targa, F.; Gonzalez-Mariscal, L.; Lopez-Bayghen, E. Cyclin D1 is transcriptionally down-regulated by ZO-2 via an E box and the transcription factor c-Myc. *Mol. Biol. Cell* **2007**, *18*, 4826–4836. [CrossRef]
78. Singh, A.B.; Sharma, A.; Smith, J.J.; Krishnan, M.; Chen, X.; Eschrich, S.; Washington, M.K.; Yeatman, T.J.; Beauchamp, R.D.; Dhawan, P. Claudin-1 up-regulates the repressor ZEB-1 to inhibit E-cadherin expression in colon cancer cells. *Gastroenterology* **2011**, *141*, 2140–2153. [CrossRef]
79. Dhawan, P.; Ahmad, R.; Chaturvedi, R.; Smith, J.J.; Midha, R.; Mittal, M.K.; Krishnan, M.; Chen, X.; Eschrich, S.; Yeatman, T.J.; et al. Claudin-2 expression increases tumorigenicity of colon cancer cells: Role of epidermal growth factor receptor activation. *Oncogene* **2011**, *30*, 3234–3247. [CrossRef]
80. Suh, Y.; Yoon, C.H.; Kim, R.K.; Lim, E.J.; Oh, Y.S.; Hwang, S.G.; An, S.; Yoon, G.; Gye, M.C.; Yi, J.M.; et al. Claudin-1 induces epithelial-mesenchymal transition through activation of the c-Abl-ERK signaling pathway in human liver cells. *Oncogene* **2013**, *32*, 4873–4882. [CrossRef]
81. Fredriksson, K.; Van Itallie, C.M.; Aponte, A.; Gucek, M.; Tietgens, A.J.; Anderson, J.M. Proteomic analysis of proteins surrounding occludin and claudin-4 reveals their proximity to signaling and trafficking networks. *PLoS ONE* **2015**, *10*, e0117074. [CrossRef] [PubMed]
82. Zona, L.; Lupberger, J.; Sidahmed-Adrar, N.; Thumann, C.; Harris, H.J.; Barnes, A.; Florentin, J.; Tawar, R.G.; Xiao, F.; Turek, M.; et al. HRas signal transduction promotes hepatitis C virus cell entry by triggering assembly of the host tetraspanin receptor complex. *Cell Host Microbe* **2013**, *13*, 302–313. [CrossRef] [PubMed]
83. De Souza, W.F.; Fortunato-Miranda, N.; Robbs, B.K.; de Araujo, W.M.; de-Freitas-Junior, J.C.; Bastos, L.G.; Viola, J.P.; Morgado-Diaz, J.A. Claudin-3 overexpression increases the malignant potential of colorectal cancer cells: Roles of ERK1/2 and PI3K-Akt as modulators of EGFR signaling. *PLoS ONE* **2013**, *8*, e74994. [CrossRef]
84. Lupberger, J.; Zeisel, M.B.; Xiao, F.; Thumann, C.; Fofana, I.; Zona, L.; Davis, C.; Mee, C.J.; Turek, M.; Gorke, S.; et al. EGFR and EphA2 are host factors for hepatitis C virus entry and possible targets for antiviral therapy. *Nat. Med.* **2011**, *17*, 589–595. [CrossRef] [PubMed]
85. Singh, A.B.; Sharma, A.; Dhawan, P. Claudin-1 expression confers resistance to anoikis in colon cancer cells in a Src-dependent manner. *Carcinogenesis* **2012**, *33*, 2538–2547. [CrossRef] [PubMed]
86. Herrero, R.; Prados, L.; Ferruelo, A.; Puig, F.; Pandolfi, R.; Guillamat-Prats, R.; Moreno, L.; Matute-Bello, G.; Artigas, A.; Esteban, A.; et al. Fas activation alters tight junction proteins in acute lung injury. *Thorax* **2019**, *74*, 69–82. [CrossRef] [PubMed]
87. Spadaro, D.; Le, S.; Laroche, T.; Mean, I.; Jond, L.; Yan, J.; Citi, S. Tension-Dependent Stretching Activates ZO-1 to Control the Junctional Localization of Its Interactors. *Curr. Biol.* **2017**, *27*, 3783–3795. [CrossRef]
88. Zhou, B.; Flodby, P.; Luo, J.; Castillo, D.R.; Liu, Y.; Yu, F.X.; McConnell, A.; Varghese, B.; Li, G.; Chimge, N.O.; et al. Claudin-18-mediated YAP activity regulates lung stem and progenitor cell homeostasis and tumorigenesis. *J. Clin. Investig.* **2018**, *128*, 970–984. [CrossRef]
89. Fujibe, M.; Chiba, H.; Kojima, T.; Soma, T.; Wada, T.; Yamashita, T.; Sawada, N. Thr203 of claudin-1, a putative phosphorylation site for MAP kinase, is required to promote the barrier function of tight junctions. *Exp. Cell Res.* **2004**, *295*, 36–47. [CrossRef]

90. Ishizaki, T.; Chiba, H.; Kojima, T.; Fujibe, M.; Soma, T.; Miyajima, H.; Nagasawa, K.; Wada, I.; Sawada, N. Cyclic AMP induces phosphorylation of claudin-5 immunoprecipitates and expression of claudin-5 gene in blood-brain-barrier endothelial cells via protein kinase A-dependent and -independent pathways. *Exp. Cell Res.* **2003**, *290*, 275–288. [CrossRef]
91. Ikari, A.; Ito, M.; Okude, C.; Sawada, H.; Harada, H.; Degawa, M.; Sakai, H.; Takahashi, T.; Sugatani, J.; Miwa, M. Claudin-16 is directly phosphorylated by protein kinase A independently of a vasodilator-stimulated phosphoprotein-mediated pathway. *J. Cell. Physiol.* **2008**, *214*, 221–229. [CrossRef]
92. Schmitt, M.; Horbach, A.; Kubitz, R.; Frilling, A.; Haussinger, D. Disruption of hepatocellular tight junctions by vascular endothelial growth factor (VEGF): A novel mechanism for tumor invasion. *J. Hepatol.* **2004**, *41*, 274–283. [CrossRef] [PubMed]
93. Ni, Y.; Teng, T.; Li, R.; Simonyi, A.; Sun, G.Y.; Lee, J.C. TNFalpha alters occludin and cerebral endothelial permeability: Role of p38MAPK. *PLoS ONE* **2017**, *12*, e0170346. [CrossRef] [PubMed]
94. Elias, B.C.; Suzuki, T.; Seth, A.; Giorgianni, F.; Kale, G.; Shen, L.; Turner, J.R.; Naren, A.; Desiderio, D.M.; Rao, R. Phosphorylation of Tyr-398 and Tyr-402 in occludin prevents its interaction with ZO-1 and destabilizes its assembly at the tight junctions. *J. Biol. Chem.* **2009**, *284*, 1559–1569. [CrossRef] [PubMed]
95. Ma, T.Y.; Iwamoto, G.K.; Hoa, N.T.; Akotia, V.; Pedram, A.; Boivin, M.A.; Said, H.M. TNF-alpha-induced increase in intestinal epithelial tight junction permeability requires NF-kappa B activation. *Am. J. Physiol. Gastrointest. Liver Physiol.* **2004**, *286*, G367–G376. [CrossRef]
96. Kalluri, R. EMT: When epithelial cells decide to become mesenchymal-like cells. *J. Clin. Investig.* **2009**, *119*, 1417–1419. [CrossRef]
97. Lamouille, S.; Xu, J.; Derynck, R. Molecular mechanisms of epithelial-mesenchymal transition. *Nat. Rev. Mol. Cell Biol.* **2014**, *15*, 178–196. [CrossRef]
98. Gissen, P.; Arias, I.M. Structural and functional hepatocyte polarity and liver disease. *J. Hepatol.* **2015**, *63*, 1023–1037. [CrossRef]
99. Tanaka, H.; Imasato, M.; Yamazaki, Y.; Matsumoto, K.; Kunimoto, K.; Delpierre, J.; Meyer, K.; Zerial, M.; Kitamura, N.; Watanabe, M.; et al. Claudin-3 regulates bile canalicular paracellular barrier and cholesterol gallstone core formation in mice. *J. Hepatol.* **2018**, *69*, 1308–1316. [CrossRef]
100. Rahner, C.; Mitic, L.L.; Anderson, J.M. Heterogeneity in expression and subcellular localization of claudins 2, 3, 4, and 5 in the rat liver, pancreas, and gut. *Gastroenterology* **2001**, *120*, 411–422. [CrossRef]
101. Somoracz, A.; Korompay, A.; Torzsok, P.; Patonai, A.; Erdelyi-Belle, B.; Lotz, G.; Schaff, Z.; Kiss, A. Tricellulin expression and its prognostic significance in primary liver carcinomas. *Pathol. Oncol. Res.* **2014**, *20*, 755–764. [CrossRef] [PubMed]
102. Orban, E.; Szabo, E.; Lotz, G.; Kupcsulik, P.; Paska, C.; Schaff, Z.; Kiss, A. Different expression of occludin and ZO-1 in primary and metastatic liver tumors. *Pathol. Oncol. Res.* **2008**, *14*, 299–306. [CrossRef] [PubMed]
103. Byass, P. The global burden of liver disease: A challenge for methods and for public health. *BMC Med.* **2014**, *12*, 159. [CrossRef] [PubMed]
104. D'Amico, G.; Morabito, A.; D'Amico, M.; Pasta, L.; Malizia, G.; Rebora, P.; Valsecchi, M.G. Clinical states of cirrhosis and competing risks. *J. Hepatol.* **2018**, *68*, 563–576. [CrossRef]
105. Marcellin, P.; Kutala, B.K. Liver diseases: A major, neglected global public health problem requiring urgent actions and large-scale screening. *Liver Int.* **2018**, *38*, 2–6. [CrossRef]
106. Pradhan-Sundd, T.; Zhou, L.; Vats, R.; Jiang, A.; Molina, L.; Singh, S.; Poddar, M.; Russell, J.; Stolz, D.B.; Oertel, M.; et al. Dual catenin loss in murine liver causes tight junctional deregulation and progressive intrahepatic cholestasis. *Hepatology* **2018**, *67*, 2320–2337. [CrossRef]
107. Pradhan-Sundd, T.; Vats, R.; Russell, J.O.; Singh, S.; Michael, A.A.; Molina, L.; Kakar, S.; Cornuet, P.; Poddar, M.; Watkins, S.C.; et al. Dysregulated Bile Transporters and Impaired Tight Junctions During Chronic Liver Injury in Mice. *Gastroenterology* **2018**, *155*, 1218–1232-e24. [CrossRef]
108. Takaki, Y.; Hirai, S.; Manabe, N.; Izumi, Y.; Hirose, T.; Nakaya, M.; Suzuki, A.; Mizuno, K.; Akimoto, K.; Tsukita, S.; et al. Dynamic changes in protein components of the tight junction during liver regeneration. *Cell Tissue Res.* **2001**, *305*, 399–409. [CrossRef]
109. Liang, T.J.; Rehermann, B.; Seeff, L.B.; Hoofnagle, J.H. Pathogenesis, natural history, treatment, and prevention of hepatitis C. *Ann. Intern. Med.* **2000**, *132*, 296–305. [CrossRef]
110. Thrift, A.P.; El-Serag, H.B.; Kanwal, F. Global epidemiology and burden of HCV infection and HCV-related disease. *Nat. Rev. Gastroenterol. Hepatol.* **2017**, *14*, 122–132. [CrossRef]

111. WHO. *Global Hepatitis Report*; WHO: Geneva, Switzerland, 2017.
112. Douam, F.; Lavillette, D.; Cosset, F.L. The mechanism of HCV entry into host cells. *Prog. Mol. Biol. Transl Sci.* **2015**, *129*, 63–107. [CrossRef] [PubMed]
113. Sourisseau, M.; Michta, M.L.; Zony, C.; Israelow, B.; Hopcraft, S.E.; Narbus, C.M.; Parra Martin, A.; Evans, M.J. Temporal analysis of hepatitis C virus cell entry with occludin directed blocking antibodies. *PLoS Pathog.* **2013**, *9*, e1003244. [CrossRef] [PubMed]
114. Shimizu, Y.; Shirasago, Y.; Kondoh, M.; Suzuki, T.; Wakita, T.; Hanada, K.; Yagi, K.; Fukasawa, M. Monoclonal Antibodies against Occludin Completely Prevented Hepatitis C Virus Infection in a Mouse Model. *J. Virol.* **2018**, *92*, e02258-17. [CrossRef] [PubMed]
115. Liu, S.; Kuo, W.; Yang, W.; Liu, W.; Gibson, G.A.; Dorko, K.; Watkins, S.C.; Strom, S.C.; Wang, T. The second extracellular loop dictates Occludin-mediated HCV entry. *Virology* **2010**, *407*, 160–170. [CrossRef] [PubMed]
116. Dorner, M.; Horwitz, J.A.; Robbins, J.B.; Barry, W.T.; Feng, Q.; Mu, K.; Jones, C.T.; Schoggins, J.W.; Catanese, M.T.; Burton, D.R.; et al. A genetically humanized mouse model for hepatitis C virus infection. *Nature* **2011**, *474*, 208–211. [CrossRef] [PubMed]
117. Dorner, M.; Horwitz, J.A.; Donovan, B.M.; Labitt, R.N.; Budell, W.C.; Friling, T.; Vogt, A.; Catanese, M.T.; Satoh, T.; Kawai, T.; et al. Completion of the entire hepatitis C virus life cycle in genetically humanized mice. *Nature* **2013**, *501*, 237–241. [CrossRef] [PubMed]
118. Ding, Q.; von Schaewen, M.; Hrebikova, G.; Heller, B.; Sandmann, L.; Plaas, M.; Ploss, A. Mice Expressing Minimally Humanized CD81 and Occludin Genes Support Hepatitis C Virus Uptake In Vivo. *J. Virol.* **2017**, *91*, e01799-16. [CrossRef]
119. Nakamuta, M.; Fujino, T.; Yada, R.; Aoyagi, Y.; Yasutake, K.; Kohjima, M.; Fukuizumi, K.; Yoshimoto, T.; Harada, N.; Yada, M.; et al. Expression profiles of genes associated with viral entry in HCV-infected human liver. *J. Med. Virol.* **2011**, *83*, 921–927. [CrossRef]
120. Mensa, L.; Crespo, G.; Gastinger, M.J.; Kabat, J.; Perez-del-Pulgar, S.; Miquel, R.; Emerson, S.U.; Purcell, R.H.; Forns, X. Hepatitis C virus receptors claudin-1 and occludin after liver transplantation and influence on early viral kinetics. *Hepatology* **2011**, *53*, 1436–1445. [CrossRef]
121. Krieger, S.E.; Zeisel, M.B.; Davis, C.; Thumann, C.; Harris, H.J.; Schnober, E.K.; Mee, C.; Soulier, E.; Royer, C.; Lambotin, M.; et al. Inhibition of hepatitis C virus infection by anti-claudin-1 antibodies is mediated by neutralization of E2-CD81-claudin-1 associations. *Hepatology* **2010**, *51*, 1144–1157. [CrossRef]
122. Fofana, I.; Krieger, S.E.; Grunert, F.; Glauben, S.; Xiao, F.; Fafi-Kremer, S.; Soulier, E.; Royer, C.; Thumann, C.; Mee, C.J.; et al. Monoclonal anti-claudin 1 antibodies prevent hepatitis C virus infection of primary human hepatocytes. *Gastroenterology* **2010**, *139*, 953–964. [CrossRef] [PubMed]
123. Colpitts, C.C.; Tawar, R.G.; Mailly, L.; Thumann, C.; Heydmann, L.; Durand, S.C.; Xiao, F.; Robinet, E.; Pessaux, P.; Zeisel, M.B.; et al. Humanisation of a claudin-1-specific monoclonal antibody for clinical prevention and cure of HCV infection without escape. *Gut* **2018**, *67*, 736–745. [CrossRef] [PubMed]
124. Mailly, L.; Xiao, F.; Lupberger, J.; Wilson, G.K.; Aubert, P.; Duong, F.H.T.; Calabrese, D.; Leboeuf, C.; Fofana, I.; Thumann, C.; et al. Clearance of persistent hepatitis C virus infection in humanized mice using a claudin-1-targeting monoclonal antibody. *Nat. Biotechnol.* **2015**, *33*, 549–554. [CrossRef] [PubMed]
125. Fofana, I.; Fafi-Kremer, S.; Carolla, P.; Fauvelle, C.; Zahid, M.N.; Turek, M.; Heydmann, L.; Cury, K.; Hayer, J.; Combet, C.; et al. Mutations that alter use of hepatitis C virus cell entry factors mediate escape from neutralizing antibodies. *Gastroenterology* **2012**, *143*, 223–233. [CrossRef]
126. Xiao, F.; Fofana, I.; Thumann, C.; Mailly, L.; Alles, R.; Robinet, E.; Meyer, N.; Schaeffer, M.; Habersetzer, F.; Doffoel, M.; et al. Synergy of entry inhibitors with direct-acting antivirals uncovers novel combinations for prevention and treatment of hepatitis C. *Gut* **2014**. [CrossRef]
127. Xiao, F.; Fofana, I.; Heydmann, L.; Barth, H.; Soulier, E.; Habersetzer, F.; Doffoel, M.; Bukh, J.; Patel, A.H.; Zeisel, M.B.; et al. Hepatitis C virus cell-cell transmission and resistance to direct-acting antiviral agents. *PLoS Pathog.* **2014**, *10*, e1004128. [CrossRef]
128. Okai, K.; Ichikawa-Tomikawa, N.; Saito, A.C.; Watabe, T.; Sugimoto, K.; Fujita, D.; Ono, C.; Fukuhara, T.; Matsuura, Y.; Ohira, H.; et al. A novel occludin-targeting monoclonal antibody prevents hepatitis C virus infection in vitro. *Oncotarget* **2018**, *9*, 16588–16598. [CrossRef]
129. Michta, M.L.; Hopcraft, S.E.; Narbus, C.M.; Kratovac, Z.; Israelow, B.; Sourisseau, M.; Evans, M.J. Species-specific regions of occludin required by hepatitis C virus for cell entry. *J. Virol.* **2010**, *84*, 11696–11708. [CrossRef]

130. Bray, F.; Ferlay, J.; Soerjomataram, I.; Siegel, R.L.; Torre, L.A.; Jemal, A. Global cancer statistics 2018, GLOBOCAN estimates of incidence and mortality worldwide for 36 cancers in 185 countries. *CA Cancer J. Clin.* **2018**, *68*, 394–424. [CrossRef]
131. Bouchagier, K.A.; Assimakopoulos, S.F.; Karavias, D.D.; Maroulis, I.; Tzelepi, V.; Kalofonos, H.; Kardamakis, D.; Scopa, C.D.; Tsamandas, A.C. Expression of claudins-1, -4, -5, -7 and occludin in hepatocellular carcinoma and their relation with classic clinicopathological features and patients' survival. *In Vivo* **2014**, *28*, 315–326.
132. Holczbauer, A.; Gyongyosi, B.; Lotz, G.; Torzsok, P.; Kaposi-Novak, P.; Szijarto, A.; Tatrai, P.; Kupcsulik, P.; Schaff, Z.; Kiss, A. Increased expression of claudin-1 and claudin-7 in liver cirrhosis and hepatocellular carcinoma. *Pathol. Oncol. Res.* **2014**, *20*, 493–502. [CrossRef]
133. Huang, G.W.; Ding, X.; Chen, S.L.; Zeng, L. Expression of claudin 10 protein in hepatocellular carcinoma: Impact on survival. *J. Cancer Res. Clin. Oncol.* **2011**, *137*, 1213–1218. [CrossRef] [PubMed]
134. Zhou, S.; Parham, D.M.; Yung, E.; Pattengale, P.; Wang, L. Quantification of glypican 3, beta-catenin and claudin-1 protein expression in hepatoblastoma and paediatric hepatocellular carcinoma by colour deconvolution. *Histopathology* **2015**, *67*, 905–913. [CrossRef] [PubMed]
135. Kim, J.H.; Kim, E.L.; Lee, Y.K.; Park, C.B.; Kim, B.W.; Wang, H.J.; Yoon, C.H.; Lee, S.J.; Yoon, G. Decreased lactate dehydrogenase B expression enhances claudin 1-mediated hepatoma cell invasiveness via mitochondrial defects. *Exp. Cell Res.* **2011**, *317*, 1108–1118. [CrossRef] [PubMed]
136. Cheung, S.T.; Leung, K.L.; Ip, Y.C.; Chen, X.; Fong, D.Y.; Ng, I.O.; Fan, S.T.; So, S. Claudin-10 expression level is associated with recurrence of primary hepatocellular carcinoma. *Clin. Cancer Res.* **2005**, *11*, 551–556. [PubMed]
137. Li, C.P.; Cai, M.Y.; Jiang, L.J.; Mai, S.J.; Chen, J.W.; Wang, F.W.; Liao, Y.J.; Chen, W.H.; Jin, X.H.; Pei, X.Q.; et al. CLDN14 is epigenetically silenced by EZH2-mediated H3K27ME3 and is a novel prognostic biomarker in hepatocellular carcinoma. *Carcinogenesis* **2016**, *37*, 557–566. [CrossRef]
138. Jiang, L.; Yang, Y.D.; Fu, L.; Xu, W.; Liu, D.; Liang, Q.; Zhang, X.; Xu, L.; Guan, X.Y.; Wu, B.; et al. CLDN3 inhibits cancer aggressiveness via Wnt-EMT signaling and is a potential prognostic biomarker for hepatocellular carcinoma. *Oncotarget* **2014**, *5*, 7663–7676. [CrossRef]
139. Gerardo-Ramirez, M.; Lazzarini-Lechuga, R.; Hernandez-Rizo, S.; Jimenez-Salazar, J.E.; Simoni-Nieves, A.; Garcia-Ruiz, C.; Fernandez-Checa, J.C.; Marquardt, J.U.; Coulouarn, C.; Gutierrez-Ruiz, M.C.; et al. GDF11 exhibits tumor suppressive properties in hepatocellular carcinoma cells by restricting clonal expansion and invasion. *Biochim. Biophys. Acta Mol. Basis Dis.* **2019**, *1865*, 1540–1554. [CrossRef]
140. Hou, X.; Yang, L.; Jiang, X.; Liu, Z.; Li, X.; Xie, S.; Li, G.; Liu, J. Role of microRNA-141-3p in the progression and metastasis of hepatocellular carcinoma cell. *Int. J. Biol. Macromol.* **2019**, *128*, 331–339. [CrossRef]
141. Wang, S.C.; Lin, X.L.; Li, J.; Zhang, T.T.; Wang, H.Y.; Shi, J.W.; Yang, S.; Zhao, W.T.; Xie, R.Y.; Wei, F.; et al. MicroRNA-122 triggers mesenchymal-epithelial transition and suppresses hepatocellular carcinoma cell motility and invasion by targeting RhoA. *PLoS ONE* **2014**, *9*, e101330. [CrossRef]
142. Nagai, T.; Arao, T.; Nishio, K.; Matsumoto, K.; Hagiwara, S.; Sakurai, T.; Minami, Y.; Ida, H.; Ueshima, K.; Nishida, N.; et al. Impact of Tight Junction Protein ZO-1 and TWIST Expression on Postoperative Survival of Patients with Hepatocellular Carcinoma. *Dig. Dis.* **2016**, *34*, 702–707. [CrossRef]
143. Bekker, V.; Chanock, S.J.; Yeager, M.; Hutchinson, A.A.; von Hahn, T.; Chen, S.; Xiao, N.; Dotrang, M.; Brown, M.; Busch, M.P.; et al. Genetic variation in CLDN1 and susceptibility to hepatitis C virus infection. *J. Viral. Hepat.* **2010**, *17*, 192–200. [CrossRef]
144. Zadori, G.; Gelley, F.; Torzsok, P.; Sarvary, E.; Doros, A.; Deak, A.P.; Nagy, P.; Schaff, Z.; Kiss, A.; Nemes, B. Examination of claudin-1 expression in patients undergoing liver transplantation owing to hepatitis C virus cirrhosis. *Transplant. Proc.* **2011**, *43*, 1267–1271. [CrossRef] [PubMed]
145. Liu, S.; Yang, W.; Shen, L.; Turner, J.R.; Coyne, C.B.; Wang, T. Tight junction proteins claudin-1 and occludin control hepatitis C virus entry and are downregulated during infection to prevent superinfection. *J. Virol.* **2009**, *83*, 2011–2014. [CrossRef]
146. De Vos, R.; Desmet, V.J. Morphologic changes of the junctional complex of the hepatocytes in rat liver after bile duct ligation. *Br. J. Exp. Pathol.* **1978**, *59*, 220–227.
147. Sakisaka, S.; Kawaguchi, T.; Taniguchi, E.; Hanada, S.; Sasatomi, K.; Koga, H.; Harada, M.; Kimura, R.; Sata, M.; Sawada, N.; et al. Alterations in tight junctions differ between primary biliary cirrhosis and primary sclerosing cholangitis. *Hepatology* **2001**, *33*, 1460–1468. [CrossRef]

148. Nemeth, Z.; Szasz, A.M.; Tatrai, P.; Nemeth, J.; Gyorffy, H.; Somoracz, A.; Szijarto, A.; Kupcsulik, P.; Kiss, A.; Schaff, Z. Claudin-1, -2, -3, -4, -7, -8, and -10 protein expression in biliary tract cancers. *J. Histochem. Cytochem.* **2009**, *57*, 113–121. [CrossRef]
149. Sambrotta, M.; Strautnieks, S.; Papouli, E.; Rushton, P.; Clark, B.E.; Parry, D.A.; Logan, C.V.; Newbury, L.J.; Kamath, B.M.; Ling, S.; et al. Mutations in TJP2 cause progressive cholestatic liver disease. *Nat. Genet.* **2014**, *46*, 326–328. [CrossRef]
150. Vitale, G.; Gitto, S.; Vukotic, R.; Raimondi, F.; Andreone, P. Familial intrahepatic cholestasis: New and wide perspectives. *Dig. Liver Dis.* **2019**, *51*, 922–933. [CrossRef]
151. Baala, L.; Hadj-Rabia, S.; Hamel-Teillac, D.; Hadchouel, M.; Prost, C.; Leal, S.M.; Jacquemin, E.; Sefiani, A.; De Prost, Y.; Courtois, G.; et al. Homozygosity mapping of a locus for a novel syndromic ichthyosis to chromosome 3q27–q28. *J. Investig. Dermatol.* **2002**, *119*, 70–76. [CrossRef]
152. Lindor, K.D.; Gershwin, M.E.; Poupon, R.; Kaplan, M.; Bergasa, N.V.; Heathcote, E.J. American Association for Study of Liver, D. Primary biliary cirrhosis. *Hepatology* **2009**, *50*, 291–308. [CrossRef] [PubMed]
153. Karlsen, T.H.; Folseraas, T.; Thorburn, D.; Vesterhus, M. Primary sclerosing cholangitis—A comprehensive review. *J. Hepatol.* **2017**, *67*, 1298–1323. [CrossRef] [PubMed]
154. Nakanuma, Y.; Tsuneyama, K.; Gershwin, M.E.; Yasoshima, M. Pathology and immunopathology of primary biliary cirrhosis with emphasis on bile duct lesions: Recent progress. *Semin. Liver Dis.* **1995**, *15*, 313–328. [CrossRef] [PubMed]
155. Feldmeyer, L.; Huber, M.; Fellmann, F.; Beckmann, J.S.; Frenk, E.; Hohl, D. Confirmation of the origin of NISCH syndrome. *Hum. Mutat.* **2006**, *27*, 408–410. [CrossRef] [PubMed]
156. Nagtzaam, I.F.; van Geel, M.; Driessen, A.; Steijlen, P.M.; van Steensel, M.A. Bile duct paucity is part of the neonatal ichthyosis-sclerosing cholangitis phenotype. *Br. J. Dermatol.* **2010**, *163*, 205–207. [CrossRef] [PubMed]
157. Shah, I.; Bhatnagar, S. NISCH syndrome with hypothyroxinemia. *Ann. Hepatol.* **2010**, *9*, 299–301. [CrossRef]
158. Kirchmeier, P.; Sayar, E.; Hotz, A.; Hausser, I.; Islek, A.; Yilmaz, A.; Artan, R.; Fischer, J. Novel mutation in the CLDN1 gene in a Turkish family with neonatal ichthyosis sclerosing cholangitis (NISCH) syndrome. *Br. J. Dermatol.* **2014**, *170*, 976–978. [CrossRef]
159. Youssefian, L.; Vahidnezhad, H.; Saeidian, A.H.; Sotoudeh, S.; Zeinali, S.; Uitto, J. Gene-Targeted Next-Generation Sequencing Identifies a Novel CLDN1 Mutation in a Consanguineous Family With NISCH Syndrome. *Am. J. Gastroenterol.* **2017**, *112*, 396–398. [CrossRef]
160. Nagtzaam, I.F.; Peeters, V.P.M.; Vreeburg, M.; Wagner, A.; Steijlen, P.M.; van Geel, M.; van Steensel, M.A.M. Novel CLDN1 mutation in ichthyosis-hypotrichosis-sclerosing cholangitis syndrome without signs of liver disease. *Br. J. Dermatol.* **2018**, *178*, e202–e203. [CrossRef]
161. Szepetowski, S.; Lacoste, C.; Mallet, S.; Roquelaure, B.; Badens, C.; Fabre, A. NISCH syndrome, a rare cause of neonatal cholestasis: A case report. *Arch. Pediatr.* **2017**, *24*, 1228–1234. [CrossRef]
162. Carlton, V.E.; Harris, B.Z.; Puffenberger, E.G.; Batta, A.K.; Knisely, A.S.; Robinson, D.L.; Strauss, K.A.; Shneider, B.L.; Lim, W.A.; Salen, G.; et al. Complex inheritance of familial hypercholanemia with associated mutations in TJP2 and BAAT. *Nat. Genet.* **2003**, *34*, 91–96. [CrossRef] [PubMed]
163. Patel, T. Increasing incidence and mortality of primary intrahepatic cholangiocarcinoma in the United States. *Hepatology* **2001**, *33*, 1353–1357. [CrossRef] [PubMed]
164. Saha, S.K.; Zhu, A.X.; Fuchs, C.S.; Brooks, G.A. Forty-Year Trends in Cholangiocarcinoma Incidence in the U.S: Intrahepatic Disease on the Rise. *Oncologist* **2016**, *21*, 594–599. [CrossRef] [PubMed]
165. Von Hahn, T.; Ciesek, S.; Wegener, G.; Plentz, R.R.; Weismuller, T.J.; Wedemeyer, H.; Manns, M.P.; Greten, T.F.; Malek, N.P. Epidemiological trends in incidence and mortality of hepatobiliary cancers in Germany. *Scand. J. Gastroenterol.* **2011**, *46*, 1092–1098. [CrossRef]
166. Ehlken, H.; Schramm, C. Primary sclerosing cholangitis and cholangiocarcinoma: Pathogenesis and modes of diagnostics. *Dig. Dis.* **2013**, *31*, 118–125. [CrossRef]
167. Shaib, Y.H.; El-Serag, H.B.; Davila, J.A.; Morgan, R.; McGlynn, K.A. Risk factors of intrahepatic cholangiocarcinoma in the United States: A case-control study. *Gastroenterology* **2005**, *128*, 620–626. [CrossRef]
168. Ralphs, S.; Khan, S.A. The role of the hepatitis viruses in cholangiocarcinoma. *J. Viral. Hepat.* **2013**, *20*, 297–305. [CrossRef]
169. Tyson, G.L.; El-Serag, H.B. Risk factors for cholangiocarcinoma. *Hepatology* **2011**, *54*, 173–184. [CrossRef]

170. Jakab, C.; Kiss, A.; Schaff, Z.; Szabo, Z.; Rusvai, M.; Galfi, P.; Szabara, A.; Sterczer, A.; Kulka, J. Claudin-7 protein differentiates canine cholangiocarcinoma from hepatocellular carcinoma. *Histol. Histopathol.* **2010**, *25*, 857–864. [CrossRef]

171. Lodi, C.; Szabo, E.; Holczbauer, A.; Batmunkh, E.; Szijarto, A.; Kupcsulik, P.; Kovalszky, I.; Paku, S.; Illyes, G.; Kiss, A.; et al. Claudin-4 differentiates biliary tract cancers from hepatocellular carcinomas. *Mod. Pathol.* **2006**, *19*, 460–469. [CrossRef]

172. Bunthot, S.; Obchoei, S.; Kraiklang, R.; Pirojkul, C.; Wongkham, S.; Wongkham, C. Overexpression of claudin-4 in cholangiocarcinoma tissues and its possible role in tumor metastasis. *Asian Pac. J. Cancer Prev.* **2012**, *13*, 71–76. [PubMed]

173. Shinozaki, A.; Shibahara, J.; Noda, N.; Tanaka, M.; Aoki, T.; Kokudo, N.; Fukayama, M. Claudin-18 in biliary neoplasms. Its significance in the classification of intrahepatic cholangiocarcinoma. *Virchows Arch.* **2011**, *459*, 73–80. [CrossRef] [PubMed]

174. Nemeth, Z.; Szasz, A.M.; Somoracz, A.; Tatrai, P.; Nemeth, J.; Gyorffy, H.; Szijarto, A.; Kupcsulik, P.; Kiss, A.; Schaff, Z. Zonula occludens-1, occludin, and E-cadherin protein expression in biliary tract cancers. *Pathol. Oncol. Res.* **2009**, *15*, 533–539. [CrossRef] [PubMed]

5

Tight Junctions in Cell Proliferation

Mónica Díaz-Coránguez, Xuwen Liu and David A. Antonetti *

Department of Ophthalmology and Visual Sciences, University of Michigan, Kellogg Eye Center, Ann Arbor, MI 48105, USA; mdiazcor@med.umich.edu (M.D.-C.); xuwen@med.umich.edu (X.L.)
* Correspondence: dantonet@med.umich.edu

Abstract: Tight junction (TJ) proteins form a continuous intercellular network creating a barrier with selective regulation of water, ion, and solutes across endothelial, epithelial, and glial tissues. TJ proteins include the claudin family that confers barrier properties, members of the MARVEL family that contribute to barrier regulation, and JAM molecules, which regulate junction organization and diapedesis. In addition, the membrane-associated proteins such as MAGUK family members, i.e., zonula occludens, form the scaffold linking the transmembrane proteins to both cell signaling molecules and the cytoskeleton. Most studies of TJ have focused on the contribution to cell-cell adhesion and tissue barrier properties. However, recent studies reveal that, similar to adherens junction proteins, TJ proteins contribute to the control of cell proliferation. In this review, we will summarize and discuss the specific role of TJ proteins in the control of epithelial and endothelial cell proliferation. In some cases, the TJ proteins act as a reservoir of critical cell cycle modulators, by binding and regulating their nuclear access, while in other cases, junctional proteins are located at cellular organelles, regulating transcription and proliferation. Collectively, these studies reveal that TJ proteins contribute to the control of cell proliferation and differentiation required for forming and maintaining a tissue barrier.

Keywords: tight junctions; cell growth; epithelia; endothelia; proliferation; migration

---

1. TJ Expression in Epithelial Differentiation

The role of TJs in growth, proliferation, and differentiation becomes evident in development during epithelial specialization in eukaryotic organisms. Studies of junction formation in early development reveal the contribution of TJs to the early differentiation process and are largely associated with barrier formation. In mice, influenced by the reproductive niche, the germ stem cells undergo the first cell proliferation cycle. The subsequent transcription and translational activity from the embryonic genome during the 2-cell and 4-cell stages is crucial for the first morphogenetic transition in the embryo that occurs during the 8-cell stage, a process that is known as compaction (Figure 1A). Compaction is characterized by the expression and organization of intercellular adhesion and polarization complexes. At a molecular level, protein kinase C isoform alpha (PKCα) and myosin light-chain kinase signaling become activated and maintain cell contact through the regulation of the adherens junction (AJ) proteins E-cadherin, nectin-2, epithin, vezatin, and β-catenin [1,2]. Meanwhile, the membrane-associated guanylate kinase (MAGUK) homolog family member zonula occludens-1 (ZO-1) α^-, the smaller splice variant of the two ZO-1α isoforms [3], arrives at the blastocyst membrane cell contact where it binds with the Rab-GTPase, Rab13 [4–6]. The junctional adhesion molecule (JAM)-A is also expressed at this stage [7] and is localized at the apical microvillous pole together with the partitioning defective (PAR) complex proteins PAR-3, PAR-6, and the atypical PKC isoform iota (aPKCι) as well as Cdc42 [8]. *Cldn3, 6, 7, 8, 10, 12,* and *15*, and *Ocln* genes [9] are also expressed in this compaction stage. *Cldn3* gene shows a particularly high expression compared to other claudins in the progenitors of the trophectoderm during late compaction [10], suggesting a role of claudin-3 protein in differentiation. The TJ protein content increases as the embryo continues to develop and shows an even distribution at the lateral domain similar to the AJs.

The first epithelial specialization differentiation takes places in cavitation, which is a process of the formation of the first cavity in the embryonic body or blastocoel. Cavitation is characterized by asymmetrical divisions and the coordinated polarization of the trophectoderm epithelial cells, a process regulated by junctional components. At the 16-cell stage, E-cadherin assembles at the membrane at cell-cell contact sites and confers the apical and basolateral cell polarization of blastomeres, which leads to a restricted localization of polarity proteins [11]. The polarity complexes Crumbs, which includes Crumbs3, a protein associated with lin seven 1 (PALS1) and a protein associated with tight junctions (PATJ); PAR, including PAR3, PAR6, and aPKCι; and Scribble, including Scribble, lethal giant larvae (LGL), and disc large (DLG), distribute in three specific domains in a mechanism orchestrated by the recruitment of JAM-A into the junctions. JAM-A is recruited to newly formed cell–cell junctions simultaneously with ZO-1. Then, PAR-3 interacts with JAM-A and also tethers to the junction. Although the mechanism is not completely understood, it has been proposed that the JAM-induced localization of the PAR protein complex controls the asymmetrical divisions of the blastocyst [12,13]. Meanwhile, cingulin and ZO-2 assemble to the TJ for the first time [2,14]. At the late 32-cell stage, ZO-1α^+ [5] is expressed and co-localizes with occludin [15] at the Golgi. Together with claudin-1 and 3, ZO-1α^+ and occludin assemble at the junctions [16], allowing the embryo to generate a barrier between trophectoderm cells and the nascent blastocoel cavity. The specific function of the ZO-1 α-domain is not completely clear, but a recent study showed that ZO-1α^+ expression correlates with the grade of Caco-2 cell differentiation. During the exponential phase of Caco-2 cell growth, cells shift in isoform expression from elevated ZO-1α^- to elevated ZO-1α^+ as cells reach confluence [17]. Further, this shift in expression pattern could be manipulated by the cell substrate and correlated with differentiation of the cells, implicating ZO-1α^+ in differentiation of junctional properties. Little is known about the molecular mechanisms that promote the proper localization of these junctional domains, but PKC isoforms have been implicated through chemical activation of PKC isoform zeta (PKCζ) in trophectoderm cells, which promotes the assembly of ZO-1α^+ [18], while PKCι interacts with PAR3/6 proteins, promoting localization and positioning at the junctions [19].

The mechanisms of epithelial polarization and differentiation during development differ between species. In addition to mammals, this process has been studied in invertebrate organisms including *C. elegans*, *Xenopus*, and *Drosophila* embryos. More detailed reviews of these species may be found in [11,20]. In contrast with mammals, the polarization of *C. elegans* blastomeres is not directly linked to cell fate specialization since at the 4-cell stage the blastomeres are already polarized but do not form junctions. In fact, the first epithelial specialization of *C. elegans* appears later during organogenesis [21]. In *Xenopus* embryos, both polarization and junction formation start together with the first cleavage, but in this case, the epithelial differentiation process occurs independently of cell adhesion [22]. Distinct from these organisms, the *Drosophila* embryo has a unique cleavage mechanism named cellularization. In this process, the embryo undergoes multiple cell divisions at the same time that are mediated through membrane invaginations. The resultant tightly packed epithelium of 13 columnar hexagonal cells, possesses cytoskeleton-based landmarks that act as localized clusters for AJ and septate junction (SJ) recruitment [23,24]. In *Drosophila*, SJ performs a similar role as TJ. These studies highlight important differences in epithelial junction specializations among species and reveal unique evolutionary resolution of epithelial differentiation.

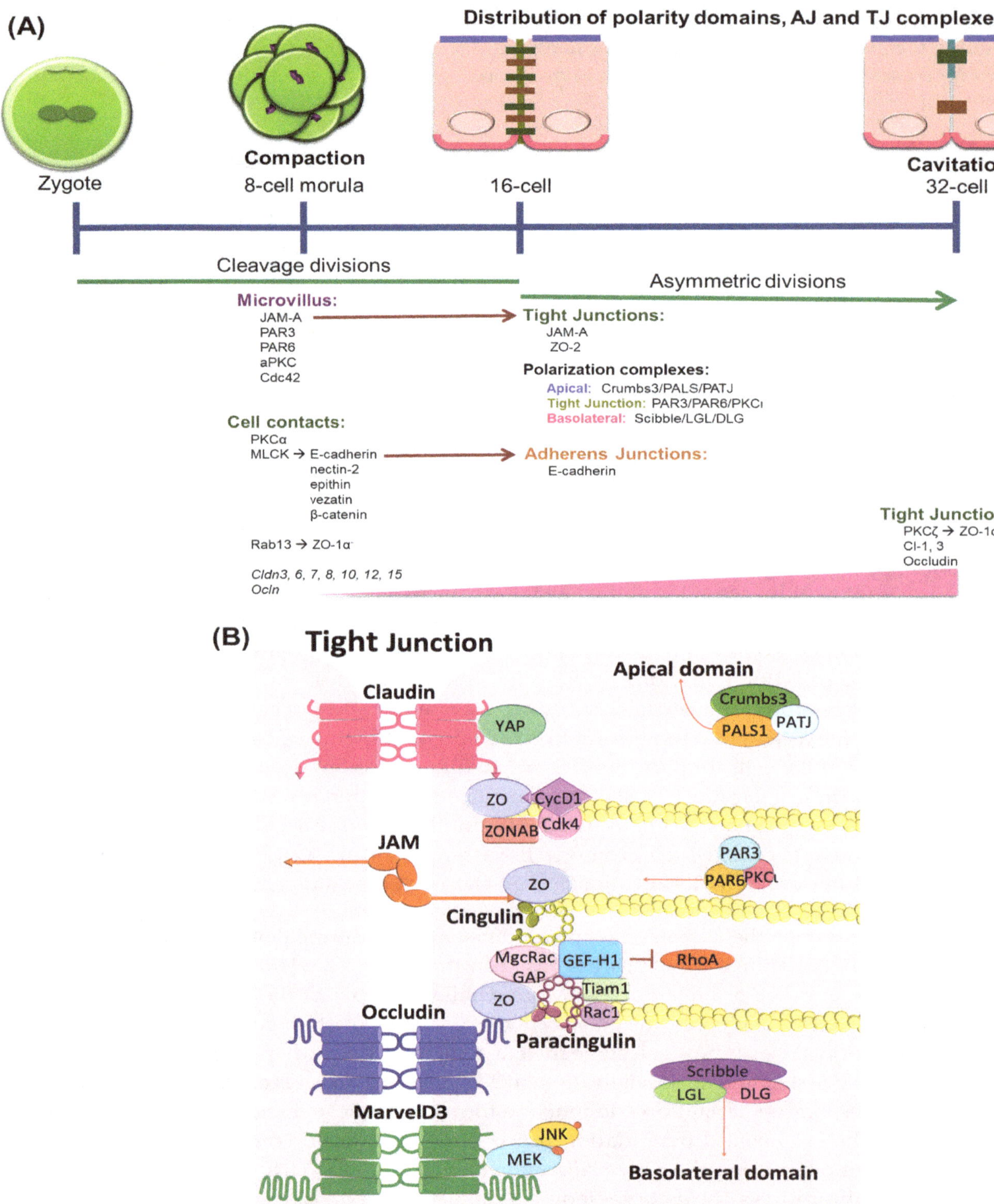

Figure 1. Tight junction proteins in the differentiation of mouse epithelial cells. (**A**) The morula is produced by a series of cleavage divisions of the early embryo, starting with a single cell zygote. Once the embryo has divided into 8 cells, it begins to form a clustered cell mass that expresses AJ and TJ proteins at cell contacts. While JAM and the PAR complex are already expressed at this stage, they localize at the microvillus. As the embryo continues to develop, JAM is recruited into the junctions and polarization complexes are tethered into three specific domains. The subsequent asymmetric divisions initiate the cavitation process at the 32-cell stage. The expression of TJ proteins occludin and claudins

increases and together with ZO-1α^+, they assemble at the junctions, correlating with the differentiation of the first epithelium specialization in the body. (**B**) In fully differentiated tissues, TJ proteins localize at the contact sites of adjacent cells and create a selective paracellular barrier. In addition, the junctional proteins bind and regulate a number of signal transduction factors including those involved in the regulation of cell proliferation. These include sequestering GEF-H1 preventing RhoA activity, the transcription factor ZONAB, and the CycD1/Cdk4 complex, all involved in G1 to S transition. The complexes Crumbs, PAR, and Scribble regulate the proper polarization of apical, TJ, and basolateral domains, respectively.

2. TJ Expression in Endothelial Differentiation

In the central nervous system (CNS), the vasculature has unique anatomic and functional properties that contribute to blood-brain (BBB) and blood-retinal (BRB) barriers. The endothelial cells are in close association with neurons, glia, and pericytes that promote the differentiation to a specialized endothelium with TJ proteins that restrict the transport of molecules into the neural tissue. The development of CNS vascularization and angiogenesis begins around embryonic (E) day 9 in rodents. Starting from the perineural vascular plexus, the endothelial network sprouts and covers the entire surface of the neural tube [25].

Endothelial differentiation and specialization in the CNS barriers result from the polarization of the endothelial network and the localization of junctional components along endothelial cell contacts. For the mouse BBB, the differentiation process takes place between E15.5 and E18.5 [26], while the BRB angiogenesis starts after birth, and the barrier properties are established simultaneously with capillary formation between postnatal day 7 and 15 [27]. In humans, both BBB and BRB develop during gestation, but it is not completely clear if the establishment of barrier properties culminate with birth [28]. Although all endothelial cells from the most primitive linage express the TJ protein claudin-5, a junctional protein required for brain barrier properties [29], the sealing of a functional barrier occurs later.

The Wnt/β-catenin signaling pathway is vital for growth and specialization of the CNS vasculature. During mouse brain development, the neuroepithelium-derived Wnt7a/Wnt7b ligands activate signaling through Frizzled-4 (Fzd4) receptor, low-density lipoprotein receptor-related protein 5/6 (Lrp5/6) co-receptor, G-protein-coupled receptor 124 (Gpr124), and reversion-inducing cysteine-rich protein (Reck) co-activators to promote β-catenin activation and angiogenesis [30–32]. This signaling is essential for neural tube vascularization because its depletion causes CNS vascular malformation and lethality. Moreover, β-catenin signaling contributes to the specialization of the CNS endothelium by inducing the expression of the glucose transporter and the TJ protein claudin-5 [31]. The non-Wnt ligand norrin activates β-catenin signaling though its receptor Fzd4, the Lrp5/6 co-receptor, and the tetraspanin-12 (Tspan12) co-activator, and this signaling is essential for proper retinal angiogenesis and barrier formation as well as cerebellum barrier formation [33]. In the retina, while Wnt3, Wnt7a, and Wnt7b ligands have only a small contribution in barrier formation [34], the Müller glia-derived norrin is the predominant β-catenin activator that regulates BRB formation and maintenance. Downstream of β-catenin, gene expression of the sex-determining region Y-related high-mobility group box factors 7, 17, and 18 (*Sox7*, *Sox17*, and *Sox18*, respectively) promotes BRB formation [35].

3. Epithelial-Mesenchymal Transition (EMT)

As part of normal development, epithelial cells transition to mesenchymal phenotype in a process called epithelial mesenchymal transition (EMT). This occurs, for example, during implantation of embryos, as part of gastrulation and neural crest formation. EMT also contributes to both wound repair and to the pathology of neoplasia as a part of tumor dedifferentiation. Under a mesenchymal state, tumor cells are highly invasive and able to migrate to distant sites, establishing metastases.

After invading a new tissue, tumors cells may differentiate towards the recovery of their epithelial characteristics in a process called mesenchymal-epithelial transition (MET) [36].

Changes in TJ protein expression and organization are a fundamental step in EMT. Indeed, the most common markers that indicate the epithelial cell transitioning to mesenchymal phenotype include the suppression of TJ protein transcription and loss of junctional complex organization and polarity. These changes coincide with additional changes such as increased synthesis of cytoskeletal components, actin stress fibers, spindle-shaped appearance, increased movement, and resistance to apoptosis [37,38].

An early step in EMT in cancer cells is loss of gene transcription for junctional components, where the high expression of the transcription regulators Snail, Slug, Twist, zinc finger E-box-binding homeobox 1 (ZEB-1), SMAD interacting protein 1 (Sip-1), and lymphoid enhancer-binding factor 1 (LEF-1) reduces *Ocln* and *Cldn* synthesis [39,40] (Figure 2). With the progression of EMT, the junction complex is disassembled via transforming growth factor beta (TGFβ) signaling. The binding of TGFβ to its receptor TGFβR2 results in its recruitment to the junctional complex where it binds to occludin and promotes phosphorylation of the polarity protein PAR6. Then, the endogenous E3 ubiquitin ligase Smurf1 redistributes to cell junctions and promotes RhoA ubiquitination and degradation, thus leading to cytoskeleton rearrangement and TJ disassembly [41]. Another example is epidermal growth factor (EGF) activation of its receptor (ERBB2), which then interacts with the PAR6-aPKC complex and causes PAR3 dissociation and ultimately TJ breakdown [42]. Other growth factors that promote EMT through their tyrosine kinase receptors include the hepatocyte growth factor (HGF) through its receptor Met; the fibroblast growth factor (FGF); and the bone morphogenetic protein (BMP) [39]. While BMP2 and BMP4 promote EMT [43,44], BMP7 induces MET [45].

While TJ proteins are often reduced in cancers of epithelial origin, in several human cancers, TJ proteins are overexpressed potentially by epigenetic regulation [46]. It is important to note in these cases that the junctional complexes have lost their membrane organization and barrier function. However, their intracellular localization suggests that they might contribute to other cellular functions. Recently, E-cadherin was found to promote metastasis in models of invasive ductal carcinomas [47]. In this work, E-cadherin gene (*Cdh1*) was depleted in MMTV-PyMT invasive ductal carcinoma cells and as expected, this resulted in increased invasion. However, *Cdh1*-null cells also exhibited increased TGFβ/SMAD2/3 and reactive oxygen species signaling, which resulted in reduced cell proliferation, lower survival, and inhibition of metastasis. These studies suggest that E-cadherin acts as a survival factor in invasive ductal carcinomas by limiting reactive oxygen-mediated apoptosis and highlight the non-barrier function of this adherens junction protein.

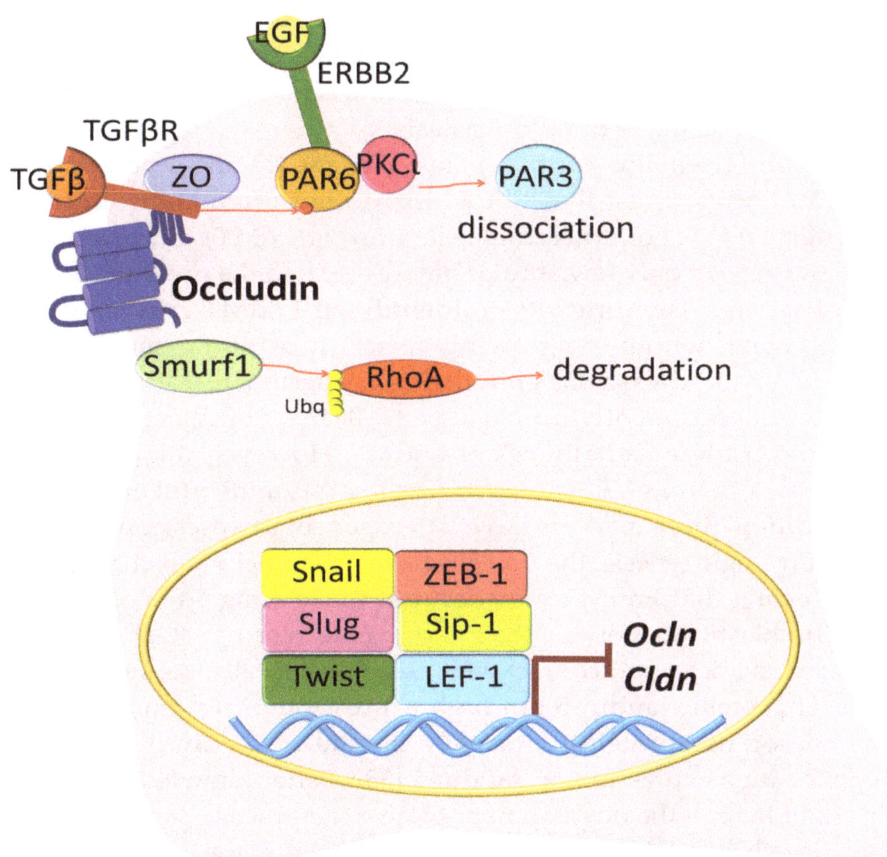

Figure 2. Tight junction proteins in EMT. As an early step in EMT, epithelial cells lose polarity and TJs are disrupted. TGFβ binds its receptor and is recruited to the junction where it interacts with ZO-1 and occludin. TGFβR activation promotes PAR6 phosphorylation. ERBB2 binds to PAR6/PKCι proteins, but PAR3 becomes dissociated from the complex, and this results in overall altered cell polarization. Smurf1 is also recruited into the TJ, where it induces RhoA ubiquitination (Ubq) and degradation. Meanwhile, during EMT, a series of nuclear transcription factors inhibit the expression of TJ genes *Ocln* and *Cldn*. Growth factors including FGF, HGF, and BMPs promote EMT, but the exact mechanism is not completely clear. Together, these molecular mechanisms promote epithelial transformation into a mesenchymal state.

4. Endothelial-Mesenchymal Transition (EndMT)

Similar to epithelial cells, endothelial tissues have the plasticity to transition to a mesenchymal phenotype (EndMT), and this transition is commonly observed in vasculogenesis. EndMT was first described by Leonard M. Eisenberg in 1995 [48], and recently, considerable attention has been paid to EndMT due to the vital role it plays in human diseases, including cerebral cavernous malformation (CCM) [49] ischemic stroke, atherosclerosis, cardiac fibrosis associated with diabetes mellitus, nephrosis, pulmonary arterial hypertension, diabetic nephropathy, alcoholic liver disease, multiple myeloma, and other cancers [50]. CCM is caused by loss of function mutations in *CCM* genes 1, 2 or 3. The *CCM* gene products bind to the endothelial adherens junction complex in the cytoplasm [51]. In CCM, increased TGFβ and BMP signaling and the consequent EndMT in *CCM1*-null endothelial cells are crucial events in the onset and progression of the disease [49]. Further, recent evidence reveals limited *CCM3*-null endothelial cell clonal expansion induces EndMT in wild-type cells [52]. Importantly, while in epithelial cells all three TGFβ isoforms can induce EMT [53,54], EndMT is primarily stimulated by the TGFβ2 isoform [55–57], by BMP2 or by BMP4 [43,44].

In cancer, it has been proposed that EndMT may contribute to metastasis. The cancer microenvironment, which is rich in growth factors secreted by stromal cells, promotes a sustained tip phenotype in most of the endothelial cells undergoing angiogenesis in tumors [58]. Moreover, recent studies suggest that TGFβ induces EndMT in brain endothelial cells [59]. In this study, cells with the mesenchymal phenotype had better adhesion to melanoma cells with increased migration potential, suggesting that EndMT might facilitate brain metastasis.

In some diseases, EndMT remains controversial. Lineage tracing studies employing multiple independent murine Cre lines suggest that fibroblasts do not originate from hematopoietic cells, endothelial cells (through EndMT) or epicardial cells (through EMT) but proliferate from resident fibroblast lineages [60]. Moreover, only few studies have described the presence of EndMT in vascular cells with endothelial barriers. The difficulty in identifying EndMT may be due to the presence of various differentiated states within the growing vessel. Loss of the endothelial markers CD31, von Willebrand factor (vWF), and vascular endothelial cadherin (VE-cadherin) and an increase in the mesenchymal markers alpha smooth muscle actin (α-SMA), fibroblast-specific protein 1 (FSP1), and vimentin have been considered a hallmark of EndMT. However, it has been shown that some level of endothelial markers such as VE-Cadherin, Tie-1, vWF, and cytokeratins is still detected in cells after EndMT. In addition, these markers have heterogeneous expression within the same vessel. In CNS vessels undergoing angiogenesis, the tip cells have a mesenchymal character that allows them to migrate [61,62], suggesting that EndMT may occur at the leading edge of the branching vessels. While tip cells have a fibroblastic phenotype, the intermediate vascular stalk cells proliferate but also have a more differentiated character and contact with neighboring cells. As such, tip cells demonstrate low expression of the TJ protein claudin-5 but high expression of the transcytosis marker PLVAP (plasmalemma vesicle-associated protein). However, the stalk cells simultaneously proliferate and form a barrier through the high expression of claudin-5 [33]. Better knowledge about the mechanisms that control EndMT might help in the development of new therapies in several diseases.

While extensive research has elaborated a role for AJ proteins in the control of epithelial (reviewed in [63]) and endothelial (reviewed in [51]) cell proliferation, growing evidence now implicates the TJ proteins in cell cycle control and regulation of proliferation.

5. Role of Tight Junction Scaffold Proteins in Controlling Cell Proliferation

5.1. *Zona Occludens (ZO)*

ZO proteins serve as links between the transmembrane TJ proteins and the cytoskeleton. ZO proteins bind to the TJ proteins claudins, occludin, and JAMs (Figure 1B) as well as other ZO proteins, promoting the polymerization of the junctional complex and binding cytoskeletal-associated proteins α-catenin and afadin (AF6) linking the TJ to the cytoskeleton [64–66]. ZO proteins are part of the MAGUK family, and there are three genes that encode these proteins: ZO-1, -2, and -3 [67–69]. Although they have redundant functions, knockout mice studies suggest a role of ZO-1 and -2 in early development. Mice deficient of ZO-1 gene (*Tjp1*) die in E10.5 due to embryonic defects mediated by apoptosis in the neural tube, the notochord, and the allantois areas, as well as extra-embryonic defects in the angiogenesis of the yolk sac, suggesting that ZO-1 regulates tissue organization and remodeling in both epithelial and endothelial tissues [70]. Similarly, mice deficient of ZO-2 gene (*Tjp2*) die shortly after implantation due to an arrest in early gastrulation [71].

Recent publications suggest that ZO proteins are able to transmit information about the degree of cell-cell contacts to the nucleus, thus maintaining a balance between proliferation and differentiation. ZO -1, -2, and -3 sequester key regulators of cell cycle progression at the junction sites [72] (Figure 1B), including ZO-1-associated nucleic acid binding protein (ZONAB). At low-density cell numbers, the Y-box transcription factor ZONAB is located at the nuclear fractions, promoting the expression of G1/S-phase transition through regulation of proliferating cell nuclear antigen (*PCNA*) gene expression (Figure 3). However, at higher density and cell contact, ZO-1 sequesters ZONAB at the TJ, reducing nuclear concentration and thus controlling cell proliferation [73,74]. Cyclin D1 (CycD1) is also sequestrated at the junctions by ZO proteins, together with its associated cyclin-dependent kinase 4

(Cdk4). CycD1 forms a complex with the ZO-3 protein, specifically through its PDZ-binding motif. This interaction regulates epithelial cell proliferation by CycD1 stabilization at the membrane during cell proliferation, since knockdown of ZO-3 gene (*TJP3*) with siRNA results in G0/G1 cell-cycle arrest [75].

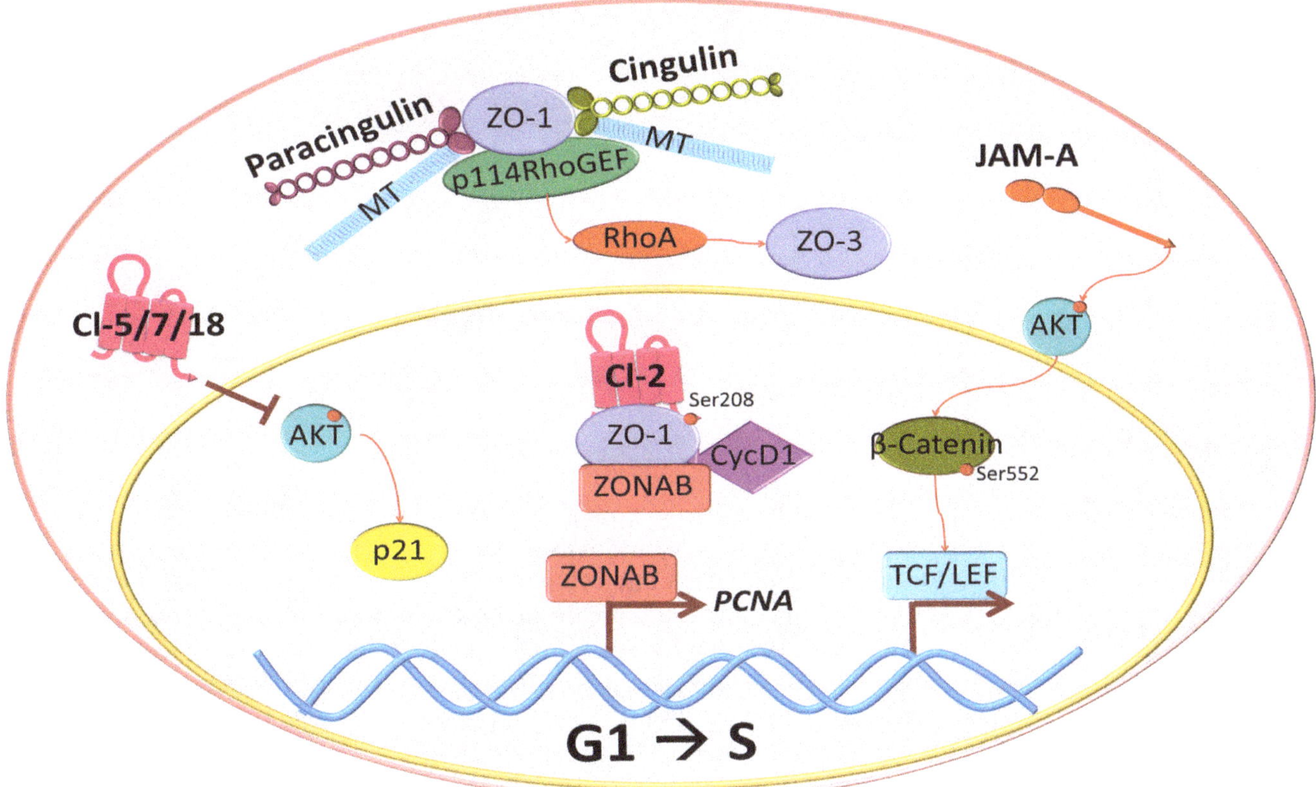

Figure 3. Tight junction proteins in G1/S transition. TJ proteins have been implicated in the control of cell cycle progression. Claudin-5, -7 or -18 (Cl-5/7/18) expression in cancer cells inhibits G1/S transition associated with AKT activation. Alternatively, cell cycle progression can be activated by cingulin and paracingulin dissociation from the junction, which results in a conformational change that allows interaction with microtubules (MT), ZO-1, and p114RhoGEF, promoting RhoA activation and ZO-3 and claudin-2 accumulation that are related to increased G1/S transition. ZO-1 also has been found in nuclear fractions and interacts with claudin-2 (Cl-2) phosphorylated at Ser208, CycD1, and ZONAB. The latter, in sparse cells, can promote *PCNA* gene expression and increase proliferation. In mice deficient of JAM-A gene (*F11r*), AKT becomes activated and phosphorylates β-catenin on Ser552. As a result, β-catenin translocates to the nucleus and activates transcription mediated by TCF/LEF.

Other studies reveal changes in ZO proteins in carcinogenesis. The ZO-1 protein has been found to be highly expressed in adenocarcinoma samples, as compared with healthy tissue [76]. Also, in pancreatic cancer cells, EGFR activation correlates with ZO-1 protein localization at the cytoplasm or at the nucleus, and inhibition of EGFR with AG1478 induces the redistribution of ZO-1 to the junctions [77] (Figure 4). In lung cancer, PKC isoform epsilon (PKCε) activation regulates the interaction between α5-integrin and ZO-1, and this correlates with poor prognosis [78]. Moreover, in EMT, receptor serine/threonine kinase TGFβR type II and I co-localize with ZO-1 at the junctions in a mechanism dependent on TGFβ stimulation and the phosphorylation of PAR6 by TGFβRII [41] (Figure 2). Collectively, these studies suggest that ZO proteins contribute to contact regulated control of cell proliferation.

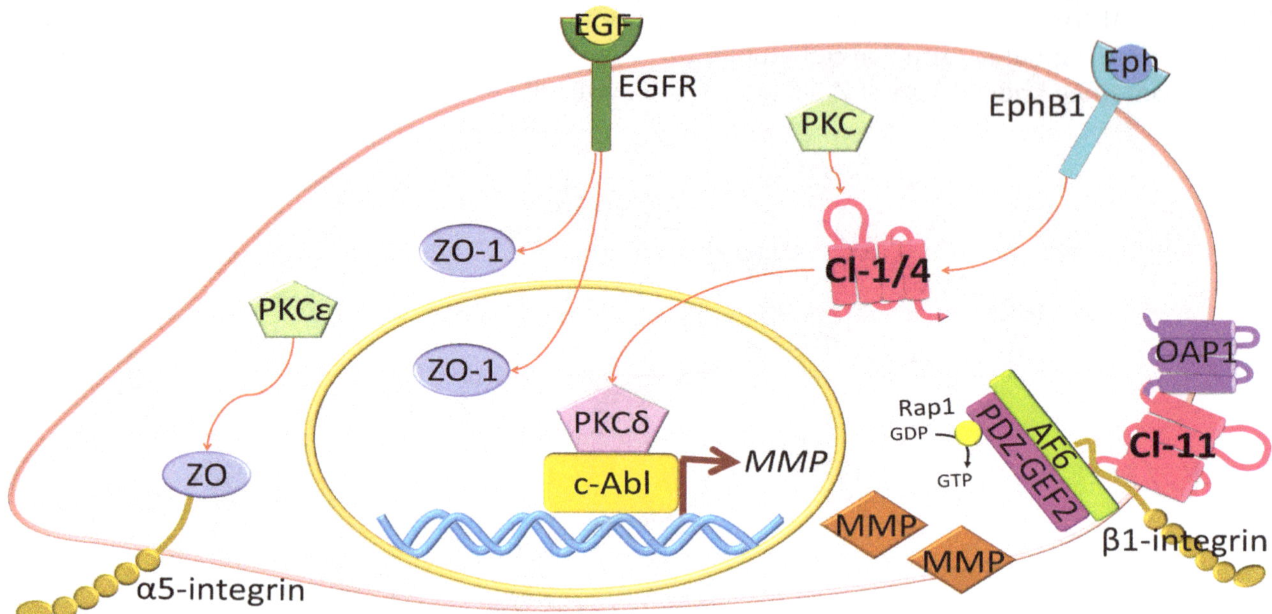

Figure 4. Tight junction proteins in migration and invasion. Overexpression of TJ proteins has been associated with the promotion of cell migration and invasion in cancer cells. EGFR expression has been correlated with ZO-1 localization at nuclear and cytoplasmic fractions, and inhibition of EGFR phosphorylation leads to relocalization of ZO-1 to cell-contacts. PKCε promotes ZO interaction with α5-integrin. Claudin-1 (Cl-1) activates PKCδ, which in turn, binds to c-Abl transcription factors and activates *MMP* transcription. MMPs are secreted and induce basal membrane degradation, increasing the invasive potential of cancer cells. Similarly, EphB1 receptor phosphorylation has been associated with claudin-4 (Cl-4) altered expression promoting MMP expression and secretion. Claudin-11 (Cl-11) interaction with OAP1 and β1-integrin increases cell migration through AF6 and PDZ-GEF2 interaction and Rap1 activation.

5.2. Cingulin

Cingulin is a cytoskeletal adaptor protein that has a crucial role in transducing the mechanical force generated by the contraction of the actin-myosin cytoskeleton into functional regulation of the epithelial and endothelial barriers [79]. Its localization at the junctions is mediated by the interaction with the TJ proteins ZO and JAMs, along with its anchoring to the actin cytoskeleton (Figure 1B).

Recent studies have demonstrated a role of cingulin in cell proliferation and migration through its ability to interact with microtubule (MT)-associated small GTPase activators of RhoA, such as the guanine nucleotide exchange factor H1 (GEF-H1) [80–83]. Knockdown of cingulin gene (*CGN*) in Mardin-Darby Canine Kidney (MDCK) cells results in increased expression of claudin-2, ZO-3, and RhoA activation (Figure 3). While the knockdown did not affect TJ protein localization or its barrier function, loss of *CGN* increased RhoA-induced G1/S phase transition through its interaction with GEF-H1 [84]. During neural tube closure, the pre-migratory neural crest cells initiate EMT by TJ disruption and cingulin-induced delamination in the neuroepithelium. Both depletion and overexpression of *CGN* increase the migratory neural crest cell population, associated with loss of basal lamina and disruption of the neural tube [85]. Moreover, cingulin participates in cell polarization in epithelial cysts by interacting with the Rab11 family interacting protein 5 (FIP5), an effector of Rab11 GTPase. Cingulin serves as the tethering factor to ensure the fidelity of apical endosome targeting the apical membrane initiation sites [86].

5.3. Paracingulin

Paracingulin, cingulin-like 1 or the junction-associated-coiled-coil protein (JACOP), is a scaffold TJ protein that maintains the integrity of the association between the MT cytoskeleton and cell junctions [87,88]. With 39% sequence homology to cingulin, paracingulin also interacts with ZO-1 at the globular head domain, promoting its localization at the junctions [89].

Similar to cingulin, paracingulin activates RhoA GTPases to promote cell proliferation. Through its interaction with Rac1, paracingulin promotes the recruitment of Tiam1 (T-cell lymphoma invasion and metastasis 1) and GEF-H1 activators at the junctions (Figure 1B). Therefore, a reduced paracingulin expression in MDCK cells increases RhoA activity and promotes G1/S phase transition [90]. However, when the Rac1 inhibitor MgcRacGAP is present, this interacts with both cingulin and paracingulin at TJs and reduces cell proliferation [91].

Recent studies on endothelial cells suggest a role of paracingulin in angiogenesis. In a primary culture of human dermal microvascular endothelial cells (HDMEC), ZO-1 promoted the recruitment of paracingulin and the RhoA activator 114RhoGEF into the junctions, and this complex increased the angiogenic potential of HDMEC [92]. Supporting this idea, the silencing of paracingulin gene (*Cgnl1*), greatly impaired tubule structure formation in 3D co-culture assays and diminished the number of vascular structures formed during vascular expansion in the developing retina, suggesting paracingulin as a defining factor in new vessel formation [93]. Furthermore, using baculovirus-infected insect cells, Vasileva et al. provided evidence for the interaction of paracingulin with MTs, which is important for the formation of both strong AJ and focal adhesions to ensure stabilization and further elongation of neovascular tubules [93,94]. Together, these results suggest that paracingulin might control endothelial cell proliferation and angiogenesis through the regulation of MT and RhoA activity.

6. Transmembrane TJ Proteins in the Control of Cell Proliferation

6.1. Claudins

Claudins are transmembrane proteins that contribute to paracellular transport by forming ion selective barriers and pores in a tissue-specific manner. As mentioned previously, a hallmark of EMT includes loss of barrier properties. Indeed, claudin dysregulation is commonly found in human carcinomas, often with decreased *CLDN* gene expression. Further, claudin overexpression can sometimes reverse the malignant phenotype. In lung squamous cell carcinoma, the transfection of claudins -5, -7, and -18 was able to suppress proliferation and inhibit G1/S transition associated with inhibition of AKT phosphorylation [95] (Figure 3).

However, some cancer studies reveal a correlation between claudin overexpression and tumor progression [96–99]. Examples include claudin-3 expression, which has been correlated with increased tubulogenesis and bromo-deoxy uridine (BrdU) incorporation in mouse inner medullary collecting duct cells (mIMCD-3) [100]. Similarly, overexpression of claudin-6, -7, or -9 enhanced invasiveness and proliferation of an adenocarcinoma cell line [96], and *CLDN4* and *18* were upregulated in pancreatic cancer tissues [101,102]. Claudin-18 isoform a2 is highly expressed in gastric, esophageal, pancreatic, lung, and ovarian cancers, and while it is considered as a putative marker, its exact role in tumor progression remains unknown [103]. Importantly, the high expression of claudins in neoplastic tissues does not coincide with border localization or barrier regulation. For example, claudin-4 is highly expressed in poorly differentiated pancreatic cancer cells and is enriched at basolateral membranes rather than the apical junctional complex [104]. Claudin-11 overexpression has been associated with proliferation and migration of oligodendrocytes in a mechanism dependent on its interaction with outer surface protein-associated protein 1 (OAP1) and β1-integrin [105] (Figure 4). Moreover, claudin-2 has been found localized at the nuclear fractions in highly proliferative lung adenocarcinoma cells [106].

The molecular consequences of claudin overexpression in cancer biology have not been clearly defined. In human liver cells, it has been suggested that the activation of the protein kinase complex c-Abl/PKCδ (PKC isoform delta) is critical for the acquisition of a malignant phenotype induced by claudin-1 overexpression (Figure 4) [99]. In this study, the overexpression of claudin-1 in normal liver hepatocytes led to an increased expression of metalloproteinsases (MMPs), promoting an invasive phenotype. Similar results were found in human melanoma cells where PKC activation by phorbol myristic acid (PMA) increased *CLDN*1 transcription and contributed to invasion [107]. Recent evidence also indicates that ephrin (Eph), through its EphB1 receptor, can control epithelial transformation through interaction with claudin-1 and -4. This interaction promotes EphB1 tyrosine phosphorylation, which in cancer cells mediates migration and invasion through the downstream exocytosis of MMPs [108]. Importantly, claudins can also be phosphorylated by Eph receptors that regulate cell-contact formation [108,109]. A role of claudin-2 in proliferation has been also suggested in studies with lung adenocarcinoma cells. In this work, it was proposed that claudin-2 could induce proliferation when it is phosphorylated at Ser208 and located at the nuclear fractions, interacting with ZO-1, ZONAB, and CycD1 [106] (Figure 3). Moreover, there was a positive correlation between claudin-4 expression, production of interleukin 8 (IL-8), and increased angiogenesis in mouse xenografts [110].

The *Cldn15* knockout mouse model suggests that alterations in barrier function might lead to homeostatic changes that activate other cellular processes including proliferation. *Cldn15* null mice have an enlarged upper small intestinal phenotype or mega-intestine due to the enhanced proliferation of the crypt cells [111]. In this knockout model, the expression and localization of claudins-1, -2, -3, -4, -7, -12, -18, -20, and -23 were not affected, but barrier properties were clearly lost. The authors concluded that the increased cellular proliferation is a consequence of altered barrier function that changes the intestinal environment [112]. This alteration in cell proliferation associated with an altered microenvironment may also be observed in metastasis studies of pancreatic carcinomas where the primary tissue has low expression of *CLDN3* while the metastatic cells that invade the liver express high levels of *CLDN3* and *4* but show low expression of *CLDN1* and *7* [113,114]. The altered TJ expression might influence the cancer microenvironment, promoting changes in cancer metastasis. Similar to *Cldn15*, *Cldn18* knockout mice show proliferation changes with lung enlargement, parenchymal expansion, and increased abundance and proliferation of known distal lung progenitors, the alveolar epithelial type II (AT2) cells, activation of Yes-associated protein (YAP), and increased organ size and tumor genesis in mice [115]. Importantly, claudin-18 and YAP were found to interact and co-localize at cell contacts in control samples (Figure 1B), and claudin-18 overexpression decreased YAP nuclear localization, suggesting that similar to ZO proteins, claudin-18 restricts cell proliferation by the retention of YAP at cell contacts.

Together, these data suggest that altered claudin expression, organization or function at the barrier may be observed as a hallmark of EMT and in a number of epithelial cancers. However, some cancers involve increased expression of specific claudins despite loss of barrier function, and overexpression of some claudins can alter the cancer cell phenotype. While loss of barrier function clearly alters the microenvironment of the cancer, the precise role of claudin in proliferation and cancer remains an area that requires additional investigation. Although claudins are promising molecular targets for diagnosis and therapy [102,116], a better understanding of the mechanisms regulated by claudins in the control of cell proliferation and metastasis is needed.

6.2. Junctional Adhesion Molecules (JAM)

JAM is the most extensively studied single-span TJ protein, and there are three genes described that encode to the proteins: JAM-A, -B, and -C. JAMs belong to the immunoglobulin superfamily because they contain at least one immunoglobulin domain at the extracellular N-terminus, while the cytoplasmic tail contains a PDZ binding sequence that interacts with the PDZ domain of ZO-1.

JAM-A regulates epithelial proliferation via canonical Wnt signaling. In control mice, JAM-A shows a gradient expression along the intestinal crypt-luminal axis, which is increased in non-proliferating luminal epithelial cells [117]. JAM-A gene (*F11r*) deletion in mice results in increased intestinal epithelial cell proliferation and activation of AKT and its downstream target β-catenin, which is phosphorylated at Ser552 (Figure 3). This in turn, promotes β-catenin accumulation and its nuclear localization that is followed by an increase in T-cell factor (TCF)/LEF-induced transcriptional activity [118]. Moreover, a recent study revealed that JAM-A also regulates the cortical localization of dynein to control planar spindle orientation during mitosis via activation of Cdc42 and phosphatidyl inositol 3-kinase (PI3K) [119]. As expected, JAM-A suppression or expression of a dimerization-deficient isoform resulted in aberrant spindle orientation.

JAM-A dimerization promotes epithelial and endothelial migration. Expression of the JAM-A dimerization-defective mutant in 293T cells and the use of JAM-A dimer-disrupting antibodies reduce cell migration. Disruption of JAM-A dimerization also correlates with β1-integrin degradation, decreased GTPase Rap1 activation, and diminished numbers of focal concentrations of phosphorylated paxillin [120]. This process is mediated by the interaction of JAM-A with AF6 and PDZGEF2 [121,122]. Similar to these studies, isolated endothelial cells from mice deficient in *F11r* have enhanced spontaneous and random motility associated with increased numbers of actin-containing protrusions, reduced MT stability, and impaired focal adhesions, which can be reversed by JAM-A expression or by using glycogen synthase kinase 3-beta (GSK3β) inhibitors [123]. Moreover, mice deficient of JAM-C gene (*Jam3*) show increased *F11r* expression and enhanced retinal vascularization [124,125], supporting a role for JAM in angiogenesis.

Together, these data indicate that JAMs contribute to cell proliferation and migration. However, further studies are clearly needed in order to determine the exact role of JAMs in these processes.

6.3. MARVEL Family Proteins

The MARVEL (for MAL and related proteins for vesicle trafficking and membrane link) family members include occludin (MarvelD1), tricellulin (MarvelD2), and MarvelD3 [126–128]. MARVEL proteins possess a putative protein-lipid-interacting motif containing four intramembrane helices creating the MARVEL domain (Figure 5A). Its function remains unknown, but MARVEL domains of non-TJ proteins, such as myelin and lymphocyte protein (MAL), have been associated with the generation and stabilization of functional membrane domains via the propensity for homo-oligomerization and the ability to attract apical membrane lipids [129]. Knockout studies of TJ MARVEL genes reveal a complex phenotype in a variety of tissues (reviewed in [130]). Many tissues are able to form morphological and functionally normal TJs, suggesting that they are not individually required for TJ formation although duplication of function has yet to be fully explored. Indeed, in vitro experiments demonstrate depletion of occludin gene (*Ocln*) resulting in the redistribution of tricellulin [131]. *Ocln* knockout mice develop hyperplasia of the gastric epithelium and testicular atrophy, while deletion of *MarvelD2* or *Ocln* in mice present degeneration of cochlear hair cells that leads to progressive hearing loss [132,133]. Moreover, Tricellulin is essential for barrier formation and the maintenance of the TJ structure in the inner ear as determined by gene deletion studies [126,134]. Additionally, MARVEL proteins have been found in membrane raft domains with a high cholesterol content, suggesting that they might participate in bending during the formation of endocytic caveolin transport vesicles similar to other proteins with MARVEL domains [135]. To our knowledge, there is no evidence for tricellulin in the control of cell proliferation, and we will focus on occludin and MarvelD3.

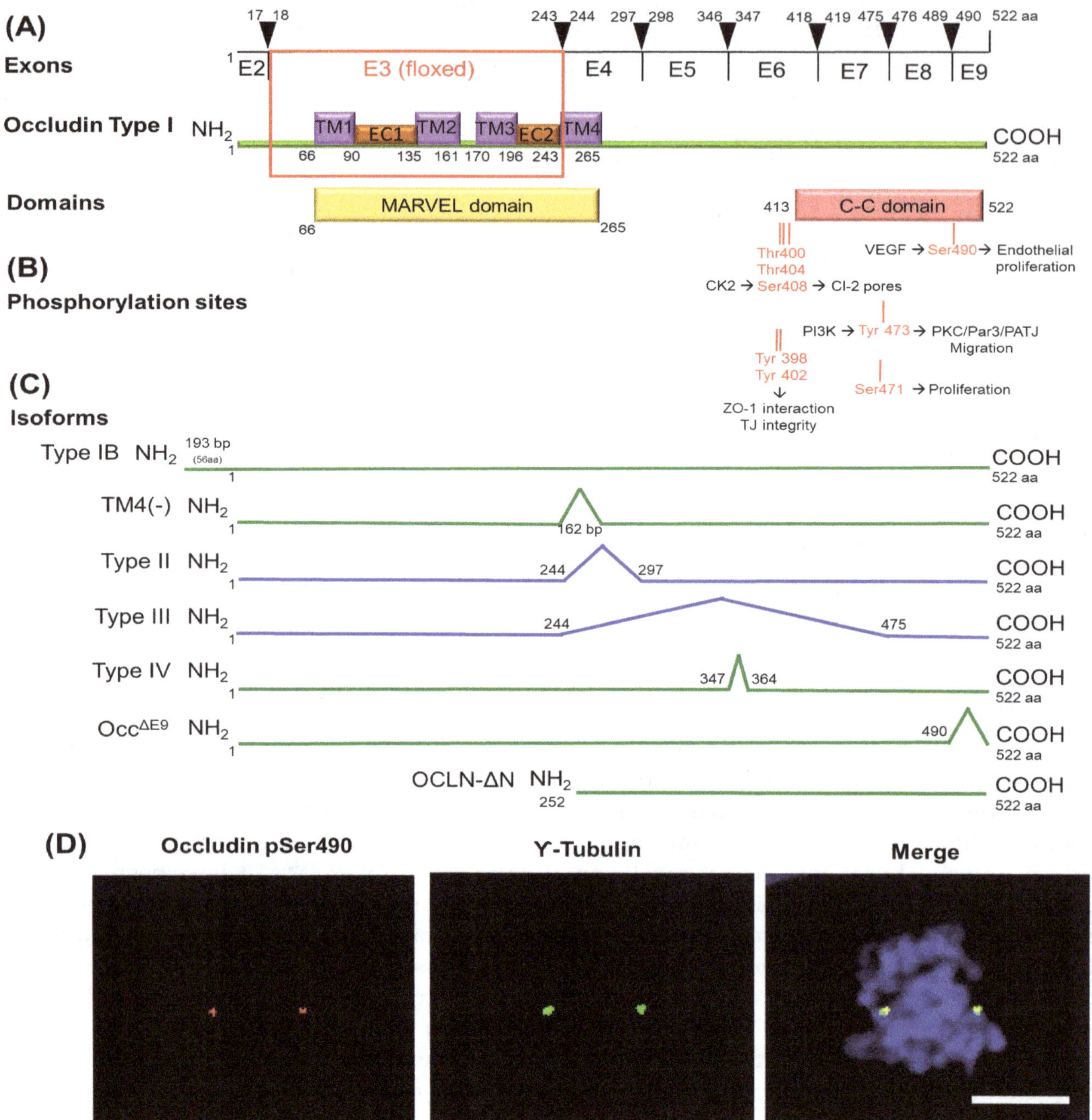

Figure 5. Occludin structure and isoforms. Occludin is a 522 amino acids protein encoded by 9 exons (**A**). Occludin full length (type I) possesses four transmembrane (TM) domains and two extracellular (EC) loops with the MARVEL domain as homology on the cytoplasmic side after each TM region. At the C-terminus (COOH) of occludin, a coiled-coiled (C-C) domain can be phosphorylated at multiple sites (**B**). Phosphorylation of occludin identified and known or implied functions. Occludin can mediate proliferation though the phosphorylation of two sites: Ser471, which regulates post contact proliferation in epithelial cells, and Ser490, which is promoted by VEGF-induced PKCβ activation and regulates both endothelial permeability and neovascularization. To date, several occludin isoforms have been described (**C**). The function of each isoform has not been fully elucidated, but most isoforms localize to the junctions except type II and III (blue lines). Interestingly, occludin deleted in exon 9 ($Occ^{\Delta E9}$) restricts cell migration. (**D**) In bovine retinal endothelial cells, occludin stained with a pS490-specific antibody (red) shows co-localization of phospho-occludin with the centrosome marker γ-tubulin (green) in pro-metaphase. Hoechst dye (blue). Scale bar = 5 μm.

Occludin was the first transmembrane protein discovered at the TJ [136]. While the *Ocln* -null mouse forms intact TJs, the animals have phenotypic alterations including growth retardation, thinning of compact bone, testicular atrophy, male infertility, loss of cytoplasmic granules in salivary epithelial cells, females are not able to lactate, brain calcification [132], and hyper proliferation of mucous epithelial cells in the intestinal lining [137]. In vitro, the silencing of *OCLN* gene has a limited effect on barrier properties, with increases in permeability to divalent organic cations and also to small molecules under hydrostatic pressure [138,139]. In ARPE-19 cells, a human retinal pigmented epithelial cell line, loss of *OCLN* increases the DNA synthesis rate and cell proliferation [139]. Moreover, during neurogenesis, occludin loss has been found in neural tubes of chicken and mouse embryos at E9 [140].

Occludin function is regulated by phosphorylation. Specially, the C-terminal region has been studied extensively (Figure 5B). The 3 sites Thr400, Thr404, and Ser408 that lie just prior to the coiled-coil (C-C) domain regulate its interaction with ZO proteins and TJ integrity [141]. While occludin phosphorylation at Thr404 regulates its localization at the junctions [142], phosphorylation of Ser408 promotes ion flux through control of claudin-2 dimerization. The complex formed by occludin/ZO-1/claudin-2 is dissociated when occludin is phosphorylated by casein kinase 2 (CK2) at Ser408. As a result, claudin-2 interacts in trans with claudin-2 from neighboring cells, forms a cation pore, and increases ion flux [143]. Other phosphorylation sites nearby including Tyr398 and 402 residues may also alter ZO-1 binding [144]. Ser490 phosphorylation controls vascular endothelial growth factor (VEGF)-induced endothelial permeability with the expression of S490A point mutants preventing VEGF-induced permeability in cell culture. In endothelial cells, this results in increased vascular permeability [145] in a PKC isoform beta (PKCβ)-dependent manner [146].

Collectively, these studies demonstrate that occludin phosphorylation contributes to regulation of barrier properties by promoting its binding to ZO proteins and the recruitment of occludin and claudins into the junctions.

Occludin phosphorylation has also been associated with the regulation of cell proliferation and migration. The phosphorylation at Tyr473 promotes directional migration of epithelial cells by the activation of PI3K signaling and by promoting the organization of the aPKC, PAR3, and PATJ polarity complex [147], suggesting a role of occludin in cell migration. In addition, previous studies in our group have identified five phosphorylation sites on the C-C domain of occludin, including Ser471 and Ser490, mentioned above, on the two turns of the C-C domain, which have been associated with the control of cell proliferation. Expression of the S471A mutant as a stable cell line has no effect on sub-confluent proliferation but inhibits proliferation and cell packing after cell contact in MDCK cells as determined by cell number and DNA synthesis, leading to enlarged cells. Inhibition of proliferation by expressing S471A point mutant occludin, inhibition of cell proliferation in cell contacted immature monolayers or inhibition of the Ser471 kinase G-protein coupled receptor kinase (GRK), all profoundly inhibit TJ formation and epithelial monolayer maturation [148]. The Hippo signaling pathway is an important determinant of cell and organ sizes, and nuclear exclusion of the co-activator YAP accompanies proliferative quiescence. The Hippo pathway elements YAP and TEAD (TEA-dependent) have been found to co-localize with occludin in pancreatic cancer cells and regulate cell proliferation [149].

The second phosphorylation site of occludin that regulates cell proliferation was identified at Ser490. This phosphorylation can be induced by VEGF and promotes occludin ubiquitination and its intracellular trafficking. In endothelial cells, this results in increased vascular permeability [145] in a PKCβ-dependent manner [146]. Moreover, studies on the phosphorylation of this site have revealed a novel function of occludin as a regulator of centrosome separation and mitosis initiation. In MDCK cells, the expression of occludin mutated at Ser490 to Ala slows cell proliferation and hindered mitotic entry due to delayed centrosome separation. Stable expression of aspartic acid phosphomimetic (S490D) in MDCK cells results in centrosomal localization of occludin and cell proliferation. However, expression of the nonphosphorylatable alanine mutation (S490A) of occludin impedes centrosome separation, delays mitotic entry, and reduces proliferation. Collectively, these studies demonstrate a novel location and function for occludin in centrosome separation and mitosis [150]. Similar results were found in endothelial cell culture and in retinal tissue, where the induction of endothelial-specific expression of the occludin S490A mutant through viral delivery completely inhibited neovascularization [151]. In this study, primary bovine retinal endothelial cells in collagen matrices responded to VEGF with increased Ser490 phosphorylation coincident with tube formation. Transfection of occludin with the S490A point mutant inhibited tube formation, proliferation, and migration compared to wild-type occludin. Western blotting revealed increased occludin phosphorylation associated with angiogenesis, and whole-mount immunofluorescent staining of the retinas revealed centrosomal occludin organization in proliferating vessels. Further, mice with doxycycline-inducible *VEGF* expression from photoreceptors were used to study occludin control of angiogenesis in vivo. Expression of the occludin S490A mutant by sub-retinal injection of adeno-associated virus significantly reduced retinal vessel growth in vivo. Importantly, in these studies, occludin was located at the centrosomes (Figure 5D), and increased occludin phosphorylation was found in dividing endothelial cells, mouse retinas with neovascularization, and human surgical samples of retinal neovessels, suggesting a novel role for occludin in regulation of endothelial proliferation and neovascularization in a phosphorylation-dependent manner

Similar to claudins, previous publications have demonstrated that oncogenic transformation of a variety of cell types is associated with altered occludin expression [152]. *OCLN* gene is downregulated in premalignant foci in kidneys from patients with germ line tumor suppressor von Hippel–Lindau (*VHL*) gene mutations [153]. At a molecular level, it has been suggested that *OCLN* is transcriptionally repressed following constitutive Raf-1 expression [154] this is mediated through a direct interaction between activated Slug and the E-box in the *Ocln* promoter [155]. Similarly, in murine melanoma cells, *Ocln* is epigenetically silenced through promoter hyper-methylation, and its forced expression also reduces tumor migration. This is supported by other studies suggesting that occludin possess anti-tumorigenic properties [156,157]. The expression of exogenous occludin suppresses tumor growth in nude mice of Raf1-transformed rat salivary gland epithelial cells [158]. Similarly, stable occludin expression in melanoma and breast cancer cells followed by injection into the craniolateral thorax and mammary fat pad, respectively, reduced the size of lung metastases [157]. In a cell culture model of uveal melanoma, blood vessel epicardial substance (BVES) protein overexpression led to an increase in ZO-1 and occludin, which correlated with decreased cell proliferation [159]. Further, occludin induces premature senescence in breast cancer cells, which can be blocked by chemical inhibition of the mitogen-activated protein kinase (MEK) pathway [160]. Conversely, occludin has been also implicated in EMT, where occludin targets the TGFβ receptor to the junctional complex and promotes efficient epithelial transformation [161] (Figure 2), a mechanism that correlates with the simultaneous repression of the genes encoding E-cadherin, claudins, and occludin [39]. Finally, *OCLN* depletion in MDCK cultures demonstrated impaired mitotic spindle orientation due to a reduced interaction with ZO-1, suggesting an alteration of the cues necessary for polarization in cell division [162].

Other studies support a role of occludin in proliferation and interaction with centrosomal proteins. *OCLN* mutations in human patients can lead to microcephaly and band-like calcifications with polymicrogyria characterized by loss of cortical convolutions, shallow or absent sulci, and multiple small gyri giving the cortex surface a roughened irregular appearance [163–168]. To date, thirteen pathogenic mutations in *OCLN* gene have been identified in thirteen families, and seven mutations are situated in exon 3 [165–168]. Primary microcephaly (MCPH, for microcephaly primary hereditary) is a disorder of brain development that results in a head circumference more than three standard deviations below the mean for age and gender. Notably, many of the causative genes for MCPH encode centrosomal proteins involved in centriole biogenesis [169].

Recently, a new isoform of occludin was discovered by Bendriem et al. with a specific function in the proliferation of human embryonic stem cells (hESCs) and in neural progenitors that alter cortex size in the developing mouse brain (Bendriem et al., accepted for publication in ELife). The original *Ocln* knockout mouse line was generated by excising exon 3 and was believed to be a null model.

While mouse full-length occludin (OCLN-FL) is no longer expressed, a truncated form that lacks its N-terminus and three of its four transmembrane domains (OCLN-ΔN) is still expressed. This is a 32-34 kDa protein that results from a shorter ΔN transcript lacking exons 2 and 3 (Figure 5C). Both OCLN-FL and OCLN-ΔN isoforms localize to the centrosomes; however, in the homozygous mutant mouse line *Ocln*$^{\Delta N/\Delta N}$, OCLN-ΔN localizes to interphase and mitotic centrosomes in the embryonic mouse cortex but not at the plasma membrane, suggesting the C-terminal domain of occludin is important for this centrosomal localization. Consistent with *OCLN* mutation in patients with microcephaly, depletion of *Ocln* in mice led to microcephaly. An increased mitotic index was found in *Ocln*$^{\Delta N/\Delta N}$ mutant mice along with a higher percentage of cells of *Ocln*$^{\Delta N/\Delta N}$ E12.5 cortices in prometaphase and metaphase compared to the wild-type. Moreover, prolonged mitosis led to a higher percentage of activated (cleaved) caspase 3 (CC3)-positive apoptotic cells in mutant embryos compared to controls prior to E14.5.

Occludin centrosomal localization was also confirmed in vitro in two hESCs lines that closely resemble the *Ocln*$^{\Delta N/\Delta N}$ mouse mutant. Mutant hESC-derived organoids displayed pronounced proliferation defects, premature differentiation, and apoptosis. Cells numbers with a reduced ratio of the basal neural progenitor marker HOPX (homeodomain-only protein homeobox) compared to the early neuronal marker NeuroD1 may be responsible for the reduced size of human organoids. Importantly, centrosomal occludin co-localized and immune-precipitated with NuMA (for Nuclear Mitotic Apparatus protein) and the small GTPase RAN (RAs-related Nuclear protein), two important proteins in mitotic spindle assembly and stabilization, and mutant hESCs exhibited impaired mitotic spindles and abnormal morphology at the spindle poles. These studies demonstrated an important role for occludin in neurogenesis through its centrosomal interactions and promotion of proper functioning neural progenitor mitotic spindles.

Along with occludin localization at the centrosomes, studies have also shown that intracellular occludin-containing vesicles move along MTs and contribute to the regulation of cell proliferation [170]. MTs interacting with plasma membranes participate in the preservation of epithelial TJ structure and function. This is regulated by the binding of MT plus end–tracking proteins at the scaffold in the AJs or may be achieved through MT minus end binding of nezha/calmodulin-regulated spectrin-associated protein (CAMSAP) and ninein to the AJs [171–175]. The junctional localization of several MT organizing center proteins indicates the crucial role of junctions as sites that orchestrate MT organization in polarized cells [94]. Glotfelty et al. [170] reported that intracellular occludin-containing vesicles move along MTs and that the rate of movement depends on intact MT networks. This suggests that dynein, a key regulator of MT-vesicle trafficking, may regulate this process in the minus-end direction. Consistent with this hypothesis, the siRNA knockdown of dynein/dynactin induced occludin accumulation in the cytosol, whereas plus-end motor kinesin knockdown did not. This model of MT-dependent TJ trafficking was further supported by the results from studies on dynein and Rab11 [176]. Rab11 utilizes MTs for trafficking and has been shown to participate in occludin trafficking by the regulation of the Rab11 FIPs (Rab11 family interacting protein) [176].

Occludin-containing vesicles may traffic bi-directionally on MTs to regulate cell proliferation. Previous studies suggest that MTs traffic likely contribute to TJ assembly by functioning as tracks for the delivery of TJ proteins through MT-associated vesicles. Particularly occludin-containing granules have been found near the tip of oolemma ingression in dividing *Xenopus* oocytes [177]. Rab13 or junction rab (JRab) is a key mediator of the endocytic recycling of occludin [178] through its binding partners Rab13-binding protein and MICAL-like protein 2 (MICAL-L2) [179,180]. VAP-33 (VAMP associated protein of 33 kDa) is implicated in vesicle docking/fusion and binds to occludin, and its overexpression promotes occludin movement along the lateral edge of the plasma membrane [181]. Identification and characterization of a homolog of VAP-33 in *Drosophila* (DVAP-33A) revealed that DVAP-33A regulates the division of boutons at the synaptic terminals by stabilizing and directing the MT cytoskeleton during budding [182]. Thus, TJ assembly mediated by occludin-MT interaction may play an important role in both cell proliferation and junction organization.

Together, these studies indicate that occludin both regulates barrier function by controlling TJ protein internalization and localizes at centrosomes, contributing to the regulation of cell proliferation.

MarvelD3 is identified as the third member of proteins with a MARVEL domain [128]. MarvelD3 is expressed as two isoforms that show a broad tissue distribution. The two isoforms represent splice variants and share the predicted N-terminal cytoplasmic domain of 198 amino acids, but differ in their C-terminal halves that contain the transmembrane domains. MarvelD3 does not have the long cytoplasmic C-terminus ending in a C-C domain found on occludin and tricellulin. RNA interference experiments in steady-state Caco-2 monolayers indicate that MarvelD3 is required to maintain epithelial integrity under osmotic stress, but this is not essential for TJ formation [183]. However, Raleigh et al. reported that knockdown of *MARVELD3* delays TJs assembly in Caco-2 cells [127]. Further, depletion of *MARVELD3* by siRNAs in the human pancreatic cancer cell line (HPAC) resulted in downregulation of barrier function as shown by decreased electrical resistance and increased permeability to fluorescent dextran tracers, whereas knockdown did not affect the fence function of TJs maintaining apical and basolateral membrane protein restriction [184]. Interestingly, in a genome-wide association study, an intergenic single nucleotide in *MARVELD3* correlated with resistance to severe malaria [185] while its specific role has not yet been discovered.

Recent studies suggest a role of MarvelD3 in EMT, cell proliferation, and migration. In tumor cells, MarvelD3 was downregulated during EMT in human pancreatic cancer cells [184]. Similar effects were also observed when MCF-7 breast cancer cells that are known to express MarvelD3 were compared to MiaPaca-2 pancreatic tumor cells, which do not express detectable MarvelD3. In this study, transfection of MarvelD3 in MiaPaca-2 cells reduced cell migration and proliferation. When they were injected in a xenograft model, MarvelD3 overexpressing MiaPaca-2 cells revealed reduced tumor volume as compared to cells with low MarvelD3 expression. Moreover, siRNA-mediated depletion of *MARVELD3* induced Caco-2 cell migration and proliferation, and the effects could be rescued by expressing mouse MarvelD3. These data indicate that MarvelD3 regulates cell proliferation and cell migration of differentiating and dedifferentiated epithelial model cell lines. The authors suggest that MarvelD3 functions as a signaling transmembrane component of TJs.

MarvelD3 activates MEKK1/JNK (mitogen-activated protein kinase kinase 1/c-Jun NH_2-terminal kinase) signaling through its N-terminal cytoplasmic domain to control cell proliferation. [183]. The interaction of the N-terminal domain of MarvelD3 with upstream component MEKK1 that regulates JNK activation was first demonstrated in glutathione S-transferase (GST) pull-down assays in Caco-2 cells. Further studies in MDCK cell lines expressing MarvelD3 fusion proteins carrying the biotin ligase at both the C-terminus and N-terminus supported this conclusion and corroborated the specific interaction of MEKK1 with the N-terminal domain of MarvelD3. Moreover, in MarvelD3 overexpressing MiaPaca-2 cells, MEKK1 was partially recruited to cell–cell contacts when MarvelD3 was expressed in MiaPaca-2 cells but not in control cells, indicating that re-expression of MarvelD3 is sufficient to stimulate membrane recruitment of MEKK1.

MarvelD3 modulates cell proliferation and migration in early embryogenesis. A recent study with *Xenopus* indicated that MarvelD3 modulates cell proliferation in early eye development and regulates cell survival during eye morphogenesis [186]. In these studies, the JNK pathway was required for proper eye morphogenesis by acting as an inhibitor of the expression of eye-field transcription factors (EFTFs). EFTFs specify a single eye-field in the most anterior region of the neural plate. In this region, inhibition of cell-cycle activators occurs to favor EFTF expression, while duration of the expression of the transcription factors is established by cell-cycle-independent factors. The single eye-field is then divided into two eye primordia under the influence of Sonic hedgehog signaling. Moreover, MarvelD3 is part of a regulatory feedback loop that coordinates JNK activity with neural crest formation. *MarvelD3* depletion enhances JNK signaling, which leads to disruption of neural crest derivative differentiation and neural crest precursor formation, as well as displacement of the neural plate border. Consistent with this model, inhibition of JNK signaling is sufficient to rescue the phenotype induced by *marvelD3* depletion, while constitutively active JNK disrupts neural crest development, supporting the importance of controlled regulation of JNK activity [187]. Together, these data present a novel role of MarvelD3 as an essential regulator of early vertebrate development and neural crest induction that relies on interplay between gene expression, cell proliferation, and cell migration.

7. Conclusions

While numerous studies clearly demonstrate the role of TJ proteins in barrier regulation, more recent research suggests a number of tight junction proteins also contribute to the control of cell proliferation and growth. The junction proteins may contribute to proliferation in a number of mechanisms. In some cases, control of barrier properties alters the microenvironment contributing to growth control. Further, TJ proteins may localize transcription factors and cell cycle control proteins at the plasma membrane, restricting nuclear access or inhibiting function. In the presence of growth factors, cell polarization is disrupted and the TJ proteins release these factors, allowing progression through the cell cycle. In addition, a number of TJ proteins demonstrate non-typical cellular localization associated with cell proliferation, revealing distinct and unique functions from their role in cell-cell contact. TJ proteins have been found localized in the nucleus at G1/S cell cycle transition, associated with integrins at the basal membrane during migration, or in mitosis interacting with microtubules or with centrosome proteins. Together, these studies reveal the coordination of epithelial and endothelial barrier formation with cell proliferation, an essential component of cellular growth and differentiation.

Author Contributions: D.A.A. and M.D.-C. designed the manuscript outline, M.D.-C., X.L., and D.A.A. wrote the manuscript.

Abbreviations

α-SMA	alpha smooth muscle actin
AF6	afadin
AJ	adherens junction
AT2	alveolar epithelial type II cells
BBB	blood-brain barrier
BMP	bone marrow morphogenetic protein
BRB	blood-retinal barrier
BrdU	bromo-deoxy uridine
BVES	blood vessel epicardial substance protein
CAMSAP	calmodulin-regulated spectrin-associated protein
C-C	coiled-coil domain
CC3	cleaved caspase 3
CCM	cerebral cavernous malformation
Cdk4	cyclin-dependent kinase 4
CK2	casein kinase 2
Cl	claudin

α-SMA	alpha smooth muscle actin
AF6	afadin
AJ	adherens junction
AT2	alveolar epithelial type II cells
BBB	blood-brain barrier
BMP	bone marrow morphogenetic protein
BRB	blood-retinal barrier
BrdU	bromo-deoxy uridine

BVES	blood vessel epicardial substance protein
CAMSAP	calmodulin-regulated spectrin-associated protein
C-C	coiled-coil domain
CC3	cleaved caspase 3
CCM	cerebral cavernous malformation
Cdk4	cyclin-dependent kinase 4
CK2	casein kinase 2
Cl	claudin
CNS	central nervous system
CycD1	cyclin-D1
DLG	disc large
EC	extracellular
EFTFs	eye-field transcription factors
EGF	epidermal growth factor
EGFR2 or ERBB2	epidermal growth factor receptor
EMT	epithelial-mesenchymal transition
EndMT	endothelial-mesenchymal transition
Eph	ephrin
EphB1	ephrin B1 receptor
FGF	fibroblast growth factor
FSP1	fibroblast-specific protein 1
FIP	family interacting protein
Fzd4	frizzled-4
GEF-H1	guanine nucleotide exchange factor H1
Gpr124GRK	G-protein-coupled receptor 124
GSK3β	G-protein coupled receptor kinase glycogen synthase kinase 3 beta
GST	glutathione S-transferase
HDMEC	human dermal microvascular endothelial cells
hESCs	human embryonic stem cells
HGF	hepatocyte growth factor
HOPX	homeodomain-only protein homeobox
HPAC	human pancreatic cancer cell line
IL-8	interleukin 8
JACOP	junction-associated-coiled-coil protein
JAM	junctional adhesion molecule
JNK	c-Jun NH_2-terminal kinase
JRAB	junction Rab
LEF	lymphoid enhancer-binding factor
LGL	lethal giant larvae
Lrp5/6	low density lipoprotein receptor-related protein 5/6
MAGUK	membrane-associated guanylate kinase homolog
MAL	myelin and lymphocyte protein
MARVEL	MAL and related proteins for vesicle trafficking and membrane link
MCPH	microcephaly primary hereditary
MDCK	Madin–Darby canine kidney
MEK	mitogen-activated protein kinase
MEKK	mitogen-activated protein kinase kinase
MET	mesenchymal-epithelial transition
MICAL-L2	MICAL-like protein 2
mIMCD-3	mouse inner medullary collecting duct cells
MMP	matrix metalloproteinase
MT	microtubules
NuMA	nuclear mitotic apparatus protein
OAP1	outer surface protein -associated protein 1
$Occ^{\Delta E9}$	occludin deleted in exon 9

OCLN-FL	occludin-full length
OCLN-ΔN	occludin N-terminal deleted
PALS1	protein associated with lin seven 1
PAR	partitioning defective
PATJ	protein associated with tight junctions
PCNA	proliferating cell nuclear antigen
PI3K	phosphatidyl inositol 3-kinase
PKC	protein kinase C
PLVAP	plasmalemma vesicle-associated protein
PMA	phorbol myristic acid
RAN	Ras-related nuclear protein
Reck	reversion-inducing cysteine-rich protein
Sip-1	SMAD interacting protein 1
SJ	septate junction
Sox	sex determining region Y-related high mobility group box factors
TCF	T-cell factor
TEAD	TEA-dependent
TGFβ	transforming growth factor-beta
TGFβR	transforming growth factor-beta receptor
Tiam1	T-cell lymphoma invasion and metastasis 1
TJ	tight junctions
TM	transmembrane
Tspan12	tetraspanin-12
Ubq	ubiquitination
VAMP	vesicle-associated membrane proteins
VAP-33	VAMP associated protein of 33 kDa
VE-cadherin	vascular endothelial cadherin
VEGF	vascular endothelial growth factor
VHL	von Hippel–Lindau
vWF	von Willebrand factor
YAP	c-yes associated protein
ZEB-1	zinc finger E-box-binding homeobox 1
ZO	zona occludens
ZONAB	ZO-1-associated nucleic acid binding protein

References

1. Pauken, C.M.; Capco, D.G. Regulation of cell adhesion during embryonic compaction of mammalian embryos: Roles for PKC and beta-catenin. *Mol. Reprod. Dev.* **1999**, *54*, 135–144. [CrossRef]
2. Fleming, T.P.; Wilkins, A.; Mears, A.; Miller, D.J.; Thomas, F.; Ghassemifar, M.R.; Fesenko, I.; Sheth, B.; Kwong, W.Y.; Eckert, J.J. Society for Reproductive Biology Founders' Lecture 2003. The making of an embryo: Short-term goals and long-term implications. *Reprod. Fertil. Dev.* **2004**, *16*, 325–337. [CrossRef] [PubMed]
3. Balda, M.S.; Anderson, J.M. Two classes of tight junctions are revealed by ZO-1 isoforms. *Am. J. Physiol.* **1993**, *264 Pt 4*, C918–C924. [CrossRef]
4. Fleming, T.P.; McConnell, J.; Johnson, M.H.; Stevenson, B.R. Development of tight junctions de novo in the mouse early embryo: Control of assembly of the tight junction-specific protein, ZO-1. *J. Cell Biol.* **1989**, *108*, 1407–1418. [CrossRef] [PubMed]
5. Sheth, B.; Fesenko, I.; Collins, J.E.; Moran, B.; Wild, A.E.; Anderson, J.M.; Fleming, T.P. Tight junction assembly during mouse blastocyst formation is regulated by late expression of ZO-1 alpha+ isoform. *Development* **1997**, *124*, 2027–2037. [PubMed]
6. Sheth, B.; Fontaine, J.J.; Ponza, E.; McCallum, A.; Page, A.; Citi, S.; Louvard, D.; Zahraoui, A.; Fleming, T.P. Differentiation of the epithelial apical junctional complex during mouse preimplantation development: A role for rab13 in the early maturation of the tight junction. *Mech. Dev.* **2000**, *97*, 93–104. [CrossRef]

7. Thomas, F.C.; Sheth, B.; Eckert, J.J.; Bazzoni, G.; Dejana, E.; Fleming, T.P. Contribution of JAM-1 to epithelial differentiation and tight-junction biogenesis in the mouse preimplantation embryo. *J. Cell Sci.* **2004**, *117 Pt 23*, 5599–5608. [CrossRef]
8. Macara, I.G. Par proteins: Partners in polarization. *Curr. Biol.* **2004**, *14*, R160–R162. [CrossRef]
9. Cui, X.S.; Li, X.Y.; Shen, X.H.; Bae, Y.J.; Kang, J.J.; Kim, N.H. Transcription profile in mouse four-cell, morula, and blastocyst: Genes implicated in compaction and blastocoel formation. *Mol. Reprod. Dev.* **2007**, *74*, 133–143. [CrossRef]
10. Hamatani, T.; Carter, M.G.; Sharov, A.A.; Ko, M.S. Dynamics of global gene expression changes during mouse preimplantation development. *Dev. Cell* **2004**, *6*, 117–131. [CrossRef]
11. St Johnston, D.; Ahringer, J. Cell polarity in eggs and epithelia: Parallels and diversity. *Cell* **2010**, *141*, 757–774. [CrossRef] [PubMed]
12. Ebnet, K.; Suzuki, A.; Horikoshi, Y.; Hirose, T.; Zu Brickwedde, M.K.M.; Ohno, S.; Vestweber, D. The cell polarity protein ASIP/PAR-3 directly associates with junctional adhesion molecule (JAM). *EMBO J.* **2001**, *20*, 3738–3748. [CrossRef] [PubMed]
13. Betschinger, J.; Mechtler, K.; Knoblich, J.A. The Par complex directs asymmetric cell division by phosphorylating the cytoskeletal protein Lgl. *Nature* **2003**, *422*, 326–330. [CrossRef] [PubMed]
14. Javed, Q.; Fleming, T.P.; Hay, M.; Citi, S. Tight junction protein cingulin is expressed by maternal and embryonic genomes during early mouse development. *Development* **1993**, *117*, 1145–1151.
15. Sheth, B.; Moran, B.; Anderson, J.M.; Fleming, T.P. Post-translational control of occludin membrane assembly in mouse trophectoderm: A mechanism to regulate timing of tight junction biogenesis and blastocyst formation. *Development* **2000**, *127*, 831–840.
16. Fleming, T.P.; Sheth, B.; Fesenko, I. Cell adhesion in the preimplantation mammalian embryo and its role in trophectoderm differentiation and blastocyst morphogenesis. *Front. Biosci.* **2001**, *6*, D1000–D1007. [CrossRef]
17. Ciana, A.; Meier, K.; Daum, N.; Gerbes, S.; Veith, M.; Lehr, C.M.; Minetti, G. A dynamic ratio of the alpha+ and alpha− isoforms of the tight junction protein ZO-1 is characteristic of Caco-2 cells and correlates with their degree of differentiation. *Cell Biol. Int.* **2010**, *34*, 669–678. [CrossRef]
18. Eckert, J.J.; McCallum, A.; Mears, A.; Rumsby, M.G.; Cameron, I.T.; Fleming, T.P. Relative contribution of cell contact pattern, specific PKC isoforms and gap junctional communication in tight junction assembly in the mouse early embryo. *Dev. Biol.* **2005**, *288*, 234–247. [CrossRef]
19. Gopalakrishnan, S.; Hallett, M.A.; Atkinson, S.J.; Marrs, J.A. aPKC-PAR complex dysfunction and tight junction disassembly in renal epithelial cells during ATP depletion. *Am. J. Physiol. Cell Physiol.* **2007**, *292*, C1094–C1102. [CrossRef]
20. Nance, J. Getting to know your neighbor: Cell polarization in early embryos. *J. Cell Biol.* **2014**, *206*, 823–832. [CrossRef]
21. Nance, J.; Priess, J.R. Cell polarity and gastrulation in *C. elegans*. *Development* **2002**, *129*, 387–397. [PubMed]
22. Cardellini, P.; Davanzo, G.; Citi, S. Tight junctions in early amphibian development: Detection of junctional cingulin from the 2-cell stage and its localization at the boundary of distinct membrane domains in dividing blastomeres in low calcium. *Dev. Dyn.* **1996**, *207*, 104–113. [CrossRef]
23. Harris, T.J.C.; Sawyer, J.K.; Peifer, M. How the Cytoskeleton Helps Build the Embryonic Body Plan: Models of Morphogenesis from *Drosophila*. *Curr. Top. Dev. Biol.* **2009**, *89*, 55–85. [PubMed]
24. Lecuit, T. Junctions and vesicular trafficking during *Drosophila* cellularization. *J. Cell Sci.* **2004**, *117 Pt 16*, 3427–3433. [CrossRef]
25. Risau, W. Mechanisms of angiogenesis. *Nature* **1997**, *386*, 671–674. [CrossRef]
26. Haddad-Tovolli, R.; Dragano, N.R.V.; Ramalho, A.F.S.; Velloso, L.A. Development and Function of the Blood-Brain Barrier in the Context of Metabolic Control. *Front. Neurosci.* **2017**, *11*, 224. [CrossRef]
27. Diaz-Coranguez, M.; Ramos, C.; Antonetti, D.A. The inner blood-retinal barrier: Cellular basis and development. *Vis. Res.* **2017**, *139*, 123–137. [CrossRef]
28. Saunders, N.R.; Dreifuss, J.J.; Dziegielewska, K.M.; Johansson, P.A.; Habgood, M.D.; Mollgard, K.; Bauer, H.C. The rights and wrongs of blood-brain barrier permeability studies: a walk through 100 years of history. *Front. Neurosci.* **2014**, *8*, 404. [CrossRef]
29. Nitta, T.; Hata, M.; Gotoh, S.; Seo, Y.; Sasaki, H.; Hashimoto, N.; Furuse, M.; Tsukita, S. Size-selective loosening of the blood-brain barrier in claudin-5-deficient mice. *J. Cell Biol.* **2003**, *161*, 653–660. [CrossRef]

30. Stenman, J.M.; Rajagopal, J.; Carroll, T.J.; Ishibashi, M.; McMahon, J.; McMahon, A.P. Canonical Wnt signaling regulates organ-specific assembly and differentiation of CNS vasculature. *Science* **2008**, *322*, 1247–1250. [CrossRef]
31. Liebner, S.; Corada, M.; Bangsow, T.; Babbage, J.; Taddei, A.; Czupalla, C.J.; Reis, M.; Felici, A.; Wolburg, H.; Fruttiger, M.; et al. Wnt/beta-catenin signaling controls development of the blood-brain barrier. *J. Cell Biol.* **2008**, *183*, 409–417. [CrossRef] [PubMed]
32. Daneman, R.; Agalliu, D.; Zhou, L.; Kuhnert, F.; Kuo, C.J.; Barres, B.A. Wnt/beta-catenin signaling is required for CNS, but not non-CNS, angiogenesis. *Proc. Natl. Acad. Sci. USA* **2009**, *106*, 641–646. [CrossRef] [PubMed]
33. Zhou, Y.; Wang, Y.; Tischfield, M.; Williams, J.; Smallwood, P.M.; Rattner, A.; Taketo, M.M.; Nathans, J. Canonical WNT signaling components in vascular development and barrier formation. *J. Clin. Investig.* **2014**, *124*, 3825–3846. [CrossRef] [PubMed]
34. Wang, Y.; Cho, C.; Williams, J.; Smallwood, P.M.; Zhang, C.; Junge, H.J.; Nathans, J. Interplay of the Norrin and Wnt7a/Wnt7b signaling systems in blood-brain barrier and blood-retina barrier development and maintenance. *Proc. Natl. Acad. Sci. USA* **2018**, *115*, E11827–E11836. [CrossRef] [PubMed]
35. Zhou, Y.; Williams, J.; Smallwood, P.M.; Nathans, J. Sox7, Sox17, and Sox18 Cooperatively Regulate Vascular Development in the Mouse Retina. *PLoS ONE* **2015**, *10*, e0143650. [CrossRef] [PubMed]
36. Morris, H.T.; Machesky, L.M. Actin cytoskeletal control during epithelial to mesenchymal transition: Focus on the pancreas and intestinal tract. *Br. J. Cancer* **2015**, *112*, 613–620. [CrossRef] [PubMed]
37. Robson, E.J.; Khaled, W.T.; Abell, K.; Watson, C.J. Epithelial-to-mesenchymal transition confers resistance to apoptosis in three murine mammary epithelial cell lines. *Differentiation* **2006**, *74*, 254–264. [CrossRef]
38. Vega, S.; Morales, A.V.; Ocana, O.H.; Valdes, F.; Fabregat, I.; Nieto, M.A. Snail blocks the cell cycle and confers resistance to cell death. *Genes Dev.* **2004**, *18*, 1131–1143. [CrossRef]
39. Ikenouchi, J.; Matsuda, M.; Furuse, M.; Tsukita, S. Regulation of tight junctions during the epithelium-mesenchyme transition: Direct repression of the gene expression of claudins/occludin by Snail. *J. Cell Sci.* **2003**, *116 Pt 10*, 1959–1967. [CrossRef]
40. Nieto, M.A. The snail superfamily of zinc-finger transcription factors. *Nat. Rev. Mol. Cell Biol.* **2002**, *3*, 155–166. [CrossRef]
41. Ozdamar, B.; Bose, R.; Barrios-Rodiles, M.; Wang, H.R.; Zhang, Y.; Wrana, J.L. Regulation of the polarity protein Par6 by TGFbeta receptors controls epithelial cell plasticity. *Science* **2005**, *307*, 1603–1609. [CrossRef] [PubMed]
42. Aranda, V.; Haire, T.; Nolan, M.E.; Calarco, J.P.; Rosenberg, A.Z.; Fawcett, J.P.; Pawson, T.; Muthuswamy, S.K. Par6-aPKC uncouples ErbB2 induced disruption of polarized epithelial organization from proliferation control. *Nat. Cell Biol.* **2006**, *8*, 1235–1245. [CrossRef] [PubMed]
43. Ma, L.; Lu, M.F.; Schwartz, R.J.; Martin, J.F. Bmp2 is essential for cardiac cushion epithelial-mesenchymal transition and myocardial patterning. *Development* **2005**, *132*, 5601–5611. [CrossRef] [PubMed]
44. McCulley, D.J.; Kang, J.O.; Martin, J.F.; Black, B.L. BMP4 is required in the anterior heart field and its derivatives for endocardial cushion remodeling, outflow tract septation, and semilunar valve development. *Dev. Dyn.* **2008**, *237*, 3200–3209. [CrossRef]
45. Zeisberg, M.; Hanai, J.; Sugimoto, H.; Mammoto, T.; Charytan, D.; Strutz, F.; Kalluri, R. BMP-7 counteracts TGF-beta1-induced epithelial-to-mesenchymal transition and reverses chronic renal injury. *Nat. Med.* **2003**, *9*, 964–968. [CrossRef]
46. Bhat, A.A.; Uppada, S.; Achkar, I.W.; Hashem, S.; Yadav, S.K.; Shanmugakonar, M.; Al-Naemi, H.A.; Haris, M.; Uddin, S. Tight Junction Proteins and Signaling Pathways in Cancer and Inflammation: A Functional Crosstalk. *Front. Physiol.* **2018**, *9*, 1942. [CrossRef]
47. Padmanaban, V.; Krol, I.; Suhail, Y.; Szczerba, B.M.; Aceto, N.; Bader, J.S.; Ewald, A.J. E-cadherin is required for metastasis in multiple models of breast cancer. *Nature* **2019**, *573*, 439–444. [CrossRef]
48. Eisenberg, L.M.; Markwald, R.R. Molecular regulation of atrioventricular valvuloseptal morphogenesis. *Circ. Res.* **1995**, *77*, 1–6. [CrossRef]
49. Maddaluno, L.; Rudini, N.; Cuttano, R.; Bravi, L.; Giampietro, C.; Corada, M.; Ferrarini, L.; Orsenigo, F.; Papa, E.; Boulday, G.; et al. EndMT contributes to the onset and progression of cerebral cavernous malformations. *Nature* **2013**, *498*, 492–496. [CrossRef]
50. Hong, L.; Du, X.; Li, W.; Mao, Y.; Sun, L.; Li, X. EndMT: A promising and controversial field. *Eur. J. Cell Biol.* **2018**, *97*, 493–500. [CrossRef]

51. Lampugnani, M.G.; Dejana, E.; Giampietro, C. Vascular Endothelial (VE)-Cadherin, Endothelial Adherens Junctions, and Vascular Disease. *Cold Spring Harb. Perspect. Biol.* **2018**, *10*, a029322. [CrossRef]
52. Malinverno, M.; Maderna, C.; Abu Taha, A.; Corada, M.; Orsenigo, F.; Valentino, M.; Pisati, F.; Fusco, C.; Graziano, P.; Giannotta, M.; et al. Endothelial cell clonal expansion in the development of cerebral cavernous malformations. *Nat. Commun.* **2019**, *10*, 2761. [CrossRef] [PubMed]
53. Akhurst, R.J.; Derynck, R. TGF-beta signaling in cancer—A double-edged sword. *Trends Cell Biol.* **2001**, *11*, S44–S51. [PubMed]
54. Medici, D.; Hay, E.D.; Olsen, B.R. Snail and Slug promote epithelial-mesenchymal transition through beta-catenin-T-cell factor-4-dependent expression of transforming growth factor-beta3. *Mol. Biol. Cell* **2008**, *19*, 4875–4887. [CrossRef]
55. Liebner, S.; Cattelino, A.; Gallini, R.; Rudini, N.; Iurlaro, M.; Piccolo, S.; Dejana, E. Beta-catenin is required for endothelial-mesenchymal transformation during heart cushion development in the mouse. *J. Cell Biol.* **2004**, *166*, 359–367. [CrossRef]
56. Kokudo, T.; Suzuki, Y.; Yoshimatsu, Y.; Yamazaki, T.; Watabe, T.; Miyazono, K. Snail is required for TGFbeta-induced endothelial-mesenchymal transition of embryonic stem cell-derived endothelial cells. *J. Cell Sci.* **2008**, *121 Pt 20*, 3317–3324. [CrossRef]
57. Medici, D.; Potenta, S.; Kalluri, R. Transforming growth factor-beta2 promotes Snail-mediated endothelial-mesenchymal transition through convergence of Smad-dependent and Smad-independent signalling. *Biochem. J.* **2011**, *437*, 515–520. [CrossRef]
58. Gerhardt, H.; Golding, M.; Fruttiger, M.; Ruhrberg, C.; Lundkvist, A.; Abramsson, A.; Jeltsch, M.; Mitchell, C.; Alitalo, K.; Shima, D.; et al. VEGF guides angiogenic sprouting utilizing endothelial tip cell filopodia. *J. Cell Biol.* **2003**, *161*, 1163–1177. [CrossRef] [PubMed]
59. Krizbai, I.A.; Gasparics, A.; Nagyoszi, P.; Fazakas, C.; Molnar, J.; Wilhelm, I.; Bencs, R.; Rosivall, L.; Sebe, A. Endothelial-mesenchymal transition of brain endothelial cells: Possible role during metastatic extravasation. *PLoS ONE* **2015**, *10*, e0123845. [CrossRef]
60. Moore-Morris, T.; Tallquist, M.D.; Evans, S.M. Sorting out where fibroblasts come from. *Circ. Res.* **2014**, *115*, 602–604. [CrossRef]
61. Davis, G.E.; Senger, D.R. Endothelial extracellular matrix: Biosynthesis, remodeling, and functions during vascular morphogenesis and neovessel stabilization. *Circ. Res.* **2005**, *97*, 1093–1107. [CrossRef]
62. Potenta, S.; Zeisberg, E.; Kalluri, R. The role of endothelial-to-mesenchymal transition in cancer progression. *Br. J. Cancer* **2008**, *99*, 1375–1379. [CrossRef]
63. Pinheiro, D.; Bellaiotache, Y. Mechanical Force-Driven Adherens Junction Remodeling and Epithelial Dynamics. *Dev. Cell* **2018**, *47*, 391. [CrossRef] [PubMed]
64. Itoh, M.; Nagafuchi, A.; Moroi, S.; Tsukita, S. Involvement of ZO-1 in cadherin-based cell adhesion through its direct binding to alpha catenin and actin filaments. *J. Cell Biol.* **1997**, *138*, 181–192. [CrossRef]
65. Wittchen, E.S.; Haskins, J.; Stevenson, B.R. Protein interactions at the tight junction. Actin has multiple binding partners, and ZO-1 forms independent complexes with ZO-2 and ZO-3. *J. Biol. Chem.* **1999**, *274*, 35179–35185. [CrossRef]
66. Yamamoto, T.; Harada, N.; Kawano, Y.; Taya, S.; Kaibuchi, K. In vivo interaction of AF-6 with activated Ras and ZO-1. *Biochem. Biophys. Res. Commun.* **1999**, *259*, 103–107. [CrossRef]
67. Gumbiner, B.; Lowenkopf, T.; Apatira, D. Identification of a 160-kDa polypeptide that binds to the tight junction protein ZO-1. *Proc. Natl. Acad. Sci. USA* **1991**, *88*, 3460–3464. [CrossRef]
68. Haskins, J.; Gu, L.; Wittchen, E.S.; Hibbard, J.; Stevenson, B.R. ZO-3, a novel member of the MAGUK protein family found at the tight junction, interacts with ZO-1 and occludin. *J. Cell Biol.* **1998**, *141*, 199–208. [CrossRef]
69. Stevenson, B.R.; Siliciano, J.D.; Mooseker, M.S.; Goodenough, D.A. Identification of ZO-1: A high molecular weight polypeptide associated with the tight junction (zonula occludens) in a variety of epithelia. *J. Cell Biol.* **1986**, *103*, 755–766. [CrossRef]
70. Katsuno, T.; Umeda, K.; Matsui, T.; Hata, M.; Tamura, A.; Itoh, M.; Takeuchi, K.; Fujimori, T.; Nabeshima, Y.; Noda, T.; et al. Deficiency of zonula occludens-1 causes embryonic lethal phenotype associated with defected yolk sac angiogenesis and apoptosis of embryonic cells. *Mol. Biol. Cell* **2008**, *19*, 2465–2475. [CrossRef]
71. Xu, J.; Kausalya, P.J.; Phua, D.C.; Ali, S.M.; Hossain, Z.; Hunziker, W. Early embryonic lethality of mice lacking ZO-2, but Not ZO-3, reveals critical and nonredundant roles for individual zonula occludens proteins in mammalian development. *Mol. Cell Biol.* **2008**, *28*, 1669–1678. [CrossRef] [PubMed]

72. Balda, M.S.; Matter, K. Tight junctions and the regulation of gene expression. *Biochim. Biophys. Acta* **2009**, *1788*, 761–767. [CrossRef]
73. Sourisseau, T.; Georgiadis, A.; Tsapara, A.; Ali, R.R.; Pestell, R.; Matter, K.; Balda, M.S. Regulation of PCNA and cyclin D1 expression and epithelial morphogenesis by the ZO-1-regulated transcription factor ZONAB/DbpA. *Mol. Cell Biol.* **2006**, *26*, 2387–2398. [CrossRef] [PubMed]
74. Lima, W.R.; Parreira, K.S.; Devuyst, O.; Caplanusi, A.; N'Kuli, F.; Marien, B.; Van Der Smissen, P.; Alves, P.M.; Verroust, P.; Christensen, E.I.; et al. ZONAB promotes proliferation and represses differentiation of proximal tubule epithelial cells. *J. Am. Soc. Nephrol.* **2010**, *21*, 478–488. [CrossRef]
75. Capaldo, C.T.; Koch, S.; Kwon, M.; Laur, O.; Parkos, C.A.; Nusrat, A. Tight function zonula occludens-3 regulates cyclin D1-dependent cell proliferation. *Mol Biol Cell* **2011**, *22*, 1677–1685. [CrossRef]
76. Kleeff, J.; Shi, X.; Bode, H.P.; Hoover, K.; Shrikhande, S.; Bryant, P.J.; Korc, M.; Buchler, M.W.; Friess, H. Altered expression and localization of the tight junction protein ZO-1 in primary and metastatic pancreatic cancer. *Pancreas* **2001**, *23*, 259–265. [CrossRef]
77. Takai, E.; Tan, X.; Tamori, Y.; Hirota, M.; Egami, H.; Ogawa, M. Correlation of translocation of tight junction protein Zonula occludens-1 and activation of epidermal growth factor receptor in the regulation of invasion of pancreatic cancer cells. *Int. J. Oncol.* **2005**, *27*, 645–651.
78. Tuomi, S.; Mai, A.; Nevo, J.; Laine, J.O.; Vilkki, V.; Ohman, T.J.; Gahmberg, C.G.; Parker, P.J.; Ivaska, J. PKCepsilon regulation of an alpha5 integrin-ZO-1 complex controls lamellae formation in migrating cancer cells. *Sci. Signal.* **2009**, *2*, ra32. [CrossRef]
79. Turner, J.R. 'Putting the squeeze' on the tight junction: Understanding cytoskeletal regulation. *Semin. Cell Dev. Biol.* **2000**, *11*, 301–308. [CrossRef]
80. Ren, Y.; Li, R.; Zheng, Y.; Busch, H. Cloning and characterization of GEF-H1, a microtubule-associated guanine nucleotide exchange factor for Rac and Rho GTPases. *J. Biol. Chem.* **1998**, *273*, 34954–34960. [CrossRef]
81. Cordenonsi, M.; D'Atri, F.; Hammar, E.; Parry, D.A.; Kendrick-Jones, J.; Shore, D.; Citi, S. Cingulin contains globular and coiled-coil domains and interacts with ZO-1, ZO-2, ZO-3, and myosin. *J. Cell Biol.* **1999**, *147*, 1569–1582. [CrossRef]
82. Citi, S.; D'Atri, F.; Parry, D.A. Human and Xenopus cingulin share a modular organization of the coiled-coil rod domain: Predictions for intra- and intermolecular assembly. *J. Struct. Biol.* **2000**, *131*, 135–145. [CrossRef]
83. Aijaz, S.; D'Atri, F.; Citi, S.; Balda, M.S.; Matter, K. Binding of GEF-H1 to the tight junction-associated adaptor cingulin results in inhibition of Rho signaling and G1/S phase transition. *Dev. Cell* **2005**, *8*, 777–786. [CrossRef]
84. Guillemot, L.; Citi, S. Cingulin regulates claudin-2 expression and cell proliferation through the small GTPase RhoA. *Mol. Biol. Cell* **2006**, *17*, 3569–3577. [CrossRef] [PubMed]
85. Wu, C.Y.; Jhingory, S.; Taneyhill, L.A. The tight junction scaffolding protein cingulin regulates neural crest cell migration. *Dev. Dyn.* **2011**, *240*, 2309–2323. [CrossRef]
86. Mangan, A.J.; Sietsema, D.V.; Li, D.; Moore, J.K.; Citi, S.; Prekeris, R. Cingulin and actin mediate midbody-dependent apical lumen formation during polarization of epithelial cells. *Nat. Commun.* **2016**, *7*, 12426. [CrossRef]
87. Ohnishi, H.; Nakahara, T.; Furuse, K.; Sasaki, H.; Tsukita, S.; Furuse, M. JACOP, a novel plaque protein localizing at the apical junctional complex with sequence similarity to cingulin. *J. Biol. Chem.* **2004**, *279*, 46014–46022. [CrossRef]
88. Paschoud, S.; Yu, D.; Pulimeno, P.; Jond, L.; Turner, J.R.; Citi, S. Cingulin and paracingulin show similar dynamic behaviour, but are recruited independently to junctions. *Mol. Membr. Biol.* **2011**, *28*, 123–135. [CrossRef]
89. Pulimeno, P.; Paschoud, S.; Citi, S. A role for ZO-1 and PLEKHA7 in recruiting paracingulin to tight and adherens junctions of epithelial cells. *J. Biol. Chem.* **2011**, *286*, 16743–16750. [CrossRef]
90. Guillemot, L.; Paschoud, S.; Jond, L.; Foglia, A.; Citi, S. Paracingulin regulates the activity of Rac1 and RhoA GTPases by recruiting Tiam1 and GEF-H1 to epithelial junctions. *Mol. Biol. Cell* **2008**, *19*, 4442–4453. [CrossRef]
91. Guillemot, L.; Guerrera, D.; Spadaro, D.; Tapia, R.; Jond, L.; Citi, S. MgcRacGAP interacts with cingulin and paracingulin to regulate Rac1 activation and development of the tight junction barrier during epithelial junction assembly. *Mol. Biol. Cell* **2014**, *25*, 1995–2005. [CrossRef]

92. Tornavaca, O.; Chia, M.; Dufton, N.; Almagro, L.O.; Conway, D.E.; Randi, A.M.; Schwartz, M.A.; Matter, K.; Balda, M.S. ZO-1 controls endothelial adherens junctions, cell-cell tension, angiogenesis, and barrier formation. *J. Cell Biol.* **2015**, *208*, 821–838. [CrossRef]
93. Chrifi, I.; Hermkens, D.; Brandt, M.M.; van Dijk, C.G.M.; Burgisser, P.E.; Haasdijk, R.; Pei, J.; van de Kamp, E.H.M.; Zhu, C.; Blonden, L.; et al. Cgnl1, an endothelial junction complex protein, regulates GTPase mediated angiogenesis. *Cardiovasc. Res.* **2017**, *113*, 1776–1788. [CrossRef]
94. Vasileva, E.; Citi, S. The role of microtubules in the regulation of epithelial junctions. *Tissue Barriers* **2018**, *6*, 1539596. [CrossRef]
95. Akizuki, R.; Shimobaba, S.; Matsunaga, T.; Endo, S.; Ikari, A. Claudin-5, -7, and -18 suppress proliferation mediated by inhibition of phosphorylation of Akt in human lung squamous cell carcinoma. *Biochim. Biophys. Acta Mol. Cell Res.* **2017**, *1864*, 293–302. [CrossRef]
96. Zavala-Zendejas, V.E.; Torres-Martinez, A.C.; Salas-Morales, B.; Fortoul, T.I.; Montano, L.F.; Rendon-Huerta, E.P. Claudin-6, 7, or 9 overexpression in the human gastric adenocarcinoma cell line AGS increases its invasiveness, migration, and proliferation rate. *Cancer Investig.* **2011**, *29*, 1–11. [CrossRef]
97. Ikari, A.; Sato, T.; Takiguchi, A.; Atomi, K.; Yamazaki, Y.; Sugatani, J. Claudin-2 knockdown decreases matrix metalloproteinase-9 activity and cell migration via suppression of nuclear Sp1 in A549 cells. *Life Sci.* **2011**, *88*, 628–633. [CrossRef]
98. Takehara, M.; Nishimura, T.; Mima, S.; Hoshino, T.; Mizushima, T. Effect of claudin expression on paracellular permeability, migration and invasion of colonic cancer cells. *Biol. Pharm. Bull.* **2009**, *32*, 825–831. [CrossRef]
99. Yoon, C.H.; Kim, M.J.; Park, M.J.; Park, I.C.; Hwang, S.G.; An, S.; Choi, Y.H.; Yoon, G.; Lee, S.J. Claudin-1 acts through c-Abl-protein kinase Cdelta (PKCdelta) signaling and has a causal role in the acquisition of invasive capacity in human liver cells. *J. Biol. Chem.* **2010**, *285*, 226–233. [CrossRef]
100. Haddad, N.; El Andalousi, J.; Khairallah, H.; Yu, M.; Ryan, A.K.; Gupta, I.R. The tight junction protein claudin-3 shows conserved expression in the nephric duct and ureteric bud and promotes tubulogenesis in vitro. *Am. J. Physiol. Ren. Physiol.* **2011**, *301*, F1057–F1065. [CrossRef]
101. Neesse, A.; Griesmann, H.; Gress, T.M.; Michl, P. Claudin-4 as therapeutic target in cancer. *Arch. Biochem. Biophys.* **2012**, *524*, 64–70. [CrossRef]
102. Karanjawala, Z.E.; Illei, P.B.; Ashfaq, R.; Infante, J.R.; Murphy, K.; Pandey, A.; Schulick, R.; Winter, J.; Sharma, R.; Maitra, A.; et al. New markers of pancreatic cancer identified through differential gene expression analyses: Claudin 18 and annexin A8. *Am. J. Surg. Pathol.* **2008**, *32*, 188–196. [CrossRef]
103. Sahin, U.; Koslowski, M.; Dhaene, K.; Usener, D.; Brandenburg, G.; Seitz, G.; Huber, C.; Tureci, O. Claudin-18 splice variant 2 is a pan-cancer target suitable for therapeutic antibody development. *Clin. Cancer Res.* **2008**, *14*, 7624–7634. [CrossRef]
104. Yamaguchi, H.; Kojima, T.; Ito, T.; Kyuno, D.; Kimura, Y.; Imamura, M.; Hirata, K.; Sawada, N. Effects of Clostridium perfringens enterotoxin via claudin-4 on normal human pancreatic duct epithelial cells and cancer cells. *Cell. Mol. Biol. Lett.* **2011**, *16*, 385–397. [CrossRef]
105. Tiwari-Woodruff, S.K.; Buznikov, A.G.; Vu, T.Q.; Micevych, P.E.; Chen, K.; Kornblum, H.I.; Bronstein, J.M. OSP/claudin-11 forms a complex with a novel member of the tetraspanin super family and beta1 integrin and regulates proliferation and migration of oligodendrocytes. *J. Cell Biol.* **2001**, *153*, 295–305. [CrossRef]
106. Ikari, A.; Watanabe, R.; Sato, T.; Taga, S.; Shimobaba, S.; Yamaguchi, M.; Yamazaki, Y.; Endo, S.; Matsunaga, T.; Sugatani, J. Nuclear distribution of claudin-2 increases cell proliferation in human lung adenocarcinoma cells. *Biochim. Biophys. Acta* **2014**, *1843*, 2079–2088. [CrossRef]
107. Leotlela, P.D.; Wade, M.S.; Duray, P.H.; Rhode, M.J.; Brown, H.F.; Rosenthal, D.T.; Dissanayake, S.K.; Earley, R.; Indig, F.E.; Nickoloff, B.J.; et al. Claudin-1 overexpression in melanoma is regulated by PKC and contributes to melanoma cell motility. *Oncogene* **2007**, *26*, 3846–3856. [CrossRef]
108. Tanaka, M.; Kamata, R.; Sakai, R. Phosphorylation of ephrin-B1 via the interaction with claudin following cell-cell contact formation. *EMBO J.* **2005**, *24*, 3700–3711. [CrossRef]
109. Tanaka, M.; Kamata, R.; Sakai, R. EphA2 phosphorylates the cytoplasmic tail of Claudin-4 and mediates paracellular permeability. *J. Biol. Chem.* **2005**, *280*, 42375–42382. [CrossRef]
110. Li, J.; Chigurupati, S.; Agarwal, R.; Mughal, M.R.; Mattson, M.P.; Becker, K.G.; Wood, W.H., 3rd; Zhang, Y.; Morin, P.J. Possible angiogenic roles for claudin-4 in ovarian cancer. *Cancer Biol. Ther.* **2009**, *8*, 1806–1814. [CrossRef]

111. Tamura, A.; Kitano, Y.; Hata, M.; Katsuno, T.; Moriwaki, K.; Sasaki, H.; Hayashi, H.; Suzuki, Y.; Noda, T.; Furuse, M.; et al. Megaintestine in claudin-15-deficient mice. *Gastroenterology* **2008**, *134*, 523–534. [CrossRef]
112. Tsukita, S.; Yamazaki, Y.; Katsuno, T.; Tamura, A.; Tsukita, S. Tight junction-based epithelial microenvironment and cell proliferation. *Oncogene* **2008**, *27*, 6930–6938. [CrossRef]
113. Borka, K.; Kaliszky, P.; Szabo, E.; Lotz, G.; Kupcsulik, P.; Schaff, Z.; Kiss, A. Claudin expression in pancreatic endocrine tumors as compared with ductal adenocarcinomas. *Virchows Arch.* **2007**, *450*, 549–557. [CrossRef]
114. Holczbauer, A.; Gyongyosi, B.; Lotz, G.; Szijarto, A.; Kupcsulik, P.; Schaff, Z.; Kiss, A. Distinct claudin expression profiles of hepatocellular carcinoma and metastatic colorectal and pancreatic carcinomas. *J. Histochem. Cytochem.* **2013**, *61*, 294–305. [CrossRef]
115. Zhou, B.; Flodby, P.; Luo, J.; Castillo, D.R.; Liu, Y.; Yu, F.X.; McConnell, A.; Varghese, B.; Li, G.; Chimge, N.O.; et al. Claudin-18-mediated YAP activity regulates lung stem and progenitor cell homeostasis and tumorigenesis. *J. Clin. Investig.* **2018**, *128*, 970–984. [CrossRef]
116. Michl, P.; Barth, C.; Buchholz, M.; Lerch, M.M.; Rolke, M.; Holzmann, K.H.; Menke, A.; Fensterer, H.; Giehl, K.; Lohr, M.; et al. Claudin-4 expression decreases invasiveness and metastatic potential of pancreatic cancer. *Cancer Res.* **2003**, *63*, 6265–6271.
117. Nava, P.; Capaldo, C.T.; Koch, S.; Kolegraff, K.; Rankin, C.R.; Farkas, A.E.; Feasel, M.E.; Li, L.; Addis, C.; Parkos, C.A.; et al. JAM-A regulates epithelial proliferation through Akt/beta-catenin signalling. *EMBO Rep.* **2011**, *12*, 314–320. [CrossRef]
118. Perry, J.M.; He, X.C.; Sugimura, R.; Grindley, J.C.; Haug, J.S.; Ding, S.; Li, L. Cooperation between both Wnt/[189]-catenin and PTEN/PI3K/Akt signaling promotes primitive hematopoietic stem cell self-renewal and expansion. *Genes Dev.* **2011**, *25*, 1928–1942. [CrossRef]
119. Tuncay, H.; Brinkmann, B.F.; Steinbacher, T.; Schurmann, A.; Gerke, V.; Iden, S.; Ebnet, K. JAM-A regulates cortical dynein localization through Cdc42 to control planar spindle orientation during mitosis. *Nat. Commun.* **2015**, *6*, 8128. [CrossRef]
120. Severson, E.A.; Jiang, L.; Ivanov, A.I.; Mandell, K.J.; Nusrat, A.; Parkos, C.A. Cis-dimerization mediates function of junctional adhesion molecule A. *Mol. Biol. Cell* **2008**, *19*, 1862–1872. [CrossRef]
121. Mandell, K.J.; Babbin, B.A.; Nusrat, A.; Parkos, C.A. Junctional adhesion molecule 1 regulates epithelial cell morphology through effects on beta1 integrins and Rap1 activity. *J. Biol. Chem.* **2005**, *280*, 11665–11674. [CrossRef]
122. Severson, E.A.; Lee, W.Y.; Capaldo, C.T.; Nusrat, A.; Parkos, C.A. Junctional adhesion molecule A interacts with Afadin and PDZ-GEF2 to activate Rap1A, regulate beta1 integrin levels, and enhance cell migration. *Mol. Biol. Cell* **2009**, *20*, 1916–1925. [CrossRef]
123. Bazzoni, G.; Tonetti, P.; Manzi, L.; Cera, M.R.; Balconi, G.; Dejana, E. Expression of junctional adhesion molecule-A prevents spontaneous and random motility. *J. Cell Sci.* **2005**, *118 Pt 3*, 623–632. [CrossRef]
124. Daniele, L.L.; Adams, R.H.; Durante, D.E.; Pugh, E.N., Jr.; Philp, N.J. Novel distribution of junctional adhesion molecule-C in the neural retina and retinal pigment epithelium. *J. Comp. Neurol.* **2007**, *505*, 166–176. [CrossRef]
125. Economopoulou, M.; Avramovic, N.; Klotzsche-von Ameln, A.; Korovina, I.; Sprott, D.; Samus, M.; Gercken, B.; Troullinaki, M.; Grossklaus, S.; Funk, R.H.; et al. Endothelial-specific deficiency of Junctional Adhesion Molecule-C promotes vessel normalisation in proliferative retinopathy. *Thromb. Haemost.* **2015**, *114*, 1241–1249.
126. Ikenouchi, J.; Furuse, M.; Furuse, K.; Sasaki, H.; Tsukita, S.; Tsukita, S. Tricellulin constitutes a novel barrier at tricellular contacts of epithelial cells. *J. Cell Biol.* **2005**, *171*, 939–945. [CrossRef]
127. Raleigh, D.R.; Marchiando, A.M.; Zhang, Y.; Shen, L.; Sasaki, H.; Wang, Y.; Long, M.; Turner, J.R. Tight junction-associated MARVEL proteins marveld3, tricellulin, and occludin have distinct but overlapping functions. *Mol. Biol. Cell* **2010**, *21*, 1200–1213. [CrossRef]
128. Steed, E.; Rodrigues, N.T.; Balda, M.S.; Matter, K. Identification of MarvelD3 as a tight junction-associated transmembrane protein of the occludin family. *BMC Cell Biol.* **2009**, *10*, 95. [CrossRef]
129. Yaffe, Y.; Shepshelovitch, J.; Nevo-Yassaf, I.; Yeheskel, A.; Shmerling, H.; Kwiatek, J.M.; Gaus, K.; Pasmanik-Chor, M.; Hirschberg, K. The MARVEL transmembrane motif of occludin mediates oligomerization and targeting to the basolateral surface in epithelia. *J. Cell Sci.* **2012**, *125 Pt 15*, 3545–3556. [CrossRef]
130. Mariano, C.; Sasaki, H.; Brites, D.; Brito, M.A. A look at tricellulin and its role in tight junction formation and maintenance. *Eur. J. Cell Biol.* **2011**, *90*, 787–796. [CrossRef]

131. Cording, J.; Berg, J.; Kading, N.; Bellmann, C.; Tscheik, C.; Westphal, J.K.; Milatz, S.; Gunzel, D.; Wolburg, H.; Piontek, J.; et al. In tight junctions, claudins regulate the interactions between occludin, tricellulin and marvelD3, which, inversely, modulate claudin oligomerization. *J. Cell Sci.* **2013**, *126 Pt 2*, 554–564. [CrossRef]
132. Saitou, M.; Furuse, M.; Sasaki, H.; Schulzke, J.D.; Fromm, M.; Takano, H.; Noda, T.; Tsukita, S. Complex phenotype of mice lacking occludin, a component of tight junction strands. *Mol. Biol. Cell* **2000**, *11*, 4131–4142. [CrossRef] [PubMed]
133. Kamitani, T.; Sakaguchi, H.; Tamura, A.; Miyashita, T.; Yamazaki, Y.; Tokumasu, R.; Inamoto, R.; Matsubara, A.; Mori, N.; Hisa, Y.; et al. Deletion of Tricellulin Causes Progressive Hearing Loss Associated with Degeneration of Cochlear Hair Cells. *Sci. Rep.* **2015**, *5*, 18402. [CrossRef] [PubMed]
134. Nayak, G.; Lee, S.I.; Yousaf, R.; Edelmann, S.E.; Trincot, C.; Van Itallie, C.M.; Sinha, G.P.; Rafeeq, M.; Jones, S.M.; Belyantseva, I.A.; et al. Tricellulin deficiency affects tight junction architecture and cochlear hair cells. *J. Clin. Investig.* **2013**, *123*, 4036–4049. [CrossRef] [PubMed]
135. Sanchez-Pulido, L.; Martin-Belmonte, F.; Valencia, A.; Alonso, M.A. MARVEL: A conserved domain involved in membrane apposition events. *Trends Biochem. Sci.* **2002**, *27*, 599–601. [CrossRef]
136. Furuse, M.; Hirase, T.; Itoh, M.; Nagafuchi, A.; Yonemura, S.; Tsukita, S.; Tsukita, S. Occludin: A novel integral membrane protein localizing at tight junctions. *J. Cell Biol.* **1993**, *123 Pt 6*, 1777–1788. [CrossRef]
137. Schulzke, J.D.; Gitter, A.H.; Mankertz, J.; Spiegel, S.; Seidler, U.; Amasheh, S.; Saitou, M.; Tsukita, S.; Fromm, M. Epithelial transport and barrier function in occludin-deficient mice. *Biochim. Biophys. Acta* **2005**, *1669*, 34–42. [CrossRef]
138. Yu, A.S.; McCarthy, K.M.; Francis, S.A.; McCormack, J.M.; Lai, J.; Rogers, R.A.; Lynch, R.D.; Schneeberger, E.E. Knockdown of occludin expression leads to diverse phenotypic alterations in epithelial cells. *Am. J. Physiol. Cell Physiol.* **2005**, *288*, C1231–C1241. [CrossRef]
139. Phillips, B.E.; Cancel, L.; Tarbell, J.M.; Antonetti, D.A. Occludin independently regulates permeability under hydrostatic pressure and cell division in retinal pigment epithelial cells. *Investig. Ophthalmol. Vis. Sci.* **2008**, *49*, 2568–2576. [CrossRef]
140. Aaku-Saraste, E.; Hellwig, A.; Huttner, W.B. Loss of occludin and functional tight junctions, but not ZO-1, during neural tube closure–remodeling of the neuroepithelium prior to neurogenesis. *Dev. Biol.* **1996**, *180*, 664–679. [CrossRef]
141. Dorfel, M.J.; Westphal, J.K.; Bellmann, C.; Krug, S.M.; Cording, J.; Mittag, S.; Tauber, R.; Fromm, M.; Blasig, I.E.; Huber, O. CK2-dependent phosphorylation of occludin regulates the interaction with ZO-proteins and tight junction integrity. *Cell Commun. Signal.* **2013**, *11*, 40. [CrossRef] [PubMed]
142. Suzuki, T.; Elias, B.C.; Seth, A.; Shen, L.; Turner, J.R.; Giorgianni, F.; Desiderio, D.; Guntaka, R.; Rao, R. PKC eta regulates occludin phosphorylation and epithelial tight junction integrity. *Proc. Natl. Acad. Sci. USA* **2009**, *106*, 61–66. [CrossRef] [PubMed]
143. Raleigh, D.R.; Boe, D.M.; Yu, D.; Weber, C.R.; Marchiando, A.M.; Bradford, E.M.; Wang, Y.; Wu, L.; Schneeberger, E.E.; Shen, L.; et al. Occludin S408 phosphorylation regulates tight junction protein interactions and barrier function. *J. Cell Biol.* **2011**, *193*, 565–582. [CrossRef] [PubMed]
144. Elias, B.C.; Suzuki, T.; Seth, A.; Giorgianni, F.; Kale, G.; Shen, L.; Turner, J.R.; Naren, A.; Desiderio, D.M.; Rao, R. Phosphorylation of Tyr-398 and Tyr-402 in occludin prevents its interaction with ZO-1 and destabilizes its assembly at the tight junctions. *J. Biol. Chem.* **2009**, *284*, 1559–1569. [CrossRef]
145. Murakami, T.; Felinski, E.A.; Antonetti, D.A. Occludin phosphorylation and ubiquitination regulate tight junction trafficking and vascular endothelial growth factor-induced permeability. *J. Biol. Chem.* **2009**, *284*, 21036–21046. [CrossRef]
146. Murakami, T.; Frey, T.; Lin, C.; Antonetti, D.A. Protein kinase C beta phosphorylates occludin regulating tight junction trafficking in vascular endothelial growth factor-induced permeability in vivo. *Diabetes* **2012**, *61*, 1573–1583. [CrossRef]
147. Du, D.; Xu, F.; Yu, L.; Zhang, C.; Lu, X.; Yuan, H.; Huang, Q.; Zhang, F.; Bao, H.; Jia, L.; et al. The tight junction protein, occludin, regulates the directional migration of epithelial cells. *Dev. Cell* **2010**, *18*, 52–63. [CrossRef]
148. Bolinger, M.T.; Ramshekar, A.; Waldschmidt, H.V.; Larsen, S.D.; Bewley, M.C.; Flanagan, J.M.; Antonetti, D.A. Occludin S471 Phosphorylation Contributes to Epithelial Monolayer Maturation. *Mol. Cell Biol.* **2016**, *36*, 2051–2066. [CrossRef]

149. Cravo, A.S.; Carter, E.; Erkan, M.; Harvey, E.; Furutani-Seiki, M.; Mrsny, R. Hippo pathway elements Co-localize with Occludin: A possible sensor system in pancreatic epithelial cells. *Tissue Barriers* **2015**, *3*, e1037948. [CrossRef]
150. Runkle, E.A.; Sundstrom, J.M.; Runkle, K.B.; Liu, X.; Antonetti, D.A. Occludin localizes to centrosomes and modifies mitotic entry. *J. Biol. Chem.* **2011**, *286*, 30847–30858. [CrossRef]
151. Liu, X.; Dreffs, A.; Diaz-Coranguez, M.; Runkle, E.A.; Gardner, T.W.; Chiodo, V.A.; Hauswirth, W.W.; Antonetti, D.A. Occludin S490 Phosphorylation Regulates Vascular Endothelial Growth Factor-Induced Retinal Neovascularization. *Am. J. Pathol.* **2016**, *186*, 2486–2499. [CrossRef] [PubMed]
152. Runkle, E.A.; Mu, D. Tight junction proteins: From barrier to tumorigenesis. *Cancer Lett.* **2013**, *337*, 41–48. [CrossRef] [PubMed]
153. Harten, S.K.; Shukla, D.; Barod, R.; Hergovich, A.; Balda, M.S.; Matter, K.; Esteban, M.A.; Maxwell, P.H. Regulation of renal epithelial tight junctions by the von Hippel-Lindau tumor suppressor gene involves occludin and claudin 1 and is independent of E-cadherin. *Mol. Biol. Cell* **2009**, *20*, 1089–1101. [CrossRef] [PubMed]
154. Li, D.; Mrsny, R.J. Oncogenic Raf-1 disrupts epithelial tight junctions via downregulation of occludin. *J. Cell Biol.* **2000**, *148*, 791–800. [CrossRef]
155. Wang, Z.; Wade, P.; Mandell, K.J.; Akyildiz, A.; Parkos, C.A.; Mrsny, R.J.; Nusrat, A. Raf 1 represses expression of the tight junction protein occludin via activation of the zinc-finger transcription factor slug. *Oncogene* **2007**, *26*, 1222–1230. [CrossRef]
156. Martin, T.A.; Mansel, R.E.; Jiang, W.G. Loss of occludin leads to the progression of human breast cancer. *Int. J. Mol. Med.* **2010**, *26*, 723–734. [CrossRef]
157. Osanai, M.; Murata, M.; Nishikiori, N.; Chiba, H.; Kojima, T.; Sawada, N. Epigenetic silencing of occludin promotes tumorigenic and metastatic properties of cancer cells via modulations of unique sets of apoptosis-associated genes. *Cancer Res.* **2006**, *66*, 9125–9133. [CrossRef]
158. Wang, Z.; Mandell, K.J.; Parkos, C.A.; Mrsny, R.J.; Nusrat, A. The second loop of occludin is required for suppression of Raf1-induced tumor growth. *Oncogene* **2005**, *24*, 4412–4420. [CrossRef]
159. Jayagopal, A.; Yang, J.L.; Haselton, F.R.; Chang, M.S. Tight junction-associated signaling pathways modulate cell proliferation in uveal melanoma. *Investig. Ophthalmol. Vis. Sci.* **2011**, *52*, 588–593. [CrossRef]
160. Osanai, M.; Murata, M.; Nishikiori, N.; Chiba, H.; Kojima, T.; Sawada, N. Occludin-mediated premature senescence is a fail-safe mechanism against tumorigenesis in breast carcinoma cells. *Cancer Sci.* **2007**, *98*, 1027–1034. [CrossRef]
161. Barrios-Rodiles, M.; Brown, K.R.; Ozdamar, B.; Bose, R.; Liu, Z.; Donovan, R.S.; Shinjo, F.; Liu, Y.; Dembowy, J.; Taylor, I.W.; et al. High-throughput mapping of a dynamic signaling network in mammalian cells. *Science* **2005**, *307*, 1621–1625. [CrossRef] [PubMed]
162. Odenwald, M.A.; Choi, W.; Buckley, A.; Shashikanth, N.; Joseph, N.E.; Wang, Y.; Warren, M.H.; Buschmann, M.M.; Pavlyuk, R.; Hildebrand, J.; et al. ZO-1 interactions with F-actin and occludin direct epithelial polarization and single lumen specification in 3D culture. *J. Cell Sci.* **2017**, *130*, 243–259. [CrossRef] [PubMed]
163. Jenkinson, E.M.; Livingston, J.H.; O'Driscoll, M.C.; Desguerre, I.; Nabbout, R.; Boddaert, N.; Soares, G.; Goncalves da Rocha, M.; D'Arrigo, S.; Rice, G.I.; et al. Comprehensive molecular screening strategy of OCLN in band-like calcification with simplified gyration and polymicrogyria. *Clin. Genet.* **2018**, *93*, 228–234. [CrossRef] [PubMed]
164. Abdel-Hamid, M.S.; Abdel-Salam, G.M.H.; Issa, M.Y.; Emam, B.A.; Zaki, M.S. Band-like calcification with simplified gyration and polymicrogyria: Report of 10 new families and identification of five novel OCLN mutations. *J. Hum. Genet.* **2017**, *62*, 553–559. [CrossRef]
165. Aggarwal, S.; Bahal, A.; Dalal, A. Renal dysfunction in sibs with band like calcification with simplified gyration and polymicrogyria: Report of a new mutation and review of literature. *Eur. J. Med. Genet.* **2016**, *59*, 5–10. [CrossRef]
166. Elsaid, M.F.; Kamel, H.; Chalhoub, N.; Aziz, N.A.; Ibrahim, K.; Ben-Omran, T.; George, B.; Al-Dous, E.; Mohamoud, Y.; Malek, J.A.; et al. Whole genome sequencing identifies a novel occludin mutation in microcephaly with band-like calcification and polymicrogyria that extends the phenotypic spectrum. *Am. J. Med. Genet. A* **2014**, *164*, 1614–1617. [CrossRef]

167. O'Driscoll, M.C.; Daly, S.B.; Urquhart, J.E.; Black, G.C.; Pilz, D.T.; Brockmann, K.; McEntagart, M.; Abdel-Salam, G.; Zaki, M.; Wolf, N.I.; et al. Recessive mutations in the gene encoding the tight junction protein occludin cause band-like calcification with simplified gyration and polymicrogyria. *Am. J. Hum. Genet.* **2010**, *87*, 354–364. [CrossRef]
168. LeBlanc, M.A.; Penney, L.S.; Gaston, D.; Shi, Y.; Aberg, E.; Nightingale, M.; Jiang, H.; Gillett, R.M.; Fahiminiya, S.; Macgillivray, C.; et al. A novel rearrangement of occludin causes brain calcification and renal dysfunction. *Hum. Genet.* **2013**, *132*, 1223–1234. [CrossRef]
169. Jayaraman, D.; Bae, B.I.; Walsh, C.A. The Genetics of Primary Microcephaly. *Annu. Rev. Genom. Hum. Genet.* **2018**, *19*, 177–200. [CrossRef]
170. Glotfelty, L.G.; Zahs, A.; Iancu, C.; Shen, L.; Hecht, G.A. Microtubules are required for efficient epithelial tight junction homeostasis and restoration. *Am. J. Physiol. Cell Physiol.* **2014**, *307*, C245–C254. [CrossRef]
171. Yano, T.; Matsui, T.; Tamura, A.; Uji, M.; Tsukita, S. The association of microtubules with tight junctions is promoted by cingulin phosphorylation by AMPK. *J. Cell Biol.* **2013**, *203*, 605–614. [CrossRef] [PubMed]
172. Moss, D.K.; Bellett, G.; Carter, J.M.; Liovic, M.; Keynton, J.; Prescott, A.R.; Lane, E.B.; Mogensen, M.M. Ninein is released from the centrosome and moves bi-directionally along microtubules. *J. Cell Sci.* **2007**, *120 Pt 17*, 3064–3074. [CrossRef]
173. Shaw, R.M.; Fay, A.J.; Puthenveedu, M.A.; von Zastrow, M.; Jan, Y.N.; Jan, L.Y. Microtubule plus-end-tracking proteins target gap junctions directly from the cell interior to adherens junctions. *Cell* **2007**, *128*, 547–560. [CrossRef]
174. Meng, W.; Mushika, Y.; Ichii, T.; Takeichi, M. Anchorage of microtubule minus ends to adherens junctions regulates epithelial cell-cell contacts. *Cell* **2008**, *135*, 948–959. [CrossRef] [PubMed]
175. Meng, W.; Takeichi, M. Adherens junction: Molecular architecture and regulation. *Cold Spring Harb. Perspect. Biol.* **2009**, *1*, a002899. [CrossRef] [PubMed]
176. Horgan, C.P.; Hanscom, S.R.; Jolly, R.S.; Futter, C.E.; McCaffrey, M.W. Rab11-FIP3 links the Rab11 GTPase and cytoplasmic dynein to mediate transport to the endosomal-recycling compartment. *J. Cell Sci.* **2010**, *123 Pt 2*, 181–191. [CrossRef]
177. Fesenko, I.; Kurth, T.; Sheth, B.; Fleming, T.P.; Citi, S.; Hausen, P. Tight junction biogenesis in the early Xenopus embryo. *Mech. Dev.* **2000**, *96*, 51–65. [CrossRef]
178. Morimoto, S.; Nishimura, N.; Terai, T.; Manabe, S.; Yamamoto, Y.; Shinahara, W.; Miyake, H.; Tashiro, S.; Shimada, M.; Sasaki, T. Rab13 mediates the continuous endocytic recycling of occludin to the cell surface. *J. Biol. Chem.* **2005**, *280*, 2220–2228. [CrossRef]
179. Terai, T.; Nishimura, N.; Kanda, I.; Yasui, N.; Sasaki, T. JRAB/MICAL-L2 is a junctional Rab13-binding protein mediating the endocytic recycling of occludin. *Mol. Biol. Cell* **2006**, *17*, 2465–2475. [CrossRef]
180. Nishimura, N.; Sasaki, T. Cell-surface biotinylation to study endocytosis and recycling of occludin. *Methods Mol. Biol.* **2008**, *440*, 89–96.
181. Lapierre, L.A.; Tuma, P.L.; Navarre, J.; Goldenring, J.R.; Anderson, J.M. VAP-33 localizes to both an intracellular vesicle population and with occludin at the tight junction. *J. Cell Sci.* **1999**, *112 Pt 21*, 3723–3732.
182. Pennetta, G.; Hiesinger, P.R.; Fabian-Fine, R.; Meinertzhagen, I.A.; Bellen, H.J. Drosophila VAP-33A directs bouton formation at neuromuscular junctions in a dosage-dependent manner. *Neuron* **2002**, *35*, 291–306. [CrossRef]
183. Steed, E.; Elbediwy, A.; Vacca, B.; Dupasquier, S.; Hemkemeyer, S.A.; Suddason, T.; Costa, A.C.; Beaudry, J.B.; Zihni, C.; Gallagher, E.; et al. MarvelD3 couples tight junctions to the MEKK1-JNK pathway to regulate cell behavior and survival. *J. Cell Biol.* **2014**, *204*, 821–838. [CrossRef] [PubMed]
184. Kojima, T.; Takasawa, A.; Kyuno, D.; Ito, T.; Yamaguchi, H.; Hirata, K.; Tsujiwaki, M.; Murata, M.; Tanaka, S.; Sawada, N. Downregulation of tight junction-associated MARVEL protein marvelD3 during epithelial-mesenchymal transition in human pancreatic cancer cells. *Exp. Cell Res.* **2011**, *317*, 2288–2298. [CrossRef] [PubMed]
185. Timmann, C.; Thye, T.; Vens, M.; Evans, J.; May, J.; Ehmen, C.; Sievertsen, J.; Muntau, B.; Ruge, G.; Loag, W.; et al. Genome-wide association study indicates two novel resistance loci for severe malaria. *Nature* **2012**, *489*, 443–446. [CrossRef]

186. Vacca, B.; Sanchez-Heras, E.; Steed, E.; Balda, M.S.; Ohnuma, S.I.; Sasai, N.; Mayor, R.; Matter, K. MarvelD3 regulates the c-Jun N-terminal kinase pathway during eye development in Xenopus. *Biol. Open* **2016**, *5*, 1631–1641. [CrossRef]
187. Vacca, B.; Sanchez-Heras, E.; Steed, E.; Busson, S.L.; Balda, M.S.; Ohnuma, S.I.; Sasai, N.; Mayor, R.; Matter, K. Control of neural crest induction by MarvelD3-mediated attenuation of JNK signalling. *Sci. Rep.* **2018**, *8*, 1204. [CrossRef]

6

Claudin-7 Modulates Cl^- and Na^+ Homeostasis and WNK4 Expression in Renal Collecting Duct Cells

Junming Fan [1,2], Rodney Tatum [1], John Hoggard [1] and Yan-Hua Chen [1,3,*]

1 Department of Anatomy and Cell Biology, Brody School of Medicine, East Carolina University, Greenville, NC 27834, USA
2 Institute of Hypoxia Medicine, School of Basic Medical Sciences, Wenzhou Medical University, Wenzhou 325035, China
3 East Carolina Diabetes and Obesity Institute, East Carolina University, Greenville, NC 27834, USA
* Correspondence: cheny@ecu.edu.

Abstract: Claudin-7 knockout ($CLDN7^{-/-}$) mice display renal salt wasting and dehydration phenotypes. To address the role of CLDN7 in kidneys, we established collecting duct (CD) cell lines from $CLDN7^{+/+}$ and $CLDN7^{-/-}$ mouse kidneys. We found that deletion of CLDN7 increased the transepithelial resistance (TER) and decreased the paracellular permeability for Cl^- and Na^+ in $CLDN7^{-/-}$ CD cells. Inhibition of transcellular Cl^- and Na^+ channels has no significant effect on TER or dilution potentials. Current-voltage curves were linear in both $CLDN7^{+/+}$ and $CLDN7^{-/-}$ CD cells, indicating that the ion flux was through the paracellular pathway. The impairment of Cl^- and Na^+ permeability phenotype can be rescued by CLDN7 re-expression. We also found that WNK4 (its mutations lead to hypertension) expression, but not WNK1, was significantly increased in $CLDN7^{-/-}$ CD cell lines as well as in primary $CLDN7^{-/-}$ CD cells, suggesting that the expression of WNK4 was modulated by CLDN7. In addition, deletion of CLDN7 upregulated the expression level of the apical epithelial sodium channel (ENaC), indicating a potential cross-talk between paracellular and transcellular transport systems. This study demonstrates that CLDN7 plays an important role in salt balance in renal CD cells and modulating WNK4 and ENaC expression levels that are vital in controlling salt-sensitive hypertension.

Keywords: Claudin-7; tight junctions; permeability; WNK4; epithelial sodium channel (ENaC), collecting duct cells

---

1. Introduction

Maintaining electrolytes and body fluids across epithelial layers in kidneys within the physiological range is of vital importance for blood pressure regulation. Chloride (Cl^-) and sodium (Na^+), two predominant extracellular ionic components in kidneys, determine the extracellular electrolyte balance and regulate the blood pressure in the segment of the collecting duct (CD) [1]. It is well known that sodium reabsorption in the CD is an active process through the apical epithelial sodium channel (ENaC) and is driven by the basolateral Na^+-K^+-ATPase. On the other hand, chloride reabsorption is driven by the lumen-negative transepithelial potential and mainly occurs through tight junctions (TJs), the gatekeeper of the paracellular pathway in CD [1–3].

TJ is a multi-molecular complex and plays an essential role in regulating ions and small molecules passing through the apical to basal compartment of the epithelial cells [4,5]. Claudins (CLDN), a family of transmembrane proteins with at least 24 members in mouse and human, are the most important structural and functional components of the TJs and the principal regulators in defining the properties of paracellular ion permeability of the epithelial cells [6,7]. There are more than ten CLDN members expressed in kidneys, and they are closely associated with their corresponding segment-specific ion reabsorption characters [6,8]. Deficiency or aberrant expression of distinct CLDNs has been reported to be associated with disturbance of electrolytes, which can lead to high blood pressure or hypertension-related diseases. For example, mutations in CLDN16 and CLDN19 in humans resulted in kidney disorders exhibiting renal magnesium wasting and hypercalciuria [9,10]. Mice with CLDN16 knockdown exhibit defects in paracellular cation selectivity and develop severe renal wasting of magnesium and calcium [11]. A study by Muto et al. [12] demonstrated that CLDN2-deficient mice show a significant decrease in net transepithelial reabsorption of Na^+ and Cl^- in proximal tubules, causing a loss of Na^+ selectivity and therefore relative Cl^- selectivity in the proximal tubule paracellular pathway. Results from Krug et al. [13] reported that Madin-Darby Canine Kidney (MDCK) C7 cells with CLDN17 overexpression show an increase in paracellular anion permeability and switch from cation-selective to anion-selective. Knockdown of CLDN17 in LLC-PK1 cells support CLDN17 as an anion channel. In addition, mice with CLDN10 deletion in the thick ascending limb show the impairment in paracellular Na^+ permeability and hypermagnesemia [14], and can rescue CLDN16-deficient mice from hypomagnesemia and hypercalciuria [15].

Pseudohypoaldosteronism type II (PHAII) is an autosomal-dominant hereditary hypertensive disease that is characterized by hyperkalemia and metabolic acidosis [16,17]. In 2001, Wilson et al. [18] found that the mutations in WNK4, a serine/threonine kinase with No K (lysine), were linked to the pathogenesis of PHAII. Since then, many studies have shown that several membrane channels and transporters are the molecular targets of WNK4 [16,19–24]. Since WNK4 is localized at the TJs of distal nephrons, TJ proteins could also serve as additional targets for WNK4. Indeed, two groups have reported that WNK4 regulates the paracellular Cl^- permeability in MDCK II cells [25,26], and the latest study documented by Chen et al. showed that mice with knockin Cl^--insensitive mutant WNK4 displayed hypertension, hyperkalemia, hyperactive NCC, and the authors concluded that WNK4 is a physiological intracellular Cl^- sensor [24]. Our previous study also showed that claudin-7 is the substrate of WNK4 and can be phosphorylated by WNK4 at serine^{206} in its COOH-terminus [27].

Although it is well characterized how transcellular channels and transports work in regulating Cl^- and Na^+ transport in the CD of the kidney [12,14], the molecular targets of the paracellular pathway responsible for Cl^- and Na^+ transport have not been fully elucidated. It has been reported that CLDN4 served as a Cl^- channel in mouse kidney CD cells that require the presence of CLDN8 [28]. We have previously reported that the first extracellular domain of CLDN7 affects paracellular Cl^- permeability [29]. In addition, overexpression of CLDN7 in LLC-PK1 cells decreased the paracellular Cl^- conductance and increased Na^+ conductance [30].

To study the role of CLDN7 in vivo, we generated a CLDN7 knockout ($CLDN7^{-/-}$) mouse model and discovered that CLDN7-deficient mice exhibited renal salt wasting, chronic dehydration, and severe intestinal defects, and that 90% of the pups died within 10 days after birth [31]. The kidney phenotypes suggest that CLDN7 plays an indispensable role in keeping salt homeostasis in distal nephrons. However, the functional role of CLDN7 in the distal nephron remains unclear. In this study, we isolated and purified CD cells from $CLDN7^{+/+}$ and $CLDN7^{-/-}$ mouse kidneys using Dolichos biflorus lectin-coated Dynabeads, and immortalized these cells into cell lines by Lenti-SV40 virus infection. These CD cells express AQP2, CLDN3, and CLDN4, but not CLDN2, a proximal tubule marker. We found that transepithelial resistance (TER) was significantly increased and paracellular Cl^- and Na^+ permeability was decreased in $CLDN7^{-/-}$ CD cells. These phenotypes can be rescued by the transfection of CLDN7 into $CLDN7^{-/-}$ CD cells. In addition, we found in this study that WNK4 expression, but not WNK1, was significantly increased, and so was the ENaC level in $CLDN7^{-/-}$ CD cells. These results suggest the potential influence of the paracellular pathway on transcellular pathway. Defects in the paracellular ion transport may affect transcellular transport systems, leading to the ionic imbalance in kidneys.

2. Results

2.1. Generation of CD Cell Lines Isolated from CLDN7$^{+/+}$ and CLDN7$^{-/-}$ Mouse Kidneys

To study the role of claudin-7 in CDs, we generated CD cell lines isolated and purified from CLDN7$^{+/+}$ and CLDN7$^{-/-}$ mouse kidneys using Dolichos biflorus lectin-coated Dynabeads, and then immortalized these cells into multiple cell lines by Lenti-SV40 virus infection. The immortalized CLDN7$^{+/+}$ and CLDN7$^{-/-}$ CD cells have an epithelial morphology as shown in Figure 1A. CLDN7$^{+/+}$ CD cells have a strong CLDN7 immunostaining signal localized at cell–cell contact area, while CLDN7 signal was absent in CLDN7$^{-/-}$ CD cells as expected (Figure 1B, top panel). CLDN3, CLDN4, and AQP2 were known to be expressed in CD cells, and their presences were confirmed by both immunofluorescence microscopy and western blot analysis (Figure 1B,C). It was observed that AQP2 signal was quite weak at the cell membrane in CLDN7$^{-/-}$ CD cells (Figure 1B); however, the protein expression level was similar between CLDN7$^{+/+}$ and CLDN7$^{-/-}$ CD cells (Figure 1C). CLDN8 signal was undetectable in both CLDN7$^{+/+}$ and CLDN7$^{-/-}$ CD cells (data unpublished).

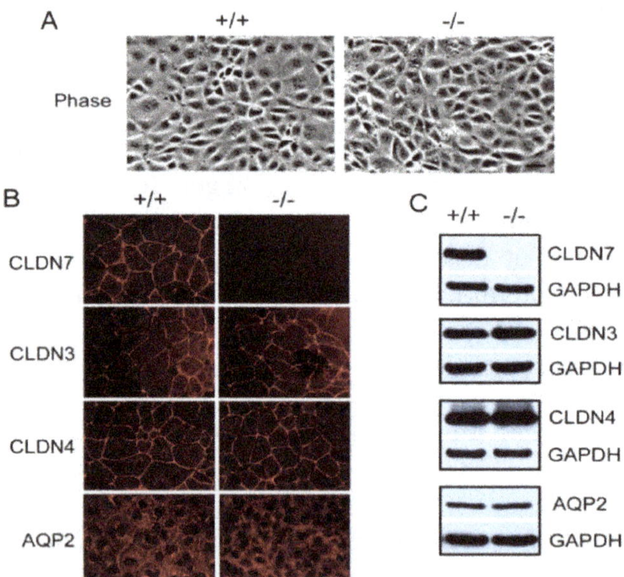

Figure 1. Characterization of immortalized mouse collecting duct (CD) cell lines. (**A**) Phase images of CLDN7$^{+/+}$ (+/+) and CLDN7$^{-/-}$ (−/−) CD cells. CD cells from 4-day old mouse CLDN7$^{+/+}$ and CLDN7$^{-/-}$ kidneys were isolated and purified by Dolichos biflorus lectin-coated Dynabeads. These purified CD cells were immortalized into cell lines using SV40 virus. The stable CD cells were grown on coverslips for 5–7 days before fixation for fluorescent light microcopy. Bar: 30 μm. (**B**) CLDN7$^{+/+}$ and CLDN7$^{-/-}$ CD cells were immunostained with antibodies against CLDN7, CLDN3, CLDN4, and AQP2 and detected by Cy3-conjugated anti-rabbit secondary antibody. CLDN7$^{+/+}$ CD cells showed a strong anti-CLDN7 immunostaining and this signal was completely absent in CLDN7$^{-/-}$ CD cells. Both CLDN3 and CLDN4 have similar staining patterns in CLDN7$^{+/+}$ and CLDN7$^{-/-}$ CD cells. AQP2 is a marker protein for CD principle cells and its membrane staining was much weaker in CLDN7$^{-/-}$ than in CLDN7$^{+/+}$ CD cells. Bar: 20 μm. (**C**) Western blot analysis confirmed the absence of CLDN7 protein in CLDN7$^{-/-}$ CD cells. CLDN3, CLDN4, and AQP2 expression levels were similar between CLDN7$^{+/+}$ and CLDN7$^{-/-}$ CD cells. GAPDH signal was used as a loading control.

2.2. Decreased Paracellular Cl^- and Na^+ Permeability in CLDN7$^{-/-}$ CD Cells

To examine CLDN7-based CD cell electrophysiology, we first measured the TER of CLDN7$^{+/+}$ and CLDN7$^{-/-}$ CD cells. The TER of CLDN7$^{+/+}$ CD cells has an average value of 510 ± 43 Ω.cm^2. Deletion of CLDN7 dramatically increased TER to 1115 ± 96 Ω.cm^2 (Figure 2A), suggesting an increase in barrier function induced by CLDN7 absence in CD cells.

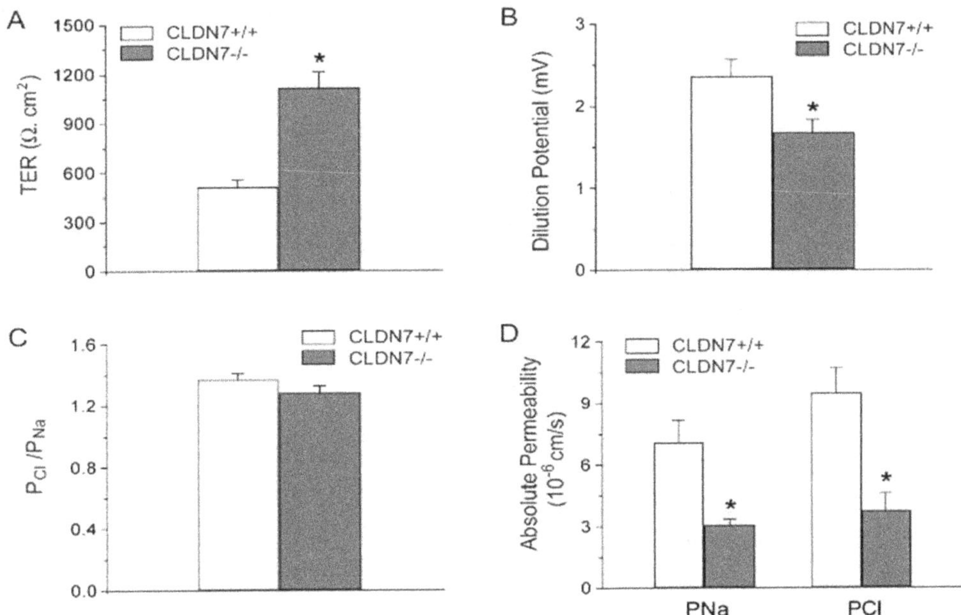

Figure 2. Deletion of CLDN7 increased transepithelial resistance (TER) and decreased paracellular Cl^- and Na^+ permeability on CD cell monolayers. (**A**) TER was measured on monolayers cultured for 7 days. (**B**) CD cells were grown on collagen-coated Snapwell filters for 7 days to reach the full confluence. The filter rings containing cell monolayers were mounted into EasyMount chambers. Both apical and basal chambers were filled with buffer containing 140 mM NaCl. Subsequently, buffer in the basal chamber was replaced by 70 mM NaCl, and dilution potentials were measured. (**C**) The ratio of the absolute permeability of Cl^- to Na^+ (P_{Cl}/P_{Na}) was calculated using the Goldman–Hodgkin–Katz equation. The ratio of P_{Cl}/P_{Na} was >1 in CLDN7$^{+/+}$, indicating that these CD cells were more permeable to Cl^- than Na^+. (**D**) The absolute permeability for P_{Cl} and P_{Na} was calculated according to the method of simplified Kimizuka and Koketsu equations. * $p < 0.05$. $n = 3$.

To further examine the ion permeability in our established CD cell lines, we performed dilution potential experiments. Our data showed that dilution potentials measured from CLDN7$^{-/-}$ CD cells were significantly reduced compared to those of CLDN7$^{+/+}$ CD cells (Figure 2B). However, the ratio of absolute permeability of Cl^- (P_{Cl}) to Na^+ (P_{Na}) was slightly decreased for CLDN7$^{-/-}$ CD cells, but without statistical significance (Figure 2C). Deletion of CLDN7 in CD cells depressed the permeation of Cl^- and Na^+ as indicated by their reduced absolute permeability values of Cl^- (P_{Cl}) and Na^+ (P_{Na}) (Figure 2D). Inhibition of epithelial Na^+ and Cl^- channels had no significant effect on TER or dilution potentials either in CLDN7$^{+/+}$ or CLDN7$^{-/-}$ CD cells, indicating that the impairment of Cl^- and Na^+ permeability in CLDN7$^{-/-}$ CD cells is through the paracellular pathway (data unpublished). Moreover, current–voltage curves were linear in both CLDN7$^{+/+}$ and CLDN7$^{-/-}$ CD cells, consistent with the conductance being attributable to the paracellular pathway for ion flux (data unpublished). Our results indicate that CLDN7 plays a vital role in NaCl reabsorption in mouse CD cells. Deletion of CLDN7 decreases paracellular permeability to Cl^- and Na^+, suggesting CLDN7 may serve as a non-selective paracellular channel in CD cells.

2.3. Increased Expression Levels of WNK4 and ENaC in CLDN7$^{-/-}$ CD Cells

We reported previously that CLDN7 was colocalized with WNK4 in kidneys and that they formed a protein complex when co-expressed in kidney epithelial cells [27]. To investigate whether CLDN7 deletion affects the expression of WNK4 and other kinases and ion channels, we performed real-time RT-PCR experiments. We found that deletion of CLDN7 significantly increased WNK4, SGK-1, and ENaC-α mRNA levels, while there were no significant changes in ROMK and AQP2 mRNA levels (Figure 3A).

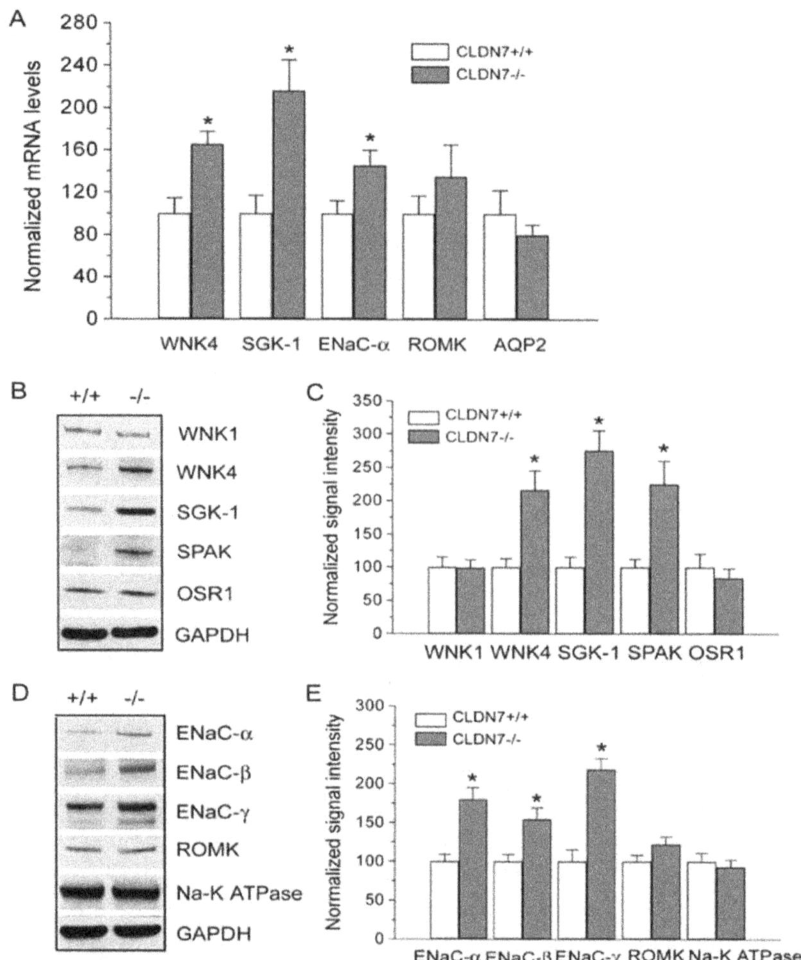

Figure 3. Deletion of CLDN7 had a significant effect on gene and protein expression levels of WNK4, SGK-1, and ENaC. (**A**) Real-time RT-PCR analysis of WNK4, SGK-1, ENaC-α, ROMK, and AQP2 mRNA levels in CLDN7$^{+/+}$ and CLDN7$^{-/-}$ CD cells. Each measurement was normalized to its β-actin level. * $p < 0.05$. $n = 3$. (**B**) Western blotting analysis of several protein kinase levels in CD cells. CLDN7$^{+/+}$ and CLDN7$^{-/-}$ CD cells were lysed in RIPA (radio-immunoprecipitation assay) buffer and a total of 30 μg protein for each lane was loaded onto the SDS NuPAGE gel. Membranes were blotted against WNK1, WNK4, SGK-1, SPARK, and OSR1. GAPDH (glyceraldehyde 3-phosphate dehydrogenase) staining was used as a loading control. (**C**) Densitometry analysis of protein expression levels shown on (**B**). Each band intensity for CLDN7$^{+/+}$ CD cells was normalized and set as a reference. * $p < 0.05$. $n = 3$. (**D**) Western blotting analysis of several ion channel levels in CD cells. Equal amounts of CLDN7$^{+/+}$ and CLDN7$^{-/-}$ CD cell lysates were loaded onto the SDS NuPAGE gel and the membranes were probed against ENaC-α, -β, -γ, ROMK, and Na^+-K^+-ATPase. (**E**) Densitometry analysis of protein expression levels shown on (**D**). Each band intensity for CLDN7$^{+/+}$ CD cells was normalized and set as a reference. * $p < 0.05$. $n = 3$.

Immunoblotting analysis also showed that the protein expression levels of WNK4, SGK-1, and SPAK were all clearly increased, but WNK1 and OSR1 levels were unchanged in CLDN7$^{-/-}$ CD cells compared to those in CLDN7$^{+/+}$ CD cells (Figure 3B,C). Interestingly, we found that the expression levels of ENaC-α, -β and −γ were all elevated with no changes in ROMK and Na-K-ATPase in CLDN7$^{-/-}$ CD cells (Figure 3D,E). We have confirmed these results in the primary CLDN7$^{+/+}$ and CLDN7$^{-/-}$ CD cells as shown in Figure 4. The phase images of primary CD cells isolated from CLDN7$^{+/+}$ and CLDN7$^{-/-}$ kidneys were shown in Figure 4A (top panel). Anti-CLDN4 and anti-AQP2 antibodies were used to stain CD cells (Figure 4A). After CD cells were removed, the remaining cells were immunostained with CLDN4 and found to be CLDN4-negative (Figure 4A, bottom panel).

Consistent with the immortalized CLDN7$^{-/-}$ CD cells, CLDN7 deletion clearly increased WNK4, SGK-1, and ENaC subunit's expression levels with no significant effects on WNK1, ROMK, or AQP2 expression levels in primary CLDN7$^{-/-}$ CD cells (Figure 4B,C).

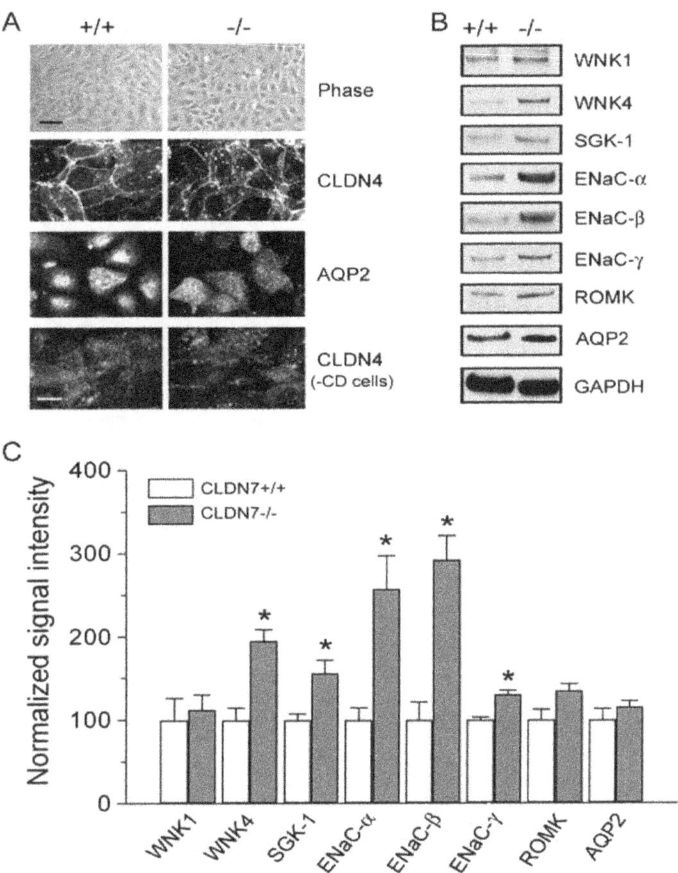

Figure 4. Deletion of CLDN7 had a significant effect on protein expression levels of representative kinases and ion channel on primary CD cells. (**A**) The establishment of primary cultures of CD cells isolated from kidneys of CLDN7$^{+/+}$ and CLDN7$^{-/-}$ pups. The top panel shows the phase images of primary CD cells isolated from kidneys of 5-day old CLDN7$^{+/+}$ and CLDN7$^{-/-}$ pups after cultured in 12-well plates for a week to form a complete monolayer. Bar: 40 μm. The cultured primary CLDN7$^{+/+}$ and CLDN7$^{-/-}$ CD cells were immunostained with anti-CLDN4 and anti-AQP2 antibodies. The last panel shows the remaining cells immunostained with anti-CLDN4 antibody after removal of CD cells. Bar: 15 μm. (**B**) The primary CD cells were lysed in RIPA buffer and a total of 30 μg protein for each lane was loaded onto the SDS NuPAGE gel. Membranes were blotted against WNK1, WNK4, SGK-1, ENaC-α, -β, and -γ, ROMK, and AQP2. GAPDH was used as a loading control. (**C**) Densitometry analysis of protein expression levels shown on (**B**). Each band intensity for CLDN7$^{+/+}$ CD cells was normalized and set as a reference. * $p < 0.05$. $n = 4$.

2.4. Rescued Function of Ion Permeability in Immortalized CLDN7$^{+/+}$ CD Cells with CLDN7 Knockdown

As we observed an increase in barrier function and a decrease in Cl^- and Na^+ permeability in CLDN7$^{-/-}$ CD cells, we tried to stably transfect CLDN7 back ('rescue') into CLDN7$^{-/-}$ CD cells to study whether CLDN7 could revert the phenotype. However, we were unable to obtain the stable cell lines after many attempts. Therefore, herein we designed specific shRNAs to knock down the expression of CLDN7 in CLDN7$^{+/+}$ CD (KD) cells and then transfected CLDN7 back into these KD cells. Immunofluorescent staining and western blot analysis confirmed the knockdown of the expression of

CLDN7 in CLDN7$^{+/+}$ CD cells by CLDN7 shRNA (Figure 5A–C). Similarly as in CLDN7$^{-/-}$ CD cells, CLDN7 KD induced an increase in WNK4 and SGK-1 expression while AQP2, CLDN3, and CLDN4 expressions were unchanged (Figure 5B,C). In addition, we found that CLDN7 KD also significantly increased the TER value by 61.2% compared with the scrambled controls (Figure 5D). Moreover, CLDN7 KD decreased dilution potential (DP) (Figure 5E) and Cl^- and Na^+ permeability (Figure 5G) as we found in CLDN7$^{-/-}$ CD cells without a significant change in PCl/PNa (Figure 5F).

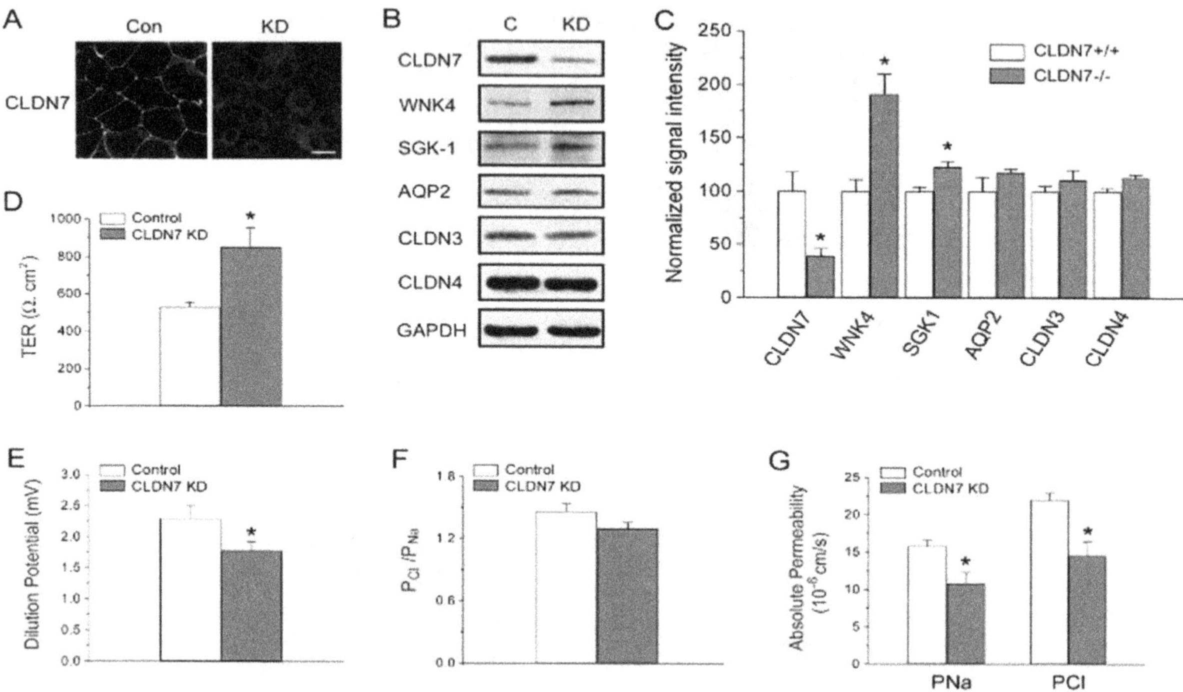

Figure 5. Knockdown of CLDN7 in CLDN7$^{+/+}$ CD cells decreased paracellular Cl^- and Na^+ permeability. (**A**) CLDN7$^{+/+}$ CD cells were transfected with either control (Con) or CLDN7 shRNA (KD) vector. The indirect immunofluorecent method shows the reduced immunostaining signal of CLDN7 in KD cells compared to that of control cells. (**B**) The control and knockdown (KD) cell lysates were subject to western blot analysis. Membranes were probed against CLDN7, WNK4, SGK-1, AQP2, CLDN3, and CLDN4. GAPDH was used as a loading control. (**C**) Densitometry analysis of protein expression levels shown on (**B**). Each band intensity for CLDN7$^{+/+}$ CD cells was normalized and set as a reference. $^* p < 0.05$. $n = 3$. (**D**) The control and KD CD cells were cultured in Transwell plates coated with collagen. TER was measured on monolayers cultured for 7 days. (**E**) The control and KD CD cells were grown on collagen-coated Snapwell filters for 7 days. The dilution potentials were measured as described in Figure 2B. (**F**) The ratio of the absolute permeability of Cl^- to Na^+ (P_{Cl}/P_{Na}) was calculated using the Goldman–Hodgkin–Katz equation. (**G**) The calculated absolute permeability for P_{Cl} and P_{Na} was significantly reduced in KD cells compared to that of control cells. $^* p < 0.05$. $n = 3$.

Transfection of CLDN7 back to the CLDN7$^{+/+}$ KD cells (KD+CLDN7) increased the protein expression of CLDN7 to 88.2% of the control cell value (Figure 6A). The TER and DP were also back to the values similar to those in control cells (Figure 6B,C). Although there was no significant difference in PCl/PNa among the control, KD, and KD+CLDN7 CD cells (Figure 6D), Cl^- and Na^+ permeability was recovered to 91.1% and 90.4% in CLDN7 rescued cells, respectively (Figure 6E).

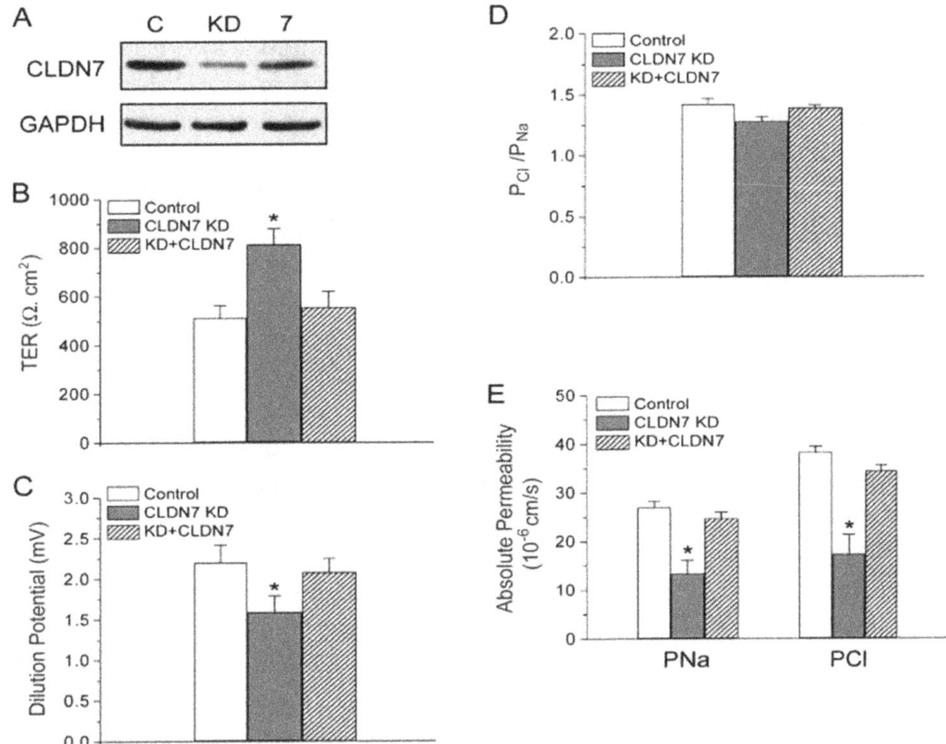

Figure 6. Re-expression ('rescue') of CLDN7 in CLDN7$^{+/+}$ CD cells with CLDN7 knockdown restored paracellular Cl^- and Na^+ permeability. (**A**) CD cells from control (C), CLDN7 knockdown (KD), KD with CLDN7 cDNA transfection (7) were lysed in RIPA buffer and subjected to western blotting. The membrane was blotted with anti-CLDN7 antibody. (**B**) TER values were measured on cell monolayers cultured for 7 days on collagen-coated Transwell plates. (**C**) Dilution potentials were measured as described in Figure 2B. (**D**) The ratio of the absolute permeability of Cl^- to Na^+ (P_{Cl}/P_{Na}), and (**E**) the absolute permeability for P_{Cl} and P_{Na} were calculated as described in Figure 2C,D, respectively. * $p < 0.05$ compared to control and KD+CLDN7. $n = 3$.

3. Discussion

In this study, we have shown that mouse renal CD cells with CLDN7 deletion exhibited significant decrease in paracellular Cl^- and Na^+ permeability. At the same time, the TER in CLDN7$^{-/-}$ CD cells was greatly increased while the dilution potential was decreased. The paracellular ion permeability in CLDN7$^{+/+}$ CD cells with CLDN7 knockdown resembled that of CLDN7$^{-/-}$ CD cells. Re-expression ('rescue') CLDN7 in CLDN7$^{+/+}$ KD cells restored the role of CLDN7 in paracellular Cl^- and Na^+ permeability. In addition, our study demonstrates that deletion of CLDN7 upregulates WNK4 expression at both mRNA and protein levels in our immortalized CD cells as well as in the primary CD cells, indicating that CLDN7 may play an important role in regulating WNK4 expression in kidneys. Interestingly, ENaC expression was also upregulated in immortalized and primary CLDN7$^{-/-}$ CD cells compared to that of CLDN7$^{+/+}$ CD cells, suggesting a potential influence of paracellular pathway on transcellular pathway.

It is known that CLDN7 is highly expressed in the distal nephron of the kidney [32]. We reported previously that CLDN7-deficient mice exhibited renal salt wasting and chronic dehydration [31]. To investigate the function of CLDN7 in distal nephron, we used the novel approach to isolate, purify, and immortalize the CD cells from CLDN7$^{+/+}$ and CLDN7$^{-/-}$ mouse kidneys. This approach allows us to study the role of CLDN7 in renal epithelial cells in a controlled environment without the stimulation of hormones and other circulating factors. We found that the TER value was around 500 $\Omega.cm^2$ for CLDN7$^{+/+}$ CD cells, which was consistent with the literature on isolated rabbit CDs [33]. However, deletion of CLDN7 increased TER value to more than 1000 $\Omega.cm^2$. The increase in TER is due to the decrease in paracellular Cl^- and Na^+ permeability. We reported previously that overexpression of CLDN7 in LLC-PK1 cells decreases the paracellular Cl^- conductance and increases paracellular Na^+ conductance [30]. It is possible that the effect of overexpression or knockdown of the same gene in different cell lines may have different functional consquences as we have observed in human lung cancer cells [34] and our unpublished data. It is known that the paracellular pathway in the CD system is mainly Cl^- selective [1,35]. However, Cl^- reabsorption must match that of Na^+ in order to maintain the homeostasis of luminal fluids and electrolytes. Our current study suggests that CLDN7 may form a non-selective paracellular channel in renal CD cells and play a critical role in Cl^- and Na^+ homeostasis in distal nephrons.

WNK4 is localized at TJs of distal nephrons and has been shown to selectively increase paracellular Cl^- permeability and phosphorylate claudins in MDCK cells [18,25,26], and many studies have revealed that mutations of WNK4 are involved in the pathogenesis of PHAII [18,24,36,37]. We previously found that CLDN7 was a substrate of WNK4, and that phosphorylation of CLDN7 by WNK4 promoted the paracellular Cl^- permeability in kidney epithelial cells [27]. Interestingly, we found in this study that deletion of CLDN7 significantly increased WNK4 expression in CD cell lines as well as in primary CD cells (Figures 3B and 4B), suggesting a previously unrecognized involvement of CLDN7 in the regulation of WNK4. In addition, ENaC expression was also increased in both CLDN7$^{-/-}$ immortalized and primary CD cells, which could be mediated through the up-regulation of SGK-1 since it has been reported that SGK-1 stimulates the membrane expression and the activity of ENaC [38–40]. It will be interesting to see in future studies whether ENaC channel activity is altered in CLDN7$^{-/-}$ CD cells.

It has been reported that CLDN4 forms a paracellular Cl^- channel in the kidney and requires CLDN8 for TJ localization [28]. However, in our CLDN7 CD cell lines, CLDN4 was well localized to the cell junction area in both CLDN7$^{+/+}$ and CLDN7$^{-/-}$ CD cells, though the CLDN8 signal was undetectable. It is possible that different renal epithelial cells may behave differently depending on the TJ components and claudin compositions.

4. Materials and methods

4.1. Antibodies and Reagents

Rabbit polyclonal anti-CLDN3 and CLDN4 antibodies were purchased from Invitrogen (Thermo Fisher Scientific, Waltham, MA, USA). Rabbit polyclonal anti-CLDN7 antibody was obtained from Immuno-Biological Laboratories (Gunma, Japan). AQP2 polyclonal antibody was purchased from CALBIOCHEM (Sigma-Aldrich, St. Louis, MO, USA). Rabbit anti-WNK4 antibody was described previously [41]. The WNK1 antibody was obtained from Novus Biologicals (Centennial, CO, USA). Anti-SGK-1 and Anti-SPAK antibodies were from Cell Signaling Technology (Danvers, MA, USA). OSR1 antibody was purchased from Abcam (Cambridge, MA, USA). All chemicals and reagents were purchased from Sigma and/or Fisher Scientific unless noted otherwise. Transwell and Snapwell plates were from Corning Costar (Corning, NY, USA).

4.2. Isolation and Immortalization of CD Cells from CLDN7$^{+/+}$ and CLDN7$^{-/-}$ Mouse Kidneys

The kidneys were removed quickly from 4–5-day-old CLDN7$^{+/+}$ and CLDN7$^{-/-}$ mice generated in this laboratory [31], minced into 1 mm^3 pieces, and digested in Hanks' Balanced Salt Solution (HBSS, Invitrogen, ThermoFisher Scientific, Waltham, MA USA) containing 0.2% collagenase A and 0.2% hyaluronidase (Sigma, St. Louis, MO, USA). After 30 min incubation at 37 °C, DNase I (100 U/mL) was added to the cell suspension to prevent cell clumping. Then the cell suspension went through a cell strainer and the collected cells were incubated with Dolichos biflorus lectin-coated Dynabeads (Invitrogen, ThermoFisher Scientific) in a tube at 8 °C for 20 min. The tube was placed in a magnet for 2 min and the supernatant was discarded. The beads-bound cells were washed with PBS, separated from the supernatant by the magnet, and re-suspended in an established renal CD medium [42]. To release the cells from the beads, DNase was added to the tube containing beads-bound cells.

The released cells and beads were separated by the magnet, and the supernatant with released cells were cultured in DMEM/Ham's F12 medium containing 5% fetal bovine serum, 2.5 μg/mL transferrin, 1 μM thyronine T_3, 30 μM sodium selenate, 2 mM L-glutamine, 15 mM HEPES, 100 U/mL penicillin, and 100 μg/mL streptomycinneomycin in a humidified 5% CO_2-air atmosphere at 37 °C. The purified CD cells were immortalized into cell lines using Lenti-SV40 virus infection according to the manufacturer's instructions (Applied Biological Materials Inc. Richmond, Canada). At least three CD cell lines were established by the above methods and used in the current study. All animal experiments were approved by the East Carolina University Animal Care and Use Committee (AUP#A172b, date of approval 16 December 2011).

4.3. Electrophysiological Measurements

Transepithelial resistant (TER) measurements: $CLDN7^{+/+}$ and $CLDN7^{-/-}$ CD cells were plated onto collagen-coated Transwell inserts at the density of 1.5×10^5 cells/cm^2 and cultured for 7–10 days. After the monolayer reached confluence, the resistance across each filter was measured by a Millicell-ERS Volt-Ohm meter (Millipore, Bedford, MA, USA). All TER values were calculated by subtracting the resistance measured in the blank insert from the resistance measured in the insert with the monolayer and then multiplied by the surface area of the membrane.

Dilution potential measurements: $CLDN7^{+/+}$ and $CLDN7^{-/-}$ CD cells were grown on Snapwell membranes coated with collagens. The apical and basal chambers were filled with P1 buffer containing (in mM): 140 NaCl, 2 $CaCl_2$, 1 $MgCl_2$, 10 HEPES, and 10 glucose (pH 7.3). To measure the dilution potential (DP), the basal chamber was switched from P1 to P2 buffer containing (in mM): 70 NaCl, 140 mannitol, 2 $CaCl_2$, 1 $MgCl_2$, 10 HEPES, and 10 glucose (pH 7.3). During the experiments, the buffer was maintained at 37 °C and bubbled constantly with 95% air and 5% CO_2. Dilution potential measurements and calculations were conducted as described in Alexandre et al. [30] and Yu et al. [43]. Briefly, the ion permeability ratio of the monolayer to Cl^- over the permeability to Na^+ ($\beta = P_{Cl}/P_{Na}$) was calculated from the dilution potential using the Goldman–Hodgkin–Katz equation. The absolute permeability values of Na^+ (P_{Na}) and Cl^- (P_{Cl}) were calculated according to the following equations, $P_{Na} = G \cdot (RT/F^2)/(\alpha(1 + \beta)$ and $P_{Cl} = P_{Na} \cdot \beta$. Here, the conductance per unit surface area (G) of the membrane can be measured by Ohm's law, α is the NaCl activity, and β is the ratio of the permeability of Cl^- to that of Na^+ as determined by the Goldman–Hodgkin–Katz equation [43]. Amiloride (100 μM, Sigma, St Louis, MO, USA), niflumic acid (NFA, 100 μM), and 4,4′-diisothiocyanatostilbene-2,2′-disulfonic acid (DIDS, 100 μM) were added to the solution to block epithelial sodium and chloride channels in CD cells.

4.4. RNA Extraction and Quantitative Real-Time PCR

The total RNA of $CLDN7^{+/+}$ and $CLDN7^{-/-}$ CD cells were isolated using a Qiagen RNeasy kit (Qiagen, Valencia, CA, USA), and the first-strand cDNA was synthesized with a Qiagen First Strand Kit according to the manufacturer's instructions. Quantitative real-time RT-PCR (qRT-PCR) was performed as previously described [44]. For each target gene, the relative gene expression was performed in triplicates and the cycle threshold (Ct) values were normalized to the internal control β-actin gene expression level and analyzed by $2^{-\Delta\Delta Ct}$ method [45].

4.5. Statistical Analysis

Statistical analysis was performed using Origin50 and VassarStats programs. The differences between two groups were analyzed using the unpaired two-tailed Student's t-test. One-way ANOVA was performed if comparisons involved more than two groups. Data are expressed as mean ± S.E.M. and *n* indicates the number of independent experiments. A *p*-value of <0.05 was considered significant.

5. Conclusion

In conclusion, our present study highlights a critical role of CLDN7 in Cl^- and Na^+ homeostasis in CD cells of kidneys and the involvement of CLDN7 in WNK4 regulation. In addition, the increased expression of ENaC in stable and primary CLDN7$^{-/-}$ CD cells suggests the influence of an altered paracellular pathway on the transcellular pathway. Future studies should involve the functional analysis of ENaC channel activity in the presence and absence of CLDN7. Therefore, our mouse CD cell lines provide a unique model for investigating the crosstalk between paracellular and transcellular pathways and how this interaction affects the ionic balance of the kidney and blood pressure in the body.

Author Contributions: Conceptualization, Y.-H.C.; Methodology, J.F., R.T., and J.H.; Validation and Data Curation, J.F., R.T., and J.H.; Funding Acquisition, Y.-H.C.; Supervision, Y.-H.C.; Writing-Original Draft Preparation, J.F.; Writing-Review and Editing, Y.-H.C.

Acknowledgments: We sincerely thank Lawrence Palmer (Weill Medical College of Cornell University) for providing ENaC antibodies and Alan Yu (University of Kansas Medical Center) for helpful discussion. We thank Beverly Jeansonne and Joani Zary Oswald for their technical assistance.

References

1. Sansom, S.C.; Weinman, E.J.; O'Neil, R.G. Microelectrode assessment of chloride-conductive properties of cortical collecting duct. *Am. J. Physiol.* **1984**, *247*, F291–F302. [CrossRef] [PubMed]
2. Schuster, V.L.; Stokes, J.B. Chloride transport by the cortical and outer medullary collecting duct. *Am. J. Physiol.* **1987**, *253*, F203–F212. [CrossRef] [PubMed]
3. Hou, J. Paracellular transport in the collecting duct. *Curr. Opin. Nephrol. Hypertens.* **2016**, *25*, 424–428. [CrossRef] [PubMed]
4. Denker, B.M.; Sabath, E. The biology of epithelial cell tight junctions in the kidney. *J. Am. Soc. Nephrol.* **2011**, *22*, 622–625. [CrossRef] [PubMed]
5. Van Itallie, C.M.; Anderson, J.M. Claudins and epithelial paracellular transport. *Annu. Rev. Physiol.* **2006**, *68*, 403–429. [CrossRef] [PubMed]
6. Balkovetz, D.F. Claudins at the gate: Determinants of renal epithelial tight junction paracellular permeability. *Am. J. Physiol. Renal Physiol.* **2006**, *290*, F572–F579. [CrossRef] [PubMed]
7. Mineta, K.; Yamamoto, Y.; Yamazaki, Y.; Tanaka, H.; Tada, Y.; Saito, K.; Tamura, A.; Igarashi, M.; Endo, T.; Takeuchi, K.; et al. Predicted expansion of the claudin multigene family. *FEBS Lett.* **2011**, *585*, 606–612. [CrossRef] [PubMed]
8. Kiuchi-Saishin, Y.; Gotoh, S.; Furuse, M.; Takasuga, A.; Tano, Y.; Tsukita, S. Differential expression patterns of claudins, tight junction membrane proteins, in mouse nephron segments. *J. Am. Soc. Nephrol.* **2002**, *13*, 875–886.
9. Konrad, M.; Schaller, A.; Seelow, D.; Pandey, A.V.; Waldegger, S.; Lesslauer, A.; Vitzthum, H.; Suzuki, Y.; Luk, J.M.; Becker, C.; et al. Mutations in the tight-junction gene claudin 19 (CLDN19) are associated with renal magnesium wasting, renal failure, and severe ocular involvement. *Am. J. Hum. Genet.* **2006**, *79*, 949–957. [CrossRef]
10. Simon, D.B.; Lu, Y.; Choate, K.A.; Velazquez, H.; Al-Sabban, E.; Praga, M.; Casari, G.; Bettinelli, A.; Colussi, G.; Rodriguez-Soriano, J.; et al. Paracellin-1, a renal tight junction protein required for paracellular Mg2+ resorption. *Science* **1999**, *285*, 103–106. [CrossRef]
11. Hou, J.; Shan, Q.; Wang, T.; Gomes, A.S.; Yan, Q.; Paul, D.L.; Bleich, M.; Goodenough, D.A. Transgenic RNAi depletion of claudin-16 and the renal handling of magnesium. *J. Biol. Chem.* **2007**, *282*, 17114–17122. [CrossRef] [PubMed]
12. Muto, S.; Hata, M.; Taniguchi, J.; Tsuruoka, S.; Moriwaki, K.; Saitou, M.; Furuse, K.; Sasaki, H.; Fujimura, A.; Imai, M.; et al. Claudin-2-deficient mice are defective in the leaky and cation-selective paracellular permeability properties of renal proximal tubules. *Proc. Natl. Acad. Sci. USA* **2010**, *107*, 8011–8016. [CrossRef] [PubMed]
13. Krug, S.M.; Gunzel, D.; Conrad, M.P.; Rosenthal, R.; Fromm, A.; Amasheh, S.; Schulzke, J.D.; Fromm, M. Claudin-17 forms tight junction channels with distinct anion selectivity. *Cell Mol. Life Sci.* **2012**, *69*, 2765–2778. [CrossRef] [PubMed]

14. Breiderhoff, T.; Himmerkus, N.; Stuiver, M.; Mutig, K.; Will, C.; Meij, I.C.; Bachmann, S.; Bleich, M.; Willnow, T.E.; Muller, D. Deletion of claudin-10 (Cldn10) in the thick ascending limb impairs paracellular sodium permeability and leads to hypermagnesemia and nephrocalcinosis. *Proc. Natl. Acad. Sci. USA* **2012**, *109*, 14241–14246. [CrossRef]
15. Breiderhoff, T.; Himmerkus, N.; Drewell, H.; Plain, A.; Gunzel, D.; Mutig, K.; Willnow, T.E.; Muller, D.; Bleich, M. Deletion of claudin-10 rescues claudin-16-deficient mice from hypomagnesemia and hypercalciuria. *Kidney Int.* **2018**, *93*, 580–588. [CrossRef]
16. Furgeson, S.B.; Linas, S. Mechanisms of type I and type II pseudohypoaldosteronism. *J. Am. Soc. Nephrol.* **2010**, *21*, 1842–1845. [CrossRef]
17. Healy, J.K. Pseudohypoaldosteronism type II: History, arguments, answers, and still some questions. *Hypertension* **2014**, *63*, 648–654. [CrossRef]
18. Wilson, F.H.; Disse-Nicodeme, S.; Choate, K.A.; Ishikawa, K.; Nelson-Williams, C.; Desitter, I.; Gunel, M.; Milford, D.V.; Lipkin, G.W.; Achard, J.M.; et al. Human hypertension caused by mutations in WNK kinases. *Science* **2001**, *293*, 1107–1112. [CrossRef]
19. Arroyo, J.P.; Gamba, G. Advances in WNK signaling of salt and potassium metabolism: Clinical implications. *Am. J. Nephrol.* **2012**, *35*, 379–386. [CrossRef]
20. Welling, P.A.; Chang, Y.P.; Delpire, E.; Wade, J.B. Multigene kinase network, kidney transport, and salt in essential hypertension. *Kidney Int.* **2010**, *77*, 1063–1069. [CrossRef]
21. Kahle, K.T.; Ring, A.M.; Lifton, R.P. Molecular physiology of the WNK kinases. *Annu. Rev. Physiol.* **2008**, *70*, 329–355. [CrossRef]
22. Wang, W.H.; Giebisch, G. Regulation of potassium (K) handling in the renal collecting duct. *Pflugers Arch.* **2009**, *458*, 157–168. [CrossRef]
23. Kahle, K.T.; Wilson, F.H.; Leng, Q.; Lalioti, M.D.; O'Connell, A.D.; Dong, K.; Rapson, A.K.; MacGregor, G.G.; Giebisch, G.; Hebert, S.C.; et al. WNK4 regulates the balance between renal NaCl reabsorption and K+ secretion. *Nat. Genet.* **2003**, *35*, 372–376. [CrossRef]
24. Chen, J.C.; Lo, Y.F.; Lin, Y.W.; Lin, S.H.; Huang, C.L.; Cheng, C.J. WNK4 kinase is a physiological intracellular chloride sensor. *Proc. Natl. Acad. Sci. USA* **2019**, *116*, 4502–4507. [CrossRef]
25. Kahle, K.T.; Macgregor, G.G.; Wilson, F.H.; Van Hoek, A.N.; Brown, D.; Ardito, T.; Kashgarian, M.; Giebisch, G.; Hebert, S.C.; Boulpaep, E.L.; et al. Paracellular Cl- permeability is regulated by WNK4 kinase: Insight into normal physiology and hypertension. *Proc. Natl. Acad. Sci. USA* **2004**, *101*, 14877–14882. [CrossRef]
26. Yamauchi, K.; Rai, T.; Kobayashi, K.; Sohara, E.; Suzuki, T.; Itoh, T.; Suda, S.; Hayama, A.; Sasaki, S.; Uchida, S. Disease-causing mutant WNK4 increases paracellular chloride permeability and phosphorylates claudins. *Proc. Natl. Acad. Sci. USA* **2004**, *101*, 4690–4694. [CrossRef]
27. Tatum, R.; Zhang, Y.; Lu, Q.; Kim, K.; Jeansonne, B.G.; Chen, Y.H. WNK4 phosphorylates ser(206) of claudin-7 and promotes paracellular Cl(-) permeability. *FEBS Lett.* **2007**, *581*, 3887–3891. [CrossRef]
28. Hou, J.; Renigunta, A.; Yang, J.; Waldegger, S. Claudin-4 forms paracellular chloride channel in the kidney and requires claudin-8 for tight junction localization. *Proc. Natl. Acad. Sci. USA* **2010**, *107*, 18010–18015. [CrossRef]
29. Alexandre, M.D.; Jeansonne, B.G.; Renegar, R.H.; Tatum, R.; Chen, Y.H. The first extracellular domain of claudin-7 affects paracellular Cl- permeability. *Biochem. Biophys. Res. Commun.* **2007**, *357*, 87–91. [CrossRef]
30. Alexandre, M.D.; Lu, Q.; Chen, Y.H. Overexpression of claudin-7 decreases the paracellular Cl- conductance and increases the paracellular Na+ conductance in LLC-PK1 cells. *J. Cell. Sci.* **2005**, *118*, 2683–2693. [CrossRef]
31. Tatum, R.; Zhang, Y.; Salleng, K.; Lu, Z.; Lin, J.J.; Lu, Q.; Jeansonne, B.G.; Ding, L.; Chen, Y.H. Renal salt wasting and chronic dehydration in claudin-7-deficient mice. *Am. J. Physiol. Renal Physiol.* **2010**, *298*, F24–34. [CrossRef]
32. Li, W.Y.; Huey, C.L.; Yu, A.S. Expression of claudin-7 and -8 along the mouse nephron. *Am. J. Physiol. Renal Physiol.* **2004**, *286*, F1063–F1071. [CrossRef]
33. Muto, S.; Yasoshima, K.; Yoshitomi, K.; Imai, M.; Asano, Y. Electrophysiological identification of alpha- and beta-intercalated cells and their distribution along the rabbit distal nephron segments. *J. Clin. Invest.* **1990**, *86*, 1829–1839. [CrossRef]
34. Lu, Z.; Ding, L.; Hong, H.; Hoggard, J.; Lu, Q.; Chen, Y.H. Claudin-7 inhibits human lung cancer cell migration and invasion through ERK/MAPK signaling pathway. *Exp. Cell Res.* **2011**, *317*, 1935–1946. [CrossRef]

35. Light, D.B.; Schwiebert, E.M.; Fejes-Toth, G.; Naray-Fejes-Toth, A.; Karlson, K.H.; McCann, F.V.; Stanton, B.A. Chloride channels in the apical membrane of cortical collecting duct cells. *Am. J. Physiol.* **1990**, *258*, F273–F280. [CrossRef]
36. Lopez-Cayuqueo, K.I.; Chavez-Canales, M.; Pillot, A.; Houillier, P.; Jayat, M.; Baraka-Vidot, J.; Trepiccione, F.; Baudrie, V.; Busst, C.; Soukaseum, C.; et al. A mouse model of pseudohypoaldosteronism type II reveals a novel mechanism of renal tubular acidosis. *Kidney Int.* **2018**, *94*, 514–523. [CrossRef]
37. Susa, K.; Sohara, E.; Rai, T.; Zeniya, M.; Mori, Y.; Mori, T.; Chiga, M.; Nomura, N.; Nishida, H.; Takahashi, D.; et al. Impaired degradation of WNK1 and WNK4 kinases causes PHAII in mutant KLHL3 knock-in mice. *Hum. Mol. Genet.* **2014**, *23*, 5052–5060. [CrossRef]
38. Alvarez de la Rosa, D.; Zhang, P.; Naray-Fejes-Toth, A.; Fejes-Toth, G.; Canessa, C.M. The serum and glucocorticoid kinase sgk increases the abundance of epithelial sodium channels in the plasma membrane of Xenopus oocytes. *J. Biol. Chem.* **1999**, *274*, 37834–37839. [CrossRef]
39. Faletti, C.J.; Perrotti, N.; Taylor, S.I.; Blazer-Yost, B.L. sgk: An essential convergence point for peptide and steroid hormone regulation of ENaC-mediated Na+ transport. *Am. J. Physiol. Cell Physiol.* **2002**, *282*, C494–500. [CrossRef]
40. Loffing, J.; Zecevic, M.; Feraille, E.; Kaissling, B.; Asher, C.; Rossier, B.C.; Firestone, G.L.; Pearce, D.; Verrey, F. Aldosterone induces rapid apical translocation of ENaC in early portion of renal collecting system: Possible role of SGK. *Am. J. Physiol. Renal. Physiol.* **2001**, *280*, F675–682. [CrossRef]
41. Jeansonne, B.; Lu, Q.; Goodenough, D.A.; Chen, Y.H. Claudin-8 interacts with multi-PDZ domain protein 1 (MUPP1) and reduces paracellular conductance in epithelial cells. *Cell Mol. Biol. (Noisy-le-grand)* **2003**, *49*, 13–21.
42. Bens, M.; Duong Van Huyen, J.P.; Cluzeaud, F.; Teulon, J.; Vandewalle, A. CFTR disruption impairs cAMP-dependent Cl(-) secretion in primary cultures of mouse cortical collecting ducts. *Am. J. Physiol. Renal Physiol.* **2001**, *281*, F434–442. [CrossRef]
43. Yu, A.S. Electrophysiological characterization of claudin ion permeability using stably transfected epithelial cell lines. *Methods Mol. Biol.* **2011**, *762*, 27–41.
44. Ding, L.; Lu, Z.; Foreman, O.; Tatum, R.; Lu, Q.; Renegar, R.; Cao, J.; Chen, Y.H. Inflammation and disruption of the mucosal architecture in claudin-7-deficient mice. *Gastroenterology* **2012**, *142*, 305–315. [CrossRef]
45. Dussault, A.A.; Pouliot, M. Rapid and simple comparison of messenger RNA levels using real-time PCR. *Biol. Proced. Online* **2006**, *8*, 1–10. [CrossRef]

7

Phosphatidylcholine Passes by Paracellular Transport to the Apical Side of the Polarized Biliary Tumor Cell Line Mz-ChA-1

Wolfgang Stremmel [1,*], Simone Staffer [2] and Ralf Weiskirchen [3]

1 Institute of Pharmacy and Molecular Biotechnology, University of Heidelberg, D-69120 Heidelberg, Germany
2 University Clinics of Heidelberg, D-69120 Heidelberg, Germany
3 Institute of Molecular Pathobiochemistry, Experimental Gene Therapy and Clinical Chemistry, RWTH University Hospital Aachen, D-52074 Aachen, Germany
* Correspondence: wolfgangstremmel@aol.com

Abstract: Phosphatidylcholine (PC) translocation into mucus of the intestine was shown to occur via a paracellular transport across the apical/lateral tight junction (TJ) barrier. In case this could also be operative in biliary epithelial cells, this may have implication for the pathogenesis of primary sclerosing cholangitis (PSC). We here evaluated the transport of PC across polarized cholangiocytes. Therefore, the biliary tumor cell line Mz-ChA-1 was grown to confluency. In transwell culture systems the translocation of PC to the apical compartment was analyzed. After 21 days in culture, polarized Mz-ChA-1 cells revealed a predominant apical translocation of choline containing phospholipids including PC with minimal intracellular accumulation. Transport was suppressed by TJ destruction employing chemical inhibitors and pretreatment with siRNA to TJ forming proteins as well as the apical transmembrane mucin 3 as PC acceptor. Apical translocation was dependent on a negative apical electrical potential created by the cystic fibrosis transmembrane conductance regulator (CFTR) and the anion exchange protein 2 (AE2). It was stimulated by apical application of secretory mucins. The results indicated the existence of a paracellular PC passage across apical/lateral TJ of the polarized biliary epithelial tumor cell line Mz-ChA-1. This has implication for the generation of a protective mucus barrier in the biliary tree.

Keywords: Mz-ChA-1 cells; biliary epithelial cells; phosphatidylcholine; mucus; tight junctions; paracellular transport

---

1. Introduction

It has been known for a long time that tight junctions (TJ) are responsible for sealing epithelial cells at their apical–lateral side. This enables stable boundaries with their respective functional implication, e.g., at the blood–brain barrier. In intestinal mucosa, they prevent the attack of microbiota. Additionally, they serve the environmental control by allowing water and electrolyte exchange. Macromolecules were not known to pass this barrier. However, in recent studies the novel pathway of phosphatidylcholine (PC) transport across lateral TJ to the luminal side of mucosal cells was described [1].

After apical translocation, PC from systemic sources (lipoproteins) is enriched within the mucus by binding to mucin 2 to establish a hydrophobic barrier against the colonic lumen. Indeed, genetic mouse models with intestinal specific TJ deletion revealed an impaired mucus barrier with lack of PC [2]. This caused an ulcerative colitis (UC) phenotype [2]. The intrinsic lack of mucus PC, the disposition for microbiota invasion and the consequent inflammation matches the situation in human UC [3,4]. The disease is often associated with primary sclerosing cholangitis (PSC), the pathogenesis of which remains obscure [5]. It was indeed postulated that the luminal lack of PC may be of etiological relevance.

When canalicular secretion of PC through the multiple drug resistance gene 2 (MDR2) was deleted in genetically modified mice, a severe cholangitis occurred [6,7]. It is believed that PC is secreted to neutralize the aggressiveness of bile acids by packing them into micelles. Thus, it prevents the attack of bile acids against the plasma membrane phospholipid bilayer of biliary epithelial cells. However, analysis in patients with PSC did not show a lack of PC in bile [8].

These studies neglected the fact that between lumen and the epithelial cells, there is a mucus layer containing secretory mucins. They bind PC to constitute a protective shield towards bile. It is unlikely that mucus PC originates from bile, containing PC-bile acid micelles with high mutual affinity. Therefore, we assume that the biliary mucus compartment is fed with PC from systemic sources as it is the case in intestinal mucosa [2]. We further assume that PC passes through lateral TJ to the apical side of the biliary epithelial cells. Indeed, a common genetic defect in intestinal and biliary epithelial cells could explain the high association of UC and PSC. To prove this concept, we first analyzed the pathway of PC from basal to the apical side employing the biliary tumor cell line Mz-ChA-1. These cells share the characteristics of physiologic biliary epithelial cells [9].

2. Results

According to our hypothesis, the transport of PC to the apical side of biliary epithelial cells requires the presence of TJ. We utilized the well-characterized biliary epithelial tumor cell line Mz-ChA-1 [9] for our experiments and grew them in transwell tissue culture systems for up to 21 days. This cell line has a highly differentiated phenotype producing large quantities of mucus. In comparison to many other human biliary tract carcinoma cell lines such as TGBC-1, TBC-51, Mz-ChA-2 and SK-ChA-1, the cell line Mz-ChA-1 is 10–1000 times less invasive through both the collagen and the basement membrane [10]. The cell line Mz-ChA-1 was originally described as an adenocarcinoma cell line. However, many subsequent studies showed that the cell line has many features of human cholangiocyte cells including TJ complexes rendering these cells as appropriate tool for respective studies [11–14].

While at day three only marginal transepithelial resistance (TER) was detectable with $120 \pm 40\ \Omega$ (non-polarized cells), it increased to $220 \pm 30\ \Omega$ at day nine and remained stable after 21 days (polarized cells) with $400 \pm 80\ \Omega$. Western blot analysis of representative TJ proteins confirmed their presence after 21 days in culture (Figure 1).

Then apical and basal PC transport was evaluated by a respective translocation of PC, provided to the opposite compartment together with taurocholate (TC) at indicated concentrations. For comparison inulin as a non-cell permeable compound was analyzed. Indeed, polarized Mz-ChA-1 cells had the capacity to translocate PC to the apical compartment, whereas basal transport diminished (Figure 2). Inulin, instead, revealed preferential transport to the basal side. These observations were in contrast to non-polarized Mz-ChA-1 cells which resulted in equilibrated PC and inulin concentrations between apical and basal compartments (Figure 2).

The intracellular accumulation was tested in non-polarized cells incubated with 100 μM for 1 h and accounted for <5% of the incubated PC, whether it was provided together with TC or albumin. Only when PC was provided with apolipoprotein B (ApoB) an uptake rate of 12.84 ± 5 nmol·mg protein^{-1}·h^{-1} was observed indicative of lipoprotein-mediated endocytosis of PC. In comparison, 1 h uptake rates for radiolabeled TC and oleate, the later complexed with TC or albumin, were significantly higher ($p < 0.01$) (Figure 3).

Apical transport of PC increased linearly with time of exposure and incubated PC concentrations, was temperature dependent with the highest rates between 25–37 °C, and a pH optimum between pH 7.0 to pH 8.0 (Figure 4).

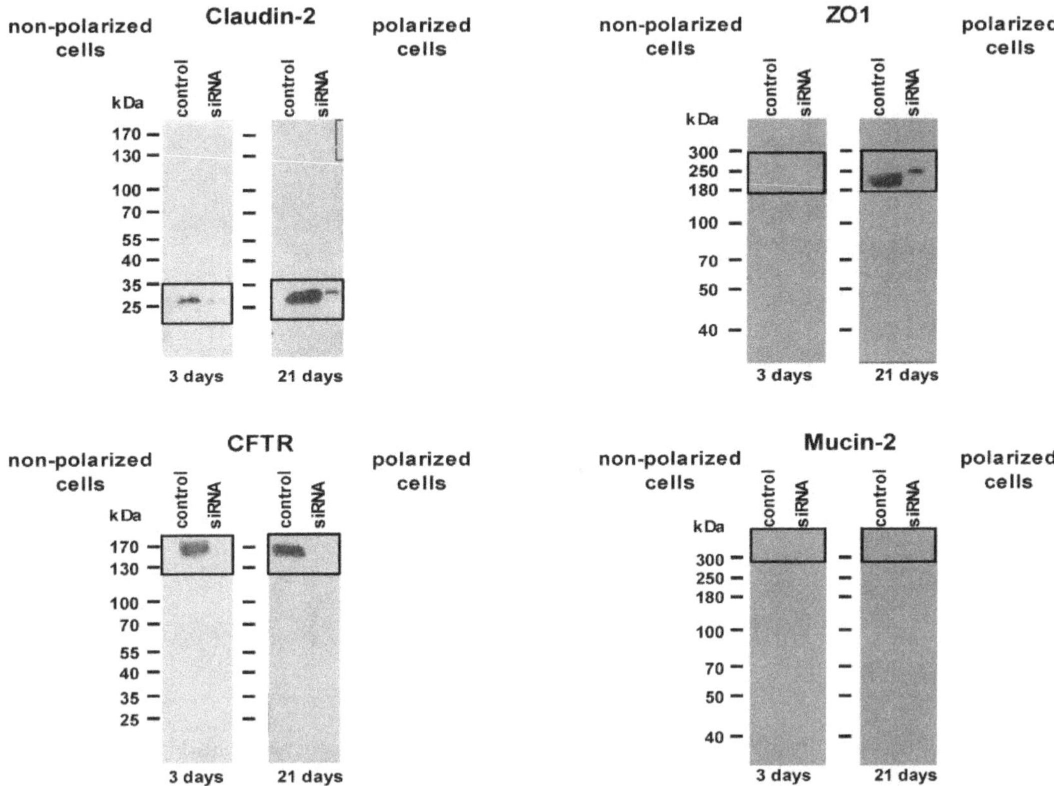

Figure 1. Western blot with non-polarized (three days cultured) and polarized (21 days cultured) MzChA-1 cells with representative proteins of the tight junction (TJ) complex (claudin-2; Zonula Occludens-1, ZO1). Cystic fibrosis transmembrane conductance regulator (CFTR) is a constitutive, non-TJ protein used as housekeeping gene, and mucin 2 as a protein that is not expressed in cholangiocytes. Applied were 30 µg of cell homogenates. The data indicate the enhancement of appearance of the TJ proteins claudin-2 and ZO1 with the polarization of the cells.

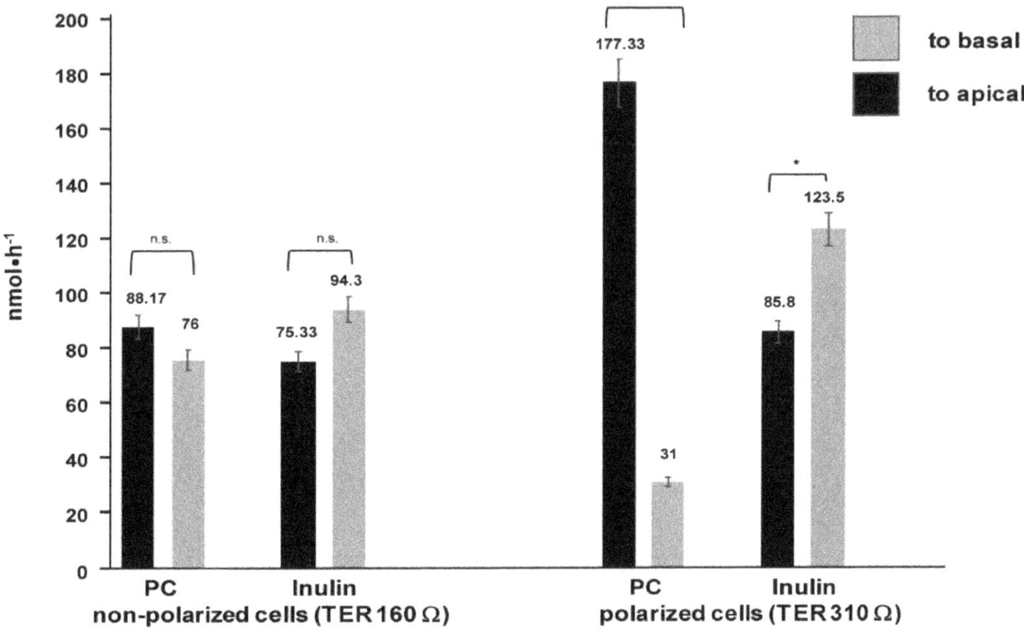

Figure 2. Apical vs. basal transport of phosphatidylcholine (PC) and inulin applied to a transwell culture system of non-polarized (three days cultured) and polarized (21 days cultured) Mz-ChA-1 cells. Means ± SD of $n = 6$; n.s. = not significant, * $p < 0.05$, *** $p < 0.001$. TER, transepithelial resistance.

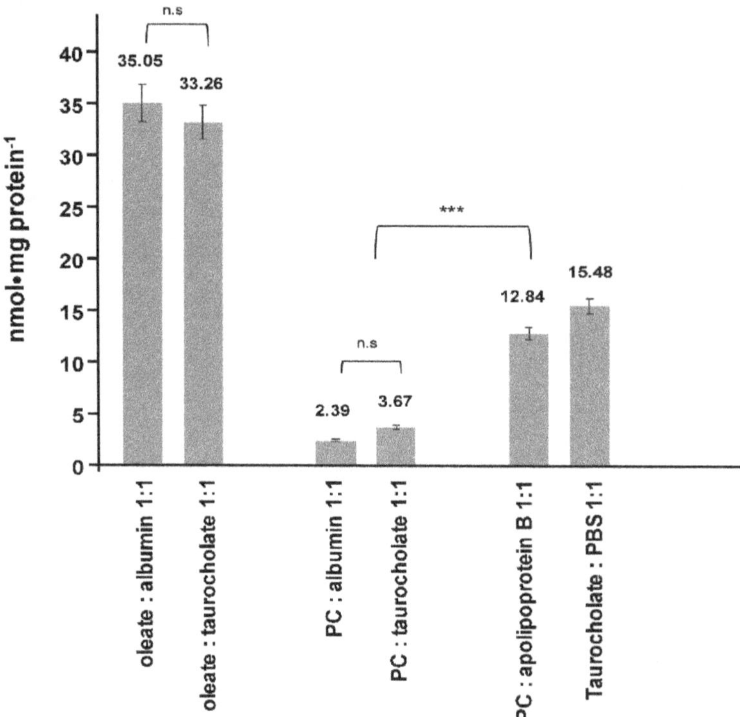

Figure 3. Intracellular accumulation of phosphatidylcholine (PC), fatty acids and taurocholate (TC) in non-polarized Mz-ChA-1 cells. After 1 h incubation of the different substrates with various binding molecules at 37 °C, the amount taken up by the cells was determined after washing the cells three times with 10 mM TC in phosphate-buffered saline (PBS). Means ± SD of $n = 6$; n.s. = not significantly different, *** $p < 0.001$.

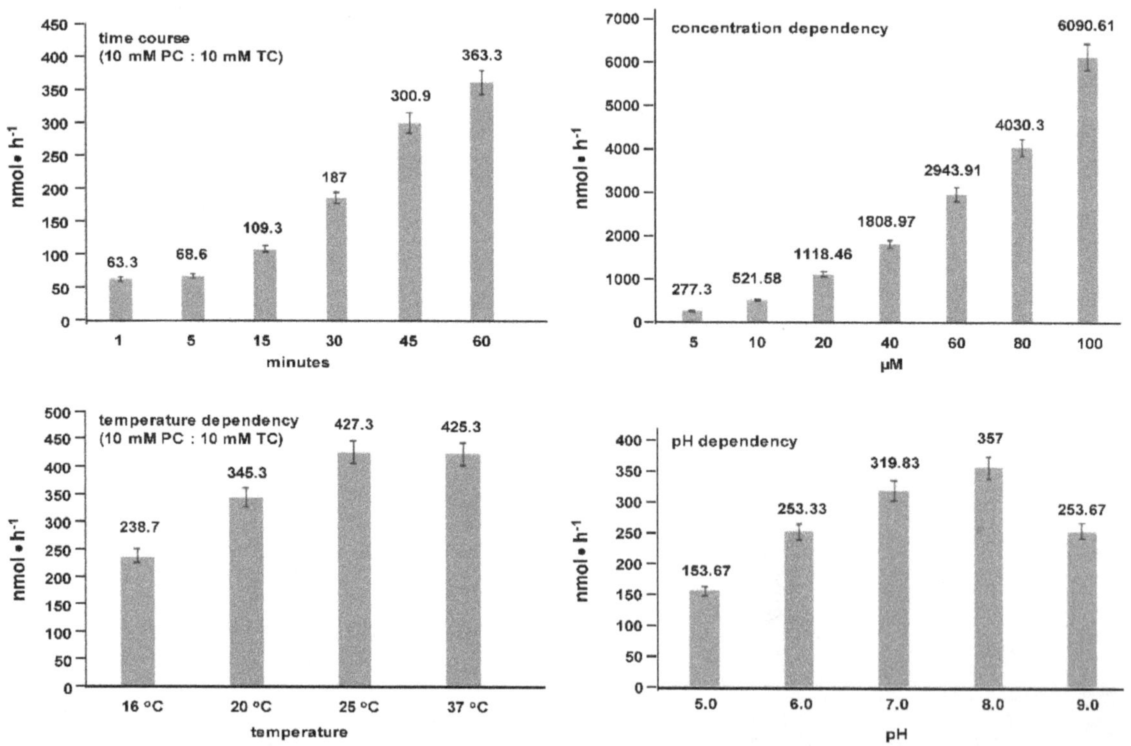

Figure 4. Transport characteristics of phosphatidylcholine (PC) to the apical side of polarized Mz-ChA-1 cells. Apical transport of PC was evaluated after exposure to the basal side. Transport was linear in regard to time and concentration (1 h incubation) with a temperature optimum between 25–37 °C and a pH optimum in the range of pH 7.0 to pH 8.0. Means ± SD of $n = 6$.

Transport was specific for the choline containing phospholipids PC, lysophosphatidylcholine (LPC) and sphingomyelin (SM) but not for other phospholipids (Figure 5).

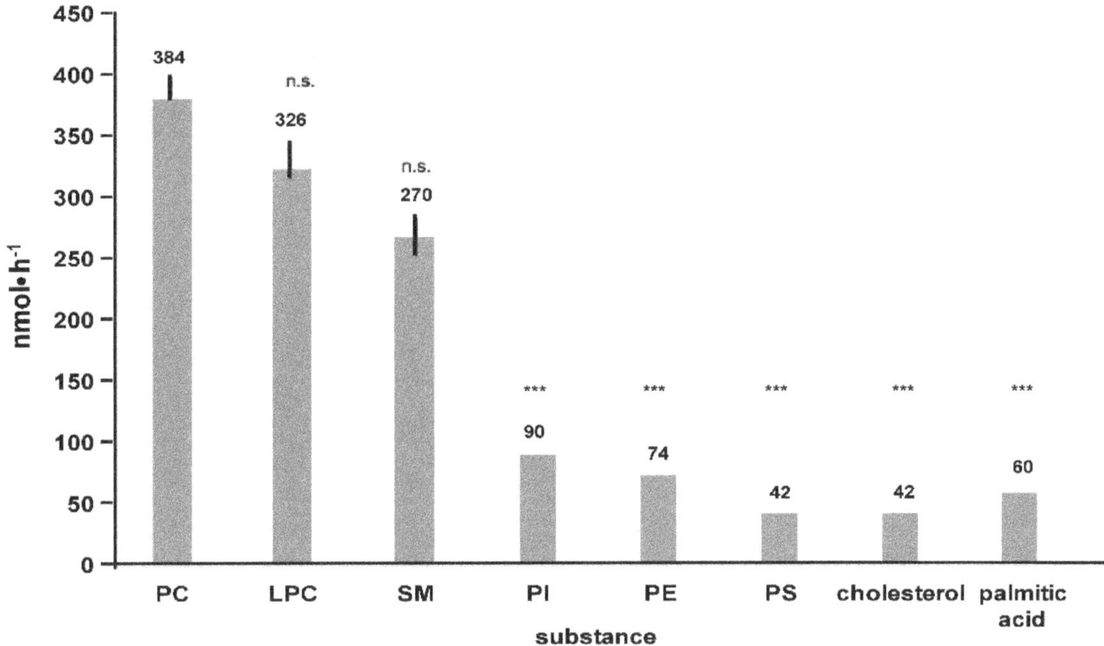

Figure 5. Specificity of apical transport for choline-containing phospholipids in polarized, nine-day cultured Mz-ChA-1 cells. Apical transport was examined for 1 h after basal application of the different substrates at 10 mM phosphatidylcholine (PC) together with 10 mM taurocholate (TC). Means ± SD of $n = 6$. Significances were calculated in relation to PC; n.s. = not significant, *** $p < 0.001$. Abbreviations used are: LPC, lysophosphatidylcholine; SM, sphingomyelin; PI, phosphatidylinositol; PE, phosphatidylethanolamine; PS, phosphatidylserine.

Apical transport was inhibited when a positive charge was generated by application of ammonium chloride (NH_4Cl), sodium thiocyanate (NaSCN) or urea, but remained stable by application of negative charge to the apical surface (Figure 6) [1]. Negative charge in vivo is generated by the cystic fibrosis transmembrane conductance regulator (CFTR) or the anion exchange protein 2 (AE2) which are both present in cholangiocytes [15].

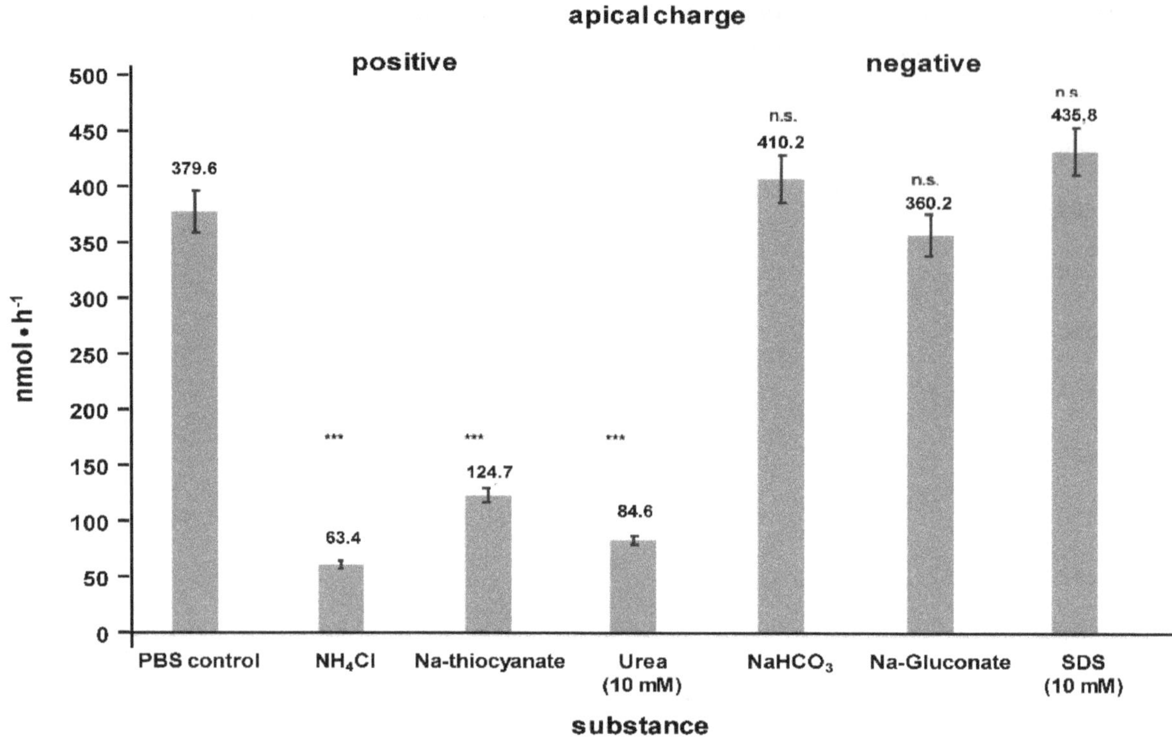

Figure 6. Ionic driving forces for apical translocation of phosphatidylcholine (PC) in polarized Mz-ChA-1 cells. Apical transport of basally applied 10 mM PC: 10 mM taurocholate was examined in polarized (21 days cultured) Mz-ChA-1 cells apically equilibrated for 1 h with indicated salts and substrates (130 mM) generating apical positive or negative charge. Means ± SD of $n = 6$. Significances were calculated in relation to the phosphate-buffered saline (PBS) control; n.s. = not significant, *** $p < 0.001$.

When CFTR or AE2 were reduced by siRNA pretreatment, PC transport was strongly inhibited (Figure 7).

The hypothesis of a TJ-mediated translocation process was supported by the observation of its inhibition by exposure of polarized Mz-ChA-1 cells to acetaldehyde (ACA) vapor or the selective peroxisome proliferator-activated receptor γ (PPARγ) inhibitors T0070907 or GW9962 (Figure 8).

Moreover, apical transport was diminished by pretreatment of polarized Mz-ChA-1 cells with siRNAs against members of the TJ complex or their upstream control proteins kindlin-1 and -2 (Figure 7). Although cholangiocytes do not synthesize mucin 1 and 2, they were shown to have mucin 3 which is a transmembrane protein serving in accepting apically transported PC as it was shown in two previous studies [1,2]. Its inhibition by siRNA reduced PC translocation significantly. From mucin 3 PC is handled to secreted mucin 1 and 2 which originates from goblet cells [16]. When it was applied to the apical surface of polarized Mz-ChA-1 cells transport was dose dependently enhanced as it was also the case when increasing concentrations of TC were apically added which favor the drainage of transported PC to the apical side by incorporation of PC into micelles (Figure 9). However, it has to be considered that Mz-ChA-1 cells as well as biliary epithelial cells in vivo do not exhibit a mucin 2 containing mucus layer.

To document a vectoral transport of PC from the basal to the apical surface, it was tested when increasing, but equal concentrations of PC in both compartments were provided. In contrast to inulin, where basal transport was higher than apical translocation, PC was overwhelmingly transported apically with only marginal appearance at the basal side indicating a predominant paracellular transport across TJ to the luminal surface (Figure 10).

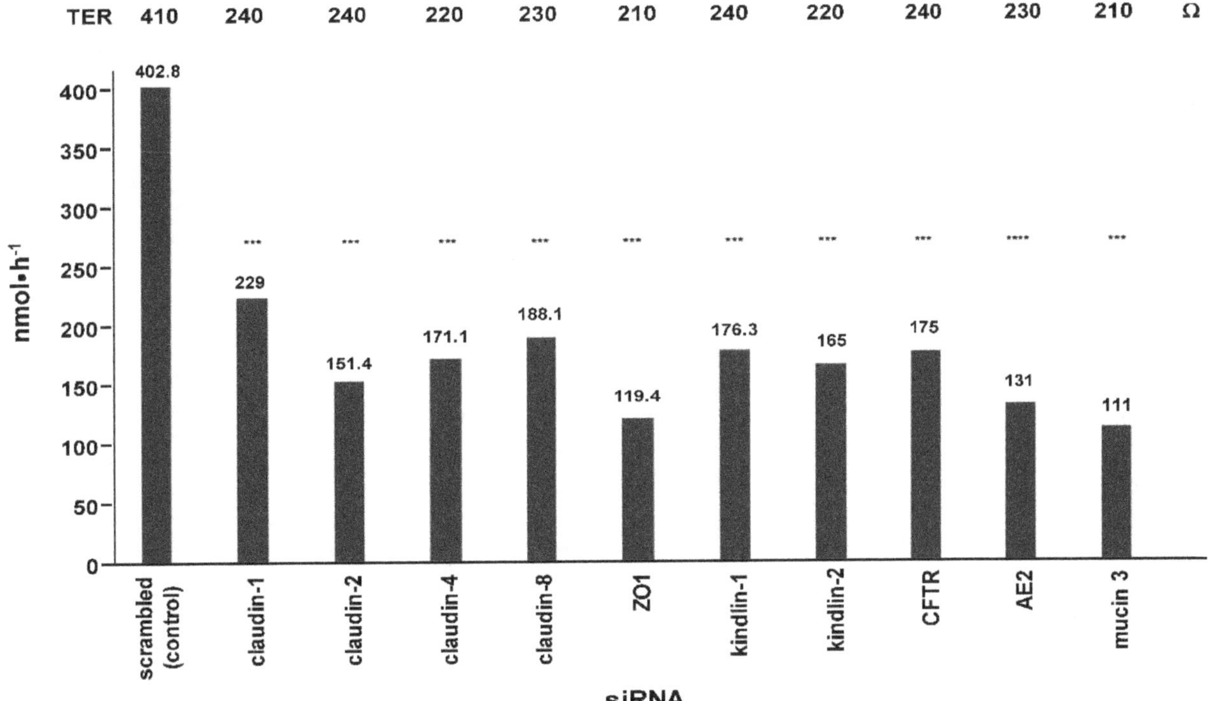

Figure 7. Effect of siRNA suppression of proteins involved in apical translocation of phosphatidylcholine (PC). Apical transport of PC after basal application of 10 mM PC: 10 mM taurocholate was examined after siRNA knockdown of indicated proteins which are all involved in tight junction-mediated luminal secretion of PC. Means ± SD of $n = 6$. Significances were calculated in relation to the control pretreated with scrambled siRNA; n.s. = not significant, *** $p < 0.001$. ZO1, Zonula Occludens-1; CFTR, cystic fibrosis transmembrane conductance regulator; AE2, anion exchange protein 2; TER, transepithelial resistance.

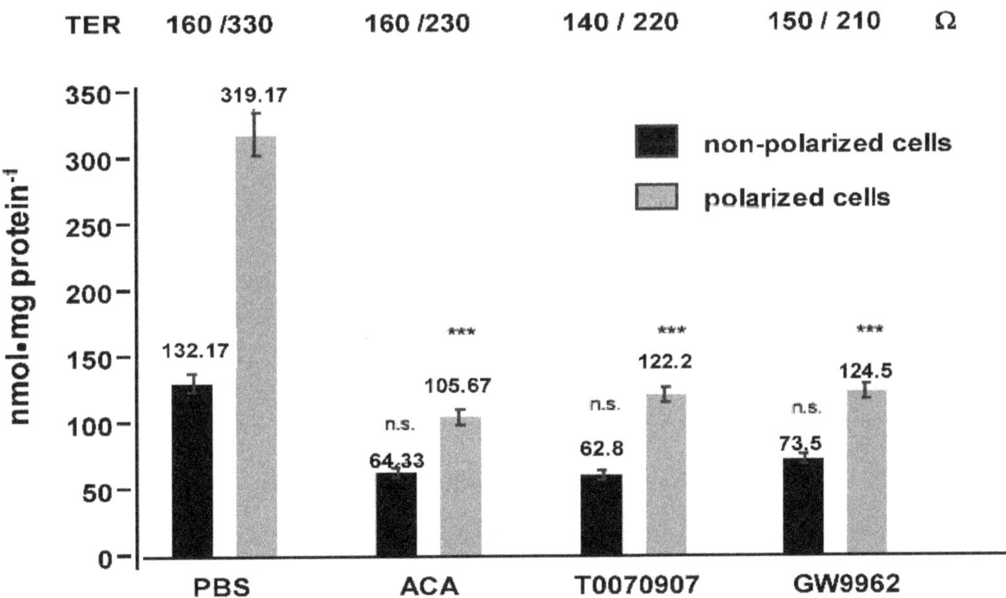

Figure 8. Effect of disruption of tight junctions (TJ) by acetaldehyde (ACA) or the indicated peroxisome proliferator-activated receptor γ (PPARγ) inhibitors on apical transport of phosphatidylcholine (PC). TJ were disrupted by exposure to ACA vapor or incubation of upper transwell chambers with the PPARγ inhibitors. Then apical transport of 10 mM PC: 10 mM taurocholate for 1 h was registered. Means ± SD of $n = 6$. Significances were calculated separately for non-polarized and polarized Mz-ChA-1 cells and compared to phosphate-buffered saline (PBS)-treated controls; n.s. = not significant, *** $p < 0.001$.

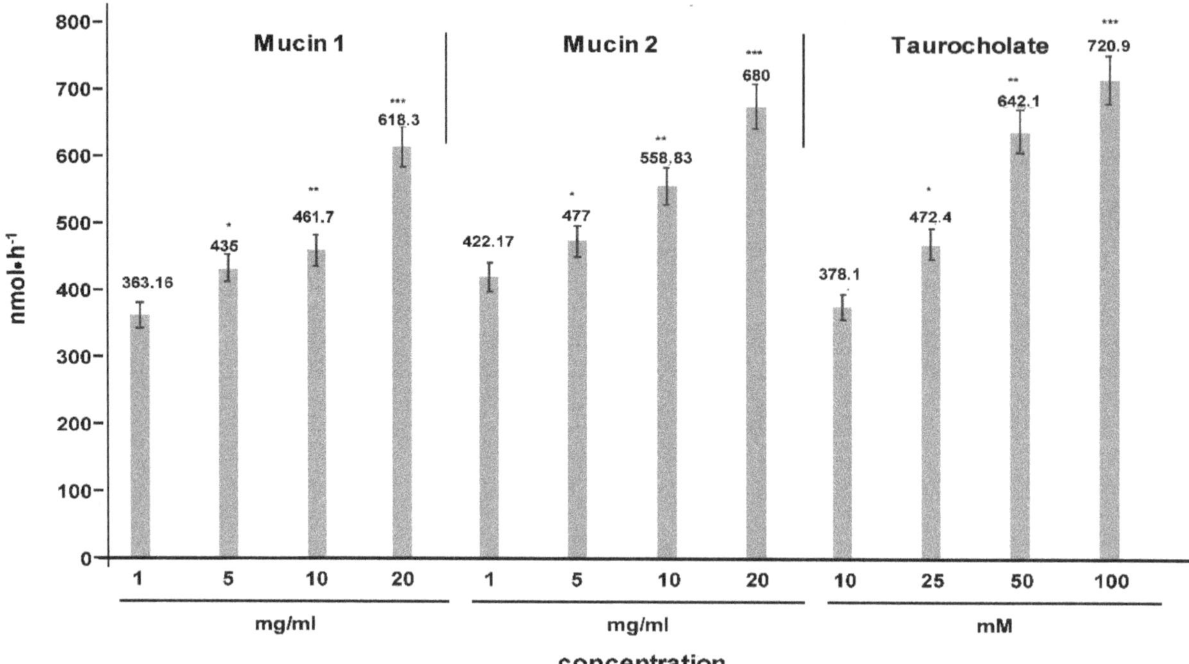

Figure 9. Enhancement of apical phosphatidylcholine (PC) transport by application of increasing concentrations of secretory mucin 1, mucin 2, and taurocholate (TC) as luminal acceptor substances. Apical transport of 10 mM PC: 10 mM TC with polarized Mz-ChA-1 cells was determined as a function of increasing concentrations of mucin 1, mucin 2 and TC in the upper culture system. Means ± SD of $n = 6$. Significances were calculated in relation to the lowest concentration of each substance added. * $p < 0.05$; ** $p < 0.01$; *** $p < 0.001$.

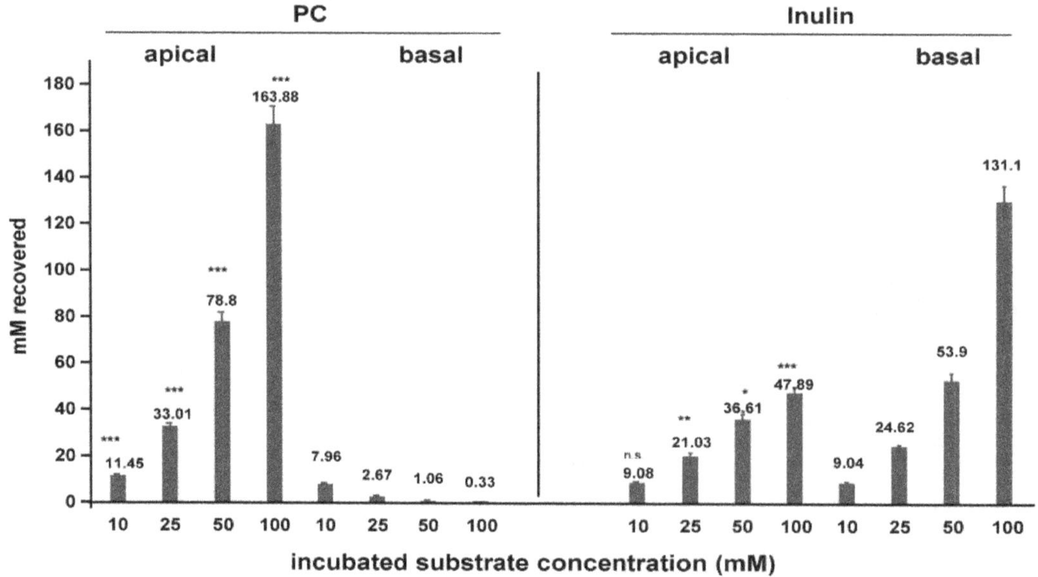

Figure 10. Equilibrium distribution of phosphatidylcholine (PC) and inulin when applied in increasing concentrations to upper and lower compartment of the transwell culture system. Increasing concentrations up to 100 mM of PC and inulin were added to the apical and basal compartments. Over 1 h observation, vectoral transport of PC was completely directed apically with decreasing PC at the luminal side (less than 1%). For inulin there was a significantly higher basal accumulation starting at 25 mM incubated. The process appeared to be less effective, at least at lower concentrations, as for the transport of PC to the apical side. Means ± SD of $n = 6$. Significances were calculated for apical versus basal transport for each in relation to the lowest concentration of each substrate concentration. n.s. = not significant; * $p < 0.05$; ** $p < 0.01$; *** $p < 0.001$.

3. Discussion

Kinetic analyses of the translocation of PC across the polarized biliary epithelial tumor cell line Mz-ChA-1 revealed a TJ-mediated process. This was as unexpected as it was for TJ-mediated PC transport in the intestinal mucosa [1,2]. This transport is strongly unidirectional to the apical side, driven by a negative apical charge generated by CFTR and AE2. The translocated PC is bound to mucin 3 from where it is handled to mucin 1 and 2 in the mucus layer of the biliary tract.

The PC in mucus serves in the biliary tree to protect the epithelial cells from the attack of bile acids, which achieve concentrations in the millimolar range within the small ducts [17]. This can be neutralized by canalicular secretion of PC via MDR2/MDR3 (ABCB4) through the formation of micelles.

However, the excessive load of bile acids could in vivo take PC from the mucus layer still maintaining a shield towards biliary epithelium, because there is a constant basal supply to the apical surface. In fact, as shown in this study, PC secretion is proportionally increased with TC in the apical (biliary) compartment. Thus, PC in cell membranes is not exposed to the luminal bile acid load. However, when, for example by a genetic defect, the apical PC secretion through the biliary epithelial cell layer is impaired, they are attacked, and a cholangitis occurs. This hypothesis will now be proven in a genetic mouse model where the biliary TJ are deleted. The observation of a disturbed TJ barrier as a potential cause of impaired secretion of PC to biliary mucus could shed light on the pathogenesis of PSC. It associates to UC with a genetically determined disruption of the TJ barrier and diminished secretion of PC to the intestinal mucus [2–4]. When this relation can also be verified in the biliary epithelium, new therapeutic strategies for PSC will be developed.

The shortcoming of this study is the use of a tissue culture model, even with a cholangiocyte-derived tumor cell line. However, biliary epithelial cells are not obtainable in high yield, in reproducible fashion and sufficient functionality. Moreover, isolated cholangiocytes do not exhibit a mucus layer. It can only be pointed out that the employed Mz-ChA-1 cell line exhibits TJ complexes and is more differentiated in comparison to other cell lines obtained from cholangiocellular carcinomas. Another shortcoming is the lack of data on the structure of the TJ under the various experimental conditions. Moreover, imaging of transport employing fluorescent PC could not be examined in these experiments. All of these shortcomings can be tackled, when a genetic mouse model with TJ deletion selective for the biliary epithelium becomes available.

4. Materials and Methods

4.1. Cell Culture Transport Studies

Transwell tissue culture for basal–apical polarization was established for the biliary epithelial tumor cell line Mz-ChA-1 obtained from Dr. Alexander Knuth (Mainz, Germany) in our laboratory [10]. In each well of the 12-well collagen-coated transwell culture dish 7.5×10^4 cells (corresponding to 80 ± 18 μg protein) were placed and cultured in Dulbecco's modified Eagle's medium containing 5% fetal calf serum ((FCS) (Life Technologies, Carlsbad, CA, USA). They were grown for up to 21 days and polarity was examined by TER [1].

Translocation of radiolabeled substrates to the apical or basal side was examined by its application to one side and its recovery in the opposite compartment. For equilibrium distribution studies substrates were located in both compartments and transport to apical and basal side was determined. Standard incubation procedures used 10 mM PC (100,000 cpm) bound to 10 mM TC in phosphate buffered saline (PBS) (pH 7.4) at 37 °C in 1 mL placed at basal side and recovery was analyzed after 1 h in the upper compartment with 1 mL 10 mM TC–PBS. As substrates we used [^{3}H]phosphatidylcholine (PC), [^{14}C]lysophosphatidylcholine (LPC), [^{3}H]sphingomyelin (SM), [^{14}C]phosphatidylethanolamine (PE), [^{3}H]phosphatidylinositol (PI), [^{3}H]palmitate (PA), [^{3}H]oleate (OA), [^{3}H]taurocholate (TC) and [^{14}C]inulin (all brought up with unlabeled substrate to desired concentrations). Transport characteristics included dependency on time, concentration, temperature and pH and was evaluated as described [1]. For driving force analysis, the 10 mM TC in the apical medium was applied in different buffers at

130 mM and pH 7.0 which generated a more positive charge in the medium (NH_4Cl, Na-thiocyanate and 10 mM urea in PBS) or a more negative charge ($NaHCO_3$, Na-gluconate and 10 mM sodium dodecylsulphate (SDS)) in PBS [1].

Radiolabeled compounds were purchased from PerkinElmer (Waltham, MA, USA). For TJ disruption apical application of 150 μM ACA for 3 h [18] or the peroxisome proliferator-activated receptor gamma inhibitors T0070907 (10 μM) or GW9662 (1 μM) for 1 h were used [19].

siRNA knockdown experiments were employed to test the functionality of proteins involved in TJ constitution and proteins of significance for PC translocation. As in earlier studies with CaCo2 cells, we used in each case 78 pmol siRNA scrambled as control and targeted to claudin-1, -2, -4 and -8, Zonula Occludens-1 (ZO1), kindlin-1 and -2, CFTR, AE2 and mucin 1, mucin 2, and mucin 3 [1]. siRNAs were apically applied for 16 h at 37 °C. After washing transport studies were initiated.

4.2. siRNA Knockdown Experiments

All sense and antisense probes except for kindlin-1 and kindlin-2 were obtained from Sigma (St. Louis, MO, USA) and are depicted in Table 1. The probes for kindlin-1 and kindlin-2 were a kind gift of Reinhard Faessler, Max Planck Institute of Biochemistry, Munich, Germany. The efficiency of siRNA knockdown was confirmed by Western blotting. Transport characteristics included dependency on time, concentration, temperature and pH and was evaluated as described [1].

Table 1. siRNAs used in knockdown Experiments.

siRNA	Sense Oligo	Antisense Oligo
scrambled	5′-gaugggaccuggccaguga-3′[dT][dT]	5′-ucacuggccaggucccauc-3′[dT][dT]
claudin-1	5′-cagucaaugccagguacga-3′[dT][dT]	5′-ucguaccuggcauugacug-3′[dT][dT]
claudin-2	5′-gacacuaccacuggaucgu-3′[dT][dT]	5′-acgauccagugguaguguc-3′[dT][dT]
claudin-4	5′-gaccaucugggagggccua-3′[dT][dT]	5′-uaggcccucccagaugguc-3′[dT][dT]
claudin-8	5′gguucaagcaucuacucuu-3′[dT][dT]	5′-aagaguagaugcuugaacc-3′[dT][dT]
mucin 2	5′-gcaacauuaccgucugcaa-3′[dT][dT]	5′-uugcagacgguaauguugc-3′[dT][dT]
mucin 3	5′-ccaaacuacucuuacuaca-3′[dT][dT]	5′-uguaguaagaguaguuugg-3′[dT][dT]
CFTR	5′-gaacacauaccuucgauau-3′[dT][dT]	5′-auaucgaagguauguguuc-3′[dT][dT]
AE2	5′-gagaucuucgccuucuuga-3′[dT][dT]	5′-ucaagaaggcgaagaucuc-3′[dT][dT]
ZO1	5′-gagaugaacgggcuacgcu-3′[dT][dT]	5′-agcguagcccguucaucuc-3′[dT][dT]
kindlin-1	5′-ggacauuacugauaucccu-3′[dT][dT]	5′-agggauaucaguaauguccc-3′[dT][dT]
kindlin-2	5′-gugugaauagaaauacugu-3′[dT][dT]	5′-acaguauuucuauucacac-3′[dT][dT]

4.3. Immunoblot Analysis

Cell homogenate samples with 30 μg protein were applied to the gel slots for electrophoretic separation and immunoblotting using a standard protocol [1]. Primary antibodies against the following human proteins were used: claudin 2 (cat. no.: TA347352; 1:500) (Acris Antibodies, Herford, Germany); ZO1 (cat. no.: ab96587; 1:200; Abcam, Cambridge, MA, USA); kindlin-2 (1:500; gift from Reinhard Faessler, Max Planck Institute of Biochemistry, Munich, Germany); mucin 2 (cat. no.: sc-59859; 1:500; Santa Cruz Biotech., Santa Cruz, CA, USA) and CFTR (sc-376683; 1:200; Santa Cruz Biotech.

4.4. Statistical Analysis

Statistical analysis was performed using Prism 4.0 software (GraphPad Software Inc., LaJolla, CA, USA). Differences between groups were evaluated using the Mann–Whitney U test. Data are presented as means ± SD and $p < 0.05$ was considered statistically significant.

5. Conclusions

The manuscript shows for the first time that phosphatidylcholine can pass by paracellular transport from systemic sources to the apical side of biliary epithelial cells. It passes the TJ barrier driven by an apical negative electrical potential, generated by CFTR and AE2. Mucin 3 is the initial apical acceptor molecule from where PC is handled to secretory mucin 1 and 2 to establish the barrier against the biliary lumen. It protects from high bile acid concentration within the biliary lumen (Figure 11).

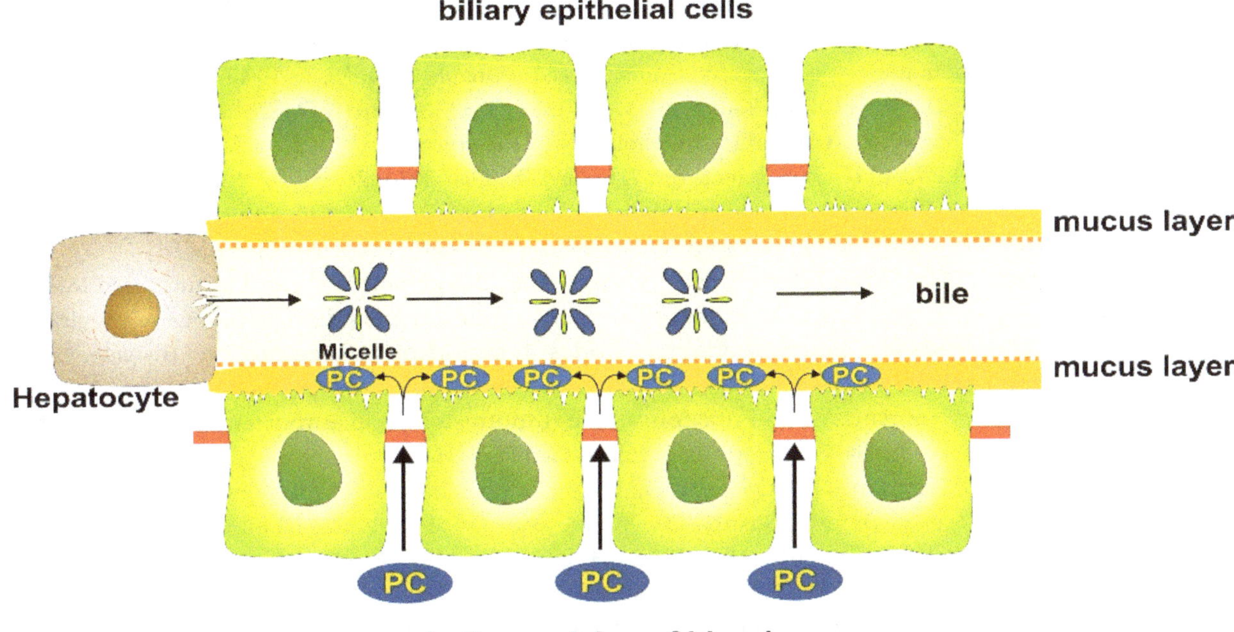

Figure 11. Scheme illustrating the paracellular transport of phosphatidylcholine (PC) across the tight junction barrier to the mucus layer at the apical side of biliary epithelium. The transport is driven by a negative electrical gradient with consequent binding to membrane-localized mucin 3 and an equilibrated shift to secretory mucin 2.

Author Contributions: Conceptualization, W.S.; formal analysis, W.S.; investigation, S.S.; validation, W.S.; visualization, W.S. and R.W.; writing—original draft preparation, W.S.; writing—review and editing, W.S. and R.W.

Abbreviations

ACA	acetaldehyde
AE2	anion exchange protein 2
ApoB	apolipoprotein B
CFTR	cystic fibrosis transmembrane conductance regulator
LPC	lysophosphatidylcholine
PBS	phosphate-buffered saline
PC	phosphatidylcholine
PSC	primary sclerosing cholangitis
SM	sphingomyelin
TER	transepithelial resistance
TC	taurocholate
TJ	tight junction(s)
UC	ulcerative colitis
ZO1	Zonula Occludens-1

References

1. Stremmel, W.; Staffer, S.; Gan-Schreier, H.; Wannhoff, A.; Bach, M.; Gauss, A. Phosphatidylcholine passes through lateral tight junctions for paracellular transport to the apical side of the polarized intestinal tumor cell-line CaCo2. *Biochim. Biophys. Acta* **2016**, *1861 Pt A*, 1161–1169. [CrossRef]

2. Stremmel, W.; Staffer, S.; Schneider, M.J.; Gan-Schreier, H.; Wannhoff, A.; Stuhrmann, N.; Gauss, A.; Wolburg, H.; Mahringer, A.; Swidsinski, A.; et al. Genetic mouse models with intestinal-specific tight junction deletion resemble an ulcerative colitis phenotype. *J. Crohns. Colitis* **2017**, *11*, 1247–1257. [CrossRef] [PubMed]
3. Ehehalt, R.; Wagenblast, J.; Erben, G.; Lehmann, W.D.; Hinz, U.; Merle, U.; Stremmel, W. Phosphatidylcholine and lysophosphatidylcholine in intestinal mucus of ulcerative colitis patients. A quantitative approach by nanoElectrospray-tandem mass spectrometry. *Scand. J. Gastroenterol.* **2004**, *39*, 737–742. [CrossRef] [PubMed]
4. Braun, A.; Schönfeld, U.; Welsch, T.; Kadmon, M.; Funke, B.; Gotthardt, D.; Zahn, A.; Autschbach, F.; Kienle, P.; Zharnikov, M.; et al. Reduced hydrophobicity of the colonic mucosal surface in ulcerative colitis as a hint at a physicochemical barrier defect. *Int. J. Colorectal. Dis.* **2011**, *26*, 989–998. [CrossRef] [PubMed]
5. Saich, R.; Chapman, R. Primary sclerosing cholangitis, autoimmune hepatitis and overlap syndromes in inflammatory bowel disease. *World J. Gastroenterol.* **2008**, *14*, 331–337. [CrossRef] [PubMed]
6. Smit, J.J.; Schinkel, A.H.; Oude Elferink, R.P.; Groen, A.K.; Wagenaar, E.; van Deemter, L.; Mol, C.A.; Ottenhoff, R.; van der Lugt, N.M.; van Roon, M.A.; et al. Homozygous disruption of the murine mdr2 P-glycoprotein gene leads to a complete absence of phospholipid from bile and to liver disease. *Cell* **1993**, *75*, 451–462. [CrossRef]
7. Oude Elferink, R.P.; Paulusma, C.C. Function and pathophysiological importance of ABCB4 (MDR3 P-glycoprotein). *Pflug. Arch.* **2007**, *453*, 601–610. [CrossRef] [PubMed]
8. Gauss, A.; Ehehalt, R.; Lehmann, W.D.; Erben, G.; Weiss, K.H.; Schaefer, Y.; Kloeters-Plachky, P.; Stiehl, A.; Stremmel, W.; Sauer, P.; et al. Biliary phosphatidylcholine and lysophosphatidylcholine profiles in sclerosing cholangitis. *World J. Gastroenterol.* **2013**, *19*, 5454–5463. [CrossRef] [PubMed]
9. Knuth, A.; Gabbert, H.; Dippold, W.; Klein, O.; Sachsse, W.; Bitter-Suermann, D.; Prellwitz, W.; Meyer zum Buschenfelde, K.H. Biliary adenocarcinoma. Characterisation of three new human tumor cell lines. *J. Hepatol.* **1985**, *1*, 579–596. [CrossRef]
10. Koike, N.; Todoroki, T.; Kawamoto, T.; Yoshida, S.; Kashiwagi, H.; Fukao, K.; Ohno, T.; Watanabe, T. The invasion potentials of human biliary tract carcinoma cell lines: Correlation between invasiveness and morphologic characteristics. *Int. J. Oncol.* **1998**, *13*, 1269–1274. [CrossRef] [PubMed]
11. Braconi, C.; Swenson, E.; Kogure, T.; Huang, N.; Patel, T. Targeting the IL-6 dependent phenotype can identify novel therapies for cholangiocarcinoma. *PLoS ONE* **2010**, *5*, e15195. [CrossRef] [PubMed]
12. Schrumpf, E.; Tan, C.; Karlsen, T.H.; Sponheim, J.; Björkström, N.K.; Sundnes, O.; Alfsnes, K.; Kaser, A.; Jefferson, D.M.; Ueno, Y.; et al. The biliary epithelium presents antigens to and activates natural killer T cells. *Hepatology* **2015**, *62*, 1249–1259. [CrossRef] [PubMed]
13. Onori, P.; Wise, C.; Gaudio, E.; Franchitto, A.; Francis, H.; Carpino, G.; Lee, V.; Lam, I.; Miller, T.; Dostal, D.E.; et al. Secretin inhibits cholangiocarcinoma growth via dysregulation of the cAMP-dependent signaling mechanisms of secretin receptor. *Int. J. Cancer* **2010**, *127*, 43–54. [CrossRef] [PubMed]
14. Zach, S.; Birgin, E.; Rückert, F. Primary cholangiocellular carcinoma cell lines. *J. Stem Cell Res. Transpl.* **2015**, *2*, 1013.
15. Tietz, P.S.; Marinelli, R.A.; Chen, X.M.; Huang, B.; Cohn, J.; Kole, J.; McNiven, M.A.; Alper, S.; LaRusso, N.F. Agonist-induced coordinated trafficking of functionally related transport proteins for water and ions in cholangiocytes. *J. Biol. Chem.* **2003**, *278*, 20413–20419. [CrossRef] [PubMed]
16. Perez-Vilar, J.; Hill, R.L. The structure and assembly of secreted mucins. *J. Biol. Chem.* **1999**, *274*, 31751–31754. [CrossRef] [PubMed]
17. Dawson, P.A.; Lan, T.; Rao, A. Bile acid transporters. *J. Lipid Res.* **2009**, *50*, 2340–2357. [CrossRef] [PubMed]
18. Dunagan, M.; Chaudhry, K.; Samak, G.; Rao, R.K. Acetaldehyde disrupts tight junctions in Caco-2 cell monolayers by a protein phosphatase 2A-dependent mechanism. *Am. J. Physiol. Gastrointest. Liver Physiol.* **2012**, *303*, G1356–G1364. [CrossRef] [PubMed]
19. Ogasawara, N.; Kojima, T.; Go, M.; Ohkuni, T.; Koizumi, J.; Kamekura, R.; Masaki, T.; Murata, M.; Tanaka, S.; Fuchimoto, J.; et al. PPARγ agonists upregulate the barrier function of tight junctions via a PKC pathway in human nasal epithelial cells. *Pharm. Res.* **2010**, *61*, 489–498. [CrossRef] [PubMed]

8

Drinking and Water Handling in the Medaka Intestine: A Possible Role of Claudin-15 in Paracellular Absorption?

Christian K. Tipsmark [1,*], Andreas M. Nielsen [2], Maryline C. Bossus [1,3], Laura V. Ellis [1], Christina Baun [4], Thomas L. Andersen [4], Jes Dreier [5], Jonathan R. Brewer [5] and Steffen S. Madsen [1,2]

1 Department of Biological Sciences, University of Arkansas, SCEN 601, Fayetteville, AR 72701, USA; maryline.bossus@lyon.edu (M.C.B.); lvellis@email.uark.edu (L.V.E.); steffen@biology.sdu.dk (S.S.M.)
2 Department of Biology, University of Southern Denmark, Campusvej 55, 5230 Odense M, Denmark; Amorck@live.dk
3 Department of Math and Sciences, Lyon College, 2300 Highland Rd, Batesville, AR 72501, USA
4 Department of Nuclear Medicine, Odense University Hospital, Sdr. Boulevard 29, 5000 Odense C, Denmark; Christina.Baun@rsyd.dk (C.B.); Thomas.Andersen@rsyd.dk (T.L.A.)
5 Department of Biochemistry and Molecular Biology, University of Southern Denmark, Campusvej 55, 5230 Odense M, Denmark; jes.dreier@cpr.ku.dk (J.D.); brewer@memphys.sdu.dk (J.R.B.)
* Correspondence: tipsmark@uark.edu.

Abstract: When euryhaline fish move between fresh water (FW) and seawater (SW), the intestine undergoes functional changes to handle imbibed SW. In Japanese medaka, the potential transcellular aquaporin-mediated conduits for water are paradoxically downregulated during SW acclimation, suggesting paracellular transport to be of principal importance in hyperosmotic conditions. In mammals, intestinal claudin-15 (CLDN15) forms paracellular channels for small cations and water, which may participate in water transport. Since two *cldn15* paralogs, *cldn15a* and *cldn15b*, have previously been identified in medaka, we examined the salinity effects on their mRNA expression and immunolocalization in the intestine. In addition, we analyzed the drinking rate and intestinal water handling by adding non-absorbable radiotracers, 51-Cr-EDTA or 99-Tc-DTPA, to the water. The drinking rate was >2-fold higher in SW than FW-acclimated fish, and radiotracer experiments showed anterior accumulation in FW and posterior buildup in SW intestines. Salinity had no effect on expression of *cldn15a*, while *cldn15b* was approximately 100-fold higher in FW than SW. Despite differences in transcript dynamics, Cldn15a and Cldn15b proteins were both similarly localized in the apical tight junctions of enterocytes, co-localizing with occludin and with no apparent difference in localization and abundance between FW and SW. The stability of the Cldn15 protein suggests a physiological role in water transport in the medaka intestine.

Keywords: aquaporin; claudin; drinking rate; epithelial fluid transport; enterocyte; occludin; osmoregulation; paracellular

---

1. Introduction

In fresh water (FW) fishes, the intestinal epithelium must limit excessive fluid absorption while securing dietary ion uptake [1]; in seawater (SW), imbibed water is absorbed in a solute-linked process [1,2]. Therefore, the functional plasticity of the enterocytic epithelium is a critical factor in euryhaline fish that are capable of going through salinity transitions. Elevated intestinal aquaporin (Aqp/*aqp*) abundance in eel [3–5] and salmonids [6] in response to SW transfer have led to propose a transcellular water path in these species [7];

however, in medaka, a consistent downregulation of several intestinal Aqp/*aqp* isoforms after SW transfer has challenged this model and suggests a major involvement of a paracellular pathway [8].

Transepithelial water transport has been suggested to be mainly transcellular via Aqps, but this matter is still under debate [9,10]. Thus, in leaky epithelia, similar to the intestine, fluid transport may primarily be paracellular as proposed based on a corneal model [9] or include both components as proposed for marine fish [2]. Taking species differences into account, it appears that the medaka intestine may be a choice comparative model to study paracellular fluid transport because a tight junction defined path seems central, as suggested by Madsen et al. [8].

Proteins belonging to the claudin (Cldn) superfamily are the main determinants of tight junction permeability properties and thus important regulators of paracellular transport [11,12]. Cldns are integral membrane proteins with 4 trans-membrane domains and two extracellular loops (ECL). The amino acid residues of the first ECL are critical for the permselectivity of the junction they create in homo- or hetero-dimeric and -tetrameric combinations [13–16]. There are 27 claudins (CLDNs) paralogs described in mammals [12]; in the teleost lineage, an extensive expansion of the *cldn* gene family due to gene duplications has led to a higher number with e.g., 56 in Fugu [17] and 54 in zebrafish [18]. The specific permselectivity has been investigated for several mammalian CLDN paralogs, and there are many examples of barrier-forming as well as specific anion- and cation-pore-forming CLDNs [12]. In addition, there are a few examples of CLDNs contributing to creating water-permeable pores. This has been convincingly demonstrated for CLDN2 [19], which has functional significance in the mammalian kidney, and most recently intestinal CLDN15 has also been assigned such a role [20] in addition to the cation-pore-forming properties of both CLDNs [12]. However, Na^+ and water fluxes through CLDN15 inhibit each other in functional contrast to CLDN2 [20]. Based on amino acid homology, especially in the first ECL, it is often assumed that fish Cldns give rise to the same permeability properties as mammalian orthologues, but only a few have been investigated thoroughly [21,22]. Mutational analysis and MD simulations [15,16] based on the crystal structure [23] have shown that especially amino acid D55 is critical for CLDN15 pore formation. In support for a similar function in medaka to the mammalian orthologue is the conservation of this amino acid in both medaka Cldn15a and Cldn15b.

A given tissue often shows the expression of several CLDN paralogs [12]. In the mammalian nephron, this is coupled to a highly segmental pattern of expression [24]. In the mammalian intestine, CLDN15 appears to be one of the most abundant CLDNs, at least in the small intestine [25–27], and it plays a critical role in the gut ontogeny of both mammals and fish [28,29]. In mice, CLDN15-mediated Na^+ back-flux into the intestinal lumen is essential for active glucose absorption through the Na^+/glucose cotransporter, safeguarding monosaccharide uptake [30]. In fishes, Cldn15a paralogs have been found to be expressed specifically in the gastrointestinal (GI) tract (salmon [31,32], zebrafish [33], medaka [34]). In medaka, we previously identified an additional new paralog, Cldn15b, which is also primarily expressed in the intestine at levels several orders of magnitude higher than any other examined organs [34].

To develop our knowledge about paracellular versus transcellular fluid transport, it will be valuable to expand our understanding of water transport and enterocyte tight junctions in medaka. Furthermore, the functional plasticity of the intestine during salinity change in this euryhaline fish is useful when seeking to understand basic principles. Therefore, the goals of this work were to first study drinking behavior and water handling in response to changes in the osmotic environment, which are unknown in adult medaka. Secondly, we examined the expression and localization of the two Cldn15 paralogs in relation to hypo- to hyperosmotic acclimation based on the assumption that intestinal Cldn15 is implicated in water transport, as seen in other models.

2. Results

2.1. Drinking Rate and Intestinal Handling of Imbibed Water

In preliminary experiments using both FW- and SW-acclimated fish, it was assured that the intestinal accumulation of radioactivity in fish continued linearly in excess of 3 h, and gut-passage time after drinking thus was well in excess of 3 h (data not shown). Therefore, the drinking rate estimation was based on a 3 h incubation in 51-Cr-EDTA containing water. The drinking rate was relatively high in FW-acclimated medaka (5 μL/g/h) but was doubled in fish acclimated to SW (Figure 1). Drinking rate measurements were based on counting radioactivity in the GI tract after incubation and a 1 h rinsing period in clean water.

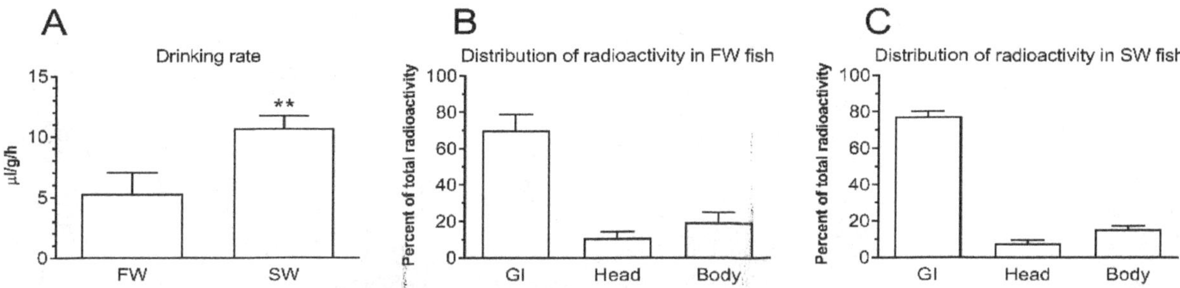

Figure 1. (**A**) Drinking rate (μL/g/h) in fresh water (FW) and seawater (SW)-acclimated medaka estimated by radioactivity in the entire gastrointestinal tract after incubation in 51-Cr-EDTA traced FW or SW for 3 h followed by rinsing in clean water for 1 h. ** $p < 0.01$. In (**B**) (FW) and (**C**) (SW), the radioactivity content of the gastrointestinal tract (GI, imbibed) is compared to radioactivity absorbed to the head and remaining body parts.

The head and body carcass where counted separately after the experiment, and the radioactivity content in these parts amounted to 8–10% and 15–20%, respectively, of the total radioactivity of the fish (Figure 1B,C). It is assumed that this is mainly due to attachment to the external mucus layer in these body parts, as Cr-EDTA has been shown to be a non-absorbable marker, which means that it does not cross the intestinal epithelium [35].

When dissected intestines were carefully fragmented into 0.5 cm segments and analyzed, it was found that radioactivity was evenly distributed but with a trend of showing higher levels at the anterior end of the FW intestine (Figure 2A). In the SW-acclimated fish, the radioactivity clearly accumulated toward the posterior end of the intestine (Figure 2B).

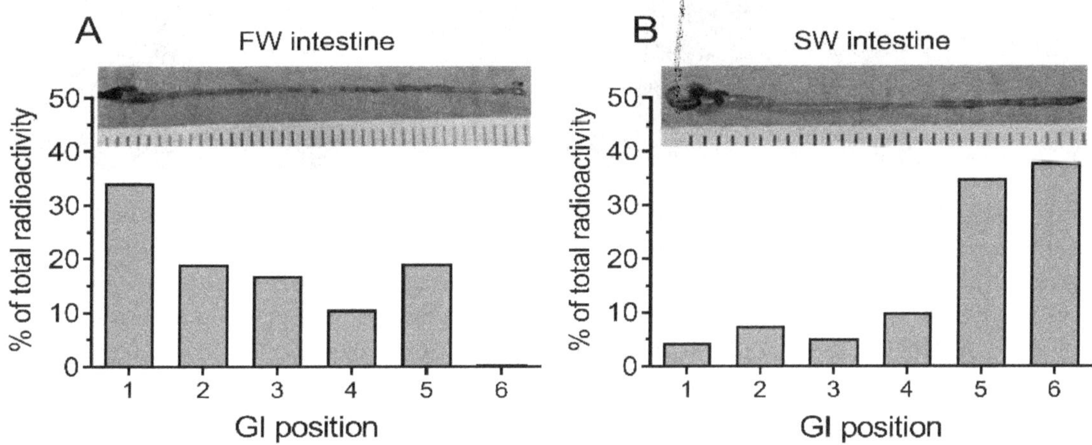

Figure 2. Distribution of radioactivity longitudinally in the gastrointestinal (GI) tract of fish allowed to drink 51-Cr-EDTA traced FW (**A**) or SW (**B**) water for 3 h. After incubation, the entire GI tract was ligatured into 0.5 cm segments and each segment was then transferred to a scintillation vial and scanned for radioactive content. The entire intestine was approximately 3 cm in length, as shown in the inserted photographs.

The progressive movement of imbibed water along the intestine was followed in a more direct way by a series of single photon emission computed tomography–computed tomography (SPECT-CT) scans of intact, euthanized fish after incubation in 99-Tc-DTPA-traced water (Figure 3). The imaging series showed an initial high intensity of tracer in the esophageal end of the GI tract with a more posterior distribution as the incubation time was increased. The precipitation of mineral salts (Mg- and Ca-carbonates) could be seen in the CT images of SW-acclimated fish ca 2/3 down the intestine (white arrow in Figure 3D).

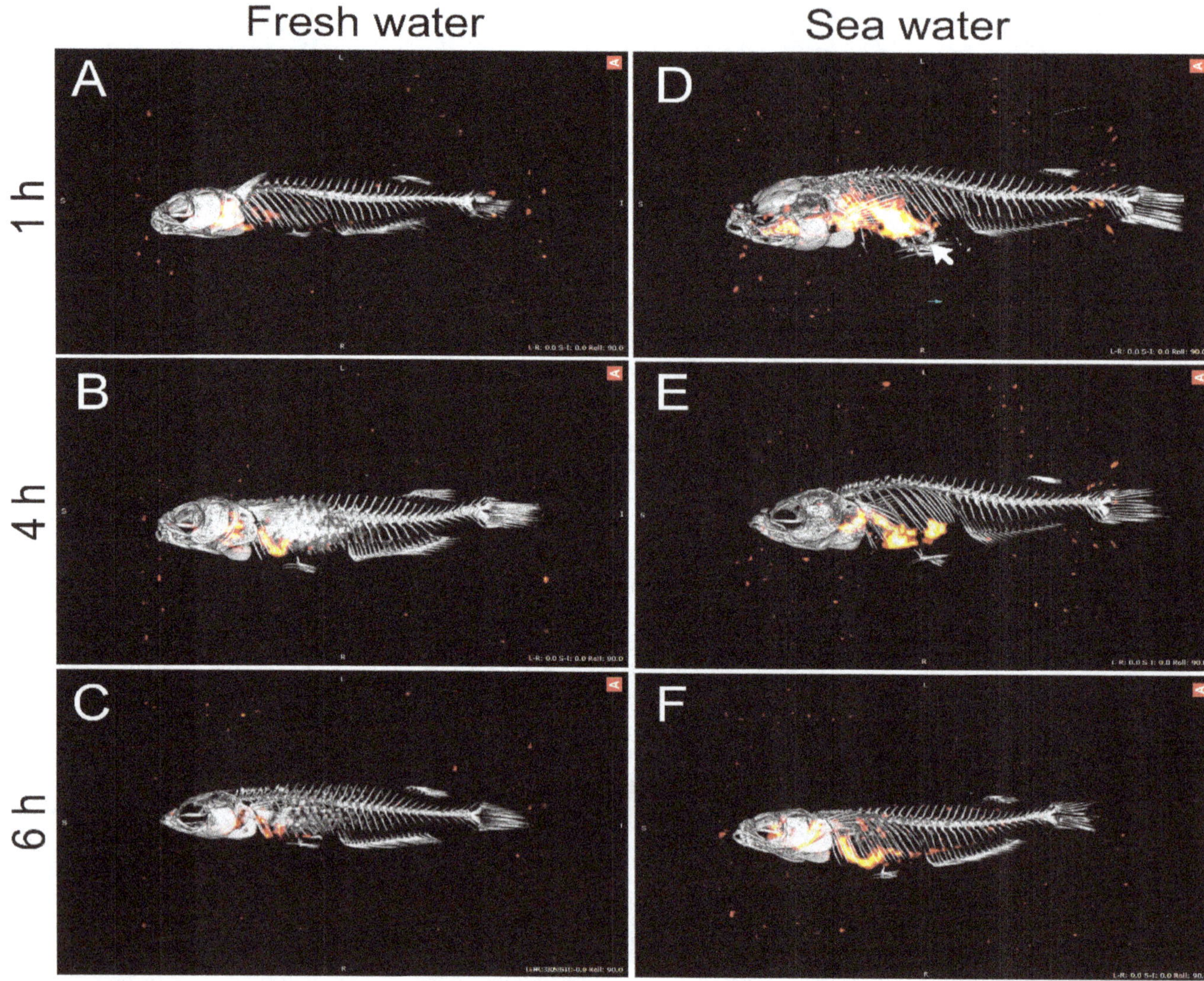

Figure 3. Visualization of the movement of imbibed water along the gastrointestinal tract of medaka acclimated to FW (**A**–**C**) or SW (**D**–**F**). Images are merged from single photon emission computed tomography (SPECT) (red intensity layer) and computed tomography (CT) (gray) scans of fish which had been incubated in 99-Tc-DTPA traced FW or SW for 1 h followed by transfer to non-radioactive FW or SW for 1 h (**A**, **D**), 4 h (**B**, **E**), or 6 h (**C**, **F**). The SPECT layer visualizes the localization (and intensity) of imbibed isotope-labeled water and the CT layer visualizes mineralized structures (skeleton and mineral precipitates in the SW intestines, white arrow in **D**). Note that 99-Tc has a short half-life (6 h), which influences the apparent intensities of the imbibed isotope. The fish was approximately 3 cm in length.

2.2. *Transcript Levels and Response to Salinity*

The transcript levels of selected targets were analyzed in intestines from medaka long-term acclimated to FW and SW (Figure 4). The absorptive Na^+, K^+,$2Cl^-$ cotransporter (*nkcc2*) level was several-fold higher in SW than FW fish, whereas *cldn15b*, *aqp1a*, and *aqp8ab* levels were significantly reduced in SW compared to FW fish. *cldn15a* was unaffected by long-term salinity acclimation. The salinity-induced changes in transcript levels observed in long-term acclimated fish were reproduced in a 7-day time course experiment (Figure 5), with *nkcc2, cldn15b, aqp1a,* and *aqp8ab* all being significantly affected by both salinity and time (two-way ANOVA). Since there was a significant interaction between the two factors on these transcripts, the effect of SW was time-dependent. Thus, the SW effect on *nkcc2* was significant at all time-points, but it was the highest after 168 h. The effect on cldn15b was significant only after 24 h and 168 h days, while both *aqps* decreased already after 6 h and 24 h but not significantly so at the 168 h time point. *cldn15a* was unaffected by salinity during the 7-day time course experiment, as observed in long-term acclimated fish.

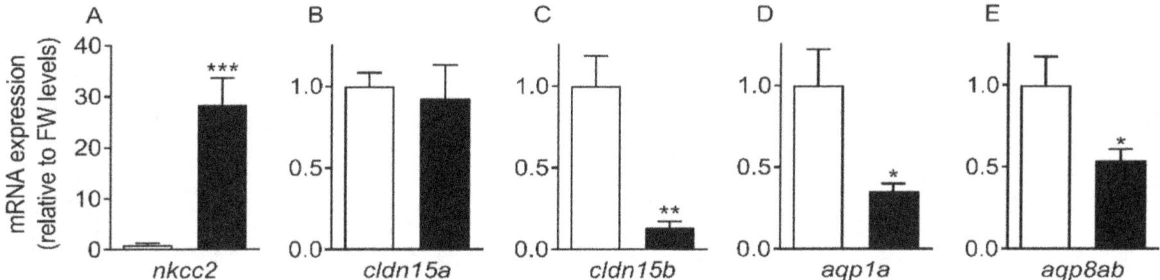

Figure 4. Transcript levels of *nkcc2* (**A**), *cldn15a* (**B**), *cldn15b* (**C**), *aqp1a* (**D**), and *aqp8ab* (**E**) in intestine from medaka acclimated to FW (white bars) or SW (black bars). Fish were acclimated to the respective salinities for over one month prior to sampling ($n = 8$). Expression levels represent the mean value ± SEM relative to FW levels. Asterisks indicate a significant difference from FW expression (* $p < 0.05$, ** $p < 0.01$, *** $p < 0.001$).

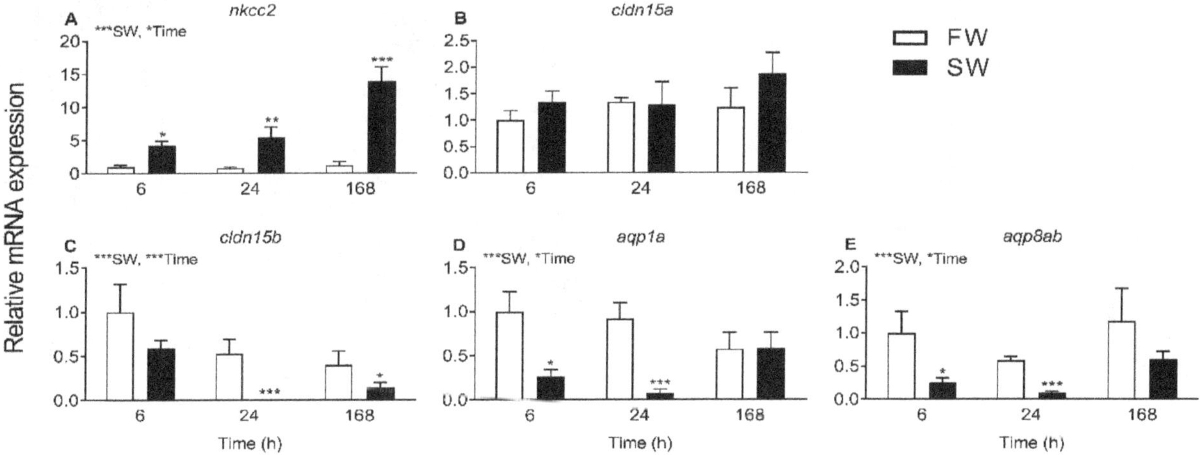

Figure 5. Effect of FW-to-SW transfer on intestinal transcript levels of *nkcc2* (**A**), *cldn15a* (**B**), *cldn15b* (**C**), *aqp1a* (**D**), and *aqp8ab* (**E**). Fish were transferred from FW-to-SW (black bars) or FW-to-FW (white bars) as a control and sampled at 6, 24, and 168 h ($n = 6$). Expression levels represent the mean value ± SEM relative to the 6 h-FW group. Asterisks next to SW and Time refers to the overall effects of a factor with two-way ANOVA. All targets with overall effects also had a significant interaction between factors, so the differences between time-matched groups were analyzed with Bonferroni multiple comparisons test (* $p < 0.05$, ** $p < 0.01$, *** $p < 0.001$) to identify the time-dependence of SW effects.

2.3. Cldn15 Localization in the Intestinal Epithelium

Cldn15a and Cldn15b showed similar localization in the intestinal epithelium. Immunoreactivity was confined to the apical area of enterocytes with distinct "hot spots" in the apical junction area between enterocytes (Figures 6 and 7). At lower magnification, these hot spots were partly masked by the non-specific staining of the brush-border area at variable intensity (Figure 6A,C).

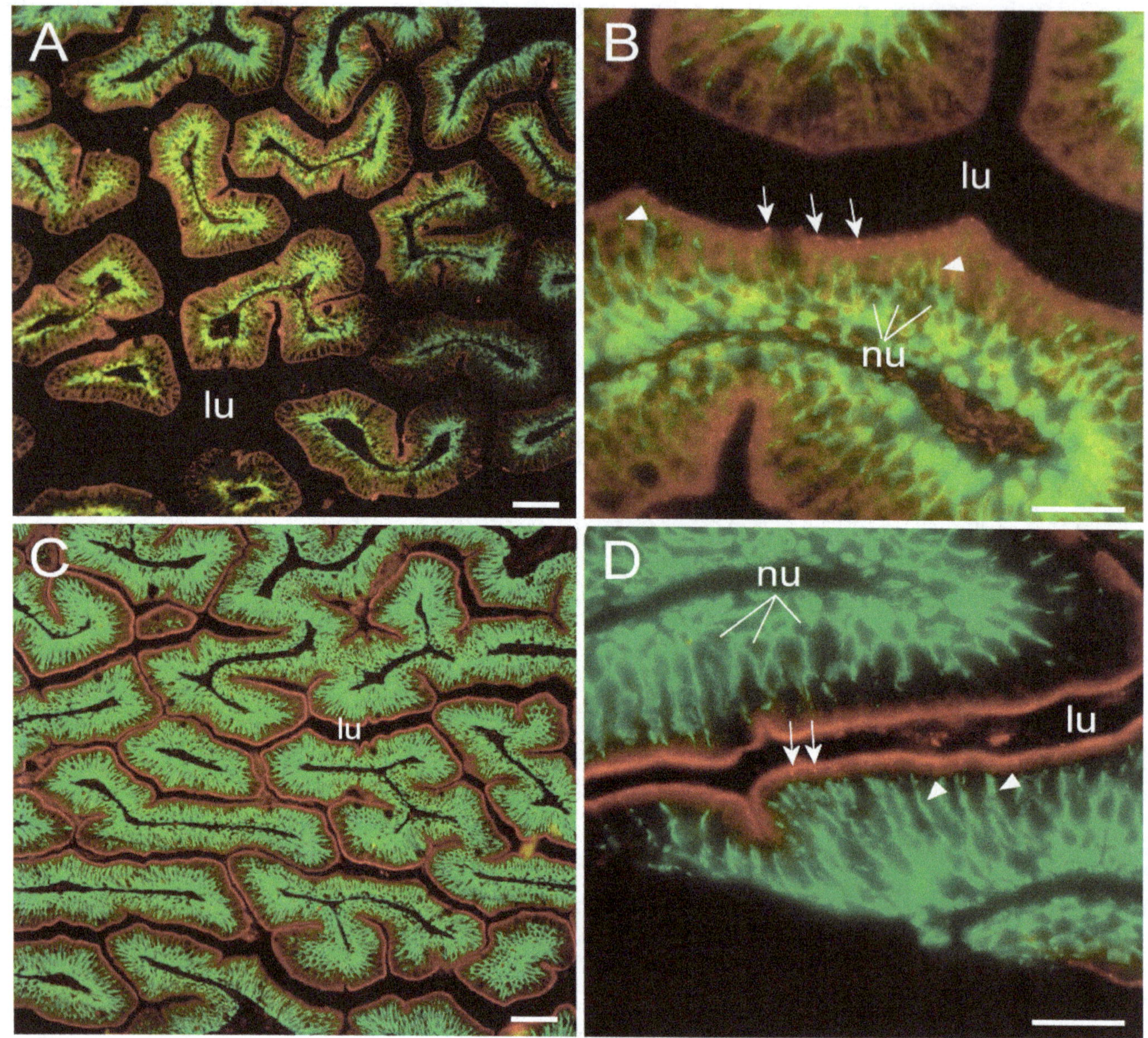

Figure 6. Immunofluorescence micrographs showing apical localization of Cldn15a (red, in **A** and **B**), and Cldn15b (red in **C** and **D**) and basolateral localization of the Na^+,K^+-ATPase alpha subunit (green) in FW-acclimated medaka middle intestine. (**A**) and (**C**) are at 200× magnification, (**B**) and (**D**) are at 1000× magnification. lu = lumen, nu = nuclei; in (**B**) and (**C**) arrows point to Cldn "hot spots" in the tight junction zones; arrowheads point to lateral membranes. Size bars indicate 50 μm (**A**, **C**) or 20 μm (**B**, **D**).

However, these "hot spots" became particularly evident at higher magnification (Figure 6B,D, Figure 7) and when inspecting the tissues with confocal and STED microscopy, which has a much narrower z-plane focus (Figures 8 and 9). Control incubation without primary Cldn15 antibodies showed a very faint general fluorescence without the distinct "hot spots" (insert in Figure 7A). Na^+,K^+-ATPase alpha subunit immunostaining revealed parallel lateral membranes, which were slightly spaced between neighboring cells, thus creating the lateral intercellular space (e.g., see Figure 6D, Figure 8A,B, marked with arrowheads). Near the basal borders, the membrane staining surrounded the nuclei, which appeared as circular dark "holes" in the images (marked "nu" in Figures 6–8). The distinct Cldn staining was apical to the Na^+,K^+-ATPase staining, i.e., at the end of an axis extrapolated from the lateral membrane area. Thus, there was no co-localization of the two antibodies. This indicates that the two Cldn15 proteins are located in the tight junction zone.

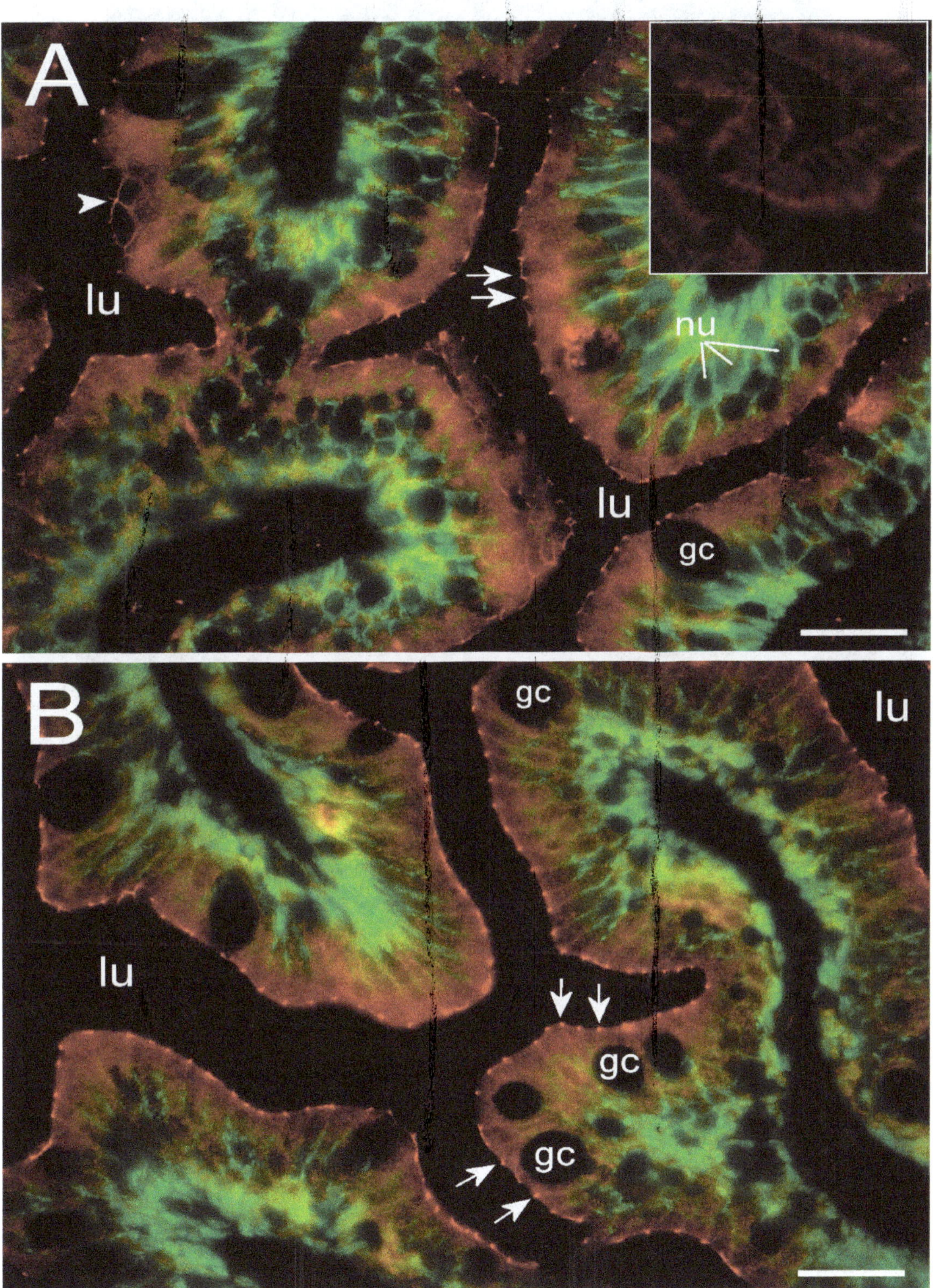

Figure 7. Immunofluorescence micrographs showing apical localization of Cldn15a (red in **A**) and Cldn15b (red in **B**) and basolateral localization of the Na^{+},K^{+}-ATPase alpha subunit (green) in SW-acclimated medaka middle intestine. The insert in the upper left corner shows control without primary antibodies. Images are at 1000× magnification. lu = lumen, gc = goblet cell, nu = nuclei; arrows point to "hot spots" in the tight junction zones; Size bars indicate 20 μm.

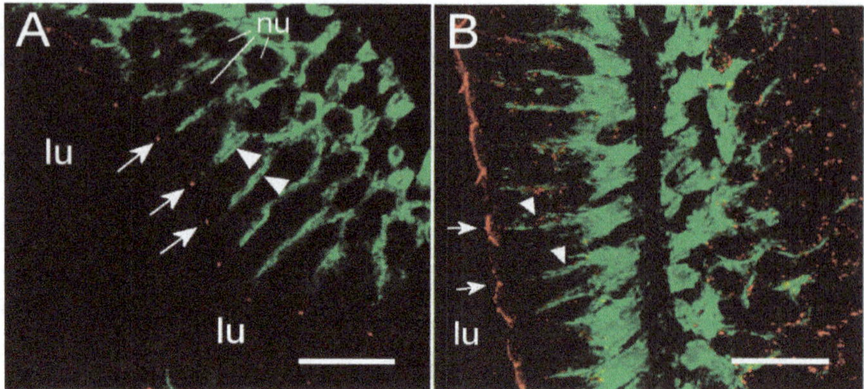

Figure 8. Confocal images showing apical localization of Cldn15a (red in **A**) and Cldn15b (red in **B**) and basolateral localization of Na^+,K^+-ATPase alpha subunit (green) in SW-acclimated medaka middle intestine. lu = lumen; arrows point to Cldn15 "hot spots" in the tight junction zones; arrowheads point to lateral membranes of enterocytes clearly separating the intercellular space. Size bars indicate 10 μm.

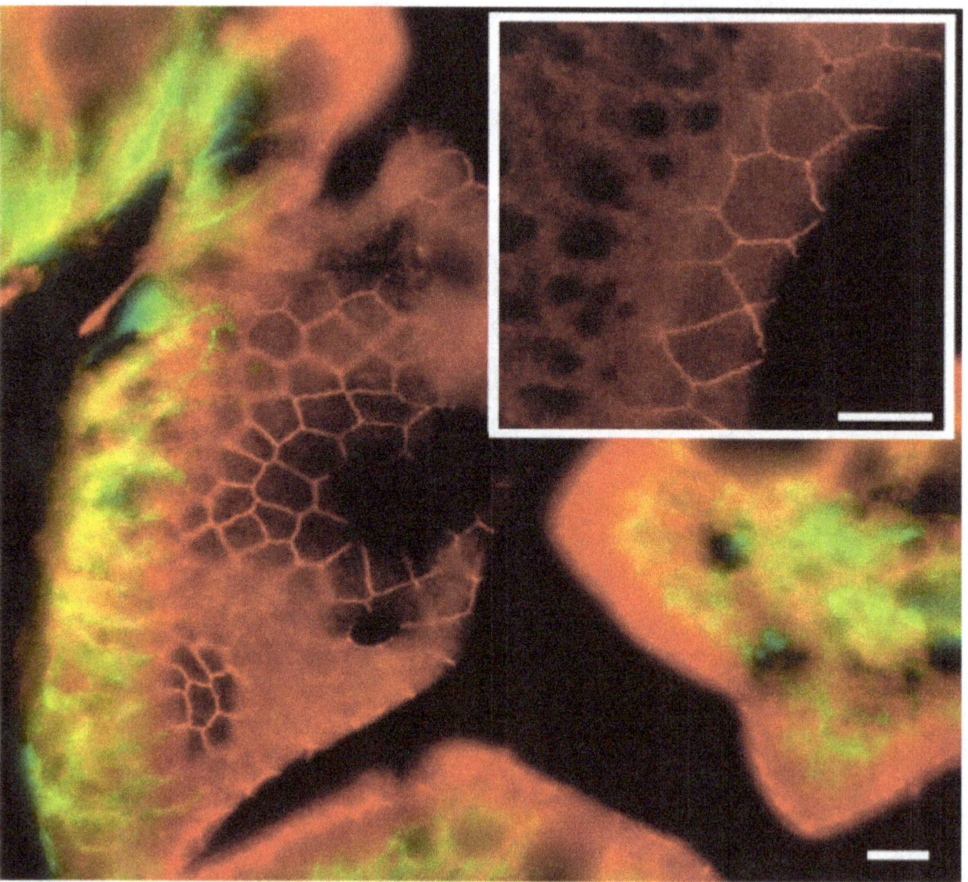

Figure 9. Confocal STED images showing the apical localization of Cldn15b (red) in SW-acclimated medaka middle intestine. Large image shows double staining with the anti-Na^+,K^+-ATPase alpha subunit (green). Na^+,K^+-ATPase is localized in basolateral membranes as shown in Figures 6–8, and it is absent in the apical area, where the tight junctions are located. Thus, the mosaic-like pattern of Cldn15b is without green overlay. The subfigure shows a subsection of the apical area focusing on the tight junction area. Size bars indicate 5 μm.

Occasionally, the section plane was slightly tilted and therefore made it possible to obtain a zoomed view of the apical junction area just below the brush border zone (Figure 9). In these cases, confocal STED microscopy showed a beautiful polygonal staining revealing the three-dimensional junction zone surrounding the individual enterocytes. These polygons varied from simple tetragons to heptagons in shape, indicating enterocytes surrounded by four to seven neighboring cells.

We also performed a double labeling with Cldn15b and occludin antibodies (Figure 10). This revealed a complete co-localization of the two proteins, thus validating that Cldn15 is indeed localized in the tight junction between enterocytes.

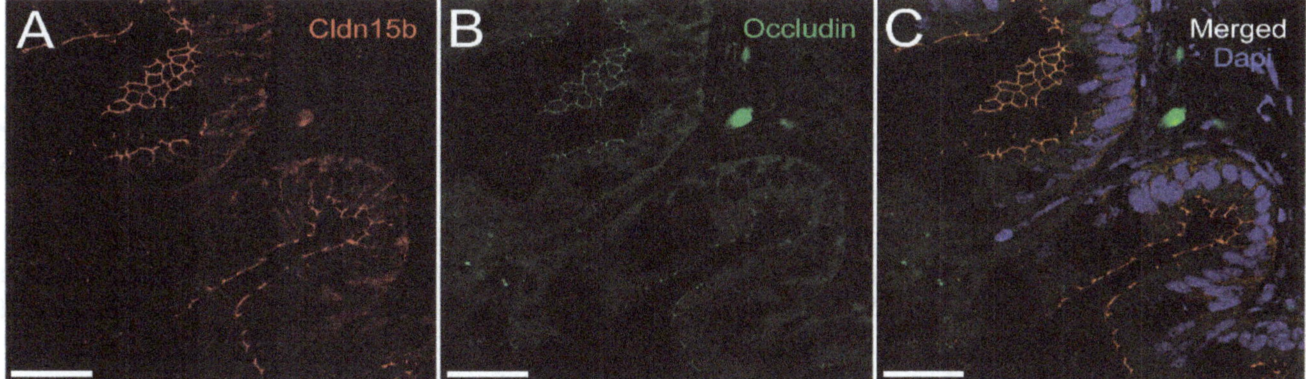

Figure 10. Confocal images showing Cldn15b (**A**, red), occludin (**B**, green) and co-localization (**C**, merged) in SW-acclimated medaka middle intestine. In (**C**), nuclei are stained blue with DAPI. Size bars indicate 20 μm.

3. Discussion

Japanese medaka can move between FW and SW while maintaining osmotic homeostasis. Based on our knowledge from several other teleosts, this requires high functional plasticity in e.g., the intestine, which in FW contributes to maintain ion balance and in SW switches to fluid absorption to compensate for dehydration [36]. Fluid absorption in fishes is driven by solute transport and is generally assumed to occur mainly through a transcellular pathway [2,7]. Accordingly, intestinal *aqp* expression is elevated during hyperosmotic exposure in order to develop the transcellular pathway [37,38]. This paradigm was challenged in previous studies, where we and others showed that in medaka *spp.*, unlike in other species studied, intestinal *aqp*/Aqp expression is downregulated at both transcript and protein levels when fish are exposed to hyperosmotic conditions [8,39]. This paradox suggests that the paracellular pathway may be of higher importance, at least in the medaka. The recent report that the mammalian tight junction CLDN15 may create intestinal water channels [20] led us to investigate the role of the medaka orthologues in relation to fluid absorption. With limited knowledge about medaka drinking behavior and intestinal water handling, we set out by examining salinity effects on drinking behavior and water handling and then addressing the specific expression of *cldn15*. If involved in paracellular fluid absorption, our expectation was that *cldn15* expression would increase after SW exposure.

3.1. Drinking Rate and Intestinal Handling of Imbibed Water

After transition to SW, drinking rates and fluid absorption in the intestine increase in most examined fish species [36]. In order to understand intestinal function using the adult medaka model, we had to describe its drinking behavior and water handling, which was until now unknown. We demonstrated that the drinking rate was 5 μL/g/h and 10 μL/g/h in FW and SW medaka, respectively (Figure 1); thus, SW-transfer doubled oral water intake, which was presumably due to the need to compensate for osmotic water loss in the concentrated environment. This is similar to what has been observed in other euryhaline fish [35,40,41] including Japanese medaka larvae [42], and rates are comparable to other studies albeit on the high side [43]. Drinking rate is inversely related to body mass [43] and probably related to surface-to-volume ratio aspects, and most fishes examined up to now were larger fish (5–800 g). The medaka used in these experiments are small (0.4–0.6 g), and the smaller SW fish examined so far have comparable drinking rates (*Pholis gunnelus*, 2–10 g: 12 μL/g/h; *Aphanius dispar*, 0.4–1 g: 10 μL/g/h; see [43]). It is often assumed that FW fish should keep oral water intake to a minimum in order to not put excessive strain on the kidney in a hypotonic environment [36]. This is certainly the case in some FW teleosts (e.g., 0.4 μL/g/h in 0.1–2.5 g *Platichthys flesus* [44]). However, there are also reports of significant drinking in FW teleosts in the μL/g/h-range [41,45,46], and this also seems to be the case in FW medaka. We do not have any physiological explanation as to why FW drinking rates were so relatively high compared to most other reported studies. Feeding events could possibly be accompanied by the swallowing of small amounts of water, but normally, including the present study, fish are unfed during drinking rate measurements. Stress is another factor that may affect drinking and water turnover, but the fish were left undisturbed during the whole experiment, so we must assume that it is negligible. It has been speculated that drinking in FW may be a source of divalent ions such as Ca^{2+} [45], but the significance of this was rejected by Lin et al. [46] based on quantitative analyses in tilapia.

When analyzing the segmental distribution of 51-Cr-EDTA-traced water in the intestine (Figure 2), our data showed that in FW fish, 51-Cr-EDTA was for the most part located anteriorly in contrast with a more posterior accumulation in SW fish. This is in perfect agreement with Kaneko and Hasegawa's [42] observations in medaka larvae, using a laser scanning technique to visualize intestinal water handling. This suggests that imbibed water in FW fish is taken up by osmosis in the anterior intestine were the tracers accumulate, and the volume regulatory problem associated must then be corrected by the kidney. While it is difficult to get reliable measurements of luminal fluid in small medaka, a study on FW tilapia showed that anterior, middle, and posterior luminal osmolality is close to plasma levels [47]. This corroborates an equilibration of the luminal fluid (FW) with plasma in the anterior intestine. In SW, after initial desalination in the esophagus, the luminal fluid osmolality in tilapia is similar all along the intestine and is higher than plasma osmolality [47]. Therefore, water uptake must rely on solute-driven transport into the lateral intercellular space in the anterior parts of the intestine [6] in combination with $CaCO_3$ precipitation in the posterior end to increase the concentration of free water molecules [2,36]. Therefore, continuous water flow along the intestine means that 51-Cr-EDTA tends to accumulate in the posterior section in SW fish. The accumulation of the non-absorbable tracer, 99-Tc-DTPA, at the posterior end indirectly supports the progressive absorption of water. This was further supported by the SPECT/CT imaging, in which the progress of water movement in the intestine was visualized directly in live fish (Figure 3). In SW fish, CT scans further revealed mineral precipitation in the posterior intestine, suggesting bicarbonate secretion, which induces Mg- and Ca-carbonate precipitation that helps drive osmotic water transport across the intestinal epithelium [2].

3.2. Transcript Levels and Response to Salinity

The progress of SW acclimation was followed by transcript analyses of a few selected targets representing intestinal NaCl uptake (*nkcc2*), which is needed to establish solute-driven water absorption and a possible transcellular water uptake pathway (*aqp1a, aqp8ab*) (Figures 4 and 5). As expected, there was a steep increase in *nkcc2* in SW, suggesting increased NaCl transport across the apical enterocytic brush border membrane. The data also confirmed the paradoxical drop in *aqp1a* and *aqp8ab* expression found in previous studies [*O. latipes*: 8; *O. dancena*: 39]. Thus, based on *aqp* dynamics, transcellular water transport is not supported in SW medaka; and it remains puzzling as to why *aqp* expression is kept higher in the FW condition. The time-course experiment showed that *cldn15b* was not significant affected by SW before 24 h and 168 h while the inhibitory effect on the two *aqps* was apparent at the 6 h and 24 h mark but not significant at the 168 h time point. Thus, while the dynamics of regulation are not straightforward, the overall inhibitory effect of SW on *cldn15b*, *aqp1a*, and *aqp8ab* observed previously was confirmed [8,34].

Based on similarities to the mammalian CLDN15 shown in Figure 11, we hypothesized that the two medaka orthologs, *cldn15a* and *cldn15b*, may share functional properties in terms of forming cation and water pores and therefore may contribute to paracellular water absorption in SW medaka. In fishes (medaka [34]; salmon [31,32]; zebrafish [33]) and mammals [12], CLDN15 orthologs are expressed especially, but not exclusively, in the GI tract. In Atlantic salmon, SW acclimation was shown to induce elevated intestinal *cldn15a* mRNA expression [32], and a different study in the same species documented higher transepithelial resistance in SW than FW intestine measured *ex vivo* [48]. Taken together, this seems counterintuitive if teleost Cldn15 paralogs such as the mammalian ortholog form cation selective pores, and thereby theoretically should *decrease* epithelial resistance rather than increase it. We did not find any effect of salinity on *cldn15a* mRNA levels in medaka. However, Na^+ and water fluxes through human CLDN15 was recently shown to inhibit each other [20], and it is possible that physiological significance depends on the local chemical conditions, which may be very different in FW and SW intestines. We found a roughly 100-fold decrease of the *cldn15b* paralog when fish are acclimated to SW, which based on an expected possible role in creating a water pore is somewhat surprising. The high FW expression level of this paralog suggests a specific role in the FW intestine, which may be related to Na^+/glucose cotransport or K^+ uptake from the diet. The interpretation of Cldn data is not straightforward, because Cldn15 may interact with other proteins and Cldn paralogs when co-expressed in enterocytes [25], and the properties and physiological significance may change depending on salinity and intestinal location. Therefore, the expression of other intestinal Cldns should be investigated in future studies.

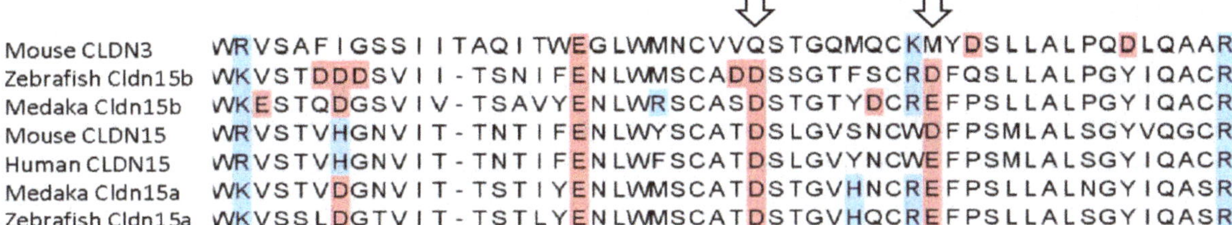

Mouse CLDN3 WRVSAFIGSSIITAQITWEGLWMNCVVQSTGQMQCKMYDSLLALPQDLQAAR
Zebrafish Cldn15b WKVSTDDDSVII-TSNIFENLWMSCADDSSGTFSCRDFQSLLALPGYIQACR
Medaka Cldn15b WKESTQDGSVIV-TSAVYENLWRSCASDSTGTYDCREFPSLLALPGYIQACR
Mouse CLDN15 WRVSTVHGNVIT-TNTIFENLWYSCATDSLGVSNCWDFPSMLALSGYVQGCR
Human CLDN15 WRVSTVHGNVIT-TNTIFENLWFSCATDSLGVYNCWEFPSMLALSGYIQACR
Medaka Cldn15a WKVSTVDGNVIT-TSTIYENLWMSCATDSTGVHNCREFPSLLALNGYIQASR
Zebrafish Cldn15a WKVSSLDGTVIT-TSTLYENLWMSCATDSTGVHQCREFPSLLALSGYIQASR

Figure 11. Alignment of the first extracellular loop of CLDN15 from human and mouse, and the orthologues from Japanese medaka and zebrafish shows that the residues critical to pore formation (D55 and D64) are found in both teleost Cldn15 paralogues. There are also differences from mammalian CLDN15; for example, both medaka Cldn15a and Cldn15b have an R63 residue and Cldn15a has an added H60. Amino acids are highlighted in red when acidic and in blue when basic. Arrows marks aspartic acids (D55 and D64), which are found to be important for cation and the water pore function of CLDN15 [11,12,15,16]. Mouse CLDN3 has been classified as a barrier protein and included as a reference. Sequences used: Human CLDN15: Acc. No. NP_001172009; Mouse CLDN15: Acc. No. NP_068365; Mouse CLDN3: NP_034032; Medaka Cldn15a: XP_004079873; Medaka Cldn15b: XP_004076514; Zebrafish Cldn15a: NP_956698; Zebrafish Cldn15b: NP_001035404.

3.3. *Cldn15 Localization in the Intestinal Epithelium*

To our knowledge, this is the first study showing enterocyte tight junction Cldn localization in a teleost fish. By using high-resolution fluorescence microscopy (Figures 6–9), we were able to demonstrate that Cldn15a and Cldn15b showed similar localization in the intestinal epithelium regardless of salinity. The use of Na^+,K^+-ATPase immunostaining to visualize basolateral membranes showed that parallel lateral membranes were slightly spaced between neighboring cells, thus creating the lateral intercellular space possibly involved in solvent drag [7]. Discrete Cldn15 immunoreaction was seen apically to Na^+,K^+-ATPase immunoreaction, demarcating apical and basolateral membranes with no co-localization of the two antibodies. The antibodies gave some apparent non-specific staining of the brush border zone, which had variable intensity between sections. It is possible that this is created by non-specific adsorption to the mucus layer in this area. Nonetheless, the specific immunoreaction of both Cldn15 antibodies was restricted to a very narrow apical-most zone, which was below the brush border and in direct extension from membranes bordering the lateral intercellular space. In cellular cross-sections at high resolution, this appeared as an apical dot-like staining, and when viewed from above in a frontal section, the pattern appeared as a circumcellular polygonal pattern, which is characteristic of epithelial tight junctions. This became particularly evident at higher magnification when using high-resolution STED microscopy. The co-localization of Cldn15b with the tight junction marker occludin confirmed a role in control of the paracellular intestinal barrier. The localization is identical to that of CLDN15 throughout the mouse intestine [49] and that of occludin in the goldfish intestine [50]. Despite their classification as tight-junction proteins, several other intestinal CLDNs (e.g., CLDN-1, -3, -4, -5, and -7) are localized further away from the apical zone in lateral and basolateral membranes in mammals (see [25]). Based on the mRNA analyses, we expected to see a significant downregulation of Cldn15b after SW-exposure, but we did not find any significant change in the localization and immunoreactivity of neither Cldn15 paralog. There was a trend that the "hot spots" of Cldn staining appeared more intense in SW specimens, but it was not possible to quantify this (compare Figure 6B,D with Figure 7A,B). Unfortunately, the antibodies did not function for Western blots, and further quantification efforts are not possible at present. Thus, we conclude that the Cldn15-based apical tight junction component is resilient to changes in salinity, suggesting that it may contribute to paracellular fluid transport.

3.4. *Conclusion and Perspectives*

The drinking rate in FW medaka is quite high though still increasing when fish are challenged with hyperosmotic conditions. This suggests that the need for fluid absorption increases as dehydration threatens osmotic homeostasis. Several Aqp isoforms are expressed in the medaka intestine [8], but paradoxically, the most abundant forms (Aqp1a, Aqp8ab, and Aqp10 [8]) are significantly downregulated in SW, in parallel with the increased demand for fluid uptake. This led us to hypothesize that in medaka, the paracellular pathway may be more important when fish move into a hyperosmotic environment. The present data do not reject this hypothesis but do not provide strong support, either. Cldn15 paralogs make a significant constituent of the apical tight junction complex and may thereby create a paracellular water (and Na^{+}) leak pathway. However, there are no signs that this is reinforced during SW acclimation.

The present study leaves behind a couple of questions: (1) What is the physiological significance of drinking in FW, and is this water really absorbed in the intestine? (2) What is the significance of uncoupled transcript and protein dynamics with regard to the Cldn15b paralog? We have attempted to analyze unidirectional water fluxes across isolated gut segments ex vivo using tritiated water as a tracer in an Ussing chamber setup, but so far, the data are inconclusive due to the fragility of the tissue. Future research in this area should pursue the functional aspects of water transport in this species by including vivo knock-down technology as well as analyses of luminal fluid chemical composition.

4. Materials and Methods

4.1. *Fish and Rearing Conditions*

The Japanese medaka (*Oryzias latipes*) used for this study came from two different sources. The fish (CAB strain) used for histological examinations, drinking rate, and drinking-related experiments were purchased from the UMS AMAGEN (Centre national de la recherche scientifique, Gif-sur-Yvette, France) and held in tanks with biofiltered FW or 30 ppt at 24–26 °C and exposed to a 12:12 light:dark photoperiod. These fish were generally fed four times per day with TetraMin® flakes (Tetra GmbH, Melle, Germany), and the food was withheld 2 days before any experimentation. The experimental procedures were approved by the Danish Animal Experiments Inspectorate in accordance with the European convention for the protection of vertebrate animals used for experiments and other scientific purposes (#86/609/EØF). Long-term acclimated fish for transcriptional analysis were obtained from Aquatic Research Organisms, Inc. (Hampton, NH, USA; CAB strain) and held in biofiltered FW or 30 ppt SW and sampled after 1 month of acclimation. They were fed daily with TetraMin tropical flakes (Tetra, United Pet Group, Blacksburg, VA, USA), and food was withheld 2 days prior to any sampling. To investigate the early response to hyperosmotic environments, 10 female and 10 male FW-acclimated medaka were transferred to both sham FW conditions and SW (30 ppt; Instant Ocean,

Spectrum Brands, Blacksburg, VA, USA; $n = 40$) and sampled after 6, 24, and 168 h ($n = 6$ per group). All handling and experimental procedures were approved by the Animal Care and Use Committee of the University of Arkansas (IACUC 17091).

4.2. Drinking Rate Measurements

A series of experiments was performed to estimate the rate of drinking in FW and SW-acclimated medaka. The gamma emitter 51-Cr-EDTA (PerkinElmer, NEZ147001MCNSA1, Waltham, MA, USA) was used as a non-absorbable marker for these experiments. The tracer (5 MBq) was added to the water (1 L of FW or 30 ppt SW), and the fish ($n = 10$–12) were then transferred to the experimental tank [35,40]. They were allowed to drink for 3 h, after which they were transferred to clean water (1 L) for 3 min and transferred to another tank with clean water (1 L) for 30 min. Then, the fish were anaesthetized in 100 mg/L MS-222 (Tricaine methanesulfonate) and killed by cervical dislocation. Before dissection, the fish were blotted by a paper towel and weighed to the nearest mg. The intestine was carefully ligatured at the anterior and posterior ends, removed from the body, and transferred to a 5 mL scintillation vial. The head was separated from the body with remaining organs and transferred to separate scintillation vials. All samples had 0.5 mL of distilled water added, and they were counted on a PerkinElmer 1480 WizardTM 3″ Automatic gamma counter. The radioactivity of a 1.0 mL water sample was measured to estimate the specific radioactivity of the drinking water. Background radioactivity was counted on a 1.0 mL non-radioactive water sample. All samples were corrected for background, and the specific drinking rate (μL/g/h) was calculated as DR = sa/(bw*time), DR = drinking rate, sa = background-corrected specific activity (counts/minute); bw = body weight; time = time in radioactive water. Prior to these experiments, the accumulation of 51-Cr radioactivity was investigated and found to be linear in excess of 3 h.

4.3. Water Passage through the GI Tract

Medaka are agastric fish, meaning that the esophagus is directly connected to the anterior part of the tube-like intestine. In order to trace the passage of imbibed water in FW and SW-acclimated fish, two fish were allowed to drink for 3 h in water to which 51-Cr-EDTA had been added as described above. After this the fish were anaesthetized in MS-222 and killed by cervical dislocation, and the complete GI tract was ligatured at both ends and removed from the fish. Then, segments of 5–6 mm were ligatured and carefully dissected into scintillation vials to estimate the longitudinal distribution of radioactivity. The counting and calculations were done according to the above methodology, and the data were graphed in percent of total radioactivity as a function of longitudinal position.

4.4. Single Photon Emission Computed Tomography (SPECT)–Computed Tomography (CT) Scanning

In order to visualize the intestinal passage of imbibed water, a series of experiments was done in which fish were allowed to drink water with added non-absorbable marker 99-Tc-DTPA (Technetium-99mTc-diethylene-triamine-pentaacetic acid). This short-lived gamma emitter ($T_{1/2}$ = 6.0067 h) is a widely used clinical radiophamaceutical for renal diagnosis and functioning. Subsequently, the fish were analyzed by SPECT-CT scanning. SPECT scanning is used for three-dimensional analysis of the radiochemical, while CT scanning creates a three-dimensional X-ray image, and when the two images are merged, a high-resolution image localizing the radiochemical to internal structures is obtained. All SPECT/CT scans were performed on a Siemens INVEON multimodality pre-clinical scanner (Siemens pre-clinical solutions, Knoxville, TN, USA).

Three fish were used for experiments in FW and SW, respectively, with imaging time-points at 1, 4, and 6 h for each group. For each salinity, two fish were transferred to a container with 100 mL water with the addition of 3.5 GBq Tc-99-DTPA, and one fish was transferred to a container with 100 mL of water with the addition of 5 GBq Tc-99-DTPA. Due to the short half-life of the 99-Tc isotope, a relatively high specific activity in the water is needed in order to obtain a good signal-to-noise ratio for visualization; thus, a higher activity was required for the late imaging group. All fish were allowed to drink in the labeled water for 1 h. Then, they were transferred to separate containers with 1 L of clean water for 5 min, followed by transfer to a second container with 1 L of clean water to rid the external surface for radioactivity. For each salinity, one fish was then euthanized in an overdose of MS-222 after a total of 1, 4, and 6 h after transfer to clean water and analyzed by SPECT-CT scanning in order to analyze the progressive movement of the imbibed isotope through the GI tract. After euthanasia, each fish was wrapped in plastic to avoid dehydration during the following imaging. The fish was placed in a lateral position on a dedicated SPECT/CT pre-clinical bed (25 mm).

CT scans were performed with the following settings; 360° rotation with 360 projections and 2×2 bin. The magnification was set at medium, yielding an isotropic pixel size of 40.00 μm and a trans-axial field view of 42 mm. The tube voltage was set to 80 kV, the current was 500 μA, and each projection was exposed for 1000 ms. CT scans were reconstructed using Feldkamp algorithm, with a Sheep–Logan filter and slight noise reduction. SPECT images were acquired using mouse high-resolution single pinhole collimators. A full 360° rotation with 60 projections and a fixed radius of 25 mm yielded a reconstructed 28 mm trans-axial field of view. A 20% energy window centered on the energy peak of 99mTc at 140 keV was used. Acquisition duration was set to 100 sec/projection. CT and SPECT images were co-registered using a transformation matrix and SPECT data was reconstructed using the Siemens MAP3D algorithm (matrix 128×128, 0.5 mm pixels, 16 iterations, and 6 subsets).

4.5. RNA Isolation, cDNA Synthesis, and qPCR

RNA isolation was conducted according to the manufacturer's protocol (TRI Reagent®; Sigma Aldrich, St. Lois, MO, USA). All samples were homogenized using a Power Max 200 rotating knife homogenizer (Advanced Homogenizing System; Manufactured by PRO Scientific for Henry Troemner LLC, Thorofare, NJ, USA). First, 500 ng of total RNA was used for cDNA synthesis using the Applied Biosystems High Capacity cDNA Reverse Transcription kit (Thermo Fisher, Waltham, MA, USA). Used primers were previously validated and published in Bossus et al. [34] and Madsen et al. [6]. Elongation Factor 1 alpha (*ef1α*), beta actin (*βact*), and ribosomal protein L7 (*rpl7*) were analyzed as normalization genes in all experiments. Quantitative PCR was run on a Bio-Rad CFX96 platform (BioRad, Hercules, CA, USA) using SYBR® Green JumpStart (Sigma Aldrich). qPCR cycling was conducted using the following protocol: a denaturation/activation step (94 °C) for 3 min, 40 cycles of a 15 s denaturation step (94 °C) followed by an annealing/elongation step for 60 s (60 °C), and finally a melting curve analysis at an interval of 5 s per degree from 55 to 94 °C. The absence of primer–dimer association was verified with no template controls (NTC). As an alternative to DNAse treatment, the absence of significant genomic DNA amplification was confirmed using total RNA samples instead of cDNA in a no reverse transcriptase control (NRT). Primer amplification efficiency was analyzed using a standard curve method with dilutions of the primers from 2 to 16 times. Amplification efficiency was used to calculate the relative copy numbers of the individual targets. Relative copy numbers were calculated by ${E_a}^{\Delta Ct}$, where Ct is the threshold cycle number and E_a is the amplification efficiency. Data were normalized to the geometric mean of the three normalization genes.

4.6. Immunofluorescence, Confocal, and Stimulated Emission Depletion (STED) Microscopy

The preparation of medaka intestines for immunofluorescence microscopy followed the procedures described previously [51]. Sections (0.5 cm) from the middle part of the intestines from medaka acclimated to FW and 30 ppt SW were sampled and fixed overnight in 4% buffered paraformaldehyde at 4 °C. After rinsing several times in 70% EtOH, the tissues were dehydrated overnight through a graded series of EtOH and xylene followed by embedding in 60 °C paraffin. Five-micron-thick transversal sections were cut on a microtome, and sections were placed on Superfrost plus (Gerhard Menzel GmbH, Braunschweig, Germany) slides before being dried overnight at 55 °C. Then, the tissue sections were hydrated through washes in xylene, 99%, 96%, and 70% EtOH and finally Na citrate (10 mM Na-citrate, pH 6.0). Antigen retrieval was performed by boiling the sections in the citrate solution for 5 min in a microwave oven and leaving them in the warm citrate solution for 30 min before being washed in 1× PBS (in mmol L^{-1}: 137 NaCl, 2.7 KCl, 1.5 KH_2PO_4, 4.3 Na_2HPO_4, pH 7.3). Then, representative sections were blocked by incubation in 2% goat serum and 2% bovine serum albumin in 1× PBS for 1 h at room temperature. This was followed by dual labeling with a cocktail of an affinity purified polyclonal rabbit antibody against medaka Cldn15a or Cldn15b, respectively, in combination with the monoclonal mouse α5 antibody, which recognizes the alpha-subunit of the Na^+,K^+-ATPase in all vertebrates (The Developmental Studies Hybridoma Bank developed under auspices of the National Institute of Child Health Development and maintained by The University of Iowa, Department of Biological Sciences, Iowa City, IA, USA). In a separate experiment, sections were dual-labeled with Cldn15b and an occludin mouse monoclonal antibody (Invitrogen, product # 33-1500) in order to verify localization in the tight junction zone. Primary antibodies were diluted in 2% goat serum and 2% bovine serum albumin in PBS and incubated overnight at 4 °C. The polyclonal Cldn antibodies were custom-made in rabbits by Genscript (Piscataway, NJ, USA) against the following epitopes near the C-termini: Japanese medaka Cldn15a PAPTRSVVASTYGR, GenBank accession XP_004079873.1; Japanese medaka Cldn15b SHAAPSNYDRNAYV, GenBank accession XP_004076514.1). They were used at the concentrations 0.5 μg/mL (Cldn15a), 0.6 μg/mL (Cldn15b), and 5 μg/mL (occludin). The α5 antibody was used at 0.2 μg/mL.

For immunofluorescence and confocal microscopy, the following secondary antibodies were used for visualization: Alexa Flour® 568 Donkey Anti-Rabbit IgG (H+L) at 1 ug/mL and Oregon Green® 488 Goat Anti-Mouse IgG (H+L) at 2ug/mL (Invitrogen™ Molecular ProbesTM, Carlsbad, CA, USA). The incubation time was 1 h at 37 °C for the secondary antibody. Then, sections were washed repeatedly in PBS, and coverslips were mounted using ProLong Gold antifade reagent (Invitrogen).

For STED microscopy, we used higher Cldn antibody concentrations: 0.7 μg/mL (Cldn15a) and 0.9 μg/mL (Cldn15b). The secondary antibodies used for STED were goat-anti-rabbit Abberior® STAR 488 and goat-anti-mouse Abberior® STAR 440SX (Sigma-Aldrich) at 1:200 and 1:1000 dilution, respectively. Negative control incubations with 2% BSA in PBS instead of primary antibodies were made routinely. The fluorescence was inspected on a Leica HC microscope (Manheim, Germany) and pictures of representative areas were captured using a Leica DC200 camera. Confocal images were taken on a Zeiss LSM510 META confocal microscope (CarlZeiss, Oberkochen, Germany) using a 63× objective with oil immersion. STED images were recorded using a Leica TSC SP8 STED setup. The excitation was done at 500 nm using a white light laser for Abberior® STAR 488 and at 458 using an Argon laser for Abberior® STAR 440SXP. The depletion laser (STED laser) was a 592 nm CW for both channels. The emission was recorded at 510–560 nm using the gated hybrid detector (0.3 ns) in counting mode for the Abberior® STAR 488 and at 500–550 nm using the non-gated hybrid detector in counting mode for the Abberior® STAR 440SXP. The images were cross-talk corrected and deconvoluted using Huygens™ (Hilversum, Netherland). The deconvolution was done to further increase the resolution of the images and decrease the background.

4.7. Statistical Analyses

All data analysis was conducted using GraphPad Prism 8.0 software (San Diego, CA, USA). Data from the salinity transfer experiments were analyzed using Bonferroni adjusted two-tailed Student's t-test in experiments with two groups and two-way ANOVA followed by Bonferroni's multiple comparisons test of time-matched groups in experiments with more groups. Drinking rates were analyzed using two-tailed Student's t-test. When required, data were log or square root transformed to meet the ANOVA assumption of homogeneity of variances as tested with Bartlett's test. Significant differences were accepted when $p < 0.05$.

Author Contributions: S.S.M., A.M.N., M.C.B. and C.K.T. conceived the idea and designed the project; A.M.N., M.C.B., and L.V.E. performed the transcript analyses; A.M.N performed the fluorescence and confocal microscopy; J.D. and J.R.B. performed the STED microscopy and image analyses; C.B. and T.L.A. performed the SPECT/CT scanning experiments and data analyses; A.M.N. and S.S.M. performed the drinking rate experiments; S.S.M. and C.K.T. wrote the manuscript. All authors have read and agreed to the published version of the manuscript.

Acknowledgments: The authors acknowledge the Danish Molecular Biomedical Imaging Center (DaMBIC, University of Southern Denmark) for the use of the bioimaging facilities.

Abbreviations

ANOVA	Analysis of variance
AQP	Aquaporin
CLDN	Claudin
CT	Computed tomography
DR	Drinking rate
DTPA	Diethylenetriamine penta-acetic acid
EDTA	2,2′,2′′,2′′′-(Ethane-1,2-diyldinitrilo)-tetraacetic acid
FW	Fresh water
GI	Gastrointestinal tract
NKCC	Sodium-potassium-chloride-cotransporter
SEM	Standard error of the mean
STED	Stimulated Emission Depletion
SPECT	Single photon emission computed tomography
SW	Seawater
PPT	Parts per thousand

References

1. Madsen, S.S.; Engelund, M.B.; Cutler, C.P. Water transport and functional dynamics of aquaporins in osmoregulatory organs of fishes. *Biol. Bull.* **2015**, *229*, 70–92. [CrossRef] [PubMed]
2. Whittamore, J.M. Osmoregulation and epithelial water transport: Lessons from the intestine of marine teleost fish. *J. Comp. Physiol.* **2012**, *182B*, 13–19. [CrossRef] [PubMed]
3. Aoki, M.; Kaneko, T.; Katoh, F.; Hasegawa, S.; Tsutsui, N.; Aida, K. Intestinal water absorption through aquaporin 1 expressed in the apical membrane of mucosal epithelial cells in seawater-adapted Japanese eel. *J. Exp. Biol.* **2003**, *206*, 3495–3505. [CrossRef] [PubMed]
4. Cutler, C.P.; Philips, C.; Hazon, N.; Cramb, G. Aquaporin 8 (AQP8) intestinal mRNA expression increases in response to salinity acclimation in yellow and silver European eels (*Anguilla anguilla*). *Comp. Biochem. Physiol.* **2009**, *153A*, S78. [CrossRef]
5. Martinez, A.S.; Cutler, C.P.; Wilson, G.D.; Phillips, C.; Hazon, N.; Cramb, G. Cloning and expression of three aquaporin homologues from the European eel (*Anguilla anguilla*): Effects of seawater acclimation and cortisol treatment on renal expression. *Biol. Cell* **2005**, *97*, 615–627. [CrossRef]
6. Madsen, S.S.; Olesen, J.H.; Bedal, K.; Engelund, M.B.; Velasco-Santamaria, Y.M.; Tipsmark, C.K. Functional characterization of water transport and cellular localization of three aquaporin paralogs in the salmonid intestine. *Front. Physiol.* **2011**, *2*, 56. [CrossRef]
7. Sundell, K.S.; Sundh, H. Intestinal fluid absorption in anadromous salmonids: Importance of tight junctions and aquaporins. *Front. Physiol.* **2012**, *3*, 388. [CrossRef]
8. Madsen, S.S.; Bujak, J.; Tipsmark, C.K. Aquaporin expression in the Japanese medaka (*Oryzias latipes*) in freshwater and seawater: Challenging the paradigm of intestinal water transport? *J. Exp. Biol.* **2014**, *217*, 3108–3121. [CrossRef]
9. Fischbarg, J. Fluid transport across leaky epithelia: Central role of the tight junction and supporting role of aquaporins. *Physiol. Rev.* **2010**, *90*, 1271–1290. [CrossRef]
10. Laforenza, U. Water channel proteins in the gastrointestinal tract. *Mol. Aspects Med.* **2012**, *33*, 642–650. [CrossRef]
11. Günzel, D.; Fromm, M. Claudins and other tight junction proteins. *Compr. Physiol.* **2012**, *2*, 1819–1852. [CrossRef] [PubMed]
12. Günzel, D.; Yu, A.S.L. Claudins and the modulation of tight junction permeability. *Physiol. Rev.* **2013**, *93*, 525–569. [CrossRef] [PubMed]

13. Angelow, S.; Yu, A.S.L. Structure-function studies of claudin extracellular domains by cysteine-scanning mutagenesis. *J. Biol. Chem.* **2009**, *284*, 29205–29217. [CrossRef] [PubMed]
14. Li, J.; Angelow, S.; Linge, A.; Zhuo, M.; Yu, A.S. Claudin-2 pore function requires an intramolecular disulfide bond between two conserved extracellular cysteines. *Am. J. Physiol.* **2013**, *305*, C190–C196. [CrossRef]
15. Samanta, P.; Wang, Y.; Fuladi, S.; Zou, J.; Li, Y.; Shen, L.; Weber, C.; Khalili-Araghi, F. Molecular determination of claudin-15 organization and channel selectivity. *J. Gen. Physiol.* **2018**, *150*, 949–968. [CrossRef]
16. Alberini, G.; Benfenati, F.; Maragliano, L. Molecular dynamics simulations of ion selectivity in a claudin-15 paracellular channel. *J. Phys. Chem. B* **2018**, *122*, 10783–10792. [CrossRef]
17. Loh, Y.H.; Christoffels, A.; Brenner, S.; Hunziker, W.; Venkatesh, B. Extensive expansion of the claudin gene family in the teleost fish, *Fugu rubripes*. *Genome Res.* **2004**, *14*, 1248–1257. [CrossRef]
18. Baltzegar, D.A.; Reading, B.J.; Brune, E.S.; Borski, R.J. Phylogenetic revision of the claudin gene family. *Mar. Genomics* **2013**, *11*, 17–26. [CrossRef]
19. Rosenthal, R.; Milatz, S.; Krug, S.M.; Oelrich, B.; Schulzke, J.-D.; Amasheh, S.; Günzel, D.; Fromm, M. Claudin-2, a component of the tight junction, forms a paracellular water channel. *J. Cell Sci.* **2010**, *123*, 1913–1921. [CrossRef]
20. Rosenthal, R.; Gunzel, D.; Piontek, J.; Krug, S.M.; Ayala-Torres, C.; Hempel, C.; Theune, D.; Fromm, M. Claudin-15 forms a water channel through the tight junction with distinct function compared to claudin-2. *Acta Physiol.* **2020**, *228*, e13334. [CrossRef]
21. Engelund, M.B.; Yu, A.S.L.; Li, J.; Madsen, S.S.; Færgeman, N.J.; Tipsmark, C.K. Functional characterization and localization of a gill-specific claudin isoform in Atlantic salmon. *Am. J. Physiol.* **2012**, *302*, R300–R322. [CrossRef] [PubMed]
22. Kwong, R.W.M.; Perry, S.F. The tight junction protein claudin-b regulates epithelial permeability and sodium handling in larval zebrafish, *Danio rerio*. *Am. J. Physiol.* **2013**, *304*, R504–R513. [CrossRef] [PubMed]
23. Suzuki, H.; Nishizawa, T.; Tani, K.; Yamazaki, Y.; Tamura, A.; Ishitani, R.; Dohmae, N.; Tsukita, S.; Nureki, O.; Fujiyoshi, Y. Crystal structure of a claudin provides insight into the architecture of tight junctions. *Science* **2014**, *344*, 304–307. [CrossRef] [PubMed]
24. Hou, J.; Rajagopal, M.; Yu, A.S.L. Claudins and the kidney. *Annu. Rev. Physiol.* **2013**, *75*, 479–501. [CrossRef]
25. Garcia-Hernandez, V.; Quiros, M.; Nusrat, A. Intestinal epithelial claudins: Expression and regulation in homeostasis and inflammation. *Ann. N. Y. Acad. Sci.* **2017**, *1397*, 66–79. [CrossRef]
26. Holmes, J.L.; Van Itallie, C.M.; Rasmussen, J.E.; Anderson, J.M. Claudin profiling in the mouse during postnatal intestinal development and along the gastrointestinal tract reveals complex expression patterns. *Gene Expr. Patterns* **2006**, *6*, 581–588. [CrossRef]
27. Lu, Z.; Ding, L.; Lu, Q.; Chen, Y.H. Claudins in intestines: Distribution and functional significance in health and diseases. *Tissue Barriers* **2013**, *1*, e24978. [CrossRef]
28. Tamura, A.; Kitano, Y.; Hatam, M.; Katsuno, T.; Moriwaki, K.; Sasaki, H.; Hayashi, H.; Suzuki, Y.; Noda, T.; Furuse, M.; et al. Megaintestine in claudin-15 deficient mice. *Gastroenterology* **2008**, *134*, 523–534. [CrossRef]
29. Bagnat, M.; Cheung, I.D.; Mostov, K.E.; Stainier, D.Y. Genetic control of single lumen formation in the zebrafish gut. *Nat. Cell Biol.* **2007**, *9*, 954–960. [CrossRef]
30. Tamura, A.; Hayashi, H.; Imasato, M.; Yamazaki, Y.; Hagiwara, A.; Wada, M.; Noda, T.; Watanabe, M.; Suzuki, Y.; Tsukita, S. Loss of claudin-15, but not claudin-2, causes Na^{+} deficiency and glucose malabsorption in mouse small intestine. *Gastroenterology* **2011**, *140*, 913–923. [CrossRef]
31. Tipsmark, C.K.; Kiilerich, P.; Nilsen, T.O.; Ebbesson, L.O.E.; Stefansson, S.O.; Madsen, S.S. Branchial expression patterns of claudin isoforms in Atlantic salmon during seawater acclimation and smoltification. *Am. J. Physiol.* **2008**, *294*, R1563–R1574. [CrossRef] [PubMed]
32. Tipsmark, C.K.; Sørensen, K.J.; Hulgard, K.; Madsen, S.S. Claudin-15 and-25b expression in the intestinal tract of Atlantic salmon in response to seawater acclimation, smoltification and hormone treatment. *Comp. Biochem. Physiol.* **2010**, *155A*, 361–370. [CrossRef] [PubMed]
33. Clelland, E.S.; Kelly, S.P. Tight junction proteins in zebrafish ovarian follicles: Stage specific mRNA abundance and response to 17beta-estradiol, human chorionic gonadotropin, and maturation inducing hormone. *Gen. Comp. Endocrinol.* **2010**, *168*, 388–400. [CrossRef] [PubMed]
34. Bossus, M.C.; Madsen, S.S.; Tipsmark, C.K. Functional dynamics of claudin expression in Japanese medaka (*Oryzias latipes*): Response to environmental salinity. *Comp. Biochem. Physiol.* **2015**, *187A*, 74–85. [CrossRef] [PubMed]

35. Usher, M.L.; Talbot, C.; Eddy, F.B. Drinking in Atlantic salmon smolts transferred to seawater and the relationship between drinking and feeding. *Aquaculture* **1988**, *73*, 237–246. [CrossRef]
36. Marshall, W.S.; Grosell, M. Ion transport, osmoregulation, and acid-base regulation. In *The Physiology of Fishes*; Evans, D.H., Clairborne, J.B., Eds.; Taylor and Francis Group: Boca Raton, FL, USA, 2006; pp. 177–210.
37. Cerdá, J.; Finn, R.N. Piscine aquaporins: An overview of recent advances. *J. Exp. Zool.* **2010**, *313A*, 623–650. [CrossRef]
38. Tipsmark, C.K.; Sørensen, K.J.; Madsen, S.S. Aquaporin expression dynamics in osmoregulatory tissues of Atlantic salmon during smoltification and seawater acclimation. *J. Exp. Biol.* **2010**, *213*, 368–379. [CrossRef]
39. Kim, Y.K.; Lee, S.Y.; Kim, B.S.; Kim, D.S.; Nam, Y.K. Isolation and mRNA expression analysis of aquaporin isoforms in marine medaka *Oryzias dancena*, a euryhaline teleost. *Comp. Biochem. Physiol.* **2014**, *171A*, 1–8. [CrossRef]
40. Perrott, M.N.; Grierson, C.E.; Hazon, N.; Balment, R.J. Drinking behaviour in sea water and fresh water teleosts, the role of the renin-angiotensin system. *Fish. Physiol. Biochem.* **1992**, *10*, 161–168. [CrossRef]
41. Fuentes, J.; Bury, N.R.; Carroll, S.; Eddy, F.B. Drinking in Atlantic salmon presmolts (*Salmo salar* L.) and juvenile rainbow trout (*Oncorhynchus mykiss* Walbaum) in response to cortisol and sea water challenge. *Aquaculture* **1996**, *141*, 129–137. [CrossRef]
42. Kaneko, T.; Hasegawa, S. Application of laser scanning microscopy to morphological observations on drinking in freshwater medaka larvae and those exposed to 80% seawater. *Fish. Sci.* **1999**, *65*, 492–493. [CrossRef]
43. Fuentes, J.; Eddy, F.B. Drinking in marine, euryhaline and freshwater teleost fish. In *Ionic Regulation in Animals: A Tribute to Professor W.T.W. Potts*; Hazon, N., Eddy, F.B., Flik, G., Eds.; Springer: Berlin/Heidelberg, Germany, 1997; pp. 136–149.
44. Hutchinson, S.; Hawkins, L.E. The influence of salinity on water balance in 0-group flounders, *Platichthys flesus* (L). *J. Fish. Biol.* **1990**, *36*, 751–764. [CrossRef]
45. Tytler, P.; Tatner, M.; Findlay, C. The ontogeny of drinking in the rainbow trout, *Oncorhychus mykiss*. (Walbaum). *J. Fish. Biol.* **1990**, *36*, 867–875. [CrossRef]
46. Lin, L.Y.; Weng, C.F.; Hwang, P.P. Regulation of drinking rate in euryhaline tilapia larvae (*Oreochromis mossambicus*) during salinity challenges. *Physiol. Biochem. Zool.* **2001**, *74*, 171–177. [CrossRef] [PubMed]
47. Grosell, M. Intestinal transport processes in marine fish osmoregulation. In *Fish. Osmoregulation*; Baldisserotto, B., Mancera, J.M., Kapoor, B.G., Eds.; Science Publishers: Enfield, NH, USA, 2007; pp. 333–357.
48. Sundell, K.; Jutfelt, F.; Agustsson, T.; Olsen, R.E.; Sandblom, E.; Hansen, T.; Bjornsson, B.T. Intestinal transport mechanisms and plasma cortisol levels during normal and out-of-season parr–smolt transformation of Atlantic salmon, *Salmo salar*. *Aquaculture* **2003**, *222*, 265–285. [CrossRef]
49. Fujita, H.; Chiba, H.; Yokozaki, H.; Sakai, N.; Sugimoto, K.; Wada, T.; Kojima, T.; Yamashita, T.; Sawada, N. Differential expression and subcellular localization of claudin-7, -8, -12, -13, and -15 along the mouse intestine. *J. Histochem. Cytochem.* **2006**, *54*, 933–944. [CrossRef] [PubMed]
50. Chasiotis, H.; Kelly, S.P. Occludin immunolocalization and protein expression in goldfish. *J. Exp. Biol.* **2008**, *211*, 1524–1534. [CrossRef]
51. Engelund, M.B.; Madsen, S.S. Tubular localization and expressional dynamics of aquaporins in the kidney of seawater-challenged Atlantic salmon. *J. Comp. Physiol. B* **2015**, *185B*, 207–223. [CrossRef]

9

Apoptotic Fragmentation of Tricellulin

Susanne Janke, Sonnhild Mittag, Juliane Reiche and Otmar Huber *

Department of Biochemistry II, Jena University Hospital, Friedrich Schiller University Jena, 07743 Jena, Germany; susanne.janke@med.uni-jena.de (S.J.); sonnhild.mittag@med.uni-jena.de (S.M.); juliane.reiche@med.uni-jena.de (J.R.)

* Correspondence: otmar.huber@med.uni-jena.de

Abstract: Apoptotic extrusion of cells from epithelial cell layers is of central importance for epithelial homeostasis. As a prerequisite cell–cell contacts between apoptotic cells and their neighbors have to be dissociated. Tricellular tight junctions (tTJs) represent specialized structures that seal polarized epithelial cells at sites where three cells meet and are characterized by the specific expression of tricellulin and angulins. Here, we specifically addressed the fate of tricellulin in apoptotic cells. Methods: Apoptosis was induced by staurosporine or camptothecin in MDCKII and RT-112 cells. The fate of tricellulin was analyzed by Western blotting and immunofluorescence microscopy. Caspase activity was inhibited by Z-VAD-FMK or Z-DEVD-FMK. Results: Induction of apoptosis induces the degradation of tricellulin with time. Aspartate residues 487 and 441 were identified as caspase cleavage-sites in the C-terminal coiled-coil domain of human tricellulin. Fragmentation of tricellulin was inhibited in the presence of caspase inhibitors or when Asp487 or Asp441 were mutated to asparagine. Deletion of the tricellulin C-terminal amino acids prevented binding to lipolysis-stimulated lipoprotein receptor (LSR)/angulin-1 and thus should impair specific localization of tricellulin to tTJs. Conclusions: Tricellulin is a substrate of caspases and its cleavage in consequence contributes to the dissolution of tTJs during apoptosis.

Keywords: tight junction; tricellulin; lipolysis-stimulated lipoprotein receptor (LSR); angulin; epithelial barrier; cell–cell contact; apoptosis; caspase

---

1. Introduction

Tight junctions (TJs) represent the most apical cell–cell contacts in polarized epithelial cell layers. They form essential barriers for solutes including nutrients, metabolites and toxins as well as ions, thereby separating luminal and external compartments from the interior of multicellular bodies. Accordingly, TJs are also important to exclude pathogens such as bacteria and viruses from organisms. Break-down of this barrier function is associated with different diseases and is involved in inflammatory bowel diseases [1] or in organ failure during sepsis [2,3]. However, it turned out that TJ components may also represent targets of specific bacteria and viruses to invade cells or tissues resulting in a deregulation of the cytoskeleton or of signaling pathways required for barrier maintenance [4]. Primarily, claudins and occludin form the apical sealing belt between two opposing cells where they are arranged in a network of TJ strands that restricts and regulates paracellular flux depending on the specific claudin composition. The situation is different at tricellular contacts where apical TJs strands of three neighboring cells meet and thus are assumed to represent weak points for the paracellular TJ network [5]. At these tricellular tight junctions, bicellular TJs converge, and TJ strands turn basally and form the so-called central sealing element [6]. These tricellular TJs (tTJs) differ from bicellular TJs (bTJs) by a specific set of proteins including tricellulin [7] and one of the three angulins [8].

Tricellulin (marvelD2) together with occludin and marvelD3 form the TJ-associated MARVEL protein (TAMP) family of 4-transmembrane-domain proteins with long N- and C-terminal domains both located in the cytosol. The tricellulin C-terminus interacts with the cytosolic adapter protein zona occludens-1 (ZO-1) and is thereby linked to the actin cytoskeleton [9–11]. Knock-down of tricellulin impairs epithelial barrier function and results in disorganized bTJs, indicating an essential role of tricellulin in maintenance of overall TJ structure and function [7]. Tricellulin at tTJs was shown to restrict macromolecular passage and strong overexpression contributes to reduced paracellular permeability [12,13]. Interestingly, in occludin knock-down cells, tricellulin is mislocated to bTJs preferentially at edges of elongating bTJs [14]. How homomeric or heteromeric tricellulin/occludin complexes may be involved in this process or in the transport of tricellulin to the TJs [15] is currently not understood. However, it is obvious that tricellulin and occludin do not form heterophilic trans-interactions within established TJs [11,16].

With the identification of LSR (lipolysis-stimulated lipoprotein receptor)/angulin-1 as a tTJ-located receptor that recruits tricellulin to tTJs, it became clear that a specific anchor–protein is responsible for the defined localization of tricellulin. Knock-down of LSR/angulin-1 impaired barrier function in reducing transepithelial resistance (TER) probably as a consequence of mislocalization of tricellulin to bTJs [17]. In addition, mislocalization of LSR was associated with cancer progression and metastasis [18]. Based on sequence homology, two LSR/angulin-1-related proteins including immunoglobulin-like domain-containing receptor (ILDR) 1/angulin-2 and ILDR 2/angulin-3 were identified. The three members show tissue-specific expression patterns and at least one family member is located at tTJs [8,19]. However, it is still open regarding how angulins are directed to tTJs.

The physiological importance of tricellulin for hair cells is emphasized by mutations in the tricellulin C-terminal domain found in patients with nonsyndromic deafness (DFNB49). All mutations result in tricellulin variants with premature stop codons and, consequently, structurally impaired C-terminal ends and limited binding to ZO-1 [20,21]. Moreover, truncation of the tricellulin C-terminal domain diminishes its correct localization to tricellular contacts [22]. Tricellulin knock-out mice show a progressive hearing loss due to apoptotic death of hair cells [23]. It is currently not clear why tricellulin knock-out mice show this limited and hair cell-specific phenotype although tricellulin was deleted in all other tissues too. Interestingly, ILDR1/angulin-2 deficiency causes a similar phenotype with progressing hearing loss by outer hair cell degeneration [19,24,25]. Whether compensatory effects or regulatory roles of tricellulin and/or angulins may play a role still has to be solved.

Little is known in respect to mechanisms involved in regulation of tricellulin function. Like occludin, tricellulin can be phosphorylated [7]. However, in occludin, a multitude of kinases and corresponding phosphosites as well as functional consequences of these modifications have been studied in detail [26]. A phosphorylation hotspot region [27] that seems to be involved in the regulation of occludin mobility, apical junctional complex dynamics and cell migration [28] has been identified. In contrast, tricellulin phosphorylation-sites have been found by mass spectrometry, but, to our knowledge, none of them has been functionally characterized. However, there is the first evidence that occludin and tricellulin are differentially targeted by specific kinases [29]. Interestingly, JNK1/JNK2-dependent phosphorylation of LSR/angulin-1 was reported to be crucial for its exclusive localization to tTJs [30]. In addition, tricellulin was shown to be targeted by the ubiquitin ligase Itch, which might affect its stability and trafficking [31].

To maintain epithelial functionality, old or damaged cells have to be removed by apoptosis and replaced by new cells. Apoptosis is induced by extrinsic and intrinsic signals. Both pathways in an initial step lead to activation of initiator caspases, which subsequently activate effector caspases. These finally hydrolyze essential structural and housekeeping proteins by cleavage after aspartate residues [32]. Caspases have multiple functions also in inflammation and immunity [33]. Extrusion of apoptotic cells from the cell layer ideally occurs without loss of barrier function and disruption of the cytoarchitecture [34]. In consequence, contacts between the dying and neighboring cells have to be disassembled. Previous studies have analyzed bTJs in this respect [35], but, to our knowledge, tTJs have not been investigated in more detail. Here, we induced apoptosis with staurosporine or camptothecin and observed a caspase-dependent cleavage of the C-terminus of tricellulin. We identified two specific caspase cleavage-sites and mutation of these sites protected tricellulin from fragmentation. In addition, we provide evidence that LSR/angulin-1, which anchors tricellulin to tTJs, is also targeted by caspases.

2. Results

2.1. Caspase-Dependent Cleavage of Tricellulin upon Apoptotic Stimuli

To investigate the fate of tricellulin during apoptosis, MDCKII cells were treated with staurosporine as a known inducer of apoptosis [36]. At different time points after addition of staurosporine, both floating and adherent cells were collected and lysed for subsequent Western blot analysis. Already three hours after induction of apoptosis, a time-dependent decline of endogenous tricellulin was detectable compared to DMSO-treated control cells (Figure 1A,B). Induction of apoptosis was confirmed by detection of cleaved poly(ADP-ribose)-polymerase (PARP), a well-known effector caspase substrate. The typical PARP cleavage fragment was detectable only in staurosporine-treated cells but not in control cells (Figure 1A). Moreover, when camptothecin was used as an alternative apoptosis inducer, the loss of endogenous tricellulin in MDCKII cells was detectable between 8 h and 24 h (Supplementary Figure S1). The produced tricellulin cleavage fragments were not detectable using the anti-tricellulin monoclonal antibody (clone 54H19L38). Immunofluorescence microscopy confirmed degradation of tricellulin. Signals for endogenous tricellulin disappeared with time (Figure 1C).

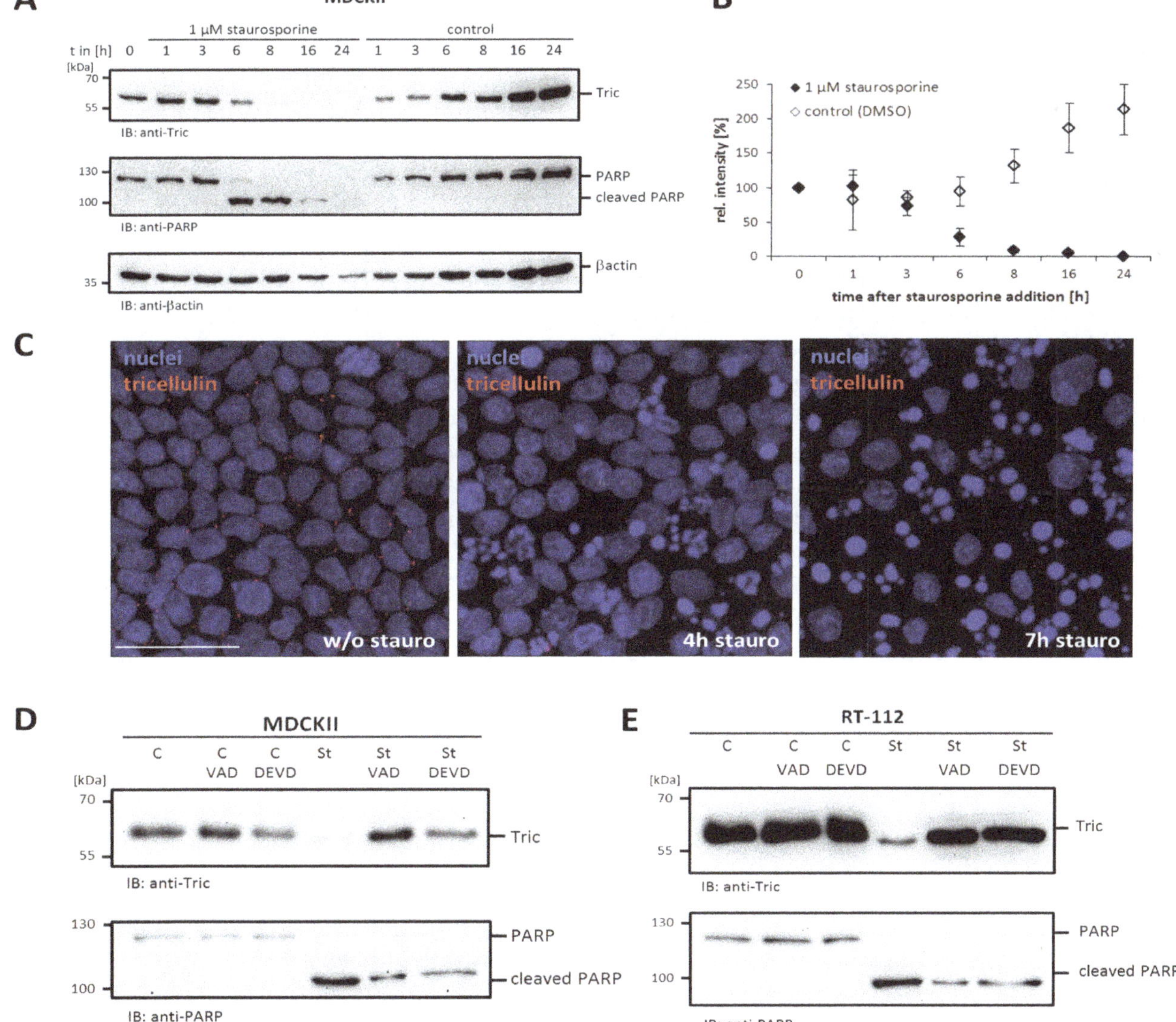

Figure 1. Tricellulin is a target of caspase cleavage in apoptotic cells. (**A**) MDCKII cells were treated with 1 μM staurosporine or DMSO as solvent control for the indicated times. Cell lysates were analyzed by Western blotting using anti-tricellulin, anti-poly[ADP-ribose] polymerase (PARP) and anti-βactin antibodies. (**B**) Quantification of tricellulin degradation by densitometric analysis. The graph represents mean values +/− SD of three independent experiments. (**C**) Loss of endogenous tricellulin detected by immunofluorescence microscopy (scale bar 30 μm). (**D**) Pre-treatment with 10 μM Z-VAD-FMK or 20 μM Z-DEVD-FMK for 1 h before stimulation with 1 μM staurosporine inhibited tricellulin degradation. Lysates were generated 6 h after induction of apoptosis. (**E**) Staurosporine-induced cleavage of tricellulin in RT-112 cells is inhibited by caspase inhibitors (10 μM Z-VAD-FMK or 20 μM Z-DEVD-FMK). All experiments are representatives of at least three independent experiments.

From our observations, we hypothesized that tricellulin like occludin [35] is a caspase target during apoptosis. To verify that the loss of tricellulin is indeed the consequence of caspase cleavage MDCKII cells were pre-treated with the pan-caspase inhibitor Z-VAD-FMK or the caspase-3 inhibitor Z-DEVD-FMK before adding staurosporine. Both caspase inhibitors significantly reduced degradation of endogenous tricellulin (Figure 1D). Inhibition of caspase activity was confirmed by the reduced formation of the PARP cleavage-fragment. A similar effect was obtained in RT-112 cells (bladder carcinoma) (Figure 1E) showing that the observed effect is not restricted to MDCKII cells.

2.2. *Mapping of Potential Caspase Cleavage-Sites in Tricellulin*

For prediction of possible caspase cleavage-sites in human tricellulin, the online software tool CaspDB [37] was applied. Two potential caspase cleavage-sites with the highest scores were predicted in the C-terminal domain of human tricellulin after amino acid aspartate 487 (D487) and/or aspartate 441 (D441) (Figure 2A). To confirm caspase-3-mediated cleavage in the C-terminal domain of tricellulin, we expressed and purified a GST-TricC fusion protein as reported previously [29,31] and applied in vitro digestion using recombinant active caspase-3. In the presence of caspase-3, two additional fragments with a molecular mass of about 40 and 35 kDa were detectable by Western blot analysis, representing potential caspase-3 cleavage-products. The molecular mass of these fragments coincides with the predicted molecular weight of GST-TricC fragments when cleaved at D487 (frag1, ~41 kDa) and D441 (frag2, ~35 kDa) (Figure 2B). Remarkably, fragment 1 is generated earlier and more efficiently than fragment 2, suggesting that D487 is the preferred caspase-3 cleavage-site and D441 appears to be used as a sequential second site after cleavage at D487. Addition of the pan-caspase inhibitor Z-VAD-FMK inhibited the generation of both GST-TricC fragments (Figure 2C). As control, purified recombinant GST alone was analyzed for caspase-3 cleavage, but no fragmentation was detectable (Figure 2D). Finally, we generated GST-TricC-D487N, GST-TricC-D441N and GST-TricC-D441N/D487N double mutant constructs and subjected them to in vitro cleavage-assays with recombinant caspase-3. Consistent with our previous results, caspase cleavage of GST-TricC-D487N only resulted in formation of fragment 2 (frag2), whereas cleavage of GST-TricC-D441N only generated fragment 1 (frag1). The GST-TricC-D441N/D487N double-mutated construct was no longer cleaved by recombinant caspase-3 (Figure 2E).

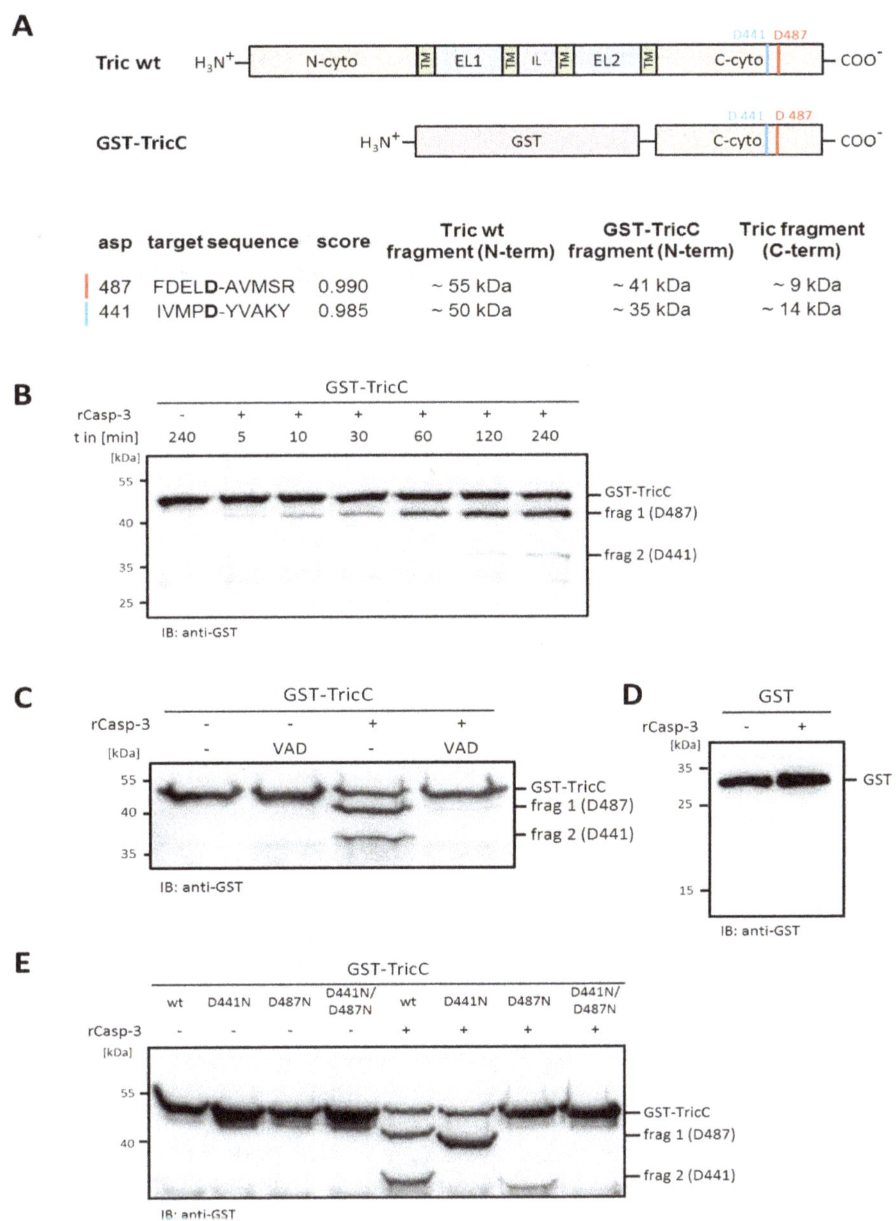

Figure 2. Identification of D487 and D441 as potential caspase-3 cleavage-sites in human tricellulin. (**A**) Schematic overview of the full-length tricellulin structure including the four transmembrane domains (TM), extracellular loops 1 and 2 (EL1, EL2), the intracellular loop (IL) and the potential caspase cleavage-sites D487 and D441. The table summarizes the molecular masses of the corresponding caspase cleavage-products expected for an in vitro caspase assay using recombinant GST-TricC fusion protein as substrate. (**B**) Time-dependent generation of the GST-TricC cleavage-products frag 1 and frag 2 after addition of recombinant caspase-3. The image is a representative of $n = 2$. (**C**) In the presence of pan-caspase inhibitor Z-VAD-FMK, fragmentation of GST-TricC is inhibited. (**D**) GST was used as a control and was not fragmented by caspase-3. (**E**) Mutation of potential caspase-3 cleavage-sites disable cleavage of GST-TricC partly (GST-TricC-D441N, GST-TricC-D487N) or completely (GST-TricC-D441N/D487N).

To validate the in vitro results, N-terminally $FLAG_3$-tagged human tricellulin was transiently transfected into MDCKII cells. Cells were subsequently treated with or without staurosporine for 6 h in the presence or absence of pan-caspase inhibitor Z-VAD-FMK and the caspase-3 inhibitor Z-DEVD-FMK, respectively. After induction of apoptosis, two bands with a molecular weight of about 55 kDa and 65 kDa were detectable (Figure 3A). Fragmentation was abrogated in the presence of each of the caspase inhibitors. In contrast to wildtype $FLAG_3$-Tric, transient transfection of a mutated $FLAG_3$-Tric-D441N construct abolished the generation of caspase-3 cleavage product frag 2 (~ 55 kDa) upon induction of apoptosis with staurosporine. Generation of cleavage-product frag 1 (~ 65 kDa) was not affected. Transfection of mutated $FLAG_3$-Tric-D487N revealed no fragment 1 and only to a very limited amount fragment 2. When the double-mutated $FLAG_3$-Tric-D441N/D487N was transfected, none of the fragments was detectable. These observations suggest that cleavage at D487 supports caspase-3-mediated cleavage at D441 (Figure 3B). Taken together, these results confirm D487 and D441 as potential caspase-sites in human tricellulin that are targeted in apoptotic cells.

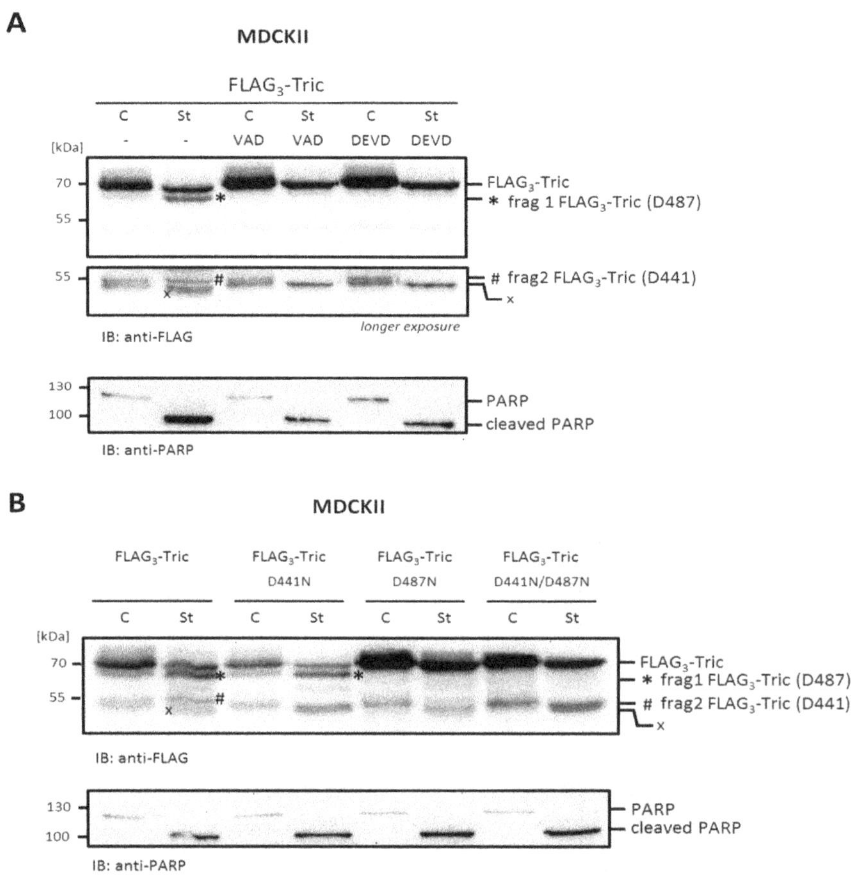

Figure 3. Caspase-3-mediated cleavage of $FLAG_3$-tricellulin in MDCKII upon apoptosis induction. (**A**) MDCKII cells were transiently transfected with p3xFLAG-CMV10-tricellulin, pre-treated with caspase inhibitors Z-VAD-FMK (VAD) or Z-DEVD-FMK (DEVD) for 1 h before induction of apoptosis with 1 μM staurosporine for 6 h. (**B**) MDCKII cells transiently transfected with $FLAG_3$-tricellulin wild-type or caspase-site mutated constructs as indicated were treated with 1 μM staurosporine for 6 h. The lower panels in (**A**) and (**B**) show Western blot detection of the typical PARP fragment generated by caspases confirming induction of apoptosis. Representative images of at least three independent experiments are shown. Bands marked with * represent caspase-dependent cleavage product frag 1 (~65 kDa) and # represents frag 2 (~55 kDa). The other bands (x) represent undefined or at least caspase-independent fragments.

2.3. The Functional Interaction of Tricellulin and LSR Is Disrupted during Apoptosis

Tricellulin is recruited to tTJs by lipolysis-stimulated lipoprotein receptor (LSR/angulin-1) in epithelial and endothelial cells [38,39]. This interaction is mediated by the cytosolic C-terminus of tricellulin [17]. In this context, the question arises if caspase-mediated cleavage within the cytosolic C-terminus of tricellulin affects its interaction with LSR. Therefore, co-immunoprecipitation experiments were performed using cell lysates obtained from HEK-293 cells transiently transfected with LSR together with either full-length tricellulin or deletion constructs lacking amino acids 487–558 ($FLAG_3$-TricΔ487–558), amino acid 441–558 ($FLAG_3$-TricΔ441–558) or complete cytosolic C-terminus ($FLAG_3$-TricΔC) (Figure 4A). Confirming literature, co-transfection of $FLAG_3$-Tric and green fluorescent protein GFP-tagged LSR in HEK-293 cells revealed an interaction of both proteins in co-immunoprecipitation experiments, whereas $FLAG_3$-TricΔC did only show a weak signal for GFP-LSR (Figure 4B). Similar to $FLAG_3$-TricΔC, an interaction between GFP-LSR and $FLAG_3$-TricΔ487–558 or $FLAG_3$-TricΔ441–558 protein lacking the cytosolic C-terminal parts released by caspases was not detectable (Figure 4B). In this context, it is interesting to note that, in a Western blot experiment, using the anti-tricellulin (clone 54H19L38) ABfinity^{TM} rabbit monoclonal antibody generated against amino acids 369–558 of human tricellulin did no longer detect neither the $FLAG_3$-TricΔ487–558 nor the $FLAG_3$-TricΔ441–558 protein in transiently transfected cells, thus suggesting that the epitope of this antibody is located between amino acids 487–558 (Supplementary Figure S2).

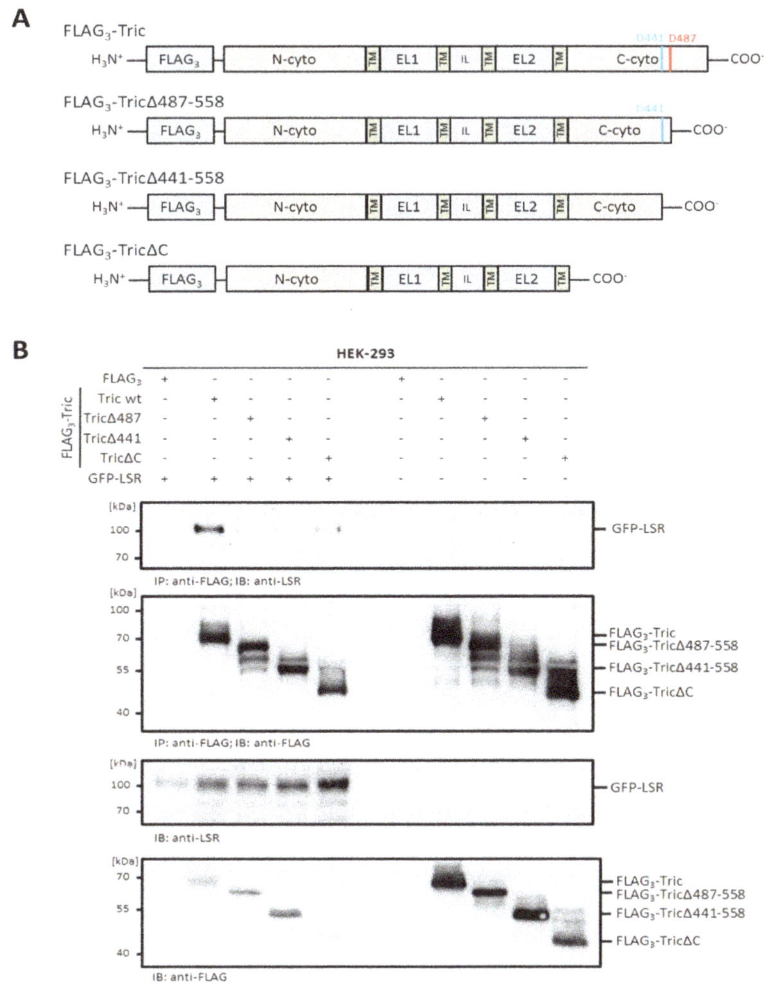

Figure 4. Deletion of the tricellulin C-terminus impairs binding to LSR/angulin-1. (**A**) Schematic representation of the $FLAG_3$-tricellulin constructs used in the transient transfection experiments. TM, transmembrane domain; EL, extracellular loop; IL, intracellular loop. (**B**) HEK-293 cells were transiently transfected with $FLAG_3$-tricellulin constructs missing the C-terminal tails corresponding to a cleavage by caspases together with GFP-LSR as indicated. Cells were lysed 48 h after transfection. Immunoprecipitaion (IP) was performed with monoclonal anti-FLAG antibody. Protein complexes from IP (upper two blots) and cell lysates (lower two blots) were analyzed by Western blotting with anti-LSR and anti-FLAG antibodies. The images are representatives of at least three independent experiments.

These results indicate that caspase-mediated cleavage of tricellulin liberates it from angulin and thus from tTJs. However, this does not exclude that LSR/angulin-1 itself is targeted during apoptosis.

Therefore, we next analyzed the fate of LSR/angulin-1 after induction of apoptosis in MDCKII cells. Indeed, staurosporine treatment led to a fragmentation of endogenous LSR/angulin-1 with time. Already 4 h after induction of apoptosis, four fragments of LSR were detectable using an anti-LSR antibody targeting the cytosolic part of LSR (Figure 5A). LSR/angulin-1 cleavage during apoptosis was inhibited by both caspase inhibitors Z-VAD-FMK and Z-DEVD-FMK. This indicates that LSR fragmentation is a consequence of caspase activation during apoptosis (Figure 5B).

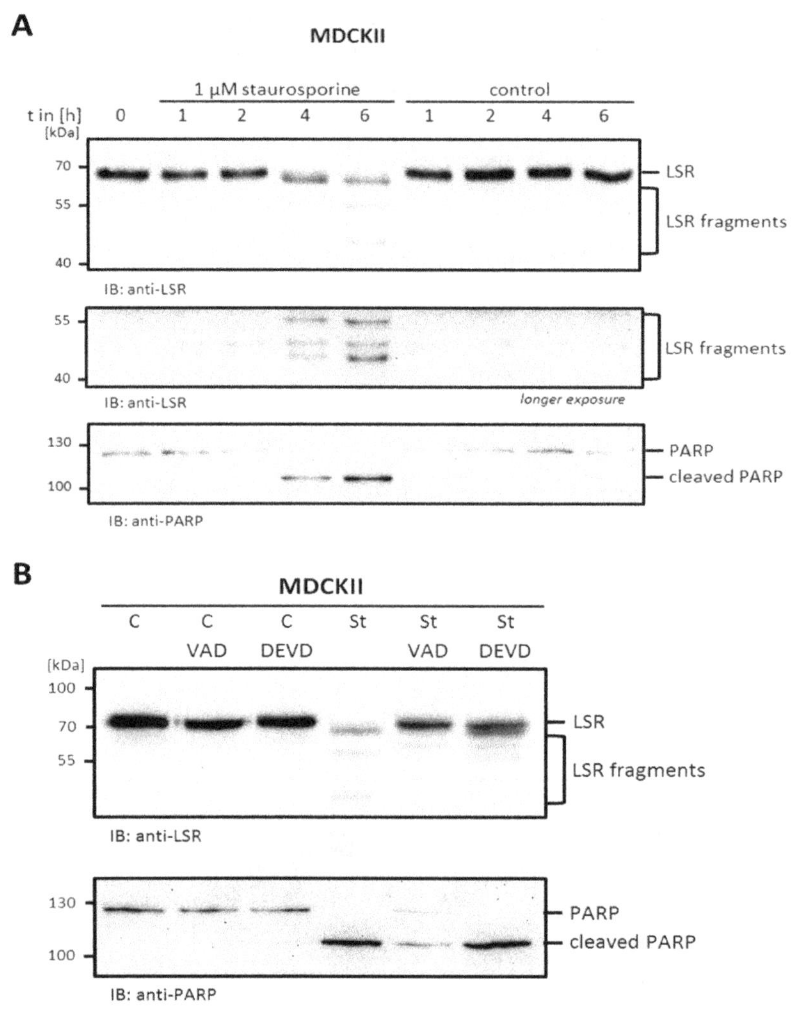

Figure 5. LSR in apoptotic cells. (**A**) Treatment of MDCKII cells with 1 μM staurosporine or solvent control for different times as indicated induced fragmentation of LSR. (**B**) Pre-treatment with caspase inhibitors (10 μM Z-VAD-FMK or 20 μM Z-DEVD-FMK for 1 h) inhibited fragmentation induced by staurosporine (1 μM for 6 h). Cell lysates were analyzed by Western blotting with anti-LSR and anti-PARP antibodies. Representative images of three independent experiments are shown.

3. Discussion

Release of apoptotic cells from epithelial and endothelial layers is critical for the maintenance of the barrier function and is often accompanied with a corresponding local decrease of transepithelial resistance [40]. In the gut, it is of importance that apoptotic cells, which are continuously extruded from the epithelium, do not generate microbial entry sites. Thus, it was not surprising that, in epithelial cell layers, there are mechanisms that efficiently break-down cell–cell contacts between apoptotic and neighboring cells. In previous studies, it was shown that components of the cadherin-catenin adhesion complex are targets of caspases. The adherens junction proteins E- and VE-cadherin are both cleaved by metalloproteinases, resulting in a release of the extracellular domain and thereby disrupting their trans-interactions with cadherin molecules on the surface of neighboring cells. Caspases are responsible for the cleavage of the cytosolic domains, thus disconnecting the interaction with the actin cytoskeleton [36,41]. Moreover, β-catenin as a linker between E-cadherin and the actin cytoskeleton is a caspase target itself [42]. Along with the adherens junctions, desmosomal cadherins and components of the cytosolic desmosomal plaque are inactivated similarly during apoptosis [43]. In addition, occludin, ZO-1 and ZO-2 were identified as caspase substrates [35]. Here, we extended this study to tricellulin and show that the C-terminal cytosolic tail of tricellulin is also targeted by caspases independent from whether apoptosis is induced by staurosporine or camptothecin and also independent from the cell system, in that case MDCKII or RT-112 cells. We identified two cleavage sites in tricellulin and mapped them C-terminal to Asp441 and Asp487. Specificity of cleavage was verified by treatment with caspase inhibitors and by mutation of the corresponding aspartate to asparagine, preventing the formation of cleavage fragments both in apoptotic cells as well as in in vitro fragmentation assays.

From our experiments, we conclude that D487 is the preferred cleavage-site. This is based on the observation that the corresponding cleavage-fragment was detectable far earlier as compared to the product generated by cleavage at D441 in in vitro cleavage-assays (Figure 2B). Moreover, it seems that initial cleavage at D487 is prerequisite for efficient cleavage at D441. This was confirmed by analyzing FLAG-tagged tricellulin where caspase cleavage-sites were mutated. The induction of apoptosis in cells transfected with wildtype $FLAG_3$-tricellulin revealed two cleavage fragments. However, in cells transfected with $FLAG_3$-tricellulin-D487N, a fragment generated by cleavage at D441 was more or less not detectable, most likely due to the lack of the initial cleavage at D487 (Figure 3B).

Caspase cleavage of tricellulin liberates a fragment of 71 C-terminal amino acids containing a region known to form a coiled-coil dimer [10]. This fragment was not detectable, even not when using high percentage SDS-PAGE gels, either because this small fragment escaped detection on Western blots or due to further degradation. Based on our observations, the antibody used in this study appears to bind to this fragment since truncated tricellulin where the C-terminus is deleted after D487 was no longer detectable. Definitively, it is worth testing by other methods if this fragment is detectable and of functional relevance. Not only due to its size can it easily translocate to the nucleus where it might be involved in transcriptional regulation, as shown for the amyloid precursor protein intracellular domain (AICD) [44] or the Notch intracellular domain (NICD) [45]. It has also been reported that nuclear localization of full-length tricellulin in pancreatic cancer promotes cell proliferation and invasiveness [46].

Based on our studies, we cannot exclude that the remaining tricellulin N-terminal part undergoes further processing steps as observed for occludin. There, the induction of apoptosis leads to caspase cleavage in the cytosolic C-terminus and to further fragmentation by metalloproteinases probably in the first extracellular loop [35]. First evidence for a potential contribution of metalloproteinases is provided by a study in Caco2 cells, where MMP-2- and MMP-3-catalyzed degradation of tricellulin is induced by N-3-(oxododecanoyl)-homoserine lactone (C12-HSL) [47].

A consequence of caspase cleavage and the concomitant release of the tricellulin C-terminal fragment is the disruption of the ZO-protein-mediated linkage of tricellulin to the actin cytoskeleton. Moreover, loss of the C-terminal amino acids might impair localization of tricellulin to tTJs by LSR/angulin-1. However, we observed that also LSR/angulin-1 is fragmented after induction of apoptosis. We conclude that, during apoptosis, tTJs are targeted at more than one level.

Taken together, here we showed that tricellulin is efficiently targeted by caspases in apoptotic cells within its C-terminal coiled-coil domain at D487 and D441. This together with concomitant cleavage of LSR/angulin-1 may contribute to a coordinated release of apoptotic cells from cell layers as it occurs continuously in the gastrointestinal tract at high frequency.

4. Materials and Methods

4.1. Cell Culture

Madine–Darby canine kidney II (MDCKII) cells were cultured in MEM with 10% (*v/v*) FBS and 1% (*v/v*) penicillin/streptomycin. Human bladder carcinoma cells (RT-112) as well as human embryonal kidney-293 cells (HEK-293) were cultivated in DMEM with 10% (*v/v*) FBS and 1% (*v/v*) penicillin/streptomycin using standard procedures as described elsewhere [31].

4.2. Reagents and Antibodies

Staurosporine was obtained from AppliChem GmbH (Darmstadt, Germany), caspase inhibitors Z-VAD-FMK and Z-DEVD-FMK as well as recombinant active caspase-3 were obtained from BD Biosciences (Heidelberg, Germany). Monoclonal anti-FLAG® M2 antibody was purchased from Sigma-Aldrich (Taufkirchen, Germany), anti-tricellulin (clone 54H19L38) ABfinity™ rabbit monoclonal antibody was from ThermoFisher Scientific (Darmstadt, Germany), anti-LSR (D3E3N) and anti-βactin (8H10D10) antibodies were obtained from Cell Signaling Technology (Frankfurt am Main, Germany) and anti-PARP (Ab-2) antibody was from Calbiochem (#AM30) (Merck KGaA, Darmstadt, Germany). Rabbit anti-GST antibody was provided by Jürgen Wienands. Goat anti-mouse-HRP and goat anti-rabbit-HRP antibodies were purchased from Sigma-Aldrich (Taufkirchen, Germany) and Alexa-Fluor™594-labeled secondary antibody was obtained from Molecular Probes (ThermoFisherScientific, Darmstadt, Germany). Dilution of the antibodies is summarized in Table 1.

Table 1. Antibodies used in the presented study including dilutions for Western blotting, immunofluorescence microscopy and immunoprecipitation.

Antibody	Dilution	Target	Species
anti-FLAG® M2	WB (1:5.000); IP 2 μg	FLAG-tag	mouse mab
anti-GST	WB (1:20.000)	glutathione-S-transferase	rabbit pab
anti-LSR (D3E3N) XP®	WB (1:1.000)	LSR	rabbit mab
anti-PARP-1 (Ab-2)	WB (1:1.000)	PARP	mouse mab
anti-tricellulin (54H19L38) ABfinity™	WB (1:1.000) IF (1 μg/mL)	tricellulin	rabbit mab
anti-βactin	WB (1:1000)	βactin	mouse
goat anti-mouse-HRP	WB (1:50.000–1:100.000)	mouse IgG	goat
goat anti-rabbit-HRP	WB (1:50.000–1:100.000)	rabbit IgG	goat
donkey anti-rabbit-Alexa594	IF (1: 1.000)	rabbit IgG	donkey

4.3. Plasmids, Site-Directed Mutagenesis and Transient Transfections

Plasmids were cloned by standard procedures and verified by sequencing. The cloning of p3xFLAG-CMV10-Tric and p3xFLAG-CMV10-TricΔC constructs was described previously [15]. To generate caspase cleavage-site mutated tricellulin variants, site-directed mutagenesis of the potential caspase cleavage-sites at positions D441 and D487 to asparagine was performed by SBOE (splicing by overlap extension)-PCR. Oligonucleotides used for the SBOE-PCR reactions are summarized in Table 2. The mutated tricellulin cDNAs were cloned into the vector p3xFLAG-CMV10 (Sigma-Aldrich, Taufkirchen, Germany) using *BamHI* restriction site. The C-terminal tricellulin deletion constructs TricΔ441–558 and TricΔ487–558 were generated by PCR using the oligonucleotides 5′-CGG ATC CTC AAA TGA TGG AAG ATC CAG-3′ as forward primer and 5′-CGG ATC CTT AGG GCA TCA CGA TAG GTT TAG-3′ or 5′-GGA TCC TTA CAG CTC ATC AAA CTT CCT CA-3′ as reverse primers, respectively, and cloned into the *BamHI* restriction site of p3xFLAG-CMV10 (Sigma-Aldrich, Taufkirchen, Germany). For recombinant expression of the tricellulin C-terminal cytosolic domain, pGEX4T1-TricC described previously [31] was used. The caspase-site mutated variants of GST-TricC were constructed by PCR employing oligonucleotides 5′-GCG GGA TCC ATG TGG AGG CAT GAG GCA GCT C-3′ and 5′-GCG GGT ACC GGA TCC TTA AGA ATA ACC TTG TAC ATC-3′ using p3xFLAG-CMV10-Tric-D441N, -D487N or -D441N/D487N as templates. After digestion with *BamHI*, the amplified products were ligated into *BamHI*-cleaved and dephosphorylated pGEX4T1 (GE Healthcare, Freiburg, Germany). pCAGGS-LSR for expression of GFP-tagged LSR was obtained from Mikio Furuse and described in [17].

MDCKII and HEK-293 cells were transiently transfected with the indicated plasmids using a DNA:PEI (poyethylenimine) ratio of 1:4 as described previously [48]. Cells were seeded in 6-well plates (2×10^5 cells for MDCKII and 5×10^5 HEK-293 cells per well one day before transfection. For transiently transfected HEK-293 or MDCKII cells, one or two 6-wells were lysed per sample. Empty vectors were transfected as a control.

Table 2. Sequences of oligonucleotides used to generate mutated tricellulin constructs. The bases marked in red represent those bases that differ from the wild-type sequence to generate the indicated mutations.

Mutation	Product	Sequence Oligonucleotides
Tric-D441N	1	5′-GCG GGT ACC GGA TCC GCC GCC ATG TCA AAT GAT GGA AGA TCC-3′ 5′-CAC ATA GTT GGG CAT CAC GAT-3′
	2	5′-ATC GTG ATG CCC AAC TAT GTG-3′ 5′-GCG GGT ACC GGA TCC TTA AGA ATA ACC TTG TAC ATC-3′
	1 + 2 (SBOE)	5′-GCG GGT ACC GGA TCC GCC GCC ATG TCA AAT GAT GGA AGA TCC-3′ 5′-GCG GGT ACC GGA TCC TTA AGA ATA ACC TTG TAC ATC-3′
Tric-D487N	1	5′-GCG GGT ACC GGA TCC GCC GCC ATG TCA AAT GAT GGA AGA TCC-3′ 5′-CAC TGC ATT CAG CTC ATC AAA-3′
	2	5′-TTT GAT GAG CTG AAT GCA GTG-3′ 5′-GCG GGT ACC GGA TCC TTA AGA ATA ACC TTG TAC ATC-3′
	1 + 2 (SBOE)	5′-GCG GGT ACC GGA TCC GCC GCC ATG TCA AAT GAT GGA AGA TCC-3′ 5′-GCG GGT ACC GGA TCC TTA AGA ATA ACC TTG TAC ATC-3′

4.4. *Induction of Apoptosis and Preparation of Cell Lysates*

MDCKII cells were plated on 6-well plates (2×10^5 cells per well for transfection; 8×10^5 cells per well without transfection) or 12-well plates (4×10^5 cells per well) and optionally transfected as described above. One day after seeding or transfection, apoptosis was induced by treatment with 1 μM staurosporine. Cells were lysed at the indicated time points. For inhibition of caspase activity, 10 μM Z-VAD-FMK or 20 μM Z-DEVD-FMK were added 1 h before induction of apoptosis for 6 h. Both floating and adherent cells from one well were harvested and pooled. Cells were lysed with 80 μL (12-well) or 100/150 μL (6-well) ice-cold modified RIPA-buffer (50 mM Tris/HCl pH 7.5, 150 mM NaCl, 0.5% (*w/v*) sodium-deoxycholate, 0.1% (*v/v*) SDS, 1% (*w/v*) Nonidet P-40 including Complete™ protease inhibitor mix (Roche Life Science, Mannheim, Germany)) for at least 15 min on ice. After sonication (15 pulses; cycle 0.5; 70% amplitude; UP100H ultrasound processor, Hilscher Ultrasound Technology, Teltow, Germany) the lysates were centrifuged (10 min, 20,800× *g*, 4 °C) to remove insoluble cell components.

4.5. *In Vitro Caspase Cleavage*

Recombinant glutathione S-transferase (GST)-tagged c-cytosolic (amino acids 366–558) domain of tricellulin wild type and mutant (D441N; D487N; D441N/D487N) proteins were expressed and purified as reported elsewhere [49]. In total, 4 μg of recombinant protein was digested with 150 ng of recombinant, active caspase-3 (BD Biosciences, Heidelberg, Germany) in 20 mM Pipes pH 7.2, 100 mM NaCl, 1 mM EDTA, 10% (*w/v*) sucrose, 10 mM DTT and 0,1% (*w/v*) CHAPS at 37 °C for the indicated incubation times. For inhibition of the in vitro caspase cleavage, 10 μM of pan-caspase inhibitor Z-VAD-FMK or DMSO as a control was added. Finally, samples were analyzed by SDS-PAGE and Western blot analysis.

4.6. Cell Lysis and Co-Immunoprecipitation

HEK-293 cells were seeded in 6-well plates and transfected (1 μg DNA each construct) with PEI as described above. Cells (two 6-wells) were harvested 48 h after transfection and lysed in 300 μL ice-cold modified RIPA-buffer including Complete™ protease inhibitor mix (Roche Life Science, Mannheim, Germany) for 20 min on ice. After lysis, samples were centrifuged (10 min, 20,800× *g*, 4 °C) to remove insoluble material. For co-immunoprecipitation experiments, 200 μL lysate was incubated with 2 μg of anti-FLAG® M2 antibody for 1 h at 4 °C under constant rotation. Subsequently, 30 μL of equilibrated Protein A Sepharose CL-4B beads (GE Healthcare, Freiburg, Germany) were added and incubated for further 30 min. Finally, beads were washed for three times with lysis buffer and protein complexes were eluted using 2 × SDS-loading buffer at 95 °C for 5 min. Samples were separated by SDS-PAGE and subsequently analyzed by Western blotting. For each experiment, 10 μL of cell lysate (5% of input) was loaded as a control.

4.7. Western Blot Analysis

Cell lysates were separated by SDS-PAGE and transferred to polyvinylidene difluoride (PVDF) membrane by tank blotting. After blocking with 5% (*w/v*) dry milk in Tris-buffered saline/Tween20 (TBS-T) for 1 h, membranes were incubated with the primary antibodies overnight at 4 °C and with horseradish peroxidase (HRP)-labeled secondary antibody at least for 1 h at room temperature (Table 1).

4.8. Immunofluorescence Microscopy

For immunofluorescence microscopy, MDCKII cells were seeded on transwell filter supports (Millicell Cell Culture Insert, 12 mm, 0.4 μm, Merck Millipore, Darmstadt, Germany) with a density of 150,000 cells/cm^2. At day 7, cells were treated with 1 μM staurosporine or DMSO for 4 h and 7 h. Subsequently, the cells were fixed with 2% paraformaledhyde in PBS and stained with 1 μg/mL anti-tricellulin (clone 54H19L38) antibody for 18 h at 4 °C followed by secondary antibody incubation (donkey anti-rabbit, Alexa-594 conjugated) for 1 h at room temperature. Nuclei were labeled using DAPI (Sigma-Aldrich, Taufkirchen, Germany). Images (z-stacks) were acquired with a laser scanning system TC/SP5 (Leica Microsystems, Wetzlar, Germany) and were maximum intensity projected.

Author Contributions: S.J., S.M. and J.R. performed experiments, analyzed and validated the data and edited the manuscript; O.H. conducted conceptual planning, discussed the data, and wrote and edited the manuscript.

Acknowledgments: We thank Mikio Furuse for the GFP-tagged LSR expression vectors. Expert technical assistance by Manuela Neumann is gratefully acknowledged.

Abbreviations

bTJ	Bicellular tight junction
CHAPS	3-[(3-cholamidopropyl)dimethylammonio]-1-propanesulfonate
DAPI	4,6-diamidino-2-phenylindole
DMEM	Dulbecco's Modified Eagle's Medium
DMSO	Dimethyl sulfoxide
EDTA	Ethylenediaminetetraacetic acid
FBS	Fetal bovine serum
GFP	Green-fluorescent protein
GST	Glutathione S-transferase
HRP	Horseradish peroxidase
LSR	Lipolysis-stimulated lipoprotein receptor
mab	Monoclonal antibody
MDCK	Madin-Darby canine kidney cells

pab	Polyclonal antibody
PARP	Poly [ADP-ribose] polymerase
PEI	Polyethlenimine
RIPA	Radioimmunoprecipitation Assay
SBOE-PCR	Splicing by overlap extension PCR
SDS	Sodium dodecyl sulfate
SDS-PAGE	Sodium dodecyl sulfate polyacrylamide gel electrophoresis
TBS-T	Tris-buffered saline/Tween 20
TER	Transepithelial resistance
TJ	Tight junction
tTJ	Tricellular tight junction
ZO-1	Zona occludens-1
Z-DEVD-FMK	N-Benzyloxycarbonyl-Asp-Glu-Val-Asp(OMe) fluoromethylketone
Z-VAD-FMK	N-Benzyloxycarbonyl-Val-Ala-Asp(OMe) fluoromethylketone

References

1. Barmeyer, C.; Schulzke, J.D.; Fromm, M. Claudin-related intestinal diseases. *Semin. Cell Dev. Biol.* **2015**, *42*, 30–38. [CrossRef] [PubMed]
2. Chawla, L.S.; Fink, M.; Goldstein, S.L.; Opal, S.; Gomez, A.; Murray, P.; Gomez, H.; Kellum, J.A.; Workgrp, A.X. The epithelium as a target in sepsis. *Shock* **2016**, *45*, 249–258. [CrossRef] [PubMed]
3. Haussner, F.; Chakraborty, S.; Halbgebauer, R.; Huber-Lang, M. Challenge to the intestinal mucosa during sepsis. *Front. Immunol.* **2019**, *10*, 891. [CrossRef] [PubMed]
4. Zihni, C.; Balda, M.S.; Matter, K. Signalling at tight junctions during epithelial differentiation and microbial pathogenesis. *J. Cell Sci.* **2014**, *127*, 3401–3413. [CrossRef] [PubMed]
5. Higashi, T.; Miller, A.L. Tricellular junctions: How to build junctions at the TRICkiest points of epithelial cells. *Mol. Biol. Cell* **2017**, *28*, 2023–2034. [CrossRef]
6. Staehelin, L.A. Further observation on the fine structure of freeze-cleaved tight junctions. *J. Cell Sci.* **1973**, *13*, 763–786. [PubMed]
7. Ikenouchi, J.; Furuse, M.; Furuse, K.; Sasaki, H.; Tsukita, S.; Tsukita, S. Tricellulin constitutes a novel barrier at tricellular contacts of epithelial cells. *J. Cell Biol.* **2005**, *171*, 939–945. [CrossRef]
8. Higashi, T.; Tokuda, S.; Kitajiri, S.; Masuda, S.; Nakamura, H.; Oda, Y.; Furuse, M. Analysis of the 'angulin' proteins LSR, ILDR1 and ILDR2-tricellulin recruitment, epithelial barrier function and implication in deafness pathogenesis. *J. Cell Sci.* **2013**, *126*, 966–977. [CrossRef]
9. Furuse, M.; Itoh, M.; Hirase, T.; Nagafuchi, A.; Yonemura, S.; Tsukita, S.; Tsukita, S. Direct association of occludin with ZO-1 and its possible involvement in the localization of occludin at tight junctions. *J. Cell Biol.* **1994**, *127*, 1617–1626. [CrossRef]
10. Schuetz, A.; Radusheva, V.; Krug, S.M.; Heinemann, U. Crystal structure of the tricellulin C-terminal coiled-coil domain reveals a unique mode of dimerization. *Ann. N. Y. Acad. Sci.* **2017**, *1405*, 147–159. [CrossRef]
11. Raleigh, D.R.; Marchiando, A.M.; Zhang, Y.; Shen, L.; Sasaki, H.; Wang, Y.; Long, M.; Turner, J.R. Tight junction-associated MARVEL proteins marveld3, tricellulin, and occludin have distinct but overlapping functions. *Mol. Biol. Cell* **2010**, *21*, 1200–1213. [CrossRef] [PubMed]
12. Krug, S.M.; Amasheh, M.; Dittmann, I.; Christoffel, I.; Fromm, M.; Amasheh, S. Sodium caprate as an enhancer of macromolecule permeation across tricellular tight junctions of intestinal cells. *Biomaterials* **2013**, *34*, 275–282. [CrossRef] [PubMed]
13. Krug, S.M.; Amasheh, S.; Richter, J.F.; Milatz, S.; Günzel, D.; Westphal, J.K.; Huber, O.; Schulzke, J.D.; Fromm, M. Tricellulin forms a barrier to macromolecules in tricellular tight junctions without affecting ion permeability. *Mol. Biol. Cell* **2009**, *20*, 3713–3724. [CrossRef] [PubMed]
14. Ikenouchi, J.; Sasaki, H.; Tsukita, S.; Furuse, M.; Tsukita, S. Loss of occludin affects tricellular localization of tricellulin. *Mol. Biol. Cell* **2008**, *19*, 4687–4693. [CrossRef] [PubMed]

15. Westphal, J.K.; Dörfel, M.J.; Krug, S.M.; Cording, J.D.; Piontek, J.; Blasig, I.E.; Tauber, R.; Fromm, M.; Huber, O. Tricellulin forms homomeric and heteromeric tight junctional complexes. *Cell. Mol. Life Sci.* **2010**, *67*, 2057–2068. [CrossRef] [PubMed]
16. Cording, J.; Berg, J.; Kading, N.; Bellmann, C.; Tscheik, C.; Westphal, J.K.; Milatz, S.; Günzel, D.; Wolburg, H.; Piontek, J.; et al. In tight junctions, claudins regulate the interactions between occludin, tricellulin and marvelD3, which, inversely, modulate claudin oligomerization. *J. Cell Sci.* **2013**, *126*, 554–564. [CrossRef]
17. Masuda, S.; Oda, Y.; Sasaki, H.; Ikenouchi, J.; Higashi, T.; Akashi, M.; Nishi, E.; Furuse, M. LSR defines cell corners for tricellular tight junction formation in epithelial cells. *J. Cell Sci.* **2011**, *124*, 548–555. [CrossRef]
18. Kohno, T.; Konno, T.; Kojima, T. Role of tricellular tight junction protein lipolysis-stimulated lipoprotein receptor (LSR) in cancer cells. *Int. J. Mol. Sci.* **2019**, *20*, 3555. [CrossRef]
19. Higashi, T.; Katsuno, T.; Kitajiri, S.; Furuse, M. Deficiency of angulin-2/ILDR1, a tricellular tight junction-associated membrane protein, causes deafness with cochlear hair cell degeneration in mice. *PLoS ONE* **2015**, *10*, e0120674. [CrossRef]
20. Riazuddin, S.; Ahmed, Z.M.; Fanning, A.S.; Lagziel, A.; Kitajiri, S.; Ramzan, K.; Khan, S.N.; Chattaraj, P.; Friedman, P.L.; Anderson, J.M.; et al. Tricellulin is a tight-junction protein necessary for hearing. *Am. J. Hum. Genet.* **2006**, *79*, 1040–1051. [CrossRef]
21. Ramzan, K.; Shaikh, R.S.; Ahmad, J.; Khan, S.N.; Riazuddin, S.; Ahmed, Z.M.; Friedman, T.B.; Wilcox, E.R.; Riazuddin, S. A new locus for nonsyndromic deafness DFNB49 maps to chromosome 5q12.3-q14.1. *Hum. Genet.* **2005**, *116*, 407–412. [CrossRef] [PubMed]
22. Nayak, G.; Lee, S.I.; Yousaf, R.; Edelmann, S.E.; Trincot, C.; Van Itallie, C.M.; Sinha, G.P.; Rafeeq, M.; Jones, S.M.; Belyantseva, I.A.; et al. Tricellulin deficiency affects tight junction architecture and cochlear hair cells. *J. Clin. Investig.* **2013**, *123*, 4036–4049. [CrossRef] [PubMed]
23. Kamitani, T.; Sakaguchi, H.; Tamura, A.; Miyashita, T.; Yamazaki, Y.; Tokumasu, R.; Inamoto, R.; Matsubara, A.; Mori, N.; Hisa, Y.; et al. Deletion of tricellulin causes progressive hearing loss associated with degeneration of cochlear hair cells. *Sci. Rep.* **2015**, *5*, 18402. [CrossRef] [PubMed]
24. Morozko, E.L.; Nishio, A.; Ingham, N.J.; Chandra, R.; Fitzgerald, T.; Martelletti, E.; Borck, G.; Wilson, E.; Riordan, G.P.; Wangemann, P.; et al. ILDR1 null mice, a model of human deafness DFNB42, show structural aberrations of tricellular tight junctions and degeneration of auditory hair cells. *Hum. Mol. Genet.* **2015**, *24*, 609–624. [CrossRef] [PubMed]
25. Sang, Q.; Li, W.; Xu, Y.; Qu, R.G.; Xu, Z.G.; Feng, R.Z.; Jin, L.; He, L.; Li, H.W.; Wang, L. ILDR1 deficiency causes degeneration of cochlear outer hair cells and disrupts the structure of the organ of Corti: A mouse model for human DFNB42. *Biol. Open* **2015**, *4*, 411–418. [CrossRef]
26. Dörfel, M.J.; Huber, O. Modulation of tight junction structure and function by kinases and phosphatases targeting occludin. *J. Biomed. Biotechnol.* **2012**, 807356. [CrossRef] [PubMed]
27. Dörfel, M.J.; Huber, O. A phosphorylation hotspot within the occludin C-terminal domain. *Ann. N. Y. Acad. Sci.* **2012**, *1257*, 38–44. [CrossRef]
28. Manda, B.; Mir, H.; Gangwar, R.; Meena, A.S.; Amin, S.; Shukla, P.K.; Dalal, K.; Suzuki, T.; Rao, R. Phosphorylation hotspot in the C-terminal domain of occludin regulates the dynamics of epithelial junctional complexes. *J. Cell Sci.* **2018**, *131*. [CrossRef]
29. Dörfel, M.J.; Westphal, J.K.; Huber, O. Differential phosphorylation of occludin and tricellulin by CK2 and CK1. *Ann. N. Y. Acad. Sci.* **2009**, *1165*, 69–73. [CrossRef]
30. Nakatsu, D.; Kano, F.; Taguchi, Y.; Sugawara, T.; Nishizono, T.; Nishikawa, K.; Oda, Y.; Furuse, M.; Murata, M. JNK1/2-dependent phosphorylation of angulin-1/LSR is required for the exclusive localization of angulin-1/LSR and tricellulin at tricellular contacts in EpH4 epithelial sheet. *Genes Cells* **2014**, *19*, 565–581. [CrossRef]
31. Jennek, S.; Mittag, S.; Reiche, J.; Westphal, J.K.; Seelk, S.; Dorfel, M.J.; Pfirrmann, T.; Friedrich, K.; Schutz, A.; Heinemann, U.; et al. Tricellulin is a target of the ubiquitin ligase Itch. *Ann. N. Y. Acad. Sci.* **2017**, *1397*, 157–168. [CrossRef] [PubMed]
32. Ramirez, M.L.G.; Salvesen, G.S. A primer on caspase mechanisms. *Semin. Cell Dev. Biol.* **2018**, *82*, 79–85. [CrossRef] [PubMed]
33. Songane, M.; Khair, M.; Saleh, M. An updated view on the functions of caspases in inflammation and immunity. *Semin. Cell Dev. Biol.* **2018**, *82*, 137–149. [CrossRef] [PubMed]

34. Gagliardi, P.A.; Primo, L. Death for life: A path from apoptotic signaling to tissue-scale effects of apoptotic epithelial extrusion. *Cell. Mol. Life Sci.* **2019**, *76*, 3571–3581. [CrossRef] [PubMed]
35. Bojarski, C.; Weiske, J.; Schöneberg, T.; Schröder, W.; Mankertz, J.; Schulzke, J.D.; Florian, P.; Fromm, M.; Tauber, R.; Huber, O. The specific fates of tight junction proteins in apoptotic epithelial cells. *J. Cell Sci.* **2004**, *117*, 2097–2107. [CrossRef] [PubMed]
36. Steinhusen, U.; Weiske, J.; Badock, V.; Tauber, R.; Bommert, K.; Huber, O. Cleavage and shedding of E-cadherin after induction of apoptosis. *J. Biol. Chem.* **2001**, *276*, 4972–4980. [CrossRef] [PubMed]
37. Kumar, S.; van Raam, B.J.; Salvesen, G.S.; Cieplak, P. Caspase cleavage sites in the human proteome: CaspDB, a database of predicted substrates. *PLoS ONE* **2014**, *9*, e110539. [CrossRef] [PubMed]
38. Iwamoto, N.; Higashi, T.; Furuse, M. Localization of angulin-1/LSR and tricellulin at tricellular contacts of brain and retinal endothelial cells in vivo. *Cell. Struct. Funct.* **2014**, *39*, 1–8. [CrossRef] [PubMed]
39. Sohet, F.; Lin, C.; Munji, R.N.; Lee, S.Y.; Ruderisch, N.; Soung, A.; Arnold, T.D.; Derugin, N.; Vexler, Z.S.; Yen, F.T.; et al. LSR/angulin-1 is a tricellular tight junction protein involved in blood-brain barrier formation. *J. Cell Biol.* **2015**, *208*, 703–711. [CrossRef] [PubMed]
40. Bojarski, C.; Gitter, A.H.; Bendfeldt, K.; Mankertz, J.; Schmitz, H.; Wagner, S.; Fromm, M.; Schulzke, J.D. Permeability of human HT-29/B6 colonic epithlium as a function of apoptosis. *J. Physiol.* **2001**, *535*, 541–552. [CrossRef] [PubMed]
41. Herren, B.; Levkau, B.; Raines, E.W.; Ross, R. Cleavage of β-catenin and plakoglobin and shedding of VE-cadherin during endothelial apoptosis: Evidence for a role for caspases and metalloproteinases. *Mol. Biol. Cell* **1998**, *9*, 1589–1601. [CrossRef] [PubMed]
42. Steinhusen, U.; Badock, V.; Bauer, A.; Behrens, J.; Wittmann-Liebold, B.; Dörken, B.; Bommert, K. Apoptosis induced cleavage of β-catenin by caspase-3 results in proteolytic fragments with reduced transactivation potential. *J. Biol. Chem.* **2000**, *275*, 16345–16353. [CrossRef] [PubMed]
43. Weiske, J.; Schöneberg, T.; Schröder, W.; Hatzfeld, M.; Tauber, R.; Huber, O. The fate of desmosomal proteins in apoptotic cells. *J. Biol. Chem.* **2001**, *276*, 41175–41181. [CrossRef] [PubMed]
44. Multhaup, G.; Huber, O.; Buee, L.; Galas, M.C. Amyloid Precursor Protein (APP) Metabolites APP intracellular fragment (AICD), Aβ42, and Tau in nuclear roles. *J. Biol. Chem.* **2015**, *290*, 23515–23522. [CrossRef] [PubMed]
45. Bray, S.J.; Gomez-Lamarca, M. Notch after cleavage. *Curr. Opin. Cell Biol.* **2018**, *51*, 103–109. [CrossRef]
46. Takasawa, A.; Murata, M.; Takasawa, K.; Ono, Y.; Osanai, M.; Tanaka, S.; Nojima, M.; Kono, T.; Hirata, K.; Kojima, T.; et al. Nuclear localization of tricellulin promotes the oncogenic property of pancreatic cancer. *Sci. Rep.* **2016**, *6*, 33582. [CrossRef]
47. Eum, S.Y.; Jaraki, D.; Bertrand, L.; Andras, I.E.; Toborek, M. Disruption of epithelial barrier by quorum-sensing N-3-(oxododecanoyl)-homoserine lactone is mediated by matrix metalloproteinases. *Am. J. Physiol. Gastrointest. Liver Physiol.* **2014**, *306*, G992–G1001. [CrossRef] [PubMed]
48. Wetzel, F.; Mittag, S.; Cano-Cortina, M.; Wagner, T.; Kramer, O.H.; Niedenthal, R.; Gonzalez-Mariscal, L.; Huber, O. SUMOylation regulates the intracellular fate of ZO-2. *Cell. Mol. Life Sci.* **2017**, *74*, 373–392. [CrossRef]
49. Mittag, S.; Valenta, T.; Weiske, J.; Bloch, L.; Klingel, S.; Gradl, D.; Wetzel, F.; Chen, Y.; Petersen, I.; Basler, K.; et al. A novel role for the tumour suppressor nitrilase1 modulating the Wnt/β-catenin signalling pathway. *Cell Discov.* **2016**, *2*. [CrossRef]

10

Intestinal Preservation Injury: A Comparison Between Rat, Porcine and Human Intestines

John Mackay Søfteland [1,2], Anna Casselbrant [3], Ali-Reza Biglarnia [4], Johan Linders [4], Mats Hellström [2], Antonio Pesce [5], Arvind Manikantan Padma [2], Lucian Petru Jiga [6], Bogdan Hoinoiu [7], Mihai Ionac [7] and Mihai Oltean [1,2,*]

1 The Transplant Institute, Sahlgrenska University Hospital, 413 45 Gothenburg, Sweden
2 Laboratory for Transplantation and Regenerative Medicine, Institute of Clinical Sciences, Sahlgrenska Academy at the University of Gothenburg, Sahlgrenska Science Park Medicinaregatan 8, 413 90 Gothenburg, Sweden
3 Department of Gastrosurgical Research and Education, Institute of Clinical Sciences, Sahlgrenska Academy at the University of Gothenburg, Sahlgrenska University Hospital, 41345 Gothenburg, Sweden
4 Department of Transplantation, Skåne University Hospital, 205 02 Malmö, Sweden
5 Department of Medical and Surgical Sciences and Advanced Technologies, University of Catania, Via Santa Sofia 86, 95123 Catania, Italy
6 Department for Plastic, Aesthetic, Reconstructive and Hand Surgery, Evangelisches Krankenhaus Oldenburg, Medical Campus University of Oldenburg, Steinweg 13–17, 26122 Oldenburg, Germany
7 Pius Branzeu Center for Laparoscopic Surgery and Microsurgery, University of Medicine and Pharmacy, P-ta. E. Murgu 2, 300041 Timisoara, Romania
* Correspondence: mihai.oltean@surgery.gu.se.

Abstract: Advanced preservation injury (PI) after intestinal transplantation has deleterious short- and long-term effects and constitutes a major research topic. Logistics and costs favor rodent studies, whereas clinical translation mandates studies in larger animals or using human material. Despite diverging reports, no direct comparison between the development of intestinal PI in rats, pigs, and humans is available. We compared the development of PI in rat, porcine, and human intestines. Intestinal procurement and cold storage (CS) using histidine–tryptophan–ketoglutarate solution was performed in rats, pigs, and humans. Tissue samples were obtained after 8, 14, and 24 h of CS), and PI was assessed morphologically and at the molecular level (cleaved caspase-3, zonula occludens, claudin-3 and 4, tricellulin, occludin, cytokeratin-8) using immunohistochemistry and Western blot. Intestinal PI developed slower in pigs compared to rats and humans. Tissue injury and apoptosis were significantly higher in rats. Tight junction proteins showed quantitative and qualitative changes differing between species. Significant interspecies differences exist between rats, pigs, and humans regarding intestinal PI progression at tissue and molecular levels. These differences should be taken into account both with regards to study design and the interpretation of findings when relating them to the clinical setting.

Keywords: tight junctions; organ preservation; intestine; transplantation; ischemia; intestinal mucosa

---

1. Introduction

Intestinal transplantation is the established therapeutic alternative in patients with complicated intestinal failure, with results continuously improving over the last two decades [1]. However, the post-transplant course is frequently marred by life-threatening complications due to ischemia–reperfusion injury (IRI), immunosuppression, and acute rejection, and patient management remains challenging [2–4]. Hence, further translational research is warranted to develop novel strategies to alleviate IRI, identify noninvasive biomarkers of rejection, and test alternative immunosuppressive strategies.

Intestinal grafts withstand the shortest cold storage (CS) period of all abdominal organs. In the clinical setting, CS is kept below ten hours due to concerns of mucosal sloughing and epithelial barrier breakdown, which may favor bacterial translocation and graft edema [5,6]. Numerous experimental approaches targeting the preservation injury have been tested in rats [7], but virtually none have been implemented clinically due to the lack of consistent evidence, including preclinical safety studies.

Rats have the advantage of simpler logistics, lower costs, and a relatively straightforward surgical procedure. Rat models have provided valuable insights into the physiology, immunology, and pathology of the transplanted intestine [8,9]. Nonetheless, anatomical, physiological, and immunological differences prevent the direct translation of many findings into clinical practice, and pigs are frequently used as a preclinical model to confirm the results of small animal studies [10–13]. Pigs share numerous anatomical and physiological similarities with humans, are easily accessible and affordable, and their use as livestock animals relieves some ethical concerns.

In spite of their use in intestinal preservation research, no direct comparison exists between rat and porcine intestines, to link the abundant data from rodents with this important preclinical model. To our knowledge, the extent to which the results obtained using porcine or rat intestines apply to the human intestine also remains unclear. Hence, it is unclear if and how the sequence and speed of development of the cellular and molecular alterations in rodents resemble the ones described in pigs and how this wealth of experimental data ultimately compares to the clinical setting. In this study, we set out to compare the development of the intestinal preservation injury in rats, pigs, and humans under similar conditions of procurement and CS.

2. Results

2.1. Histology

Intestinal CS induced the typical subepithelial lifting and edema in the vast majority of samples irrespective of species but the extent and speed of development of the subepithelial cleft revealed differences between species. Rat intestines developed significant subepithelial edema and even epithelial shedding (median Chiu/Park score 3) already after eight hours of CS, whereas the porcine intestines showed significantly lower injury score and mild edema or even normal histology (median Chiu/Park score 1) ($p < 0.001$). At the same time-point, human intestines exhibited mild or moderate subepithelial edema (median score 3)—a lesser injury than in rats ($p = 0.04$), but higher than in pigs ($p = 0.02$) (Figure 1A).

At both latter time-points, rat intestines showed a significantly more severe mucosal injury compared to porcine intestines; human intestines revealed significantly worse morphology compared to pigs after 24 h ($p < 0.01$). A particular feature in the porcine and human intestines was the significant lifting of the mucosa from the muscular layer (submucosal edema), a feature not present in the rat intestines.

Throughout the study, goblet cell (GC) counts indicated different patterns between the three species. Control rat intestine had significantly more GC than pig ileum (148 ± 28 vs 71 ± 23, $p < 0.05$). In rat and human intestines, CS induced a decrease in mucus-filled GC on the villi, which became significant after 14 h, whereas the amount of GC in porcine intestines did not differ significantly from the baseline throughout the entire experiment (Figure 1B).

Normal pig intestines had significantly higher enterocyte density and significantly fewer polymorphonuclear neutrophils (PMN) in the villi compared to rat and human intestines (Figure 2).

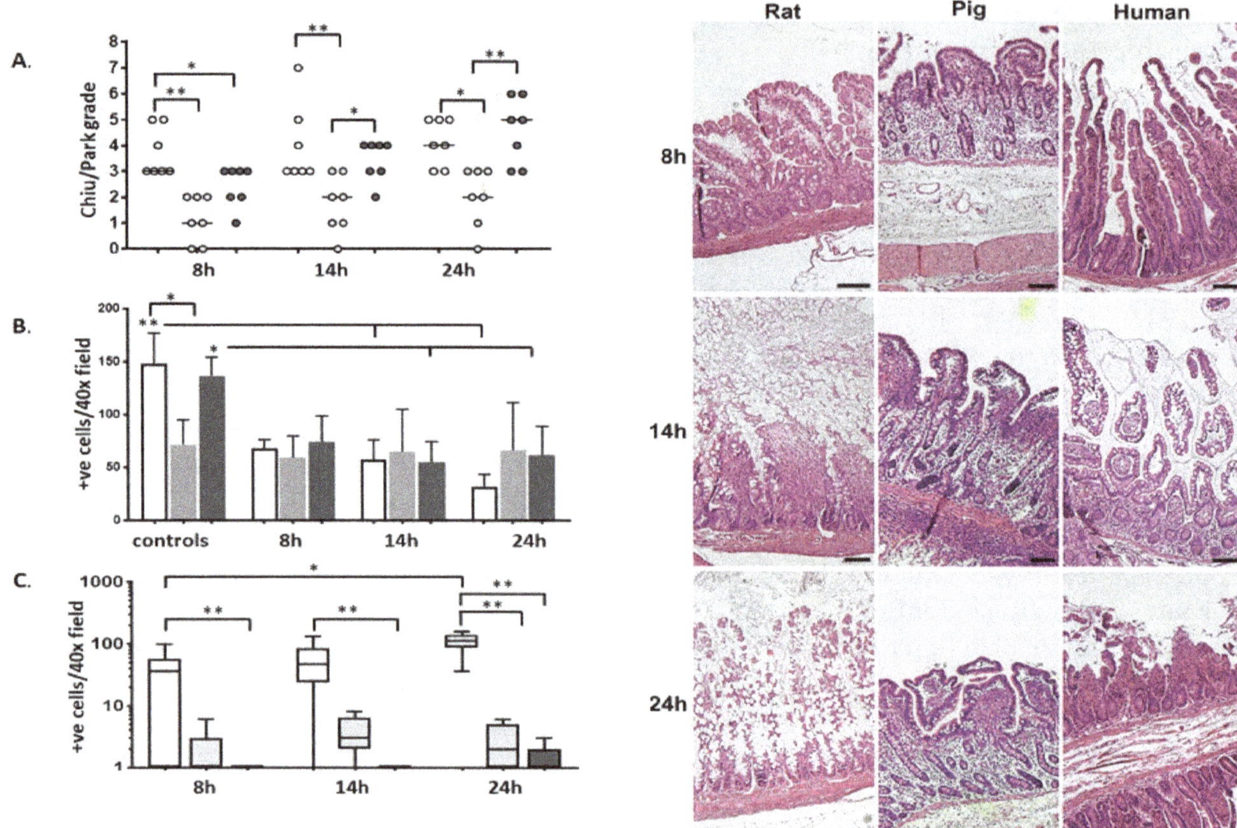

Figure 1. Light microscopy of rat (white), pig (light grey), and human (dark grey) intestines after different periods of cold storage (CS). (**A**) Summary of the tissue injury (Chiu score) induced by CS with each dot representing one individual ($n = 7$) and the bar showing the median value; (**B**) goblet cell count; (**C**) enterocyte apoptosis quantified by caspase-3 positive cells (box plot showing the median, 5–95th percentile, and lowest and highest values at each time point). * $p < 0.05$, ** $p < 0.01$. A large number of apoptotic enterocytes (positive for active caspase-3) were found in rat intestines after 8 h of CS. Rat intestines had more apoptotic enterocytes than human intestines at all time points (Figure 1C). Right: representative microphotographs from each species at each of the three time-points (hematoxillin eosin stain, original magnification ×100, scale bar 100 microns).

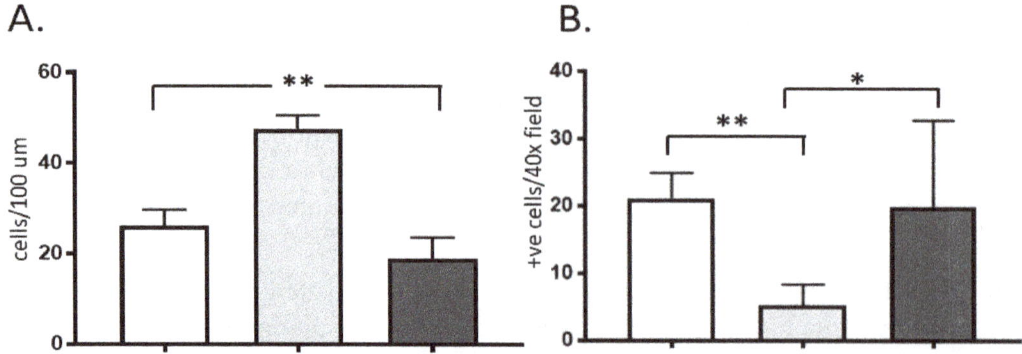

Figure 2. Enterocyte (**A**) and polymorphonuclear (PMN) leukocyte (**B**) counts in rat, (white bar) pig (light grey), and human (dark grey) intestines ($n = 7$). Enterocytes were counted using 4′,6-diamidino-2-phenylindole (DAPI) staining on the sides of longitudinally oriented villi on several 100 μm segments; PMNs were counted in villi on ten random fields at high magnification (×400) (data shown as mean ± SD). * $p < 0.05$, ** $p < 0.01$.

2.2. Immunohistochemistry

In all three species, zonula occludens (ZO)-1 was detected as an intense, thin fluorescent signal at the apical tips of the basolateral membrane, from the crypts to the tip of the villi. Whereas ZO-1 staining in rats and humans frequently appeared like a line or large dots, ZO-1 staining in pigs often had the appearance of a dotted line or small dots (probably due to a narrower apical membrane and higher cellularity). Claudin-3 was visualized as a thin, reticular signal along the entire basolateral membrane. Claudin-3 frequently colocalized with ZO-1 in an area corresponding to the apical edge of the basolateral membrane (data not shown).

After eight hours of CS, ZO-1 staining became absent or discontinuous at the tip of some villi in rat intestines but overall it was maintained along the entire villus (Figure 3). Pig and human intestines revealed a strong immunosignal along the entire contour of the villi. In rats, claudin-3 staining was found between enterocytes but showed a widespread de-colocalization from ZO-1 as well as some cytoplasmic staining. Both pig and human showed strong claudin-3 staining as a sharp, reticular fluorescence signal along the entire basolateral membrane (Figure S1).

In rats, fourteen hours of CS led to a marked decrease in the ZO-1 immunostaining, which was preserved only towards the base of the villi and in the crypts. Porcine and human intestines continued, however, to show unchanged, well-preserved ZO-1 expression along the villus. Claudin-3 staining between enterocytes became more diffuse while cytoplasmic staining was also noted. Overall, the staining pattern remained thin and fibrillar but with a tendency towards less sharp, diffuse membrane staining and cytoplasmic staining. Stronger subjunctional intensity was also noted in some samples.

After 24 h, all rat intestines completely lacked villus staining for ZO-1, while both porcine and human continued to show immunofluorescent staining frequently reaching villus tips. In both pig and human intestines, claudin-3 revealed more diffuse, discontinuous staining along the basolateral membrane with an obvious subjunctional staining gradient.

2.3. Western Blot Analysis

All proteins analyzed by Western blot were detected in rat, pig, and human samples. Generally, all proteins studied were found to have the lowest expression in the rat small intestine. After eight hours of CS, the expression of claudin-3, claudin-4, tricellulin, and ZO-1 was significantly higher in human samples compared to rat samples. This difference persisted after fourteen and 24 h for claudin-4 but subsided for claudin-3, tricellulin, and ZO-1.

In four out of six tight junction (TJ) proteins studied (occludin, tricellulin, claudin-3, ZO-1) no differences between species were found after 14 h and 24 h of CS.

Pig tissue expressed more occludin at eight hours as well as more Ck8 protein at all time points compared to rats (Figure 4).

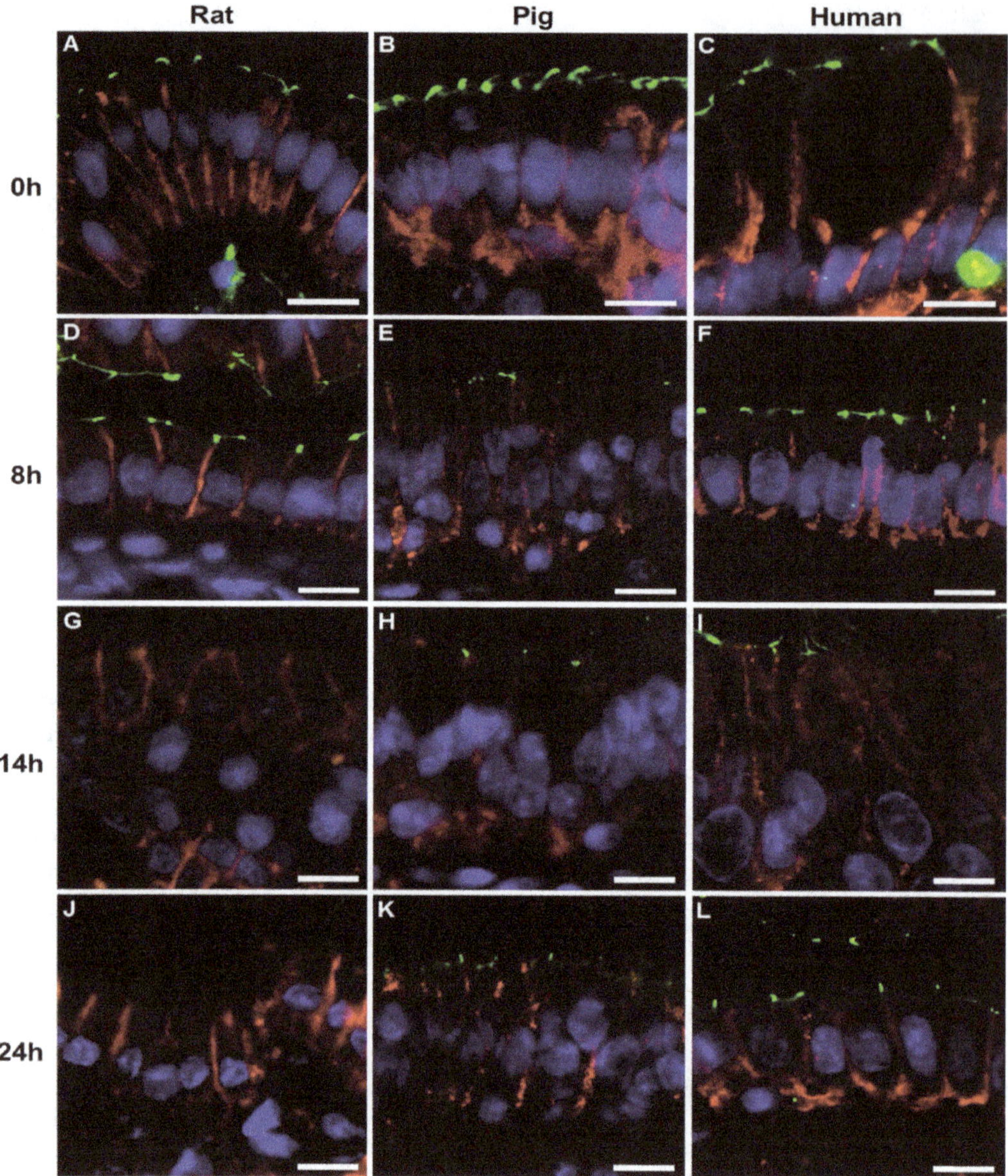

Figure 3. Immunofluorescent staining for zonula occludens (ZO)-1 (green) and claudin-3 (red) after various periods of cold storage (CS); strong immunofluorescent signal for both proteins after 8 h CS in rat (**A**), pig (**B**), and human intestine (**C**); after 14 h, ZO-1 signal was lost in rat (**D**) but not pig (**E**) or human (**F**) intestinal mucosa; after 24 h (**J–L**) of CS, ZO-1 staining was absent and claudin-3 revealed diffuse membrane staining and cytoplasmic staining in rat intestines (**G**), while in pig (**H**) and human (**I**) intestines, ZO-1 signal was maintained and claudin-3 stained more diffuse, stronger on the subjunctional basolateral membrane, together with some cytoplasmic staining. Nuclei were stained blue using 4′,6-diamidino-2-phenylindole (DAPI). Images were acquired from areas where enterocytes still remained attached to the lamina propria and as close to the villus tip as possible. Original magnification ×400, scale bar, 10 μm.

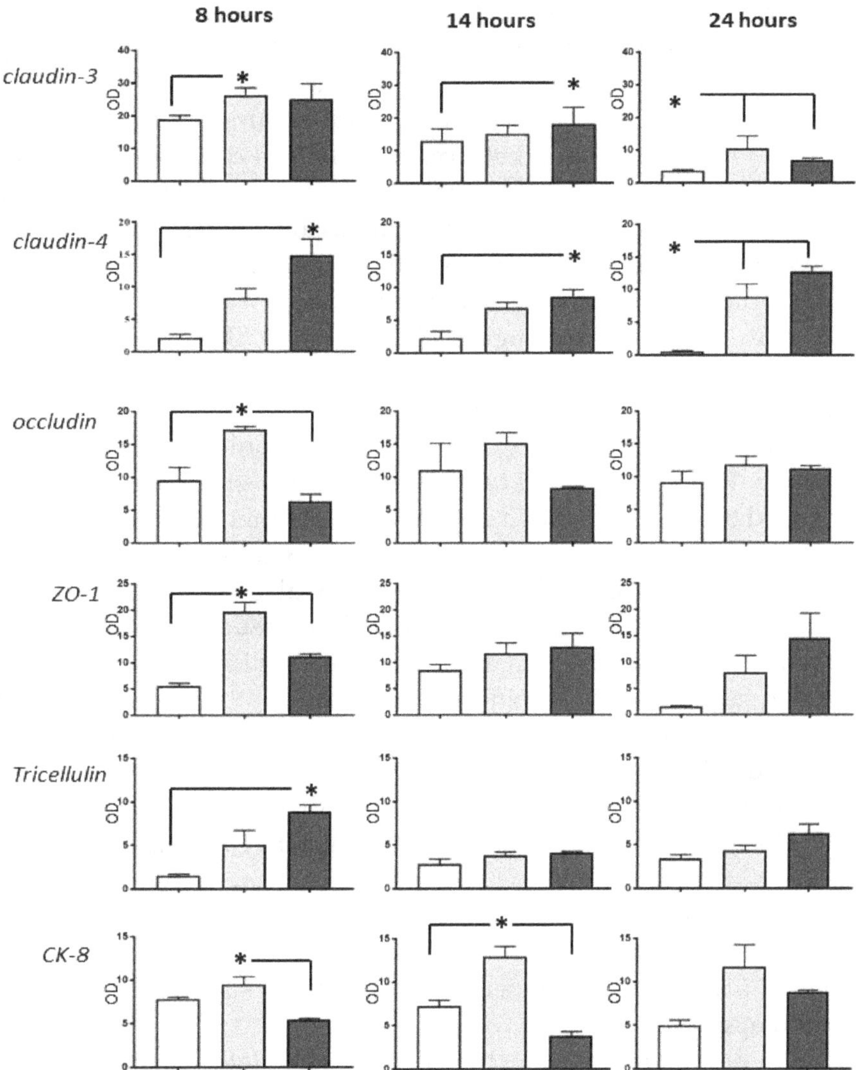

Figure 4. Western blot analysis of several junctional proteins and cytokeratin (CK)-8 in rat (white bars), pig (light grey bars), and human (dark grey bars) intestines after different periods of cold storage. Samples (15 μg) from the three species from the same time-point were run simultaneously (n = 3–4). Results (mean ± standard error) were normalized to glyceraldehyde-3-phosphate dehydrogenase and presented as semiquantitative results (optical density, OD); * $p < 0.05$.

3. Discussion

Ischemia–reperfusion injury remains a major concern after intestinal transplantation as tissue damage may favor bacterial translocation and sepsis, anastomotic leaks, and intestinal graft edema with risk for abdominal compartment syndrome [2]. Moreover, an advanced ischemic injury may promote the later occurrence of graft fibrosis and graft dysmotility [14]. The susceptibility of the intestine to ischemic injury and the life-threatening complications it may lead to, continues to mandate a search for protective and therapeutic interventions.

Although some studies infer a higher resilience of porcine intestines towards intestinal ischemia as compared to rats [15–17], to our knowledge this is the first direct comparison between these species. The sequential evaluation of the intestinal preservation injury in this study found a substantially different pattern of changes in human intestines compared with rats and pigs, both regarding the time course and the type of tissue damage. Rat and human intestines developed significant mucosal changes already after eight hours of CS whereas porcine intestines revealed near-normal epithelium after the same time span. Conversely, both the human and porcine intestines developed significant submucosal edema, a feature not observed in rodents.

Goblet cells are critical for the integrity and repair of the intestinal epithelium and are considered a good marker of intestinal health [18]. GC mucus depletion occurs rapidly after the onset of intestinal ischemia [19]. In rats and humans, goblet cell count decreased during the CS compared to normal tissue, whereas this phenomenon was absent in the porcine ileum, which revealed a stable GC number throughout the experiment. Interestingly, control porcine tissue had less GC compared to both rats and humans. It is unclear whether this finding is intrinsic to (juvenile) pigs or it is the result of preoperative fasting as a decrease in amount and mucus content of GC count have been reported early and after fasting or weaning and in malnourished piglets [20]. Notably, the initial GC count seemed higher in the human intestines, while it is likely that the human organ donors did not receive any enteral nutrition during the day preceding the organ procurement either.

Pigs developed mucosal alterations at the later time points compared to humans and rats. For example, the epithelial lesion recorded after 24 h (massive epithelial lifting, grade 3) in the pig intestines usually occurred between 8 and 14 h of CS in rat and human intestines. Part of the explanation for these interspecies differences may be the higher mucosal cellularity in pigs, leading to a higher TJ density. The lesser amount of tissue PMNs in pigs may also play a role as the hypoxic, stressed leukocytes could release its lytic enzymes in the surrounding tissue already before reperfusion [21,22]. The practical consequence of this finding is that preservation studies using pig intestines would require longer CS periods than in rats to attain a significant tissue injury. When designing experimental studies that mimic the human situation it may be prudent to adjust the ischemia time required to create a comparable injury to reflect the interspecies differences.

Caspase-3 activation and increased apoptosis has been reported earlier following CS of rat kidneys, livers, and intestines [23–26]. Similarly, we found abundant active caspase-3 in rat intestines; however, caspase-3 positive pig or human enterocytes were significantly fewer. This intriguing finding is difficult to explain considering that humans had very few caspase-3 positive enterocytes but it may also reflect interspecies differences. In another study, under similar conditions (60 min of intestinal ischemia and 120 min of reperfusion) 75% of rat Paneth cells entered apoptosis, whereas only 25% of the human Paneth cells were found apoptotic [27].

Ischemia and ATP depletion disrupt the actin cytoskeleton in various types of cells and tissues [28,29]. The actin cytoskeleton plays essential roles in the functional and structural integrity of the cells, including the structure and function of tight junctions. Thus, one suggested mechanism behind the TJ dysfunction is the strain on the TJs by the neighboring, contracting cells. Internalization of TJ proteins and TJ disassembly have been shown to occur rapidly after various stimuli, followed by TJ dysfunction and increased permeability [30]. Earlier studies revealed quantitative TJ protein changes during intestinal ischemia [31] yet the qualitative changes (i.e., cytoplasmic shift, altered membrane staining pattern) revealed by the immunofluorescence claudin-3 staining may be equally relevant. Herein, the progress of injury seems to have coincided with the occurrence of significant qualitative and quantitative alterations in claudin expression (particularly the TJ-sealing protein, claudin-3), whereas the expression of ZO-1 seems to have limited importance for injury development. In addition, it is tempting to speculate that the rapid TJ-protein alterations in the non-fasted rats, particularly tricellulin and ZO-1, may be also due to the mucosal exposure to the aggressive intestinal chyle containing bile acids, pancreatic enzymes combined with the depletion of the protective CG and mucus layer.

An advantage of the current study is the systematic use of distal small intestine. The different digestive functions of the various intestinal segments are also reflected in the varying content of different junctional proteins [32]. Hence, we minimized the differences between different anatomical areas as responsible for the differences noted between various junctional proteins in different species. An inherent drawback of the study was the use of human intestines from brain dead organ donors. This setting was both necessary and relevant as it mirrored the clinical situation of intestinal procurement and preservation for transplantation. Nonetheless, this may have induced additional changes and differences compared to the young, healthy animals, as brain death induces both local and systemic inflammatory processes [33,34]. The relatively short period of brain death customarily encountered in Sweden may have limited the effect of donor inflammation on the intestine. Besides brain death, organ donors may also have been subjected to hemodynamic instability, cardiac arrest, trauma, or medical interventions (vasopressors, fluid resuscitation) that may potentially have affected the intestine. Whereas some of these factors were present in our human tissue donors, these factors not always preclude intestinal donation [35]. Discordant age between study subjects (young animals vs middle-aged organ donors) could also be regarded as another limitation. Though the current practice usually restricts the use of intestinal donors older than 50 [36], the impact of age on the development of intestinal preservation injury is yet unknown. Last but not least, the antibodies used in the study may have different affinity to different species. This implies that a higher protein expression detected on the Western blot does not always reflect its real tissue expression but rather shows the antibody–antigen affinity, which may differ between species.

In conclusion, this report provides the first direct comparison of the development of intestinal preservation injury in the rat, pig, and human at histological and molecular levels. The current results suggest that porcine intestines have a slower development of the tissue injury compared to human intestines, while rat intestines appear to have a faster injury development. These differences should be taken into account when designing experimental studies to allow meaningful endpoints and results.

4. Materials and Methods

4.1. Animals, Surgery, and Sampling

Male, Sprague–Dawley rats ($n = 7$) aged around 3 months were purchased from Charles River (Sulzfeld, Germany), housed in the University animal quarters, and acclimatized for one week. The rats received rat chow and water ad libitum and were not fasted before surgery. The study followed the regulations outlined by the European Union (2010/63/EU) and was reviewed and approved by the Gothenburg committee of the Swedish Animal Welfare Agency (#135/07). Under 2.5% isoflurane anesthesia, the small intestine was perfused with and stored in ice-cold histidine–tryptophan–ketoglutarate solution (HTK, Custodiol®, Fresenius Kohler Chemie GmbH, Alsbach-Hähnlein, Germany) as described earlier [15]. The distal half (ileum) was resected and its ends were tightly ligated using silk 3/0. After 8 h, 14 h, and 24 h of CS 3 cm segments of ileum were sampled and stored in 4% buffered formalin or snap frozen.

Landrace pigs of either sex ($n = 7$), weighing around 30 kg were purchased from a commercial supplier and housed individually at the Pius Branzeu Center in Timisoara. Animals were acclimatized for one week, fed once daily with standard pig diet and provided with water ad libitum. Food was withdrawn 24 h before surgery but animals' unrestricted access to water was maintained. All experiments were reviewed and approved by the Ethics and Deontology Committee for Research on Animals of the University of Medicine and Pharmacy, Timisoara, Romania (13008/9 May 2013). Following premedication with ketamine (20 mg/kg; Pfizer Pharma GmbH, Germany), xylazine (2 mg/kg), and atropine (0.05 mg/kg), pigs were intubated and ventilated using a mixture of isoflurane and oxygen. Using a previously described approach [16] and following perfusion with 1.5 L HTK solution, the complete small intestine was then excised. In an ice basin on a backtable, the last meter of the ileum was resected, placed in ice-cold HTK solution, and sampled after 8 h, 14 h, and 24 h of CS. Samples were stored in 4% buffered formalin or snap frozen.

4.2. Human Organ Donors

Ileal segments were obtained from seven deceased brain dead (DBD) multiorgan donors with an intensive care unit stay of less than four days. Donor median age was 48 years (range 14–63) (additional donor information is provided in Table S1). The donors (or next of kin) previously consented for tissue use for medical research. The use of human tissue in the study was reviewed and approved by the regional ethical review committee (Dnr 204-17).

Organ retrieval was performed in the standard fashion using retrograde aortic perfusion with HTK solution and venous venting through the inferior vena cava. One meter of the distal small intestine (ileum) was resected immediately after the organ perfusion with HTK (3–8 L) and before any other organ was removed. Bowel ends were stapled off and the specimen was placed in an organ bag with cold HTK on ice. After 8 h, 14 h, and 24 h of CS a 10–15 cm ileal segment was removed with a stapler and samples were either placed in 4% formalin or snap frozen.

4.3. Histology

4.3.1. Light Microscopy

Formalin-fixed tissue was paraffinized, embedded, and cut into five-micron sections. Sections were stained with hematoxylin and eosin, and intestinal preservation injury was scored blinded by two experienced observers using the Chiu/Park score [37] on seven fields from three different sections.

Mucus-filled goblet cells (GCs) in the intestinal villi were stained using Alcian Blue staining and counted in ten random fields at high magnification (×400) by a single observer blinded to the study design.

Apoptosis was studied on paraffin sections using immunostaining for active (cleaved) caspase-3 using a Warp Red Chromogen kit (Bio-Care Medical, Concord, CA) according to the manufacturer's instructions. Briefly, after deparaffinization, rehydration, and antigen retrieval using citrate buffer (10 mM, pH 6.0), sections were blocked and then incubated with primary rabbit antibody against cleaved caspase-3 (1:100, #D175; Cell Signaling Technology, Danvers, MA) for 1 h at room temperature followed by incubation with an anti-rabbit probe, a rabbit alkaline phosphatase polymer and warp red chromogen. Nuclei were stained using Myers hematoxylin. Positively labeled enterocytes were counted on ten random fields at high magnification (×400) by a single observer. Polymorphonuclear neutrophils (PMN) were stained using the Naphtol AS-D chloroacetate esterase kit (Sigma Chemicals, St Louis, Mo) and counted in the villi on ten random fields at high magnification (×400).

4.3.2. Immunofluorescence

Paraffin sections were deparaffinized and rehydrated, then antigen retrieval was performed (citrate buffer). After species-specific blocking, slides were incubated overnight at 4 °C with antibodies against zonula occludens (ZO-1; 1:100, Invitrogen AB, Lidingö, Sweden) and claudin-3 (1:100; Abcam, UK). Thereafter, slides were incubated with secondary antibody conjugated with Alexa 488 and Alexa 594 (1:200; Invitrogen). The sections were counterstained with 4′6′-diamidino-2-phenylindole, mounted with aqueous mounting medium (Vector Laboratories, Burlingame, CA, USA), and examined by fluorescence microscopy (Leica). Nuclei on the villi were also counted. Image acquisition and processing were performed using the Leica LAS software.

4.3.3. Western Blot Analyses of Intestinal Mucosa

Western blot protein analysis was performed using whole tissue frozen specimens as described earlier [15). In brief, after electrophoresis and protein transfer on poly-vinyl-difluoride membranes, the membranes were blocked, then incubated overnight at 4 °C with primary antibody against claudin-3 (34-1700, Invitrogen AB, Lidingö, Sweden), claudin-4 (32-9400, Invitrogen AB), tricellulin (48-8400, Invitrogen AB), cytokeratin-8 (ab53708, Abcam, Cambridge, UK), ZO-1 (33-9100, Invitrogen AB), occludin (71-1500, Invitrogen AB), and the loading control glyceraldehyde-3-phosphate dehydrogenase (GAPDH, IMG-5143A, Imgenex, San Diego, CA). After repeated washings, a secondary antibody was applied for 1 h at room temperature and visualization was carried out using the chemoluminescent enzyme substrate CDP-Star (Tropix, Bedford, MA). The signal intensities of specific bands were detected and analyzed using a Chemidox XRS cooled charge-couple device camera and Quantity One software (BioRad Laboratories, Hercules, CA). GAPDH was used as loading control. For each sample, the optical density of primary antibody was normalized to GAPDH. Before re-probing with a new primary antibody, the membranes were incubated with stripping buffer (Re-Blot Plus Mild Solution 10×, Millipore, Temecula, CA, USA).

4.4. Statistical Analysis

Nonparametric methods were used for statistical comparisons. Statistical differences between independent groups were calculated using the Kruskal–Wallis test corrected for multiple comparisons using the Tukey test, followed by the Mann–Whitney U test (GraphPad Prism6; GraphPad Software, La Jolla, CA). Data are presented as median (range) unless otherwise stated. Results were considered as statistically significant at $p < 0.05$.

Author Contributions: Conceptualization: J.M.S., A.C., and M.O.; Methodology: J.M.S., A.C., M.H., A.P., A.-R.B., J.L., L.P.J., B.H., M.I., A.M.P., and M.O.; Formal Analysis: J.M.S., A.C., A.P., and M.O.; Writing—Original Draft Preparation: J.M.S., A.C., M.H., and M.O.; Writing—Review and Editing: A.P., A.-R.B., J.L., L.P.J., B.H., M.I., and A.M.P.; Funding Acquisition—M.O. and B.H.

References

1. Abu-Elmagd, K.M.; Costa, G.; Bond, G.J.; Soltys, K.; Sindhi, R.; Wu, T.; Koritsky, D.A.; Schuster, B.; Martin, L.; Cruz, R.J.; et al. Five hundred intestinal and multivisceral transplantations at a single center: Major advances with new challenges. *Ann. Surg.* **2009**, *250*, 567–581. [CrossRef] [PubMed]
2. Clouse, J.W.; Kubal, C.A.; Fridell, J.A.; Mangus, R.S. Posttransplant complications in adult recipients of intestine grafts without bowel decontamination. *J. Surg. Res.* **2018**, *225*, 125–130. [CrossRef] [PubMed]
3. Huard, G.; Schiano, T.D.; Moon, J.; Iyer, K. Severe acute cellular rejection after intestinal transplantation is associated with poor patient and graft survival. *Clin. Transplant.* **2017**, *31*. [CrossRef] [PubMed]
4. Varkey, J.; Simrén, M.; Jalanko, H.; Oltean, M.; Saalman, R.; Gudjonsdottir, A.; Gäbel, M.; Borg, H.; Edenholm, M.; Bentdal, O.; et al. Fifteen years' experience of intestinal and multivisceral transplantation in the Nordic countries. *Scand J. Gastroenterol.* **2015**, *50*, 278–290. [CrossRef] [PubMed]
5. Tesi, R.J.; Jaffe, B.M.; McBride, V.; Haque, S. Histopathologic changes in human small intestine during storage in Viaspan organ preservation solution. *Arch. Pathol. Lab. Med.* **1997**, *121*, 714. [PubMed]
6. deRoover, A.; de Leval, L.; Gilmaire, J.; Detry, O.; Boniver, J.; Honoré, P.; Meurisse, M. A new model for human intestinal preservation: Comparison of University of Wisconsin and Celsior preservation solutions. *Transplant. Proc.* **2004**, *36*, 270–272. [CrossRef] [PubMed]
7. Oltean, M.; Churchill, T.A. Organ-specific solutions and strategies for the intestinal preservation. *Int. Rev. Immunol.* **2014**, *33*, 234–244. [CrossRef] [PubMed]
8. Grant, D.; Hurlbut, D.; Zhong, R.; Wang, P.Z.; Chen, H.F.; Garcia, B.; Behme, R.; Stiller, C.; Duff, J. Intestinal permeability and bacterial translocation following small bowel transplantation in the rat. *Transplantation* **1991**, *52*, 221–224. [CrossRef] [PubMed]
9. Oltean, M.; Zhu, C.; Mera, S.; Pullerits, R.; Mattsby-Baltzer, I.; Mölne, J.; Hallberg, E.; Blomgren, K.; Olausson, M. Reduced liver injury and cytokine release after transplantation of preconditioned intestines. *J. Surg. Res.* **2009**, *154*, 30–37. [CrossRef] [PubMed]
10. Grant, D.; Duff, J.; Zhong, R.; Garcia, B.; Lipohar, C.; Keown, P.; Stiller, C. Successful intestinal transplantation in pigs treated with cyclosporine. *Transplantation* **1988**, *45*, 279–284. [CrossRef]
11. Pirenne, J.; Benedetti, E.; Gruessner, A.; Moon, C.; Hakim, N.; Fryer, J.P.; Troppmann, C.; Nakhleh, R.E.; Gruessner, R.W. Combined transplantation of small and large bowel. FK506 versus cyclosporine A in a porcine model. *Transplantation* **1996**, *61*, 1685–1694. [CrossRef] [PubMed]
12. Pakarinen, M.P.; Kuusanmäki, P.; Lauronen, J.; Paavonen, T.; Halttunen, J. Effects of ileum transplantation and chronic rejection on absorption and synthesis of cholesterol in pigs. *Pediatr. Surg. Int.* **2003**, *19*, 656–661. [CrossRef] [PubMed]
13. Weih, S.; Kessler, M.; Fonouni, H.; Golriz, M.; Nickkholgh, A.; Schmidt, J.; Holland-Cunz, S.; Mehrabi, A. Review of various techniques of small bowel transplantation in pigs. *J. Surg. Res.* **2011**, *171*, 709–718. [CrossRef] [PubMed]

14. Schaefer, N.; Tahara, K.; Schuchtrup, S.; Websky, M.V.; Overhaus, M.; Schmidt, J.; Wirz, S.; Abu-Elmagd, K.M.; Kalff, J.C.; Hirner, A.; et al. Perioperative glycine treatment attenuates ischemia/reperfusion injury and ameliorates smooth muscle dysfunction in intestinal transplantation. *Transplantation* **2008**, *85*, 1300–1310. [CrossRef] [PubMed]
15. Oltean, M.; Joshi, M.; Björkman, E.; Oltean, S.; Casselbrant, A.; Herlenius, G.; Olausson, M. Intraluminal polyethylene glycol stabilizes tight junctions and improves intestinal preservation in the rat. *Am. J. Transplant* **2012**, *12*, 2044–2051. [CrossRef]
16. Oltean, M.; Jiga, L.; Hellström, M.; Söfteland, J.; Papurica, M.; Hoinoiu, T.; Ionac, M.; Casselbrant, A. A sequential assessment of the preservation injury in porcine intestines. *J. Surg. Res.* **2017**, *216*, 149–157. [CrossRef] [PubMed]
17. Blikslager, A.T.; Roberts, M.C.; Rhoads, J.M.; Argenzio, R.A. Is reperfusion injury an important cause of mucosal damage after porcine intestinal ischemia? *Surgery* **1997**, *121*, 526–534. [CrossRef]
18. Ikeda, H.; Yang, C.L.; Tong, J.; Nishimaki, H.; Masuda, K.; Takeo, T.; Kasai, K.; Itoh, G. Rat small intestinal goblet cell kinetics in the process of restitution of surface epithelium subjected to ischemia-reperfusion injury. *Dig. Dis. Sci.* **2002**, *47*, 590–601. [CrossRef] [PubMed]
19. Grootjans, J.; Hundscheid, I.H.; Lenaerts, K.; Boonen, B.; Renes, I.B.; Verheyen, F.K.; Dejong, C.H.; von Meyenfeldt, M.F.; Beets, G.L.; Buurman, W.A. Ischaemia-induced mucus barrier loss and bacterial penetration are rapidly counteracted by increased goblet cell secretory activity in human and rat colon. *Gut* **2013**, *62*, 250–258. [CrossRef]
20. Lopez-Pedrosa, J.M.; Torres, M.I.; Fernández, M.I.; Ríos, A.; Gil, A. Severe malnutrition alters lipid composition and fatty acid profile of small intestine in newborn piglets. *J Nutr.* **1998**, *128*, 224–233. [CrossRef]
21. Kubes, P.; Hunter, J.; Granger, D.N. Ischemia/reperfusion-induced feline intestinal dysfunction: Importance of granulocyte recruitment. *Gastroenterology* **1992**, *103*, 807–812. [CrossRef]
22. Dabrowska, D.; Jablonska, E.; Iwaniuk, A.; Garley, M. Many ways–one destination: Different types of neutrophils death. *Int. Rev. Immunol.* **2019**, *38*, 18–32. [CrossRef] [PubMed]
23. Kohli, V.; Selzner, M.; Madden, J.F.; Bentley, R.C.; Clavien, P.A. Endothelial cell and hepatocyte deaths occur by apoptosis after ischemia-reperfusion injury in the rat liver. *Transplantation* **1999**, *67*, 1099–1105. [CrossRef] [PubMed]
24. Jani, A.; Ljubanovic, D.; Faubel, S.; Kim, J.; Mischak, R.; Edelstein, C.L. Caspase inhibition prevents the increase in caspase-3, -2, -8 and -9 activity and apoptosis in the cold ischemic mouse kidney. *Am. J. Transplant* **2004**, *4*, 1246–1254. [CrossRef] [PubMed]
25. Oltean, M.; Hellström, M.; Ciuce, C.; Zhu, C.; Casselbrant, A. Luminal solutions protect mucosal barrier during extended preservation. *J. Surg. Res.* **2015**, *194*, 289–296. [CrossRef] [PubMed]
26. Casselbrant, A.; Söfteland, J.M.; Hellström, M.; Malinauskas, M.; Oltean, M. Luminal polyethylene glycol alleviates intestinal preservation injury irrespective of molecular size. *J. Pharmacol. Exp. Ther.* **2018**, *366*, 29–36. [CrossRef] [PubMed]
27. Grootjans, J.; Hodin, C.M.; de Haan, J.J.; Derikx, J.P.; Rouschop, K.M.; Verheyen, F.K.; van Dam, R.M.; Dejong, C.H.; Buurman, W.A.; Lenaerts, K. Level of activation of the unfolded protein response correlates with Paneth cell apoptosis in human small intestine exposed to ischemia/reperfusion. *Gastroenterology* **2011**, *140*, 529–539. [CrossRef]
28. Kwon, O.; Phillips, C.L.; Molitoris, B.A. Ischemia induces alterations in actin filaments in renal vascular smooth muscle cells. *Am. J. Physiol. Renal. Physiol.* **2002**, *282*, F1012-9. [CrossRef]
29. Shi, T.; Moulton, V.R.; Lapchak, P.H.; Deng, G.M.; Dalle Lucca, J.J.; Tsokos, G.C. Ischemia-mediated aggregation of the actin cytoskeleton is one of the major initial events resulting in ischemia-reperfusion injury. *Am. J. Physiol. Gastrointest. Liver Physiol.* **2009**, *296*, G339–G347. [CrossRef]
30. Turner, J.R. Molecular basis of epithelial barrier regulation: From basic mechanisms to clinical application. *Am. J. Pathol.* **2006**, *169*, 1901–1909. [CrossRef]
31. Takizawa, Y.; Kishimoto, H.; Kitazato, T.; Tomita, M.; Hayashi, M. Changes in protein and mRNA expression levels of claudin family after mucosal lesion by intestinal ischemia/reperfusion. *Int. J. Pharm.* **2012**, *426*, 82–89. [CrossRef] [PubMed]
32. Fujita, H.; Chiba, H.; Yokozaki, H.; Sakai, N.; Sugimoto, K.; Wada, T.; Kojima, T.; Yamashita, T.; Sawada, N. Differential expression and subcellular localization of claudin-7, -8, -12, -13, and -15 along the mouse intestine. *J. Histochem. Cytochem.* **2006**, *54*, 933–944. [CrossRef] [PubMed]

33. Koudstaal, L.G.; 't Hart, N.A.; Ottens, P.J.; van den Berg, A.; Ploeg, R.J.; van Goor, H.; Leuvenink, H.G. Brain death induces inflammation in the donor intestine. *Transplantation* **2008**, *86*, 148–154. [CrossRef] [PubMed]
34. Pullerits, R.; Oltean, S.; Flodén, A.; Oltean, M. Circulating resistin levels are early and significantly increased in deceased brain dead organ donors, correlate with inflammatory cytokine response and remain unaffected by steroid treatment. *J. Transl. Med.* **2015**, *13*, 201. [CrossRef] [PubMed]
35. Matsumoto, C.S.; Kaufman, S.S.; Girlanda, R.; Little, C.M.; Rekhtman, Y.; Raofi, V.; Laurin, J.M.; Shetty, K.; Fennelly, E.M.; Johnson, L.B.; et al. Utilization of donors who have suffered cardiopulmonary arrest and resuscitation in intestinal transplantation. *Transplantation* **2008**, *86*, 941–946. [CrossRef] [PubMed]
36. Fischer-Fröhlich Königsrainer, A.; Schaffer, R.; Schaub, F.; Pratschke, J.; Pascher, A.; Steurer, W.; Nadalin, S. Organ donation: When should we consider intestinal donation. *Transpl. Int.* **2012**, *25*, 1229–1240. [CrossRef] [PubMed]
37. Park, P.O.; Haglund, U.; Bulkley, G.B.; Falt, K. The sequence of development of intestinal tissue injury after strangulation ischemia and reperfusion. *Surgery* **1990**, *107*, 574. [PubMed]

11

Celiac Disease Monocytes Induce a Barrier Defect in Intestinal Epithelial Cells

Deborah Delbue, Danielle Cardoso-Silva, Federica Branchi, Alice Itzlinger, Marilena Letizia, Britta Siegmund and Michael Schumann *

Department of Gastroenterology, Infectious Diseases and Rheumatology, Campus Benjamin Franklin, Charité – Universitätsmedizin Berlin, 12203 Berlin, Germany; deborah.delbue-da-silva@charite.de (D.D.); danielle.cardoso-da-silva@charite.de (D.C.-S.); federica.branchi@charite.de (F.B.); alice.itzlinger@charite.de (A.I.); marilena.letizia@charite.de (M.L.); britta.siegmund@charite.de (B.S.)

* Correspondence: michael.schumann@charite.de

Abstract: Intestinal epithelial barrier function in celiac disease (CeD) patients is altered. However, the mechanism underlying this effect is not fully understood. The aim of the current study was to evaluate the role of monocytes in eliciting the epithelial barrier defect in CeD. For this purpose, human monocytes were isolated from peripheral blood mononuclear cells (PBMCs) from active and inactive CeD patients and healthy controls. PBMCs were sorted for expression of CD14 and co-cultured with intestinal epithelial cells (IECs, Caco2BBe). Barrier function, as well as tight junctional alterations, were determined. Monocytes were characterized by profiling of cytokines and surface marker expression. Transepithelial resistance was found to be decreased only in IECs that had been exposed to celiac monocytes. In line with this, tight junctional alterations were found by confocal laser scanning microscopy and Western blotting of ZO-1, occludin, and claudin-5. Analysis of cytokine concentrations in monocyte supernatants revealed higher expression of interleukin-6 and MCP-1 in celiac monocytes. However, surface marker expression, as analyzed by FACS analysis after immunostaining, did not reveal significant alterations in celiac monocytes. In conclusion, CeD peripheral monocytes reveal an intrinsically elevated pro-inflammatory cytokine pattern that is associated with the potential of peripheral monocytes to affect barrier function by altering TJ composition.

Keywords: barrier function; tight junction assembly; monocytes; celiac disease

---

1. Introduction

Celiac disease (CeD) is an autoimmune enteropathy triggered by the ingestion of gluten, affecting approximately 1% of the population in Western countries [1]. Current understanding of CeD immune pathology focusses on activation of gluten-specific TH1-cells secondary to presentation of DQ2- or DQ8-restricted gliadin peptides as the cause of the small intestinal inflammatory response. As a consequence, inflammation leads to villous atrophy and crypt hyperplasia, thereby causing the typical clinical features of intestinal malabsorption of nutrients [2,3].

A so-far unresolved issue of CeD is the nature of the associated intestinal epithelial barrier defect. It not only occurs secondary to the inflammatory process located in the lamina propria in active disease, but appears to be primary, since it is verifiable in inactive CeD patients on a gluten-free diet (GFD) and in relatives of CeD patients who do not suffer from CeD [4,5]. Moreover, Kumar et al. identified barrier-defining genes associated with CeD, thereby providing genetic proof for the relevance of barrier function in CeD pathogenesis. Importantly, the barrier function is maintained by a protein-protein complex interaction, where the main structure is the tight junction (TJ) proteins. TJs are the most apical contact between enterocytes formed by integral membrane proteins, including occludin, claudins, and scaffolding proteins as ZO-1 [6]. Although structural changes in celiac barrier function can be allocated to enterocyte TJ composition and epithelial transcytosis of gliadin peptides, research aiming to clarify how these changes arise is scarce [7–10].

Interestingly, monocytes are strongly involved in the regulation of intestinal barrier function, either by secretion of cytokines or by direct interaction with intestinal epithelial cells [11–13]. After extravasation, monocytes infiltrate the lamina propria, differentiating into macrophages and producing inflammatory mediators to combat pathogens [14]. For CeD, it has been previously described that monocytes isolated from CeD patients produced substantial amounts of TNF-α and interleukin-8 (IL-8) in a gluten-dependent manner [15]. Moreover, a gliadin-stimulated monocytic cell line was shown to have the potential to modulate intestinal epithelial barrier function [16]. Furthermore, monocytes isolated from healthy individuals that were stimulated with IL-15 as a celiac-mimicking cytokine milieu were capable of secreting pro-inflammatory cytokines that are known to induce barrier defects [17].

In the current work, we aimed to analyze the potency of celiac monocytes to perturb intestinal barrier function. For this purpose, we isolated monocytes from CeD patients and co-cultured these cells with intestinal epithelial cells to analyze epithelial barrier function.

2. Results

2.1. Monocytes Derived from Celiac Patients Induce a Barrier Defect in Intestinal Epithelial Cells

To evaluate whether monocytes exert an effect on epithelial barrier function, intestinal epithelial cells (IECs) were co-cultured with human primary monocytes. Peripheral blood mononuclear cells (PBMCs) were isolated from peripheral blood of CeD patients with different disease status and of healthy donors and then sorted for CD14 expression. IECs were grown on transwell filters, where they reached confluence and built up a stable barrier function, whereas the monocytes were placed underneath the filters. Thus, interaction between monocytes and IECs was possible mostly through soluble factors that pass through the filter membranes. Interestingly, transepithelial resistance (TER) of IECs measured after 48 h of co-culture was reduced when they were co-cultured with celiac monocytes, compared to the co-culture with healthy controls. This effect added up to approx. 70% of the control level and was irrespective of CeD activity (Figure 1).

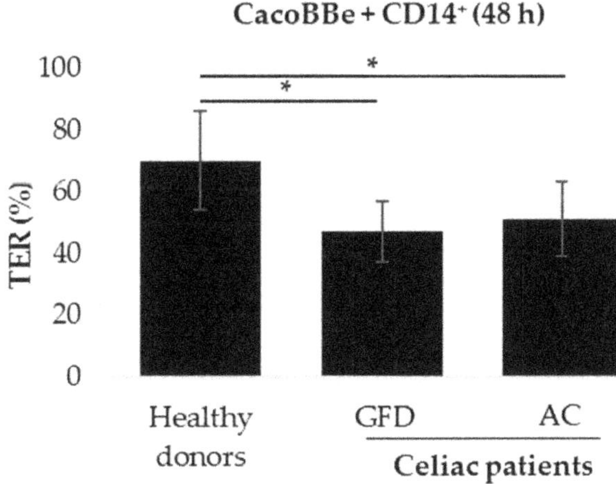

Figure 1. Epithelial barrier function after co-culture of intestinal epithelial cells (IECs) with monocytes derived from celiac disease patients. After peripheral blood mononuclear cell (PBMC) isolation and CD14+ cell-sorting, epithelial cells were co-cultured with monocytes from healthy donors. celiac patients on a gluten-free diet (GFD), or AC (active celiac disease (CeD)) patients. Subsequently, the TER was measured after 48 h of co-culture (% of TER prior to addition of monocytes). Mean of $n = 36$ (healthy donors), $n = 15$ (GFD), and $n = 20$ (AC) individual filters measurements. Monocytes used for these experiments were isolated from $n = 8$ (healthy donors), $n = 4$ (GFD) and $n = 5$ (active CeD). Mann-Whitney U * $p < 0.05$, comparison between co-cultures with monocytes from healthy donors and CeD patients.

Co-culturing IECs with unsorted (i.e., total) PBMCs caused a similar decrease in TER (Figure S1). To exclude possible direct effects of gliadin or IL-15 on the epithelium, CacoBBe cells were exposed to IL-15/Tglia alone, with monocytes or PBMCs. In IECs alone, we did not observe a decrease in TER with only IL-15/Tglia addition. Nevertheless, in the cells exposed to monocytes, TER decreased at the same levels as with monocytes plus IL-15/Tglia. (Figure S2). These results showed that effects observed in TER are independent of IL-15/Tglia stimulation and that this is rather directly associated with monocytes. In summary, this experiment uncovered the potential of celiac monocytes to alter epithelial barrier function.

2.2. Celiac Monocytes Alter IEC-TJ Structure

As a next step, we aimed to evaluate whether $CD14^+$ cells alter IEC barrier function through changes in tight junction (TJ) integrity. First, IECs that had completed 48 h of co-culture with $CD14^+$ monocytes were immunostained for TJ proteins ZO-1 and occludin. As shown in Figure 2, no effect was found regarding TJ localization or expression in IEC layers that had been co-cultured with $CD14^+$ monocytes isolated from healthy donors. However, for IECs co-cultured with monocytes derived from CeD patients, lower levels of occludin and a mosaic expression pattern of ZO-1 was found, with ZO-1 being significantly reduced in expression in some, but not all, regions of the filter. Moreover, TJs appeared to be irregular regarding loop-like linings, which was not observed when IECs were exposed to monocytes of healthy donors (Figure 2A). Additionally to the aforementioned reduction in expression level of ZO-1, XZ-projections revealed an uneven structure of the apical membrane in IECs that were co-cultured with CeD monocytes (Figure 2B). Furthermore, protein levels of occludin and the TJ-sealing claudin-5, which has previously been implicated in the CeD barrier defect, were analyzed (Figure 2C) [9]. IEC protein levels of occludin and claudin-5, after exposure to celiac monocytes, were reduced compared to protein levels of the respective healthy control monocytes. These data show evidence that CeD monocytes exert effects on the TJ structure of co-cultured IECs.

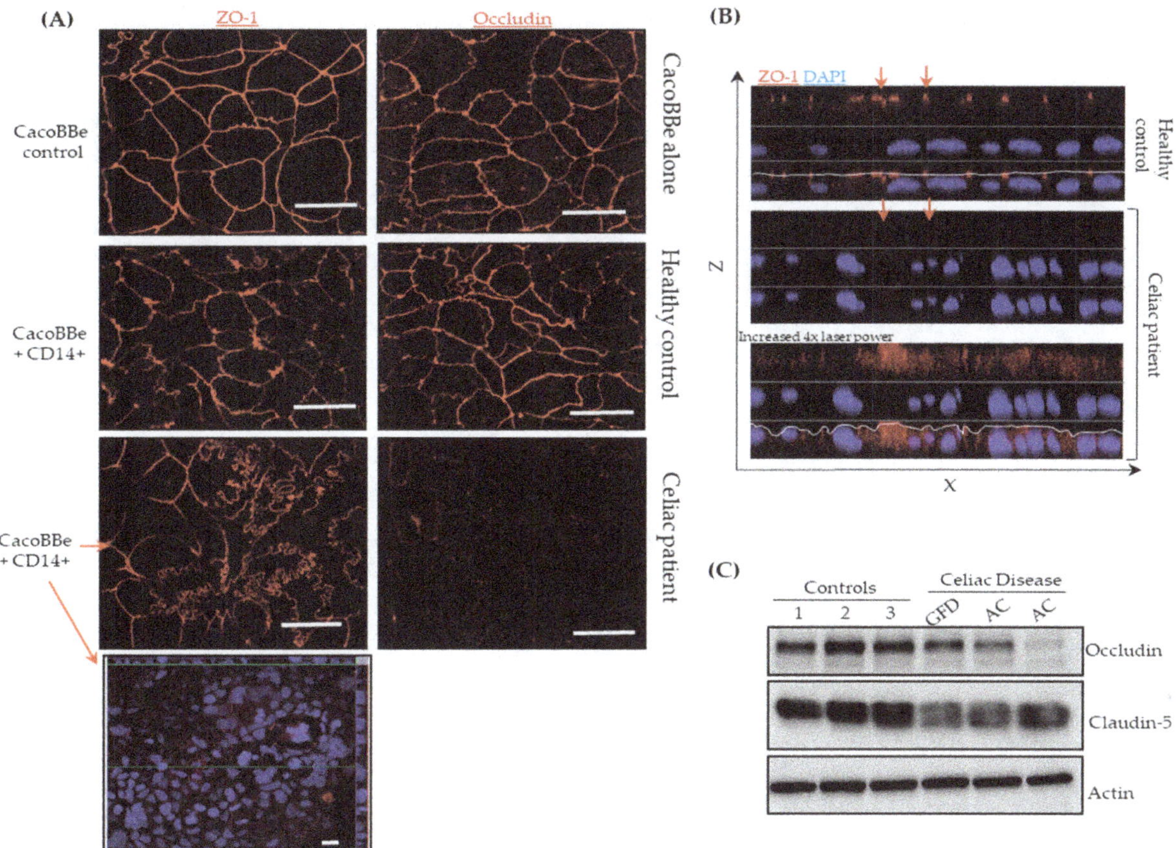

Figure 2. Tight junction (TJ) structure and protein composition after co-culture with monocytes derived from celiac disease patients. (**A**) Cellular localization of occludin and ZO-1 were investigated using confocal laser scanning microscopy after immunostaining. Representative images from n = 5 (healthy donors), n = 3 (CeD on GFD) and n = 3 (Active CeD patients). Scale bar: 50 µM. The two effects of CD14 co-culture in ZO-1 expression is pointed out by red arrows (**B**) Collapsed XZ-projections. ZO-1 staining reveals apical junctional complexes at approx. identical Z heights, as illustrated by the white lining in the merged image. When CacoBBe cells co-cultured with celiac monocytes were immunostained, lining was comparably irregular, and ZO-1 level was reduced. (**C**) IEC protein levels by Western blotting of occludin and claudin-5 after co-culture with monocytes.

2.3. Monocytes Derived from Celiac Disease Patients Present Higher Levels of Proinflammatory Cytokine Production

Next, we characterized isolated human monocytes that had previously been sorted for CD14 to uncover potential differences regarding cytokine and surface marker expression between celiac and healthy control monocytes. First, we analyzed the expression of surface markers that are characteristic of classically and non-classically activated macrophages. Surface marker expression was analyzed after CD14-sorting (Figure S3) and 24 h of culture—media including Granulocyte macrophage colony stimulating factor (GM-CSF)—by immunostaining. Using a gating strategy revealed in Figure 3A, monocyte populations were detected, doublets were excluded, and the viable population (DAPI-negative cells) was analyzed. Then, frequency of positivity for CD11b, CD80, HLA-DR, CD163, and CD16 was evaluated. No significant differences in the frequency of any of the examined surface markers were found (Figure 3B-G; Figure S3). The frequency of cells revealing a double positivity for CD80 and HLA-DR (i.e., expression of both inflammatory markers within the same cell) was also analyzed but turned out to be not significantly different. Although CD16 frequency was not significantly increased, a tendency toward higher frequencies of CD16-positive cells was observed. This points to an increased fraction of intermediate/non-classical monocytes in the peripheral blood of CeD patients.

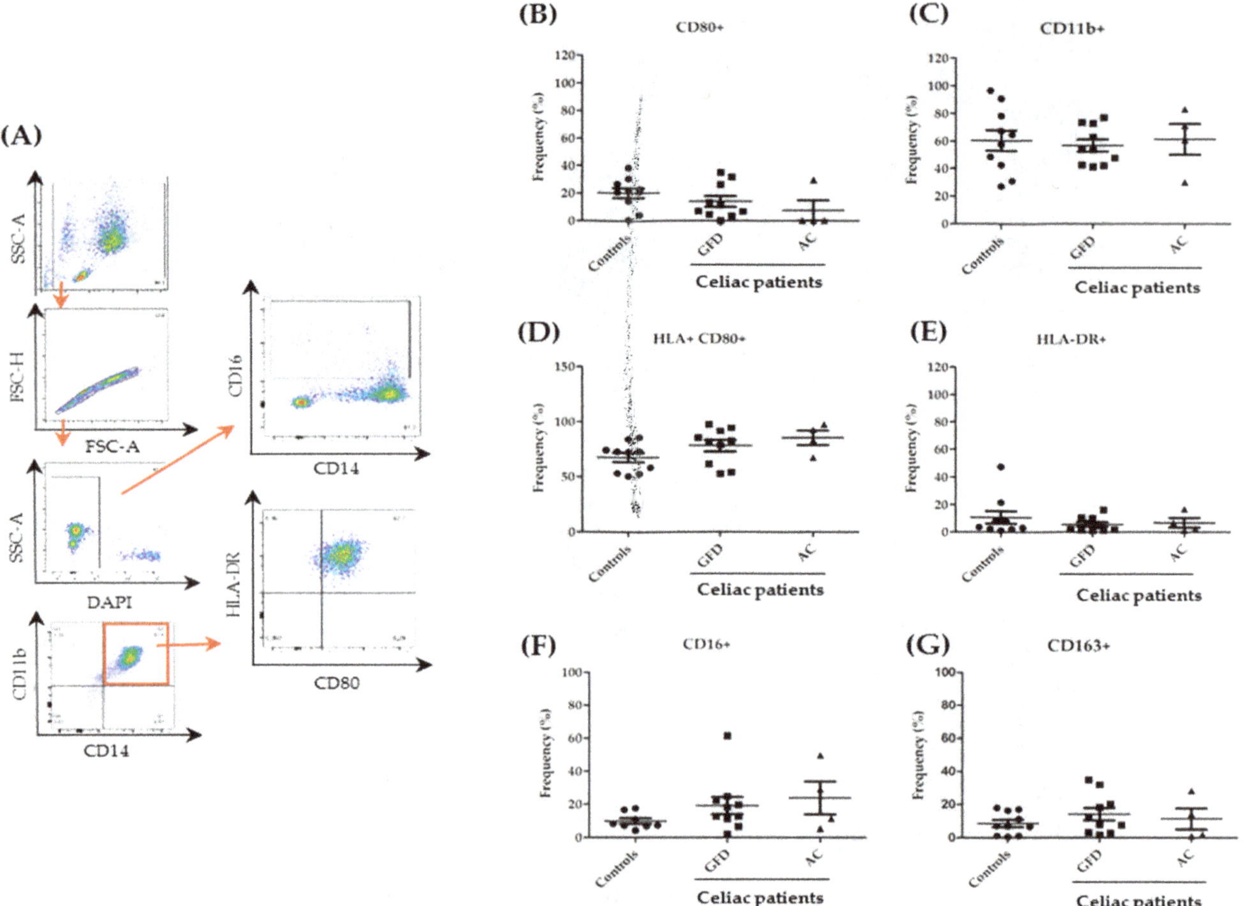

Figure 3. Expression of surface markers in peripheral monocytes from celiac patients. PBMCs were isolated from 10 controls, 10 GFD and 4 active CeD (AC) patients and sorted for CD14. Subsequently, cells were cultured in the presence of granulocyte macrophage colony stimulating factor (GM-CSF; 10 ng/ml) for 24 h and evaluated by flow cytometry. (**A**) Gating strategy used to determine surface marker expression. The stepwise gating approach is highlighted by various steps of analysis that are interconnected by red arrows. Representative plots from a healthy control are shown. (**B–G**) Results for the expression of single surface markers are shown. Each dot represents the surface marker expression result of a single patient. Mean values ± standard error of the mean (SEM) are shown. The Mann–Whitney U test revealed no significant differences.

Subsequently, we determined the concentration of a set of cytokines within the supernatant of the monocytes at the end of the 24 h time period following isolation and CD14-sorting. To guarantee survival, they were cultured at this time in the presence of GM-CSF (10 mg/ml). As illustrated in Figure 4A, concentrations of IFN-α2, IFN-λ, IL-19, IL-12p70, IL-17A, IL-18, IL-23, and IL-33 did not reach the detection level of the assay, neither in the control nor the CeD group. More interestingly, levels of IL-1β, TNF-α, IL-8, IL-10, IL-6, and MCP-1 were found at detectable levels. Although concentrations of IL-1β, TNF-α, IL-8, and IL-10 were not significantly different between the groups of patients and healthy controls, a tendency for higher levels of the pro-inflammatory cytokines was observed in CeD supernatants (Figure 4B–E). Interestingly, IL-6 and MCP-1 (monocyte chemotactic protein-1, synonymous: CC-chemokine ligand-2, CCL2) were found at significantly higher levels in the supernatant of monocytes from CeD patients who had received a GFD, compared to monocytes from healthy controls (Figure 4E,F). In summary, these data reveal that CD14$^+$ monocytes isolated from CeD patients carry a more pro-inflammatory phenotype, an effect that appears to be independent of disease activity.

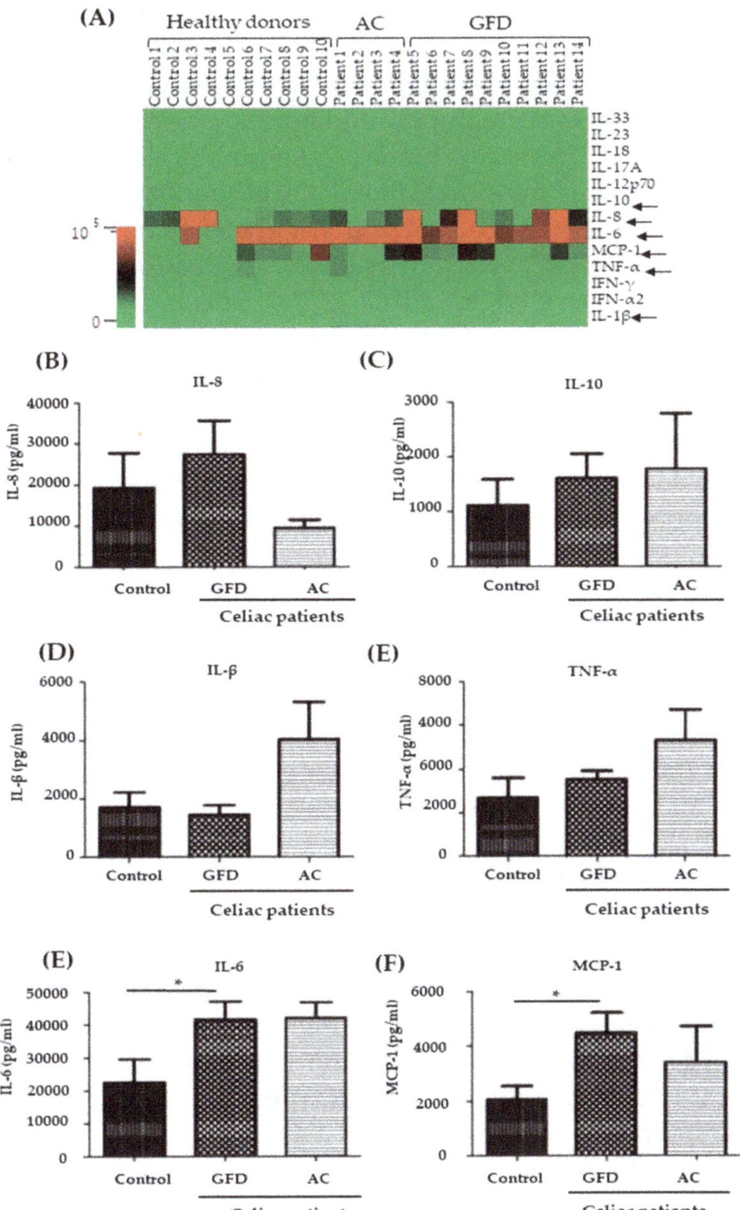

Figure 4. Increased levels of pro-inflammatory cytokines in the supernatant of monocytes derived from celiac disease patients. (**A**) PBMCs were isolated from 10 controls, 10 GFD, and 4 active CeD patients, and monocytes were sorted using CD14 MACS MicroBeads. Subsequently, cells were kept in the incubator for 24 h in the presence of GM-CSF (10 ng/mL). Supernatants of monocyte cultures were collected and cytokine levels were determined. Data are illustrated as a heat map revealing color-coded concentrations of cytokines (green: low concentration; red: high concentration). (**B**–**F**) Results for individual cytokine measurements are shown: Interleukin- (IL-)8, IL-10, IL-1β, TNF-α, IL-6, and MCP-1. Mean values ± SEM are shown. Mann–Whitney test: *, $p < 0.05$.

3. Discussion

CeD is an autoimmune enteropathy triggered by the ingestion of dietary gluten [3]. Although the central role for on the one hand gluten-specific, and on the other hand tissue-resident, cytotoxic T-cells is undisputed, results from genome-wide association studies and from functional data collected in non-affected relatives of CeD patients point to a primary defect of the epithelial barrier as a discrete pathophysiologic entity within the overall immune pathology of CeD [2,5,18]. Nevertheless, it is mostly unclear how the barrier defect in CeD is triggered. Since monocytes are major mediators in mucosal barrier defects, we aimed to elucidate the potential of peripheral human monocytes to alter intestinal epithelial barrier function. The rationale for using peripheral blood monocytes comes from early data that convincingly showed that in intestinal inflammation, monocytes infiltrate the mucosa and then further differentiate to macrophages secreting pro-inflammatory cytokines, rather than showing that resident tissue-macrophages re-differentiate into pro-inflammatory macrophages [19].

$CD14^+$ monocytes derived from peripheral blood of CeD patients revealed an intrinsically higher secretion of IL-6 and MCP-1. Moreover, a non-significant tendency for increased expression of TNF-α and IL-1β was observed. A similar pattern of cytokine expression was previously described for intestinal (i.e., not peripheral) macrophages in IBD [20,21]. Specifically, Kamada et al. described the CD14+ macrophage population secondary to their cytokine expression (e.g., TNF-α, IL-6, IL-12/23p40, and IL-23p19) as pro-inflammatory and non-resident, compared to $CD14^-$ resident macrophages that are found primarily in healthy gut and do not express pro-inflammatory cytokines. Phenotypically similar macrophages were shown to elicit a barrier defect on IECs in a co-culture model, similar to that used in the current work [11]. However, in the current work, expression analysis of IL-1β and TNF-α, which were previously shown by Lissner et al. to be mostly responsible for the IEC barrier defect in the M1- and M0-polarized macrophage model, only revealed a non-significant tendency towards higher levels of these cytokines. On the other hand, IL-6 and MCP-1 were significantly increased. MCP-1 (synonymous CCL-2) is secreted by various cells, including monocytes, as a chemoattractant for monocytes, T-cells, and dendritic cells, and enables via the CCL2-CCR2 axis the extravasation of monocytes into the lamina propria [14,22]. Our data on MCP-1 are in line with data from Italy that revealed a higher expression of CCL-2 by PBMCs after stimulation with the p31-43 gliadin peptide [23]. IL-6, a pro-inflammatory cytokine also secreted by intestinal macrophages, has previously been described to induce an intestinal barrier defect by increasing expression of pore-forming claudins, including claudin-2 [24]. Interestingly, IL-6 secreted by macrophages as the cause for a reduced intestinal barrier function was also discussed as a mechanism for the epithelial barrier defect found in liver cirrhosis [25].

Although a difference in cytokine production was found between celiac monocytes and monocytes isolated from healthy controls, which was in line with the functional barrier data revealing a defect only inducible by peripheral monocytes from CeD patients, we did not observe significant differences in monocyte surface marker expression. CD16 as a marker for alternatively activated macrophages showed a non-significant tendency towards higher expression in celiac monocytes. However, this aspect of our work remains somewhat inconclusive.

Monocytes isolated from CeD patients induced a functional barrier defect on IECs. The reduced barrier function was measured by TER and was associated with an altered TJ morphology on confocal laser scanning microscopy (LSM) as well as TJ protein expression in Western blotting. Celiac barrier defects that are localized to the TJ have been described previously [9,26]. In those studies, the alteration of the celiac TJ was complex, involving reduced expression of occludin, claudin-3, -5, and -7, and altered phosphorylation of ZO-1. This is mostly in line with our work, since celiac monocytes induced a reduction of occludin and claudin-5 expression, and confocal LSM revealed alterations in cellular ZO-1 distribution. Nevertheless, we should mention that we and others have shown that apoptosis of IECs, which was not analyzed in the current study, might also contribute to a defective epithelial barrier and that induction of (apoptotic or non-apoptotic) cell death might be a conceivable fate of an IEC that is exposed to monocytes [9,27].

Taken together, our results suggest that in CeD patients, peripheral blood monocytes have the potential to induce an epithelial defect of the intestinal mucosa. This IEC reaction to monocyte exposure is presumably related to action of monocytic IL-6 on IECs. Tight junctional alterations in the intestinal epithelial cell layer are at least partially responsible for the functional barrier defect. These effects might be primary in nature. However, work herein is not sufficient to prove the primary nature. In the future, this could be approached by analyzing monocytes of first-degree relatives to CeD patients to determine if these cells also show an impact on barrier function.

4. Materials and Methods

4.1. Human Material

The study was approved by the Ethics Committees of the Charité – Universitätsmedizin Berlin, Germany (protocol number EA4/116/18, accepted on Jan 22nd, 2019). Heparinized whole blood samples were collected from healthy individuals and CeD patients. Inactive (GFD) patients received a GFD for >1 year. All patients declared their informed consent (signed consent form). Healthy controls were individuals without a history of enteropathy and without clinical signals of CeD or other autoimmune diseases. For further characteristics of CeD patients, refer to Table 1.

Table 1. Characteristics of CeD patients.

Number of Subjects	17
Female/Male	14/3
Age at enrolment, median (range)	46 (23-83)
Age at CeD diagnosis, median (range)	32 (6-73)
Marsh Grade at enrolment, *n* (%)	
0	5 (29)
1	2 (12)
2	0 (0)
3a	1 (6)
3b	3 (18)
3c	0 (0)
not available	6 (35)
tTG at enrolment, *n* (%)	
positive	6 (35)
negative	2 (12)
not available	9 (53)
HLA-DQ status, *n* (%)	
DQ2+	11 (65)
DQ8+	0 (0)
not available	6 (35)
GFD status, *n* (%)	
Active CeD	6 (35)
New CeD diagnosis	4
CeD, non-compliant to GFD	2
CeD on GFD	11 (65)

CeD: celiac disease; tTG: transglutaminase antibodies; GFD: gluten-free diet.

4.2. Cell Line

The human colorectal cell line, CacoBBe (C2BBe1 [clone of Caco-2] ATCC®CRL-2102™), was maintained in Dulbecco's Modified Eagle's Medium (DMEM) + GlutaMAX (Gibco), with 10% fetal bovine serum (Gibco), 1% penicillin and streptomycin (Corning), 10μM HEPES-buffer and 1M non-essential amino acids (Merck Millipore). Cells were kept at 37 °C in a 5% CO_2 environment. Culture medium was changed three times per week.

4.3. PBMCs Isolation and CD14+ Sorting

Peripheral blood mononuclear cells (PBMCs) were isolated from peripheral blood of healthy donors and celiac disease patients by Biocoll (Merk Millipore) separating solution and centrifugation, as previously described [28]. Subsequently, PBMCs were sorted using CD14 MACS MicroBeads (Miltenyi Biotech, Bergisch Gladbach, Germany). As determined by flow cytometry, preparations contained >90% CD14+ cells. Monocytes were plated in 24-well dishes with RPMI-1640 as media (Gibco), supplemented with 10% of fetal bovine serum (Gibco) and 1% of penicillin and streptomycin (Corning).

CD14+ cells received human granulocyte macrophage colony stimulating factor (10ng/mL; GM-CSF) for 24 h prior to co-culture. Cell culture supernatants were collected for flow cytometry analysis.

4.4. Co-Culture and TER Measurement

Intestinal epithelial cells (IECs) were plated on permeable transwell polycarbonate filter supports (0.4 μm; 0.6 cm^2, Merck Millipore) and kept at 37 °C in a 5% CO_2 environment. Culture medium was changed three times per week. On days 10 to 16 after plating, filters were transferred to 24-well dishes containing $CD14^+$ cells (5×10^5 cells per well). In addition, IL-15/Tglia (10 mg/mL) was added (gift by W. Dieterich; LPS-free) to filters with IECs. Subsequently, the transepithelial resistance (TER) was measured, as described previously [11], for 48 h of co-culture with monocytes from CeD patients or healthy donors (controls). For immunofluorescence experiments, filters were fixed using 1% paraformaldehyde for 15 minutes at room temperature.

4.5. Immunofluorescence

Epithelial cell layers were stained using the following primary antibodies: ZO-1 (1:100; BD Biosciences, NJ, USA). The secondary antibodies used were Alexa Fluor 488 goat anti-mouse or rabbit IgG, and Alexa Fluor 594 goat anti-mouse or rabbit IgG (1:500; Thermo Fisher Scientific, MA, USA). To determine occludin expression and cellular distribution, an occludin mouse monoclonal antibody (OC-3F10) was used as an Alexa Fluor®594 Conjugate (Thermofischer). Nuclei were stained using DAPI (4′,6-Diamidin-2-phenylindol, conc. 1:2000). Immunofluorescence staining was analyzed by confocal laser scanning microscopy (LSM 780, Carl Zeiss, Jena, Germany) as previously described [9,11].

4.6. Western Blotting

For Western blotting analysis, the protocol was followed as previously described [9]. The primary antibodies used were occludin (1:1000; Sigma Aldrich, St. Louis, MO, USA), claudin-5 (1:1000; Invitrogen, Carlsbad, CA, USA) and actin (1:1000; Sigma Aldrich, St. Louis, MO, USA). The peroxidase-conjugated secondary antibodies used were goat anti-rabbit IgG or goat anti-mouse IgG (Jackson ImmunoResearch, Ely, UK). SuperSignal West Pico PLUS Stable Peroxide Solution (Thermo Scientific, Waltham, MA, USA) was used for protein detection, and Fusion FX7 imaging system (Vilber Lourmat Deutschland GmbH, Eberhardzell, Germany) was used to detect protein signal levels.

4.7. Flow Cytometric Assessment—Surface Markers and Cytokine Expression

$CD14^+$ cells were washed twice with PBS, and the surface markers were checked. The following fluorochrome-coupled antibodies were applied: anti-CD14 (61D3) from BD Biosciences, anti- anti-CD16 (3G8), anti-CD11b (ICRF44), anti-CD163 (GHI/61) from Biolegend, CD80 (2D10.4), and anti-HLA-DR (LN3) from eBioscence. Dead cells were excluded by DAPI staining. Samples were assessed by flow cytometry using a FACSCanto II and the FACS Diva software (version 6; BD Biosciences). Supernatants of the cultures after 24 h of culture with GM-CSF (10 ng/ml) were tested for cytokine expression (IL-1β, IFN-α2, IFN-λ, TNF-α, MCP-1, IL-6, IL-8, IL-19, IL-12p70, IL-17A, IL-18, IL-23, and IL-33) using the LEGENDplex Multi-Analyte Flow Assay kit–Human Inflammation Panel (13-plex) (Biolegend) according to the manufacturer's protocol. FACS data were analyzed using FlowJo (v10.6.1) and LEGENDplex v8.0 software (BioLegend, San Diego, CA, USA).

4.8. Statistical Analysis

Statistical analysis was performed using GraphPad Prism software (GraphPad Software, La Jolla, CA) by a non-parametric Mann–Whitney test to analyze differences between the control and the CeD patients. One-way ANOVA was used to compare differences in all groups analyzed. All data are expressed as mean values ± standard error of the mean (SEM). A $p < 0.05$ was considered significant.

Author Contributions: D.D. planned and carried out the experiments and wrote the manuscript. D.C.-S. performed cell culture and immunostainings. F.B. identified patients and characterized them further. M.L. performed the FACS analysis. A.I. partially performed the co-culture and the immunostainings. B.S. planned the experiments and wrote the manuscript. M.S. planned experiments, identified patients, and wrote the manuscript.

Acknowledgments: We acknowledge the very skilled support given by Claudia Heldt and the helpful support with Tglia given by Walburga Dieterich (Erlangen, Germany).

Abbreviations

AC	Active celiac
CeD	Celiac disease
CCL2	CC-chemokine ligand-2
GFD	Gluten-free diet
GM-CSF	Granulocyte macrophage colony stimulating factor
IL-1β	Interleukin-1β
IL-6	Interleukin-6
IL-8	Interleukin-8
IL-10	Interleukin-10
IL-12p70	Interleukin-12p70
IL-15	Interleukin-15
IL-17A	Interleukin-17A
IL-18	Interleukin-18
IL-23	Interleukin-23
IL-33	Interleukin-33
IFN-γ	Interferon-γ
IFNα2	Interferon-α2
MACS	Magnetic cell sorting
MCP-1	Monocyte chemotactic protein-1
PBMCs	Peripheral blood mononuclear colony stimulating factor
TER	Transepithelial resistance
Tglia	Trypsinized gliadin
TJ	Tight junctions

References

1. Mustalahti, K.; Catassi, C.; Reunanen, A.; Fabiani, E.; Heier, M.; McMillan, S.; Murray, L.; Metzger, M.H.; Gasparin, M.; Bravi, E.; et al. The prevalence of celiac disease in Europe: results of a centralized, international mass screening project. *Ann. Med.* **2010**, *42*, 587–595. [CrossRef] [PubMed]
2. Jabri, B.; Sollid, L.M. T Cells in Celiac Disease. *J. Immunol.* **2017**, *198*, 3005–3014. [CrossRef] [PubMed]
3. Ludvigsson, J.F.; Leffler, D.A.; Bai, J.C.; Biagi, F.; Fasano, A.; Green, P.H.; Hadjivassiliou, M.; Kaukinen, K.; Kelly, C.P.; Leonard, J.N.; et al. The Oslo definitions for coeliac disease and related terms. *Gut* **2013**, *62*, 43–52. [CrossRef]
4. Schumann, M.; Siegmund, B.; Schulzke, J.D.; Fromm, M. Celiac Disease: Role of the Epithelial Barrier. *Cell Mol. Gastroenterol. Hepatol.* **2017**, *3*, 150–162. [CrossRef]
5. van Elburg, R.M.; Uil, J.J.; Mulder, C.J.; Heymans, H.S. Intestinal permeability in patients with coeliac disease and relatives of patients with coeliac disease. *Gut* **1993**, *34*, 354–357. [CrossRef]
6. Cardoso-Silva, D.; Delbue, D.; Itzlinger, A.; Moerkens, R.; Withoff, S.; Branchi, F.; Schumann, M. Intestinal Barrier Function in Gluten-Related Disorders. *Nutrients* **2019**, *11*, 2325. [CrossRef] [PubMed]

7. Matysiak-Budnik, T.; Moura, I.C.; Arcos-Fajardo, M.; Lebreton, C.; Menard, S.; Candalh, C.; Ben-Khalifa, K.; Dugave, C.; Tamouza, H.; van Niel, G.; et al. Secretory IgA mediates retrotranscytosis of intact gliadin peptides via the transferrin receptor in celiac disease. *J. Exp. Med.* **2008**, *205*, 143–154. [CrossRef]
8. Menard, S.; Lebreton, C.; Schumann, M.; Matysiak-Budnik, T.; Dugave, C.; Bouhnik, Y.; Malamut, G.; Cellier, C.; Allez, M.; Crenn, P.; et al. Paracellular versus transcellular intestinal permeability to gliadin peptides in active celiac disease. *Am. J. Pathol.* **2012**, *180*, 608–615. [CrossRef]
9. Schumann, M.; Gunzel, D.; Buergel, N.; Richter, J.F.; Troeger, H.; May, C.; Fromm, A.; Sorgenfrei, D.; Daum, S.; Bojarski, C.; et al. Cell polarity-determining proteins Par-3 and PP-1 are involved in epithelial tight junction defects in coeliac disease. *Gut* **2012**, *61*, 220–228. [CrossRef]
10. Schumann, M.; Richter, J.F.; Wedell, I.; Moos, V.; Zimmermann-Kordmann, M.; Schneider, T.; Daum, S.; Zeitz, M.; Fromm, M.; Schulzke, J.D. Mechanisms of epithelial translocation of the alpha(2)-gliadin-33mer in coeliac sprue. *Gut* **2008**, *57*, 747–754. [CrossRef]
11. Lissner, D.; Schumann, M.; Batra, A.; Kredel, L.I.; Kuhl, A.A.; Erben, U.; May, C.; Schulzke, J.D.; Siegmund, B. Monocyte and M1 Macrophage-induced Barrier Defect Contributes to Chronic Intestinal Inflammation in IBD. *Inflamm. Bowel. Dis.* **2015**, *21*, 1297–1305. [CrossRef] [PubMed]
12. Managlia, E.; Liu, S.X.L.; Yan, X.; Tan, X.D.; Chou, P.M.; Barrett, T.A.; De Plaen, I.G. Blocking NF-kappaB Activation in Ly6c(+) Monocytes Attenuates Necrotizing Enterocolitis. *Am. J. Pathol.* **2019**, *189*, 604–618. [CrossRef]
13. Morhardt, T.L.; Hayashi, A.; Ochi, T.; Quiros, M.; Kitamoto, S.; Nagao-Kitamoto, H.; Kuffa, P.; Atarashi, K.; Honda, K.; Kao, J.Y.; et al. IL-10 produced by macrophages regulates epithelial integrity in the small intestine. *Sci. Rep.* **2019**, *9*, 1223. [CrossRef] [PubMed]
14. Bain, C.C.; Schridde, A. Origin, Differentiation, and Function of Intestinal Macrophages. *Front Immunol.* **2018**, *9*, 2733. [CrossRef] [PubMed]
15. Cinova, J.; Palova-Jelinkova, L.; Smythies, L.E.; Cerna, M.; Pecharova, B.; Dvorak, M.; Fruhauf, P.; Tlaskalova-Hogenova, H.; Smith, P.D.; Tuckova, L. Gliadin peptides activate blood monocytes from patients with celiac disease. *J. Clin Immunol.* **2007**, *27*, 201–209. [CrossRef] [PubMed]
16. Barilli, A.; Rotoli, B.M.; Visigalli, R.; Ingoglia, F.; Cirlini, M.; Prandi, B.; Dall'Asta, V. Gliadin-mediated production of polyamines by RAW264.7 macrophages modulates intestinal epithelial permeability in vitro. *Biochim. Biophys. Acta* **2015**, *1852*, 1779–1786. [CrossRef]
17. Harris, K.M.; Fasano, A.; Mann, D.L. Monocytes differentiated with IL-15 support Th17 and Th1 responses to wheat gliadin: implications for celiac disease. *Clin Immunol.* **2010**, *135*, 430–439. [CrossRef]
18. Kumar, V.; Gutierrez-Achury, J.; Kanduri, K.; Almeida, R.; Hrdlickova, B.; Zhernakova, D.V.; Westra, H.J.; Karjalainen, J.; Ricano-Ponce, I.; Li, Y.; et al. Systematic annotation of celiac disease loci refines pathological pathways and suggests a genetic explanation for increased interferon-gamma levels. *Hum Mol. Genet* **2015**, *24*, 397–409. [CrossRef]
19. Grimm, M.C.; Pullman, W.E.; Bennett, G.M.; Sullivan, P.J.; Pavli, P.; Doe, W.F. Direct evidence of monocyte recruitment to inflammatory bowel disease mucosa. *J. Gastroenterol. Hepatol.* **1995**, *10*, 387–395. [CrossRef]
20. Kamada, N.; Hisamatsu, T.; Okamoto, S.; Chinen, H.; Kobayashi, T.; Sato, T.; Sakuraba, A.; Kitazume, M.T.; Sugita, A.; Koganei, K.; et al. Unique CD14 intestinal macrophages contribute to the pathogenesis of Crohn disease via IL-23/IFN-gamma axis. *J. Clin Invest.* **2008**, *118*, 2269–2280.
21. Lampinen, M.; Waddell, A.; Ahrens, R.; Carlson, M.; Hogan, S.P. CD14+CD33+ myeloid cell-CCL11-eosinophil signature in ulcerative colitis. *J. Leukoc Biol.* **2013**, *94*, 1061–1070. [CrossRef]
22. Smith, P.D.; Smythies, L.E.; Shen, R.; Greenwell-Wild, T.; Gliozzi, M.; Wahl, S.M. Intestinal macrophages and response to microbial encroachment. *Mucosal Immunol.* **2011**, *4*, 31–42. [CrossRef]
23. Vincentini, O.; Maialetti, F.; Gonnelli, E.; Silano, M. Gliadin-dependent cytokine production in a bidimensional cellular model of celiac intestinal mucosa. *Clin Exp. Med.* **2015**, *15*, 447–454. [CrossRef]
24. Suzuki, T.; Yoshinaga, N.; Tanabe, S. Interleukin-6 (IL-6) regulates claudin-2 expression and tight junction permeability in intestinal epithelium. *J. Biol. Chem.* **2011**, *286*, 31263–31271. [CrossRef]
25. Du Plessis, J.; Vanheel, H.; Janssen, C.E.; Roos, L.; Slavik, T.; Stivaktas, P.I.; Nieuwoudt, M.; van Wyk, S.G.; Vieira, W.; Pretorius, E.; et al. Activated intestinal macrophages in patients with cirrhosis release NO and IL-6 that may disrupt intestinal barrier function. *J. Hepatol.* **2013**, *58*, 1125–1132. [CrossRef]

26. Ciccocioppo, R.; Finamore, A.; Ara, C.; Di Sabatino, A.; Mengheri, E.; Corazza, G.R. Altered expression, localization, and phosphorylation of epithelial junctional proteins in celiac disease. *Am. J. Clin. Pathol.* **2006**, *125*, 502–511. [CrossRef]
27. Bojarski, C.; Gitter, A.H.; Bendfeldt, K.; Mankertz, J.; Schmitz, H.; Wagner, S.; Fromm, M.; Schulzke, J.D. Permeability of human HT-29/B6 colonic epithelium as a function of apoptosis. *J. Physiol.* **2001**, *535 Pt 2*, 541–552. [CrossRef]
28. Kredel, L.I.; Batra, A.; Stroh, T.; Kuhl, A.A.; Zeitz, M.; Erben, U.; Siegmund, B. Adipokines from local fat cells shape the macrophage compartment of the creeping fat in Crohn's disease. *Gut* **2013**, *62*, 852–862. [CrossRef]

12

Cerebral Cavernous Malformation Proteins in Barrier Maintenance and Regulation

Shu Wei [1,2,†], Ye Li [2,†], Sean P. Polster [3], Christopher R. Weber [2], Issam A. Awad [3] and Le Shen [2,3,*]

1 Graduate Program in Public Health and Preventive Medicine, Wuhan University of Science and Technology, Wuhan 430081, China; shuwza@gmail.com
2 Department of Pathology, The University of Chicago, Chicago, IL 60615, USA; yli170@bsd.uchicago.edu (Y.L.); cweber@bsd.uchicago.edu (C.R.W.)
3 Section of Neurosurgery, Department of Surgery, The University of Chicago, Chicago, IL 60615, USA; spolster@uchicago.edu (S.P.P.); iawad@uchicago.edu (I.A.A.)
* Correspondence: leshen@uchicago.edu
† These authors contributed equally.

Abstract: Cerebral cavernous malformation (CCM) is a disease characterized by mulberry shaped clusters of dilated microvessels, primarily in the central nervous system. Such lesions can cause seizures, headaches, and stroke from brain bleeding. Loss-of-function germline and somatic mutations of a group of genes, called *CCM* genes, have been attributed to disease pathogenesis. In this review, we discuss the impact of *CCM* gene encoded proteins on cellular signaling, barrier function of endothelium and epithelium, and their contribution to CCM and potentially other diseases.

Keywords: cerebral cavernous malformation; endothelial barrier; epithelial barrier; Rho; ROCK; MEKK3

1. Introduction

One of the key functions of endothelial and epithelial cells is to create a barrier that separates different tissue compartments, and in the case of skin, epithelial cells separate body and outer environment. Compromised barrier function leads to abnormal mixing of different tissue components, which can contribute to pathogenesis of many diseases. In this review, we focus on a group of proteins that participates in the development of a neurovascular disease, cerebral cavernous malformation (CCM), and examine their impact on cellular signaling and barrier function.

2. Clinical Features of CCM

CCM (also known as cavernous angioma) disease is characterized by the development of abnormally dilated capillaries, primarily in the central nervous system (Figure 1) [1]. Grossly, these lesions appear to be blood filled, mulberry shaped clusters of thin-walled small vessels. Histologically, the nested microvessels have little supporting tissue and intervening parenchyma, and the dilated vessels are lined by a single layer of dysmorphic endothelial cells. Thrombi frequently form in these vessels, and hemosiderin deposits can be seen adjacent to these capillaries, indicating chronic bleeding (Figure 2). CCM patients are mostly diagnosed by magnetic resonance imaging initiated due to neurological changes, including headache, seizures, and other neurological deficits, such as nausea or vomiting, weakness or numbness, slurred speech, and altered vision. About 25% to 50% of CCM patients do not have clinical symptoms, and only a small fraction of these patients is identified incidentally [2,3]. The prevalence of CCM is about 0.5% in the general population [4,5], and about 70% to 80% of CCM patients have one lesion, and the other 20% to 30% of CCM patients have more than one lesion [6,7]. Most of the patients with one lesion have the sporadic form of the disease without a family history, while the majority of the patients who have more than one lesion have a family history with autosomal dominant Mendelian inheritance.

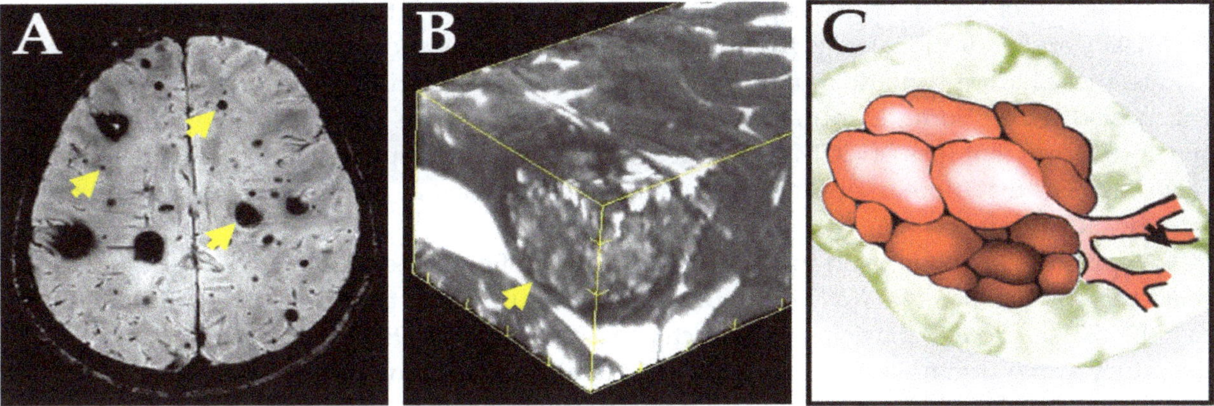

Figure 1. Radiological presentation of CCM. (**A**) MRI image of the brain of a familial CCM patient. Susceptibility weighted imaging showed multiple dark CCM lesions with various sizes. Arrows indicate representative lesions. (**B**) 3D reconstruction of T2 weighted imaging of a CCM lesion. It shows the lesion is not uniform, but with popcorn appearance. The arrow indicates the location of the lesion. (**C**) Schematic presentation of a CCM lesion showing it is composed of nested dilated microvessels.

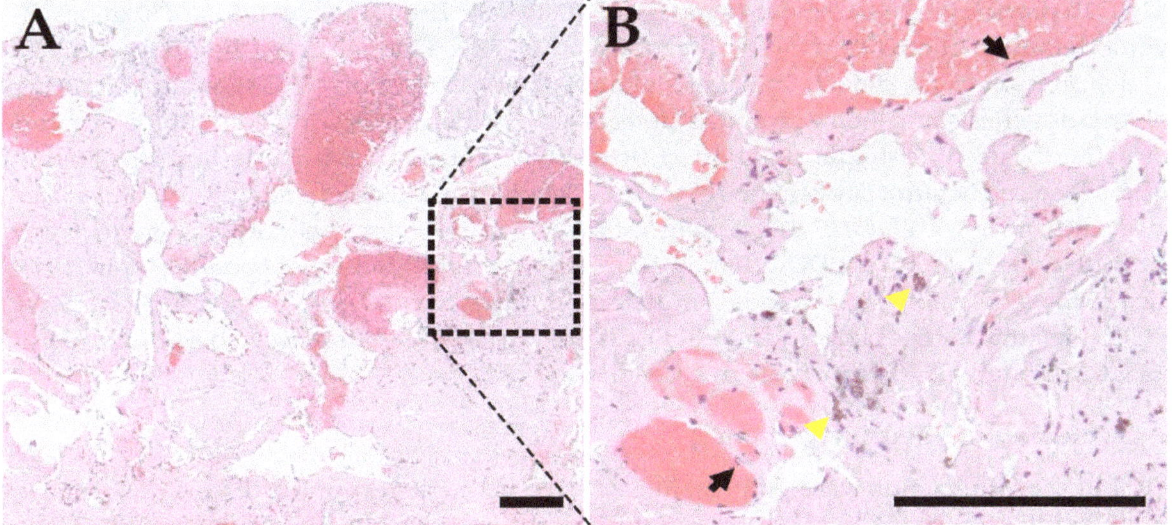

Figure 2. Histopathological presentation of CCM. (**A**) H&E staining of a surgically resected CCM lesion. It is composed of clusters of thin walled dilated microvessels with no supporting smooth muscle cells beneath the endothelial cell layer and no intervening brain parenchyma. Thrombi are present within the lumen of capillaries within the CCM lesion. (**B**) High power image of the boxed region of panel A. Black arrows point to individual endothelial cells lining the inner surface of dilated capillaries, and yellow arrowheads point to hemosiderin deposition adjacent to the capillaries, a sign of chronic bleeding. Bar = 200 μm.

3. Genetics of CCM

Based on linkage analyses, three gene loci (*CCM1* [7q21-22], *CCM2* [7p15-p13], and *CCM3* [3q25.2-q27]) have been identified in the germ-line of familial cases [8–10]. Subsequently, the genes within these loci are identified to be *CCM1/KRIT1*, *CCM2/MGC4607*, and *CCM3/PDCD10* [11–16]. Of all familial CCM patients, ~60% have *CCM1* mutations, ~20% have *CCM2* mutations, ~10% have *CCM3* mutations, and a minority of familial CCM patients do not have mutations in these three genes [17]. Although mutations of *KRIT1*, *CCM2*, and *PDCD10* genes are all associated with histologically identical CCM lesions, patients with *PDCD10* mutations have the most severe phenotype, with earlier symptomatic onset [18,19]. A large fraction of mutations identified in patients are located in the coding region of *CCM* genes and are loss-of-function mutations [20]. DNA sequencing of lesional tissue and endothelial cells from familial CCM patients showed that in addition to germ-line mutations, these harbor somatic mutations of *CCM* genes, suggesting a two-hit mechanism for CCM pathogenesis [21,22]. Somatic mutations in the same *CCM* genes have been identified in sporadic lesions, indicating that loss of CCM function also contributes to sporadic disease development [23]. This also suggests that biomarkers and therapeutic targets aimed at the familial disease will also apply in sporadic CCM cases.

CCM proteins are conserved molecules. Orthologs of all three *CCM* genes have been identified in *Caenorhabditis elegans*. The *KRIT1* ortholog *kri-1* germline affects animal longevity and germ cell survival [24,25], and *ccm-2* participates in such processes [26]. The *PDCD10* ortholog *ccm-3* is required for excretory canal organization and germline tube development through affecting a large array of cellular events, including actomyosin organization, cell polarity and endocytic recycling [26–28]. In zebrafish, *krit1* and *ccm2* loss leads to dilation of major vessels, with spreading of endothelial cells [29], and a C-terminally truncated PDCD10 causes a similar phenotype [30]. Although *Ccm* heterozygous knockout mice have little or no potential to develop CCM–like lesions in the brain, when they are on a genetically instable background ($Msh2^{-/-}$ or $Trp53^{-/-}$), these mice have a significantly higher lesion burden [19,31,32]. These findings demonstrate that loss of heterozygosity is likely an important driving force for CCM pathogenesis. Mouse studies further show that KRIT1, CCM2, and PDCD10 participate in CCM pathogenesis. Deletion of *Krit1*, *Ccm2*, and *Pdcd10* genes all cause embryonic lethality due to cardiovascular defects [33–35]. Conventional homozygous *Krit1* and *Ccm2* deletion both cause defects in branchial arch artery formation [33,34], while *Pdcd10* deletion causes vasculogenesis and hematopoiesis defects [35]. When embryonic lethality is circumvented by tamoxifen-induced postnatal deletion of floxed *Ccm* genes, CCM-like lesion formation ensues, primarily in the cerebellum, suggesting they are CCM disease causing genes [36–39]. Consistent with human studies, mice with *Pdcd10* deletion also showed a more severe phenotype than mice with *Krit1* or *Ccm2* mutations, indicating PDCD10 may affect KRIT1 and CCM2-independent events [19]. Recent evidence reveals that clonally expanded mutated endothelial cells only comprise a fraction of cells lining CCM lesions, suggesting endothelial cells with *CCM* deletions may co-opt endothelial cells without *CCM* mutations to participate in CCM disease [40].

4. CCM Proteins and Their Interactions

KRIT1 (Krev interaction trapped protein-1, CCM1) is the largest of the three CCM proteins, with 529 amino acid residues [41]. It was first identified through its binding to the small GTPase Rap1 (also called Krev-1), and it is comprised of an N-terminal Nudix domain, three NPxY/F motifs, an ankyrin-repeat region, and a C-terminal FERM (band 4.1, ezrin, radixin, moesin) domain (Figure 3). Through its N-terminal Nudix domain and NPxY/F motif containing region, KRIT1 interacts with the β1-integrin binding protein ICAP1 to limit β1-integrin activation [42,43]. The KRIT1 FERM domain binds to a transmembrane protein Heg1 and the small GTPase Rap1 and is important for KRIT1 to localize to the plasma membrane [41,44–46]. Consistent with its role in cytoskeletal regulation, KRIT1 also directly associates with microtubules [47].

CCM2 is a 444 amino acid residue protein, with a phosphotyrosine-binding domain (PTB) at its N-terminus and a C-terminal harmonin-homology domain (HHD) [13,48]. It was first characterized as an osmosensing scaffolding protein that binds to small GTPase Rac1 and protein kinases MEKK3 and MKK3 [49]. CCM2 is central to the CCM protein complex organization, as it can bind to both KRIT1 and PDCD10 (programmed cell death 10, CCM3) (Figure 3) [50–52]. The CCM2 PTB domain binds directly with the KRIT1 NPxY/F motif, and LD-like motif of CCM2 (within the linker region between the PTB and HHD domains) binds to the focal adhesion targeting (FAT) homology domain of PDCD10 [51–54].

Binding between KRIT1 and CCM2 is important for CCM2 localization [51,54], while the interaction between CCM2 and PDCD10 controls CCM2 and PDCD10 protein stability, as CCM2 depletion decreases cellular PDCD10 protein content, and PDCD10 depletion reduces CCM2 protein abundance [53]. CCM2 also associates with F-actin, bringing the actin regulating small GTPase Rac1 to the proximity of the actin cytoskeleton [55]. A paralog of CCM2, CCM2L, also exists [56]. Although CCM2L can bind to KRIT1 and compete with CCM2 for KRIT1 binding, it does not bind to PDCD10 [56]. Similar to CCM2, CCM2L also interacts with MEKK3 [57], but the significance of CCM2L for CCM disease pathogenesis and its effect on CCM protein complex organization and function remains poorly defined [58].

PDCD10 (CCM3) protein has 212 amino acid residues and consists of an N-terminal dimerization domain and a C-terminus FAT homology domain (Figure 3) [59]. It was first discovered as a gene upregulated during myeloid cell apoptosis [60]. In addition to binding to CCM2 [50], PDCD10 can bind to components of another protein complex, the striatin interacting phosphatase and kinase (STRIPAK) complex, through its dimerization domain. These proteins include striatin itself and germinal center kinase (GCK) III group of serine/threonine protein kinases MST4/MASK, MST3/STK24, and STK25/YSK1/SOK1 and other STRIPAK complex components, including STRIP1/FAM40A and STRIP2/FAM40B [61–67]. Although PDCD10 can bind to CCM2, PDCD10 primarily resides within the STRIPAK complex, rather than the CCM protein complex, in cells [63,64]. Furthermore, PDCD10 can bind to an array of other proteins, including paxillin, PTPN13, UNC13D [50,67–70]. Similar to KRIT1 and CCM2, PDCD10 also interacts with cytoskeletal regulating small GTPases. Cdc42 can co-immunoprecipitate with PDCD10, and Cdc42 deletion causes a CCM-like phenotype, suggesting Cdc42 and PDCD10 resides in the same CCM pathogenic pathway [71]. In addition, PDCD10 can directly bind to RIPOR1/FAM65A, a RhoA associated protein, providing a link between PDCD10 and RhoA signaling [72].

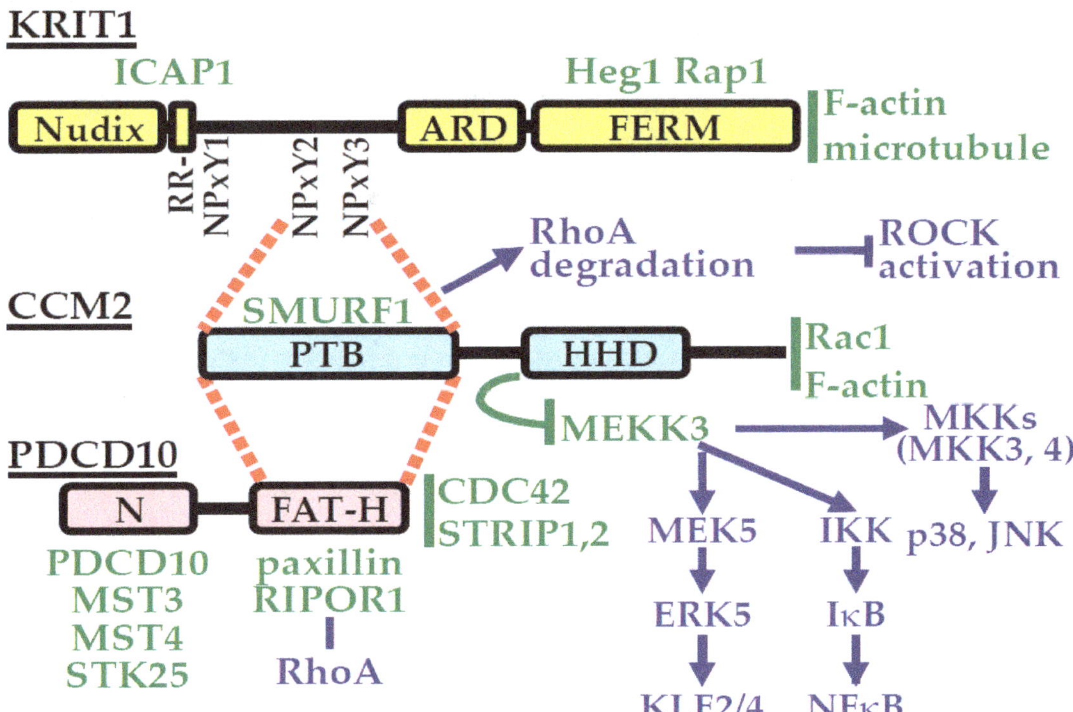

Figure 3. CCM protein domain organization and protein interactions. CCM protein domain organizations are presented schematically. Direct interaction partners are shown in green letters. Locations of the letters indicate rough interaction sites for these binding proteins. If a binding site is unknown, the binding partner is listed to the right of each CCM protein. Key pathways affected by CCM protein and their interaction partners are shown in blue letters. Dashed red lines indicate interaction sites between individual CCM proteins.

5. CCM Proteins and Cellular Signaling

Because each of the CCM proteins has a multitude of interaction partners, it is not surprising that these proteins can impact multiple signaling pathways and cellular processes, including endothelial to mesenchymal transition, autophagy, exocytosis, and Golgi complex organization [63,69,72–75]. One of the best understood CCM controlled signaling pathways is the RhoA-Rho-associated coiled-coil kinase (ROCK) signaling. Decreased expression of any of the CCM proteins leads to increased RhoA and ROCK activity [19,30,34,54,76,77], which in turn increases myosin regulatory light chain (MLC) phosphorylation, causing actomyosin contraction that affects cell migration and intercellular junction integrity. Through its PTB domain, CCM2 can bind to the E3 ubiquitin ligase Smad ubiquitin regulatory factor 1 (Smurf1) [76,78], which ubiquitinates RhoA to promote its degradation [78]. In the absence of CCM2, Smurf1-mediated RhoA degradation is reduced, leading to RhoA accumulation and increased ROCK activity [78]. Depletion of PDCD10 and its binding partners STK25, STRIP1/FAM40A, STRIP2/FAM40B, and RIPTOR/FAM65A all increase MLC phosphorylation, indicating PDCD10 may affect RhoA-ROCK activity through these proteins [19,30,65]. The enhanced RhoA-ROCK signaling is a critical component of CCM pathogenesis, which is further detailed below.

Another relatively well understood CCM-regulated pathway is the MEKK3 signaling. As discussed above, CCM2 directly interacts with MEKK3 [49]. Both *Krit1* and *Ccm2* deletion leads to activation of the MEKK3-MEK5-ERK5-KLF2/4 signaling cascade, causing increased Adamts4/5 expression [57,79,80]. These changes disrupt both embryonic cardiac development and promote CCM-like lesion formation in neonatal mice [79,80]. Consistent with the findings that CCM2 negatively regulates MEKK3, and MEKK3 is required for immune related receptor signaling [81–84], MEKK3 activating ligands lipopolysaccharide (LPS), IL-1β, and pI:pC can all promote CCM-like lesion formation [85]. There is some evidence that aberrantly activated MEKK3 signaling can lead to increased RhoA-ROCK signaling, but the exact mechanism for this potential crosstalk and its contribution to CCM disease need to be further elucidated [79,80,85,86].

CCM proteins have also been implicated in cell death regulation. The *C. elegans KRIT1* ortholog *kri-1* is required to promote irradiation-induced germ cell death through a cell-nonautonomous fashion [25], while in neuroblastoma cells, CCM2 is critical for the TrkA receptor tyrosine kinase to induce tumor cell death [87,88]. As its name suggests, PDCD10 has also been associated with apoptosis regulation. In endothelial cells, overexpression of PDCD10 promotes endothelial apoptosis, and in cardiomyocytes, PDCD10 expression is required for ischemic reperfusion injury-induced cell death [89,90]. However, the exact effect of PDCD10 on apoptosis is still under debate. For example, PDCD10 is up-regulated during oxidative stress, but one report suggested such upregulation promotes tumor cell survival, while another report suggested such upregulation enhances apoptosis [91,92]. Thus, how CCM proteins affect cell death and proliferation to impact human health and disease remains to be further explored.

6. CCM Proteins Participate in Endothelial Barrier Maintenance and Regulation

Early morphological studies showed that CCM lesions are lined by altered endothelial cells with disrupted cell–cell connections, including tight junctions [93,94]. Using MRI based in vivo permeability measurements, it is now clear that CCM lesions have increased vascular permeability [95,96]. In white matter regions away from CCM lesions, patients with familial CCM disease (harboring a germline mutation) have higher baseline permeability than patients with sporadic disease, indicating *CCM* mutations globally affect blood–brain barrier function [95,96]. Furthermore, baseline brain white matter vascular permeability can be used to distinguish familial CCM patients with non-aggressive and aggressive disease, and between stable and non-stable CCM disease [95,96]. These data suggest blood–brain barrier defect regulates CCM clinical disease presentation.

Consistent with patient-based studies, cell culture and mouse studies demonstrated how CCM proteins may affect endothelial barrier function. All three CCM proteins can limit RhoA-ROCK signaling in endothelial cells, although PDCD10 may use a mechanism distinctive of KRIT1 and CCM2 [19,54,65]. The small GTPase RhoA and the other two CCM protein binding small GTPases Rac1 and Cdc42 are cytoskeletal regulators that control barrier function [97–99]. In the case of RhoA, its effector ROCK can either directly or indirectly induce MLC phosphorylation, leading to perijunctional actomyosin contraction, which in turn causes intercellular junction remodeling to increase paracellular permeability [100]. Indeed, decreased KRIT1, CCM2 and PDCD10 expression all promote MLC phosphorylation, stress fiber formation, junctional protein redistribution, and barrier dysfunction in endothelial cells [19,34,54,101].

In addition to maintaining baseline endothelial barrier, KRIT1 also participates in endothelial barrier regulation. While tumor necrosis factor (TNF) increased arteriole and venule permeability in wild type mice, it failed to induce barrier loss in *Krit1* heterozygous knockout mice [101,102]. In contrast, histamine-induced vascular permeability increase occurred normally in *Krit1* heterozygous knockout mice [101,102]. However, another report suggests KRIT1 is required for preservation of endothelial barrier following stimuli [103]. KRIT1 depletion in cultured endothelial cells attenuated prostacyclin-induced perijunctional F-actin accumulation and tightening of endothelial barrier and enhanced cyclic stretch-induced Rho activation and endothelial barrier disruption [103]. In vivo studies further showed that *Krit1* heterozygous knockout exacerbated barrier loss induced by combined treatment of high tidal volume mechanical ventilation and TRAP6, a thrombin receptor activating peptide. This treatment also increased protein and cell content of bronchoalveolar lavage fluid, indicating partial KRIT1 loss participates in lung damage [103]. These data suggest KRIT1 may participate in endothelial barrier regulation in a stimulus-dependent manner and contribute to endothelial dysfunction-related diseases.

Because of the robust ROCK activation in CCM depleted endothelial cells, ROCK became a leading target for novel CCM therapy. ROCK inhibition not only reverses CCM depletion-induced stress fiber formation and barrier loss in vitro but also limits *Ccm* deletion-induced loss of endothelial barrier function in vivo [19,34,54,101]. Pharmacological studies further show that ROCK inhibition by fasudil, atorvastatin, and a newly identified ROCK2 specific inhibitor limits CCM-like lesion formation in multiple mouse models of CCM [104–106], highlighting ROCK inhibition may be a valid therapy for CCM disease. This proof of concept is currently being tested in a clinical trial (NCT02603328) [107].

Besides RhoA-ROCK signaling, additional cellular processes have been implicated for CCM proteins to regulate endothelial barrier. Vascular endothelial growth factor (VEGF) not only promotes endothelial growth, but also increases endothelial permeability [108]. It has been demonstrated that loss of KRIT1 and PDCD10, but not CCM2, increases VEGF production in endothelial cells, and VEGF in turn acts on VEGFR2 to increase endothelial permeability [109]. However, existing evidence also suggests that PDCD10 is required for proper VEGFR2 signaling [35], indicating the relationship between CCM proteins and VEGF and its signaling may be complex. In KRIT1 depleted cells or heterozygous knockout mice, endothelial reactive oxygen species (ROS) production is elevated, at least partially, through upregulated NAPDH oxidase expression [102,110]. When an endothelial targeting ROS scavenger was used, the increased vascular permeability was reduced in KRIT1 deficient mice, demonstrating ROS production also plays a role for KRIT1 to regulate endothelial barrier [102]. However, the molecular mechanisms for ROS to affect barrier function in endothelium, in the presence or absence of KRIT1, remain to be elucidated.

7. Tight Junctions and CCM Disease

One of the major determinants of the endothelial barrier is the tight junction. In contrast to well demarcated tight junction, adherens junction, and gap junction domains within the apical junctional complex of epithelial cells, these domains are frequently mixed at cell–cell contact sites between endothelial cells [111]. Such junctions can vary significantly in endothelial cells of different origins. Microvascular endothelial cell bodies can have a thickness of 0.3 μm, with cell–cell junction depth of ~0.5–0.9 μm, while endothelial cells from arteries and high endothelial venules the cell–cell contact sites may reach 3–10 μm in height [111]. In the brain, the endothelial cells, pericytes at the abluminal side of endothelial cells, and astrocyte end feet together form the neurovascular unit to create the highly impermeable blood–brain barrier [112]. At the endothelial junctional complex, the adherens junction component VE-cadherin provides adhesive force at the cell–cell junctions, and the tight junction proteins are critical for limiting permeability between individual endothelial cells.

The tight junction seals the paracellular space between individual cavity lining cells and is created and maintained by a large number of transmembrane proteins. The four-transmembrane-domain-containing claudin family consists of more than 25 members in mammals. Some of the claudins (including claudin-1, -3, -5) are barrier forming, while some claudin family members are forming size and charge selective pores that allow charged ions and small molecules to pass (including claudin-2, -10, -15) [113]. In the brain microvascular endothelium, the most dominantly expressed claudin is the barrier forming claudin-5 [114,115]. Although claudin-5 is not required for brain microvascular endothelial tight junction organization, its knockout increased brain microvascular permeability, leading to neonatal death [116]. The four-transmembrane domain-containing tight-junction-associated MARVEL protein (TAMP) family contains occludin, tricellulin, and marveld3 [117], and these proteins generally impact macromolecular permeability [118,119]. Occludin knockout itself does not disrupt normal epithelial tight junction organization, but causes brain calcification, particularly around small vessels [120]. Patients with homozygous recessive occludin mutations have a more severe brain phenotype, with band-like calcification with simplified gyration and polymicrogyria [121]. This suggests occludin plays a critical role in brain development, likely through affecting brain endothelial function. Additional tight junction proteins belong to the immunoglobulin superfamily of proteins with a single transmembrane domain (e.g., junctional adhesion molecule A, JAM-A and Coxsackie and adenovirus receptor, CAR) and popeye family of proteins with three transmembrane domains (Popdc1/Bves). In the intestine, JAM-A maintains proper epithelial macromolecular barrier function and limits intestinal inflammation [122,123], and endothelial JAM-A promotes leukocyte transmigration [73,124]. Similarly, CAR participates in epithelial barrier maintenance [125], and CAR affects shear stress induced endothelial immune response [126].

Multiple plaque proteins concentrate at the cytoplasmic side of the tight junction. These proteins typically bind to multiple transmembrane tight junction proteins, other tight junction plaque proteins, and the cytoskeleton, thus stabilize tight junction organization. Zonula occludens (ZO) family proteins (ZO-1, -2, -3) is a well-studied family of tight junction plaque proteins [127]. They can bind to almost all transmembrane tight junction proteins, heterodimerize among different ZO proteins, and associate with the actin cytoskeleton [127]. ZO-1 knockout mice are embryonic lethal, with defects in vascular endothelial cells [128], a finding supported by in vitro endothelial cell studies [129]. Cingulin family is another group of tight junction plaque proteins (cingulin, paracingulin/cingulin-like/JACOP) that can interact with occludin, JAM-A, ZO proteins, myosin and actin filaments, which are also required for proper endothelial function, including brain endothelial barrier function [130,131].

Many CCM affected signaling events can regulate the tight junction. As discussed above, Rho-ROCK signaling increases MLC phosphorylation to impact actomyosin contraction, which in turn regulates tight junction protein expression and localization [132–135]. In addition, ROCK can directly phosphorylate occludin and claudin-5, and such phosphorylation events are associated with blood brain barrier dysfunction [136]. Interaction between endothelial cells and basement membrane induces β1-integrin engagement, increases MLC phosphorylation in an MLC kinase and ROCK -dependent fashion to promote claudin-5, occludin, and ZO-1 reorganization at the cell–cell junction [137]. This pathway is likely affected by CCM proteins through KRIT1 binding to the β1-integrin signaling inhibitor ICAP-1, a protein that can also bind to ROCK [138–140]. The KRIT1 binding small GTPase Rap1 enhances tight junction protein localization at endothelial cell–cell contact sites and promotes endothelial barrier function [141]. Consistent with this, the Rap1 activating guanine-nucleotide-exchange factor EPAC also maintains endothelial barrier, prevents VEGF and TNF-induced endothelial permeability increase, and limits claudin-5, occludin, and ZO-1 disorganization at the cell–cell junctions [142]. Another small GTPase, Rasip1, is an effector of Rap1, which down-regulates RhoA activity through ArhGAP29 [143–145]. Rasip1 can also interact with the KRIT1 interacting transmembrane protein Heg1 [146], thus KRIT1 can bring Rasip1 and Rap1 close to one another through KRIT1-Heg1 interaction. Furthermore, engagement between individual JAM-A molecules at intercellular junctions can activate Rap1 to preserve epithelial and endothelial barrier functions through JAM-A interaction with the tight junction protein ZO-2, the adherens junction protein AF-6, and PDZ-GEF1/2 [147,148]. These data provide a complex signaling network for the tight junction proteins (JAM-A and ZO-2) and other cell surface adhesion molecules (β1-integrin and Heg1) to affect CCM-dependent cellular signaling pathways to impact tight junction barrier.

Consistent with such findings, resected CCM lesions have reduced occludin, claudin-5, and ZO-1 staining, and decreased tight junction protein expression has between associated with the tendency for local bleeding and edema [149,150]. In *Krit1* deleted brain microvascular endothelial cells, loss of claudin-5 and ZO-1 protein can be readily observed by immunofluorescent staining and western blot [151], and PDCD10 depletion in brain microvascular endothelial cells decreases claudin-5, occludin, and ZO-1 protein abundance, likely through an ERK1/2 and cortactin-dependent process [152]. A recent study suggests PDCD10 depletion in brain endothelial cells upregulates gap junction protein connexin 43 expression and increases gap junction communication, a phenomenon only minimally seen in KRIT1 or CCM2 depleted cells [153]. Such changes are associated with redistribution of tight junction proteins to gap junctions, and the connexin 43 gap junction inhibitor GAP27 can reverse tight junction disorganization and decrease endothelial barrier permeability in PDCD10 depleted cells [153]. These indicate increased gap junction function participates in tight junction disruption in *CCM3* disease. With such findings, it is likely that tight junction protein disorganization downstream of RhoA-ROCK signaling and gap junction is a key effector driving CCM pathogenesis, and it is possible that normalizing tight junction protein expression and localization at the cell–cell junctions can limit CCM development or lesional bleeding. However, the specific roles of tight junction proteins in CCM initiation and progression remain to be formally tested, likely by using transgenic or knockout mice.

8. CCM Proteins Impact Intestinal Homeostasis

In contrast to a plethora of studies on the function of CCM proteins in endothelial cells, we just start to appreciate their roles in epithelial cells. By investigating the effects of KRIT1 in β-catenin signaling, Glading and Ginsberg revealed KRIT1 depletion increases β-catenin transcriptional activity in both endothelial and epithelial cells [154]. This is functionally significant, as *Apc* mutation induced more intestinal polyp formation in *Krit1* heterozygous knockout mice with increased intestinal epithelial nuclear β-catenin accumulation [154]. A recent *C. elegans* study suggested KRIT1 can also form a complex with CCM2 to promote zinc transporter expression to cause Zn^{2+} storage in the intestinal granules, indicating KRIT1 may also impact intestinal epithelial transport [26].

Despite these findings, it was not known if CCM proteins can regulate barrier function in epithelium. Our group addressed this question by studying KRIT1 function in intestinal epithelial Caco-2 cells, a well characterized model to study intestinal epithelial barrier maintenance and regulation [155]. In this model, KRIT1 depletion caused a reduction of epithelial barrier function, characterized by selectively increased relative permeability of small cations, including Na^+, to the anion Cl^- [155]. Such a change is consistent with decreased expression of claudin-1, a tight junction protein that limits small ion permeability, in KRIT1-depleted Caco-2 cells [155,156]. In contrast to the effect of KRIT1 on endothelial cells, intestinal epithelial KRIT1 depletion does not induce MLC phosphorylation, and ROCK inhibition does not reverse KRIT1 depletion-induced barrier loss [155]. This indicates that KRIT1 regulates epithelial and endothelial barrier function through distinct mechanisms. In Caco-2 monolayers, decreasing actomyosin contractility by inhibiting either ROCK or myosin ATPase activity both reduced epithelial barrier function, along with elevated permeability to both small and large cations. These changes are inhibited in KRIT1-depleted Caco-2 monolayers, indicating KRIT1 also participates in actomyosin contraction-induced barrier regulation [155]. Furthermore, KRIT1-depleted epithelial monolayers are resistant to osmotic stress and enteric pathogen *Salmonella typhimurium*-induced epithelial barrier regulation (Figure 4), suggesting KRIT1 may impact gastrointestinal pathophysiology. With the above data, it is surprising to find that KRIT1 depletion exacerbates TNF-induced epithelial barrier loss. Mechanistic studies suggest this loss is due to aberrantly activated apoptosis in KRIT1-depleted monolayers, but we currently do not know how this occurs [155]. Nevertheless, these data suggest KRIT1 regulates epithelial barrier function through at least two distinct pathways: one is actomyosin and tight junction-dependent barrier maintenance and regulation, and the other is tight junction-independent epithelial apoptosis. Such findings not only point to a role for KRIT1 to mediate the crosstalk between distinctive epithelial barrier regulation pathways, but also suggest KRIT1 may coordinate tight junction barrier maintenance, regulation, and epithelial apoptosis to impact intestinal disease development.

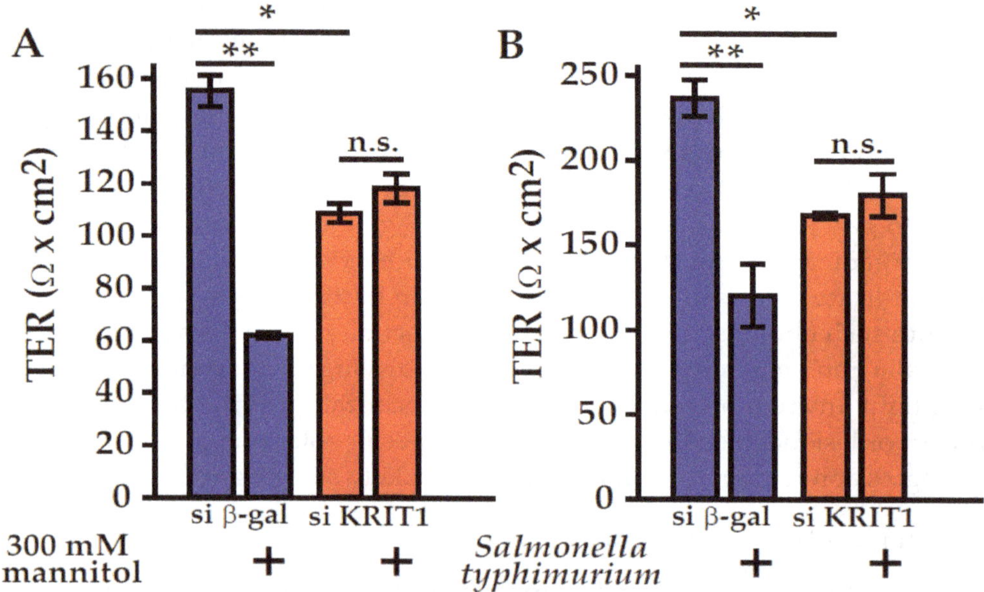

Figure 4. KRIT1 depletion limits pathophysiological stimuli-induced epithelial barrier dysfunction. KRIT1 depletion by stable transfection of a siRNA expressing plasmid decreased epithelial barrier function (A-B, assessed by transepithelial resistant (TER) measurements) in differentiated Caco-2 intestinal epithelial monolayers grown on semi-permeable Transwell inserts [155]. (**A**) Hyperosmotic stress induced by including 300 mM mannitol in Hank's balanced salt solution caused barrier loss in control (siRNA against β-galactosidase transfected, blue bars) Caco-2 monolayers. In contrast, no barrier loss was induced in KRIT1 depleted (siRNA against KRIT1 transfected, red bars) Caco-2 monolayers. (**B**) *Salmonella typhimurium* (strain ATCC 14028) infection by including bacteria in apical culture media caused barrier loss in control, but not KRIT1 depleted Caco-2 monolayers. Mean with standard error (triplicate samples) are shown. One-way ANOVA analysis with Bonferroni correction was used (* $p < 0.05$, ** $p < 0.01$).

An understanding of the potential contribution of the gastrointestinal tract to CCM disease development was stemmed from the surprising finding that neonatal mice with the same induced endothelial specific *Ccm* deletion can have drastically different CCM-like lesion burdens when they were raised in different animal facilities [85]. Fecal microbiome analysis showed that mice susceptible to *Ccm* deletion-induced lesion formation have a Gram-negative bacteria rich microbiome relative to resistant mice. Such a fecal microbiome provides the cell wall product LPS as the ligand to activate the endothelial TLR4-MEKK3-KLR2/4 signaling pathway to promote CCM development [85]. This view is further supported by the finding that germ-free mice and mice treated with antibiotics have lower lesion burden [85]. Because familial CCM patients have genetic mutations of *CCM* genes in all organs and cell types, this study also raised the possibility that CCM could function in the gastrointestinal tract to influence CCM disease development. Indeed, when *Pdcd10* was deleted in the intestinal epithelium, it promoted endothelial *Pdcd10* deletion-induced lesion formation [157]. In contrast, intestinal epithelial specific deletion of *Krit1* does not impact endothelial *Krit1* deletion-induced lesion formation [157]. This finding may at least partially explain why patients with *PDCD10* mutations have a more aggressive CCM disease than patients with *KRIT1* mutations. In addition to impacting CCM-like disease development, constitutive intestinal epithelial specific *Pdcd10* deletion alone shortened mouse life span, reduced intestinal mucus layer thickness, enlarged goblet cells, and caused intestinal inflammation [157]. These findings indicate PDCD10 is required for intestinal homeostasis and may impact intestinal disease development, which needs to be further investigated.

9. Conclusions and Future Directions

Through their many binding partners, CCM proteins impact many cellular events. The most prominent effect of CCM proteins on cellular signaling is their ability to limit RhoA-ROCK activity and MEKK3-MEK5-ERK5-KLF signaling, events that are important for endothelial function and CCM lesion formation. Despite such detailed understanding, we are just starting to grasp the full spectrum of CCM protein functions. Understanding how CCM proteins affect endothelial function through a variety of pathways to impact CCM disease and identifying therapies to preserve and promote normal CCM function in the brain remain top priorities for CCM research. With the finding that CCM proteins also function in intestinal epithelial cells, it becomes pressing to understand CCM protein functions in the gut, in the context of both CCM disease and other intestinal disorders. It also points to a need to understand CCM protein signaling in other cell types and organs. Such studies will not only advance our understanding of CCM protein biology, but also provide targets to modulate cellular functions to benefit human health.

Author Contributions: Initial draft, S.W., Y.L., and L.S.; contribution of figures, Y.L., S.P.P., C.R.W., and L.S.; draft revision, S.W., Y.L., S.P.P., C.R.W., I.A.A., and L.S. All authors have read and agreed to the published version of the manuscript.

Acknowledgments: We thank Wenli Dai, Li Dong, and Jimian Yu for comments and proofreading of the manuscript.

Abbreviations

CCM	cerebral cavernous malformation
EPAC	Rap1 guanine-nucleotide-exchange factor
FAT	focal adhesion targeting
FERM	band 4.1, ezrin, radixin, moesin
GCK	germinal center kinase
HHD	harmonin-homology domain
JAM	junctional adhesion molecule
KRIT1	Krev interaction trapped protein-1
LPS	lipopolysaccharide
MLC	myosin regulatory light chain
PDCD10	programmed cell death 10
PTB	phosphotyrosine-binding domain
ROCK	Rho-associated coiled-coil kinase
ROS	reactive oxygen species
Smurf1	Smad ubiquitin regulatory factor 1
STRIPAK	striatin interacting phosphatase and kinase
TER	transepithelial resistance
TNF	tumor necrosis factor
VEGF	vascular endothelial growth factor
ZO	zonula occludens

References

1. Awad, I.A.; Polster, S.P. Cavernous angiomas: Deconstructing a neurosurgical disease. *J. Neurosurg.* **2019**, *131*, 1–13. [CrossRef]
2. Morris, Z.; Whiteley, W.N.; Longstreth, W.T., Jr.; Weber, F.; Lee, Y.C.; Tsushima, Y.; Alphs, H.; Ladd, S.C.; Warlow, C.; Wardlaw, J.M.; et al. Incidental findings on brain magnetic resonance imaging: Systematic review and meta-analysis. *Br. Med. J.* **2009**, *339*, b3016. [CrossRef]
3. Moore, S.A.; Brown, R.D., Jr.; Christianson, T.J.; Flemming, K.D. Long-term natural history of incidentally discovered cavernous malformations in a single-center cohort. *J. Neurosurg.* **2014**, *120*, 1188–1192. [CrossRef]
4. Flemming, K.D.; Graff-Radford, J.; Aakre, J.; Kantarci, K.; Lanzino, G.; Brown, R.D., Jr.; Mielke, M.M.; Roberts, R.O.; Kremers, W.; Knopman, D.S.; et al. Population-Based Prevalence of Cerebral Cavernous Malformations in Older Adults: Mayo Clinic Study of Aging. *JAMA Neurol.* **2017**, *74*, 801–805. [CrossRef] [PubMed]
5. Otten, P.; Pizzolato, G.P.; Rilliet, B.; Berney, J. [131 cases of cavernous angioma (cavernomas) of the CNS, discovered by retrospective analysis of 24,535 autopsies]. *Neurochirurgie* **1989**, *35*, 82–83, 128–131.
6. Flemming, K.D.; Link, M.J.; Christianson, T.J.; Brown, R.D., Jr. Prospective hemorrhage risk of intracerebral cavernous malformations. *Neurology* **2012**, *78*, 632–636. [CrossRef] [PubMed]
7. Al-Shahi Salman, R.; Berg, M.J.; Morrison, L.; Awad, I.A. Hemorrhage from cavernous malformations of the brain: Definition and reporting standards. Angioma Alliance Scientific Advisory Board. *Stroke* **2008**, *39*, 3222–3230. [CrossRef] [PubMed]
8. Gunel, M.; Awad, I.A.; Anson, J.; Lifton, R.P. Mapping a gene causing cerebral cavernous malformation to 7q11.2-q21. *Proc. Natl. Acad. Sci. USA* **1995**, *92*, 6620–6624. [CrossRef] [PubMed]
9. Marchuk, D.A.; Gallione, C.J.; Morrison, L.A.; Clericuzio, C.L.; Hart, B.L.; Kosofsky, B.E.; Louis, D.N.; Gusella, J.F.; Davis, L.E.; Prenger, V.L. A locus for cerebral cavernous malformations maps to chromosome 7q in two families. *Genomics* **1995**, *28*, 311–314. [CrossRef] [PubMed]
10. Craig, H.D.; Gunel, M.; Cepeda, O.; Johnson, E.W.; Ptacek, L.; Steinberg, G.K.; Ogilvy, C.S.; Berg, M.J.; Crawford, S.C.; Scott, R.M.; et al. Multilocus linkage identifies two new loci for a mendelian form of stroke, cerebral cavernous malformation, at 7p15-13 and 3q25.2-27. *Hum. Mol. Genet.* **1998**, *7*, 1851–1858. [CrossRef]
11. Laberge-le Couteulx, S.; Jung, H.H.; Labauge, P.; Houtteville, J.P.; Lescoat, C.; Cecillon, M.; Marechal, E.; Joutel, A.; Bach, J.F.; Tournier-Lasserve, E. Truncating mutations in *CCM1*, encoding *KRIT1*, cause hereditary cavernous angiomas. *Nat. Genet.* **1999**, *23*, 189–193. [CrossRef] [PubMed]
12. Sahoo, T.; Johnson, E.W.; Thomas, J.W.; Kuehl, P.M.; Jones, T.L.; Dokken, C.G.; Touchman, J.W.; Gallione, C.J.; Lee-Lin, S.Q.; Kosofsky, B.; et al. Mutations in the gene encoding *KRIT1*, a Krev-1/rap1a binding protein, cause cerebral cavernous malformations (*CCM1*). *Hum. Mol. Genet.* **1999**, *8*, 2325–2333. [CrossRef] [PubMed]
13. Liquori, C.L.; Berg, M.J.; Siegel, A.M.; Huang, E.; Zawistowski, J.S.; Stoffer, T.; Verlaan, D.; Balogun, F.; Hughes, L.; Leedom, T.P.; et al. Mutations in a gene encoding a novel protein containing a phosphotyrosine-binding domain cause type 2 cerebral cavernous malformations. *Am. J. Hum. Genet.* **2003**, *73*, 1459–1464. [CrossRef] [PubMed]
14. Denier, C.; Goutagny, S.; Labauge, P.; Krivosic, V.; Arnoult, M.; Cousin, A.; Benabid, A.L.; Comoy, J.; Frerebeau, P.; Gilbert, B.; et al. Mutations within the MGC4607 gene cause cerebral cavernous malformations. *Am. J. Hum. Genet.* **2004**, *74*, 326–337. [CrossRef] [PubMed]
15. Guclu, B.; Ozturk, A.K.; Pricola, K.L.; Bilguvar, K.; Shin, D.; O'Roak, B.J.; Gunel, M. Mutations in apoptosis-related gene, *PDCD10*, cause cerebral cavernous malformation 3. *Neurosurgery* **2005**, *57*, 1008–1013. [CrossRef] [PubMed]
16. Bergametti, F.; Denier, C.; Labauge, P.; Arnoult, M.; Boetto, S.; Clanet, M.; Coubes, P.; Echenne, B.; Ibrahim, R.; Irthum, B.; et al. Mutations within the programmed cell death 10 gene cause cerebral cavernous malformations. *Am. J. Hum. Genet.* **2005**, *76*, 42–51. [CrossRef]
17. Zafar, A.; Quadri, S.A.; Farooqui, M.; Ikram, A.; Robinson, M.; Hart, B.L.; Mabray, M.C.; Vigil, C.; Tang, A.T.; Kahn, M.L.; et al. Familial Cerebral Cavernous Malformations. *Stroke* **2019**, *50*, 1294–1301. [CrossRef]
18. Denier, C.; Labauge, P.; Bergametti, F.; Marchelli, F.; Riant, F.; Arnoult, M.; Maciazek, J.; Vicaut, E.; Brunereau, L.; Tournier-Lasserve, E.; et al. Genotype-phenotype correlations in cerebral cavernous malformations patients. *Ann. Neurol.* **2006**, *60*, 550–556. [CrossRef]

19. Shenkar, R.; Shi, C.; Rebeiz, T.; Stockton, R.A.; McDonald, D.A.; Mikati, A.G.; Zhang, L.; Austin, C.; Akers, A.L.; Gallione, C.J.; et al. Exceptional aggressiveness of cerebral cavernous malformation disease associated with *PDCD10* mutations. *Genet. Med. Off. J. Am. Coll. Med Genet.* **2015**, *17*, 188–196. [CrossRef]
20. Spiegler, S.; Rath, M.; Paperlein, C.; Felbor, U. Cerebral Cavernous Malformations: An Update on Prevalence, Molecular Genetic Analyses, and Genetic Counselling. *Mol. Syndromol.* **2018**, *9*, 60–69. [CrossRef]
21. Gault, J.; Shenkar, R.; Recksiek, P.; Awad, I.A. Biallelic somatic and germ line *CCM1* truncating mutations in a cerebral cavernous malformation lesion. *Stroke* **2005**, *36*, 872–874. [CrossRef] [PubMed]
22. Akers, A.L.; Johnson, E.; Steinberg, G.K.; Zabramski, J.M.; Marchuk, D.A. Biallelic somatic and germline mutations in cerebral cavernous malformations (CCMs): Evidence for a two-hit mechanism of CCM pathogenesis. *Hum. Mol. Genet.* **2009**, *18*, 919–930. [CrossRef] [PubMed]
23. McDonald, D.A.; Shi, C.; Shenkar, R.; Gallione, C.J.; Akers, A.L.; Li, S.; De Castro, N.; Berg, M.J.; Corcoran, D.L.; Awad, I.A.; et al. Lesions from patients with sporadic cerebral cavernous malformations harbor somatic mutations in the *CCM* genes: Evidence for a common biochemical pathway for CCM pathogenesis. *Hum. Mol. Genet.* **2014**, *23*, 4357–4370. [CrossRef]
24. Berman, J.R.; Kenyon, C. Germ-cell loss extends C-elegans life span through regulation of DAF-16 by kri-1 and lipophilic-hormone signaling. *Cell* **2006**, *124*, 1055–1068. [CrossRef] [PubMed]
25. Ito, S.; Greiss, S.; Gartner, A.; Derry, W.B. Cell-Nonautonomous Regulation of *C. elegans* Germ Cell Death by kri-1. *Curr. Biol.* **2010**, *20*, 333–338. [CrossRef] [PubMed]
26. Chapman, E.M.; Lant, B.; Ohashi, Y.; Yu, B.; Schertzberg, M.; Go, C.; Dogra, D.; Koskimaki, J.; Girard, R.; Li, Y.; et al. A conserved CCM complex promotes apoptosis non-autonomously by regulating zinc homeostasis. *Nat. Commun.* **2019**, *10*, 1791. [CrossRef]
27. Pal, S.; Lant, B.; Yu, B.; Tian, R.L.; Tong, J.F.; Krieger, J.R.; Moran, M.F.; Gingras, A.C.; Derry, W.B. CCM-3 Promotes *C. elegans* Germline Development by Regulating Vesicle Trafficking Cytokinesis and Polarity. *Curr. Biol.* **2017**, *27*, 868–876. [CrossRef]
28. Lant, B.; Yu, B.; Goudreault, M.; Holmyard, D.; Knight, J.D.R.; Xu, P.; Zhao, L.; Chin, K.; Wallace, E.; Zhen, M.; et al. CCM-3/STRIPAK promotes seamless tube extension through endocytic recycling. *Nat. Commun.* **2015**, *6*, 6449. [CrossRef]
29. Hogan, B.M.; Bussmann, J.; Wolburg, H.; Schulte-Merker, S. *Ccm1* cell autonomously regulates endothelial cellular morphogenesis and vascular tubulogenesis in zebrafish. *Hum. Mol. Genet.* **2008**, *17*, 2424–2432. [CrossRef]
30. Zheng, X.J.; Xu, C.; Di Lorenzo, A.; Kleaveland, B.; Zou, Z.Y.; Seiler, C.; Chen, M.; Cheng, L.; Xiao, J.P.; He, J.; et al. CCM3 signaling through sterile 20-like kinases plays an essential role during zebrafish cardiovascular development and cerebral cavernous malformations. *J. Clin. Investig.* **2010**, *120*, 2795–2804. [CrossRef]
31. Plummer, N.W.; Gallione, C.J.; Srinivasan, S.; Zawistowski, J.S.; Louis, D.N.; Marchuk, D.A. Loss of p53 sensitizes mice with a mutation in *Ccm1* (*KRIT1*) to development of cerebral vascular malformations. *Am. J. Pathol.* **2004**, *165*, 1509–1518. [CrossRef]
32. McDonald, D.A.; Shenkar, R.; Shi, C.; Stockton, R.A.; Akers, A.L.; Kucherlapati, M.H.; Kucherlapati, R.; Brainer, J.; Ginsberg, M.H.; Awad, I.A.; et al. A novel mouse model of cerebral cavernous malformations based on the two-hit mutation hypothesis recapitulates the human disease. *Hum. Mol. Genet.* **2011**, *20*, 211–222. [CrossRef] [PubMed]
33. Whitehead, K.J.; Plummer, N.W.; Adams, J.A.; Marchuk, D.A.; Li, D.Y. *Ccm1* is required for arterial morphogenesis: Implications for the etiology of human cavernous malformations. *Development* **2004**, *131*, 1437–1448. [CrossRef]
34. Whitehead, K.J.; Chan, A.C.; Navankasattusas, S.; Koh, W.; London, N.R.; Ling, J.; Mayo, A.H.; Drakos, S.G.; Jones, C.A.; Zhu, W.; et al. The cerebral cavernous malformation signaling pathway promotes vascular integrity via Rho GTPases. *Nat. Med.* **2009**, *15*, 177–184. [CrossRef]
35. He, Y.; Zhang, H.; Yu, L.; Gunel, M.; Boggon, T.J.; Chen, H.; Min, W. Stabilization of VEGFR2 signaling by cerebral cavernous malformation 3 is critical for vascular development. *Sci. Signal.* **2010**, *3*, ra26. [CrossRef] [PubMed]
36. Boulday, G.; Rudini, N.; Maddaluno, L.; Blecon, A.; Arnould, M.; Gaudric, A.; Chapon, F.; Adams, R.H.; Dejana, E.; Tournier-Lasserve, E. Developmental timing of CCM2 loss influences cerebral cavernous malformations in mice. *J. Exp. Med.* **2011**, *208*, 1835–1847. [CrossRef]

37. Mleynek, T.M.; Chan, A.C.; Redd, M.; Gibson, C.C.; Davis, C.T.; Shi, D.S.; Chen, T.; Carter, K.L.; Ling, J.; Blanco, R.; et al. Lack of CCM1 induces hypersprouting and impairs response to flow. *Hum. Mol. Genet.* **2014**, *23*, 6223–6234. [CrossRef]
38. Chan, A.C.; Drakos, S.G.; Ruiz, O.E.; Smith, A.C.; Gibson, C.C.; Ling, J.; Passi, S.F.; Stratman, A.N.; Sacharidou, A.; Revelo, M.P.; et al. Mutations in 2 distinct genetic pathways result in cerebral cavernous malformations in mice. *J. Clin. Investig.* **2011**, *121*, 1871–1881. [CrossRef]
39. Cunningham, K.; Uchida, Y.; O'Donnell, E.; Claudio, E.; Li, W.; Soneji, K.; Wang, H.; Mukouyama, Y.S.; Siebenlist, U. Conditional deletion of *Ccm2* causes hemorrhage in the adult brain: A mouse model of human cerebral cavernous malformations. *Hum. Mol. Genet.* **2011**, *20*, 3198–3206. [CrossRef]
40. Detter, M.R.; Snellings, D.A.; Marchuk, D.A. Cerebral Cavernous Malformations Develop Through Clonal Expansion of Mutant Endothelial Cells. *Circ. Res.* **2018**, *123*, 1143–1151. [CrossRef]
41. Serebriiskii, I.; Estojak, J.; Sonoda, G.; Testa, J.R.; Golemis, E.A. Association of Krev-1/rap1a with Krit1, a novel ankyrin repeat-containing protein encoded by a gene mapping to 7q21-22. *Oncogene* **1997**, *15*, 1043–1049. [CrossRef] [PubMed]
42. Liu, W.; Draheim, K.M.; Zhang, R.; Calderwood, D.A.; Boggon, T.J. Mechanism for KRIT1 release of ICAP1-mediated suppression of integrin activation. *Mol. Cell* **2013**, *49*, 719–729. [CrossRef] [PubMed]
43. Zawistowski, J.S.; Serebriiskii, I.G.; Lee, M.F.; Golemis, E.A.; Marchuk, D.A. KRIT1 association with the integrin-binding protein ICAP-1: A new direction in the elucidation of cerebral cavernous malformations (CCM1) pathogenesis. *Hum. Mol. Genet.* **2002**, *11*, 389–396. [CrossRef]
44. Kleaveland, B.; Zheng, X.; Liu, J.J.; Blum, Y.; Tung, J.J.; Zou, Z.; Sweeney, S.M.; Chen, M.; Guo, L.; Lu, M.M.; et al. Regulation of cardiovascular development and integrity by the heart of glass-cerebral cavernous malformation protein pathway. *Nat. Med.* **2009**, *15*, 169–176. [CrossRef] [PubMed]
45. Gingras, A.R.; Liu, J.J.; Ginsberg, M.H. Structural basis of the junctional anchorage of the cerebral cavernous malformations complex. *J. Cell Biol.* **2012**, *199*, 39–48. [CrossRef]
46. Gingras, A.R.; Puzon-McLaughlin, W.; Ginsberg, M.H. The structure of the ternary complex of Krev interaction trapped 1 (KRIT1) bound to both the Rap1 GTPase and the heart of glass (HEG1) cytoplasmic tail. *J. Biol. Chem.* **2013**, *288*, 23639–23649. [CrossRef] [PubMed]
47. Gunel, M.; Laurans, M.S.; Shin, D.; DiLuna, M.L.; Voorhees, J.; Choate, K.; Nelson-Williams, C.; Lifton, R.P. *KRIT1*, a gene mutated in cerebral cavernous malformation, encodes a microtubule-associated protein. *Proc. Natl. Acad. Sci. USA* **2002**, *99*, 10677–10682. [CrossRef]
48. Fisher, O.S.; Zhang, R.; Li, X.; Murphy, J.W.; Demeler, B.; Boggon, T.J. Structural studies of cerebral cavernous malformations 2 (CCM2) reveal a folded helical domain at its C-terminus. *FEBS Lett.* **2013**, *587*, 272–277. [CrossRef]
49. Uhlik, M.T.; Abell, A.N.; Johnson, N.L.; Sun, W.; Cuevas, B.D.; Lobel-Rice, K.E.; Horne, E.A.; Dell'Acqua, M.L.; Johnson, G.L. Rac-MEKK3-MKK3 scaffolding for p38 MAPK activation during hyperosmotic shock. *Nat. Cell Biol.* **2003**, *5*, 1104–1110. [CrossRef]
50. Hilder, T.L.; Malone, M.H.; Bencharit, S.; Colicelli, J.; Haystead, T.A.; Johnson, G.L.; Wu, C.C. Proteomic identification of the cerebral cavernous malformation signaling complex. *J. Proteome Res.* **2007**, *6*, 4343–4355. [CrossRef]
51. Zhang, J.; Rigamonti, D.; Dietz, H.C.; Clatterbuck, R.E. Interaction between krit1 and malcavernin: Implications for the pathogenesis of cerebral cavernous malformations. *Neurosurgery* **2007**, *60*, 353–359. [CrossRef] [PubMed]
52. Zawistowski, J.S.; Stalheim, L.; Uhlik, M.T.; Abell, A.N.; Ancrile, B.B.; Johnson, G.L.; Marchuk, D.A. CCM1 and CCM2 protein interactions in cell signaling: Implications for cerebral cavernous malformations pathogenesis. *Hum. Mol. Genet.* **2005**, *14*, 2521–2531. [CrossRef] [PubMed]
53. Draheim, K.M.; Li, X.; Zhang, R.; Fisher, O.S.; Villari, G.; Boggon, T.J.; Calderwood, D.A. CCM2-CCM3 interaction stabilizes their protein expression and permits endothelial network formation. *J. Cell Biol.* **2015**, *208*, 987–1001. [CrossRef] [PubMed]
54. Stockton, R.A.; Shenkar, R.; Awad, I.A.; Ginsberg, M.H. Cerebral cavernous malformations proteins inhibit Rho kinase to stabilize vascular integrity. *J. Exp. Med.* **2010**, *207*, 881–896. [CrossRef]
55. Hilder, T.L.; Malone, M.H.; Johnson, G.L. Hyperosmotic induction of mitogen-activated protein kinase scaffolding. *Methods Enzym.* **2007**, *428*, 297–312. [CrossRef]

56. Zheng, X.; Xu, C.; Smith, A.O.; Stratman, A.N.; Zou, Z.; Kleaveland, B.; Yuan, L.; Didiku, C.; Sen, A.; Liu, X.; et al. Dynamic regulation of the cerebral cavernous malformation pathway controls vascular stability and growth. *Dev. Cell* **2012**, *23*, 342–355. [CrossRef]
57. Cullere, X.; Plovie, E.; Bennett, P.M.; MacRae, C.A.; Mayadas, T.N. The cerebral cavernous malformation proteins CCM2L and CCM2 prevent the activation of the MAP kinase MEKK3. *Proc. Natl. Acad. Sci. USA* **2015**, *112*, 14284–14289. [CrossRef]
58. Rosen, J.N.; Sogah, V.M.; Ye, L.Y.; Mably, J.D. Ccm2-like is required for cardiovascular development as a novel component of the Heg-CCM pathway. *Dev. Biol.* **2013**, *376*, 74–85. [CrossRef]
59. Li, X.; Zhang, R.; Zhang, H.; He, Y.; Ji, W.; Min, W.; Boggon, T.J. Crystal structure of CCM3, a cerebral cavernous malformation protein critical for vascular integrity. *J. Biol. Chem.* **2010**, *285*, 24099–24107. [CrossRef]
60. Wang, Y.; Liu, H.; Zhang, Y.; Ma, D. cDNA cloning and expression of an apoptosis-related gene, humanTFAR15 gene. *Sci. China C Life Sci.* **1999**, *42*, 323–329. [CrossRef]
61. Ma, X.; Zhao, H.; Shan, J.; Long, F.; Chen, Y.; Zhang, Y.; Han, X.; Ma, D. PDCD10 interacts with Ste20-related kinase MST4 to promote cell growth and transformation via modulation of the ERK pathway. *Mol. Biol. Cell* **2007**, *18*, 1965–1978. [CrossRef] [PubMed]
62. Ceccarelli, D.F.; Laister, R.C.; Mulligan, V.K.; Kean, M.J.; Goudreault, M.; Scott, I.C.; Derry, W.B.; Chakrabartty, A.; Gingras, A.C.; Sicheri, F. CCM3/PDCD10 heterodimerizes with germinal center kinase III (GCKIII) proteins using a mechanism analogous to CCM3 homodimerization. *J. Biol. Chem.* **2011**, *286*, 25056–25064. [CrossRef] [PubMed]
63. Fidalgo, M.; Fraile, M.; Pires, A.; Force, T.; Pombo, C.; Zalvide, J. CCM3/PDCD10 stabilizes GCKIII proteins to promote Golgi assembly and cell orientation. *J. Cell Sci.* **2010**, *123*, 1274–1284. [CrossRef]
64. Goudreault, M.; D'Ambrosio, L.M.; Kean, M.J.; Mullin, M.J.; Larsen, B.G.; Sanchez, A.; Chaudhry, S.; Chen, G.I.; Sicheri, F.; Nesvizhskii, A.I.; et al. A PP2A phosphatase high density interaction network identifies a novel striatin-interacting phosphatase and kinase complex linked to the cerebral cavernous malformation 3 (CCM3) protein. *Mol. Cell Proteom.* **2009**, *8*, 157–171. [CrossRef] [PubMed]
65. Suryavanshi, N.; Furmston, J.; Ridley, A.J. The STRIPAK complex components FAM40A and FAM40B regulate endothelial cell contractility via ROCKs. *BMC Cell Biol.* **2018**, *19*, 26. [CrossRef] [PubMed]
66. Gordon, J.; Hwang, J.; Carrier, K.J.; Jones, C.A.; Kern, Q.L.; Moreno, C.S.; Karas, R.H.; Pallas, D.C. Protein phosphatase 2a (PP2A) binds within the oligomerization domain of striatin and regulates the phosphorylation and activation of the mammalian Ste20-Like kinase Mst3. *BMC Biochem.* **2011**, *12*, 54. [CrossRef] [PubMed]
67. Voss, K.; Stahl, S.; Schleider, E.; Ullrich, S.; Nickel, J.; Mueller, T.D.; Felbor, U. CCM3 interacts with CCM2 indicating common pathogenesis for cerebral cavernous malformations. *Neurogenetics* **2007**, *8*, 249–256. [CrossRef]
68. Li, X.; Ji, W.; Zhang, R.; Folta-Stogniew, E.; Min, W.; Boggon, T.J. Molecular recognition of leucine-aspartate repeat (LD) motifs by the focal adhesion targeting homology domain of cerebral cavernous malformation 3 (CCM3). *J. Biol. Chem.* **2011**, *286*, 26138–26147. [CrossRef]
69. Zhang, Y.; Tang, W.; Zhang, H.; Niu, X.; Xu, Y.; Zhang, J.; Gao, K.; Pan, W.; Boggon, T.J.; Toomre, D.; et al. A network of interactions enables CCM3 and STK24 to coordinate UNC13D-driven vesicle exocytosis in neutrophils. *Dev. Cell* **2013**, *27*, 215–226. [CrossRef]
70. Dibble, C.F.; Horst, J.A.; Malone, M.H.; Park, K.; Temple, B.; Cheeseman, H.; Barbaro, J.R.; Johnson, G.L.; Bencharit, S. Defining the Functional Domain of Programmed Cell Death 10 through Its Interactions with Phosphatidylinositol-3,4,5-Trisphosphate. *PLoS ONE* **2010**, *5*, e11740. [CrossRef]
71. Castro, M.; Lavina, B.; Ando, K.; Alvarez-Aznar, A.; Abu Taha, A.; Brakebusch, C.; Dejana, E.; Betsholtz, C.; Gaengel, K. CDC42 Deletion Elicits Cerebral Vascular Malformations via Increased MEKK3-Dependent KLF4 Expression. *Circ. Res.* **2019**, *124*, 1240–1252. [CrossRef] [PubMed]
72. Mardakheh, F.K.; Self, A.; Marshall, C.J. RHO binding to FAM65A regulates Golgi reorientation during cell migration. *J. Cell Sci.* **2016**, *129*, 4466–4479. [CrossRef] [PubMed]
73. Maddaluno, L.; Rudini, N.; Cuttano, R.; Bravi, L.; Giampietro, C.; Corada, M.; Ferrarini, L.; Orsenigo, F.; Papa, E.; Boulday, G.; et al. EndMT contributes to the onset and progression of cerebral cavernous malformations. *Nature* **2013**, *498*, 492–496. [CrossRef] [PubMed]

74. Marchi, S.; Corricelli, M.; Trapani, E.; Bravi, L.; Pittaro, A.; Delle Monache, S.; Ferroni, L.; Patergnani, S.; Missiroli, S.; Goitre, L.; et al. Defective autophagy is a key feature of cerebral cavernous malformations. *EMBO Mol. Med.* **2015**, *7*, 1403–1417. [CrossRef]
75. Zhou, H.J.; Qin, L.; Zhang, H.; Tang, W.; Ji, W.; He, Y.; Liang, X.; Wang, Z.; Yuan, Q.; Vortmeyer, A.; et al. Endothelial exocytosis of angiopoietin-2 resulting from CCM3 deficiency contributes to cerebral cavernous malformation. *Nat. Med.* **2016**, *22*, 1033–1042. [CrossRef]
76. Crose, L.E.; Hilder, T.L.; Sciaky, N.; Johnson, G.L. Cerebral cavernous malformation 2 protein promotes smad ubiquitin regulatory factor 1-mediated RhoA degradation in endothelial cells. *J. Biol. Chem.* **2009**, *284*, 13301–13305. [CrossRef]
77. Glading, A.; Han, J.; Stockton, R.A.; Ginsberg, M.H. KRIT-1/CCM1 is a Rap1 effector that regulates endothelial cell cell junctions. *J. Cell Biol.* **2007**, *179*, 247–254. [CrossRef]
78. Wang, H.R.; Zhang, Y.; Ozdamar, B.; Ogunjimi, A.A.; Alexandrova, E.; Thomsen, G.H.; Wrana, J.L. Regulation of cell polarity and protrusion formation by targeting RhoA for degradation. *Science* **2003**, *302*, 1775–1779. [CrossRef]
79. Zhou, Z.; Rawnsley, D.R.; Goddard, L.M.; Pan, W.; Cao, X.J.; Jakus, Z.; Zheng, H.; Yang, J.; Arthur, J.S.; Whitehead, K.J.; et al. The cerebral cavernous malformation pathway controls cardiac development via regulation of endocardial MEKK3 signaling and KLF expression. *Dev. Cell* **2015**, *32*, 168–180. [CrossRef]
80. Zhou, Z.; Tang, A.T.; Wong, W.Y.; Bamezai, S.; Goddard, L.M.; Shenkar, R.; Zhou, S.; Yang, J.; Wright, A.C.; Foley, M.; et al. Cerebral cavernous malformations arise from endothelial gain of MEKK3-KLF2/4 signalling. *Nature* **2016**, *532*, 122–126. [CrossRef]
81. Huang, Q.; Yang, J.; Lin, Y.; Walker, C.; Cheng, J.; Liu, Z.G.; Su, B. Differential regulation of interleukin 1 receptor and Toll-like receptor signaling by MEKK3. *Nat. Immunol.* **2004**, *5*, 98–103. [CrossRef] [PubMed]
82. Li, K.; Wang, M.; Hu, Y.; Xu, N.; Yu, Q.; Wang, Q. TAK1 knockdown enhances lipopolysaccharide-induced secretion of proinflammatory cytokines in myeloid cells via unleashing MEKK3 activity. *Cell. Immunol.* **2016**, *310*, 193–198. [CrossRef] [PubMed]
83. Samanta, A.K.; Huang, H.J.; Bast, R.C., Jr.; Liao, W.S. Overexpression of MEKK3 confers resistance to apoptosis through activation of NFkappaB. *J. Biol. Chem.* **2004**, *279*, 7576–7583. [CrossRef] [PubMed]
84. Yang, J.; Lin, Y.; Guo, Z.; Cheng, J.; Huang, J.; Deng, L.; Liao, W.; Chen, Z.; Liu, Z.; Su, B. The essential role of MEKK3 in TNF-induced NF-kappaB activation. *Nat. Immunol.* **2001**, *2*, 620–624. [CrossRef] [PubMed]
85. Tang, A.T.; Choi, J.P.; Kotzin, J.J.; Yang, Y.; Hong, C.C.; Hobson, N.; Girard, R.; Zeineddine, H.A.; Lightle, R.; Moore, T.; et al. Endothelial TLR4 and the microbiome drive cerebral cavernous malformations. *Nature* **2017**, *545*, 305–310. [CrossRef]
86. Fisher, O.S.; Deng, H.; Liu, D.; Zhang, Y.; Wei, R.; Deng, Y.; Zhang, F.; Louvi, A.; Turk, B.E.; Boggon, T.J.; et al. Structure and vascular function of MEKK3-cerebral cavernous malformations 2 complex. *Nat. Commun.* **2015**, *6*, 7937. [CrossRef]
87. Harel, L.; Costa, B.; Tcherpakov, M.; Zapatka, M.; Oberthuer, A.; Hansford, L.M.; Vojvodic, M.; Levy, Z.; Chen, Z.Y.; Lee, F.S.; et al. CCM2 mediates death signaling by the TrkA receptor tyrosine kinase. *Neuron* **2009**, *63*, 585–591. [CrossRef]
88. Costa, B.; Kean, M.J.; Ast, V.; Knight, J.D.; Mett, A.; Levy, Z.; Ceccarelli, D.F.; Badillo, B.G.; Eils, R.; Konig, R.; et al. STK25 protein mediates TrkA and CCM2 protein-dependent death in pediatric tumor cells of neural origin. *J. Biol. Chem.* **2012**, *287*, 29285–29289. [CrossRef]
89. Wu, Z.; Qi, Y.; Guo, Z.; Li, P.; Zhou, D. miR-613 suppresses ischemia-reperfusion-induced cardiomyocyte apoptosis by targeting the programmed cell death 10 gene. *Biosci. Trends* **2016**, *10*, 251–257. [CrossRef]
90. Chen, L.; Tanriover, G.; Yano, H.; Friedlander, R.; Louvi, A.; Gunel, M. Apoptotic functions of PDCD10/CCM3, the gene mutated in cerebral cavernous malformation 3. *Stroke* **2009**, *40*, 1474–1481. [CrossRef]
91. Zhang, H.; Ma, X.; Deng, X.; Chen, Y.; Mo, X.; Zhang, Y.; Zhao, H.; Ma, D. PDCD10 interacts with STK25 to accelerate cell apoptosis under oxidative stress. *Front. Biosci.* **2012**, *17*, 2295–2305. [CrossRef] [PubMed]
92. Fidalgo, M.; Guerrero, A.; Fraile, M.; Iglesias, C.; Pombo, C.M.; Zalvide, J. Adaptor protein cerebral cavernous malformation 3 (CCM3) mediates phosphorylation of the cytoskeletal proteins ezrin/radixin/moesin by mammalian Ste20-4 to protect cells from oxidative stress. *J. Biol. Chem.* **2012**, *287*, 11556–11565. [CrossRef] [PubMed]
93. Wong, J.H.; Awad, I.A.; Kim, J.H. Ultrastructural pathological features of cerebrovascular malformations: A preliminary report. *Neurosurgery* **2000**, *46*, 1454–1459. [CrossRef] [PubMed]

94. Tu, J.; Stoodley, M.A.; Morgan, M.K.; Storer, K.P. Ultrastructural characteristics of hemorrhagic, nonhemorrhagic, and recurrent cavernous malformations. *J. Neurosurg.* **2005**, *103*, 903–909. [CrossRef]
95. Girard, R.; Fam, M.D.; Zeineddine, H.A.; Tan, H.; Mikati, A.G.; Shi, C.; Jesselson, M.; Shenkar, R.; Wu, M.; Cao, Y.; et al. Vascular permeability and iron deposition biomarkers in longitudinal follow-up of cerebral cavernous malformations. *J. Neurosurg.* **2017**, *127*, 102–110. [CrossRef]
96. Mikati, A.G.; Khanna, O.; Zhang, L.; Girard, R.; Shenkar, R.; Guo, X.; Shah, A.; Larsson, H.B.; Tan, H.; Li, L.; et al. Vascular permeability in cerebral cavernous malformations. *J. Cereb. Blood Flow Metab.* **2015**, *35*, 1632–1639. [CrossRef]
97. Murali, A.; Rajalingam, K. Small Rho GTPases in the control of cell shape and mobility. *Cell. Mol. Life Sci.* **2014**, *71*, 1703–1721. [CrossRef]
98. Citalan-Madrid, A.F.; Garcia-Ponce, A.; Vargas-Robles, H.; Betanzos, A.; Schnoor, M. Small GTPases of the Ras superfamily regulate intestinal epithelial homeostasis and barrier function via common and unique mechanisms. *Tissue Barriers* **2013**, *1*, e26938. [CrossRef]
99. van Buul, J.D.; Geerts, D.; Huveneers, S. Rho GAPs and GEFs: Controling switches in endothelial cell adhesion. *Cell Adhes. Migr.* **2014**, *8*, 108–124. [CrossRef]
100. Shen, Q.; Wu, M.H.; Yuan, S.Y. Endothelial contractile cytoskeleton and microvascular permeability. *Cell Health Cytoskelet.* **2009**, *2009*, 43–50. [CrossRef]
101. Corr, M.; Lerman, I.; Keubel, J.M.; Ronacher, L.; Misra, R.; Lund, F.; Sarelius, I.H.; Glading, A.J. Decreased Krev interaction-trapped 1 expression leads to increased vascular permeability and modifies inflammatory responses *in vivo*. *Arter. Thromb. Vasc. Biol.* **2012**, *32*, 2702–2710. [CrossRef] [PubMed]
102. Goitre, L.; DiStefano, P.V.; Moglia, A.; Nobiletti, N.; Baldini, E.; Trabalzini, L.; Keubel, J.; Trapani, E.; Shuvaev, V.V.; Muzykantov, V.R.; et al. Up-regulation of NADPH oxidase-mediated redox signaling contributes to the loss of barrier function in KRIT1 deficient endothelium. *Sci. Rep.* **2017**, *7*, 8296. [CrossRef] [PubMed]
103. Meliton, A.; Meng, F.; Tian, Y.; Shah, A.A.; Birukova, A.A.; Birukov, K.G. Role of Krev Interaction Trapped-1 in Prostacyclin-Induced Protection against Lung Vascular Permeability Induced by Excessive Mechanical Forces and Thrombin Receptor Activating Peptide 6. *Am. J. Respir. Cell Mol. Biol.* **2015**, *53*, 834–843. [CrossRef] [PubMed]
104. McDonald, D.A.; Shi, C.; Shenkar, R.; Stockton, R.A.; Liu, F.; Ginsberg, M.H.; Marchuk, D.A.; Awad, I.A. Fasudil decreases lesion burden in a murine model of cerebral cavernous malformation disease. *Stroke* **2012**, *43*, 571–574. [CrossRef] [PubMed]
105. Shenkar, R.; Shi, C.; Austin, C.; Moore, T.; Lightle, R.; Cao, Y.; Zhang, L.; Wu, M.; Zeineddine, H.A.; Girard, R.; et al. RhoA Kinase Inhibition with Fasudil Versus Simvastatin in Murine Models of Cerebral Cavernous Malformations. *Stroke* **2017**, *48*, 187–194. [CrossRef] [PubMed]
106. Shenkar, R.; Peiper, A.; Pardo, H.; Moore, T.; Lightle, R.; Girard, R.; Hobson, N.; Polster, S.P.; Koskimaki, J.; Zhang, D.; et al. Rho Kinase Inhibition Blunts Lesion Development and Hemorrhage in Murine Models of Aggressive *Pdcd10/Ccm3* Disease. *Stroke* **2019**, *50*, 738–744. [CrossRef] [PubMed]
107. Polster, S.P.; Stadnik, A.; Akers, A.L.; Cao, Y.; Christoforidis, G.A.; Fam, M.D.; Flemming, K.D.; Girard, R.; Hobson, N.; Koenig, J.I.; et al. Atorvastatin Treatment of Cavernous Angiomas with Symptomatic Hemorrhage Exploratory Proof of Concept (AT CASH EPOC) Trial. *Neurosurgery* **2019**, *85*, 843–853. [CrossRef]
108. Lange, C.; Storkebaum, E.; de Almodovar, C.R.; Dewerchin, M.; Carmeliet, P. Vascular endothelial growth factor: A neurovascular target in neurological diseases. *Nat. Rev. Neurol.* **2016**, *12*, 439–454. [CrossRef]
109. DiStefano, P.V.; Kuebel, J.M.; Sarelius, I.H.; Glading, A.J. KRIT1 protein depletion modifies endothelial cell behavior via increased vascular endothelial growth factor (VEGF) signaling. *J. Biol. Chem.* **2014**, *289*, 33054–33065. [CrossRef]
110. Goitre, L.; Balzac, F.; Degani, S.; Degan, P.; Marchi, S.; Pinton, P.; Retta, S.F. KRIT1 regulates the homeostasis of intracellular reactive oxygen species. *PLoS ONE* **2010**, *5*, e11786. [CrossRef]
111. Wallez, Y.; Huber, P. Endothelial adherens and tight junctions in vascular homeostasis, inflammation and angiogenesis. *Biochim. Biophys. Acta* **2008**, *1778*, 794–809. [CrossRef] [PubMed]
112. Daneman, R.; Prat, A. The blood-brain barrier. *Cold Spring Harb. Perspect. Biol.* **2015**, *7*, a020412. [CrossRef] [PubMed]
113. Gunzel, D.; Yu, A.S. Claudins and the modulation of tight junction permeability. *Physiol. Rev.* **2013**, *93*, 525–569. [CrossRef] [PubMed]

114. Castro Dias, M.; Coisne, C.; Lazarevic, I.; Baden, P.; Hata, M.; Iwamoto, N.; Francisco, D.M.F.; Vanlandewijck, M.; He, L.; Baier, F.A.; et al. Claudin-3-deficient C57BL/6J mice display intact brain barriers. *Sci. Rep.* **2019**, *9*, 203. [CrossRef] [PubMed]
115. Ohtsuki, S.; Yamaguchi, H.; Katsukura, Y.; Asashima, T.; Terasaki, T. mRNA expression levels of tight junction protein genes in mouse brain capillary endothelial cells highly purified by magnetic cell sorting. *J. Neurochem.* **2008**, *104*, 147–154. [CrossRef]
116. Nitta, T.; Hata, M.; Gotoh, S.; Seo, Y.; Sasaki, H.; Hashimoto, N.; Furuse, M.; Tsukita, S. Size-selective loosening of the blood-brain barrier in claudin-5-deficient mice. *J. Cell Biol.* **2003**, *161*, 653–660. [CrossRef]
117. Raleigh, D.R.; Marchiando, A.M.; Zhang, Y.; Shen, L.; Sasaki, H.; Wang, Y.; Long, M.; Turner, J.R. Tight junction-associated MARVEL proteins marveld3, tricellulin, and occludin have distinct but overlapping functions. *Mol. Biol. Cell* **2010**, *21*, 1200–1213. [CrossRef]
118. Buschmann, M.M.; Shen, L.; Rajapakse, H.; Raleigh, D.R.; Wang, Y.; Lingaraju, A.; Zha, J.; Abbott, E.; McAuley, E.M.; Breskin, L.A.; et al. Occludin OCEL-domain interactions are required for maintenance and regulation of the tight junction barrier to macromolecular flux. *Mol. Biol. Cell* **2013**, *24*, 3056–3068. [CrossRef]
119. Krug, S.M.; Amasheh, S.; Richter, J.F.; Milatz, S.; Gunzel, D.; Westphal, J.K.; Huber, O.; Schulzke, J.D.; Fromm, M. Tricellulin Forms a Barrier to Macromolecules in Tricellular Tight Junctions without Affecting Ion Permeability. *Mol. Biol. Cell* **2009**, *20*, 3713–3724. [CrossRef]
120. Saitou, M.; Furuse, M.; Sasaki, H.; Schulzke, J.D.; Fromm, M.; Takano, H.; Noda, T.; Tsukita, S. Complex phenotype of mice lacking occludin, a component of tight junction strands. *Mol. Biol. Cell* **2000**, *11*, 4131–4142. [CrossRef]
121. O'Driscoll, M.C.; Daly, S.B.; Urquhart, J.E.; Black, G.C.; Pilz, D.T.; Brockmann, K.; McEntagart, M.; Abdel-Salam, G.; Zaki, M.; Wolf, N.I.; et al. Recessive mutations in the gene encoding the tight junction protein occludin cause band-like calcification with simplified gyration and polymicrogyria. *Am. J. Hum. Genet.* **2010**, *87*, 354–364. [CrossRef]
122. Laukoetter, M.G.; Nava, P.; Lee, W.Y.; Severson, E.A.; Capaldo, C.T.; Babbin, B.A.; Williams, I.R.; Koval, M.; Peatman, E.; Campbell, J.A.; et al. JAM-A regulates permeability and inflammation in the intestine *in vivo*. *J. Exp. Med.* **2007**, *204*, 3067–3076. [CrossRef] [PubMed]
123. Otani, T.; Nguyen, T.P.; Tokuda, S.; Sugihara, K.; Sugawara, T.; Furuse, K.; Miura, T.; Ebnet, K.; Furuse, M. Claudins and JAM-A coordinately regulate tight junction formation and epithelial polarity. *J. Cell Biol.* **2019**, *218*, 3372–3396. [CrossRef] [PubMed]
124. Schmitt, M.M.N.; Megens, R.T.A.; Zernecke, A.; Bidzhekov, K.; van den Akker, N.M.; Rademakers, T.; van Zandvoort, M.A.; Hackeng, T.M.; Koenen, R.R.; Weber, C. Endothelial Junctional Adhesion Molecule-A Guides Monocytes into Flow-Dependent Predilection Sites of Atherosclerosis. *Circulation* **2014**, *129*, 66–76. [CrossRef] [PubMed]
125. Cohen, C.J.; Shieh, J.T.; Pickles, R.J.; Okegawa, T.; Hsieh, J.T.; Bergelson, J.M. The coxsackievirus and adenovirus receptor is a transmembrane component of the tight junction. *Proc. Natl. Acad. Sci. USA* **2001**, *98*, 15191–15196. [CrossRef] [PubMed]
126. Chung, J.; Kim, K.H.; An, S.H.; Lee, S.; Lim, B.K.; Kang, S.W.; Kwon, K. Coxsackievirus and adenovirus receptor mediates the responses of endothelial cells to fluid shear stress. *Exp. Mol. Med.* **2019**, *51*, 1–15. [CrossRef]
127. Fanning, A.S.; Anderson, J.M. Zonula Occludens-1 and-2 Are Cytosolic Scaffolds That Regulate the Assembly of Cellular Junctions. *Ann. N. Y. Acad. Sci.* **2009**, *1165*, 113–120. [CrossRef]
128. Katsuno, T.; Umeda, K.; Matsui, T.; Hata, M.; Tamura, A.; Itoh, M.; Takeuchi, K.; Fujimori, T.; Nabeshima, Y.; Noda, T.; et al. Deficiency of zonula occludens-1 causes embryonic lethal phenotype associated with defected yolk sac angiogenesis and apoptosis of embryonic cells. *Mol. Biol. Cell* **2008**, *19*, 2465–2475. [CrossRef]
129. Tornavaca, O.; Chia, M.; Dufton, N.; Almagro, L.O.; Conway, D.E.; Randi, A.M.; Schwartz, M.A.; Matter, K.; Balda, M.S. ZO-1 controls endothelial adherens junctions, cell-cell tension, angiogenesis, and barrier formation. *J. Cell Biol.* **2015**, *208*, 821–838. [CrossRef] [PubMed]
130. Schossleitner, K.; Rauscher, S.; Groger, M.; Friedl, H.P.; Finsterwalder, R.; Habertheuer, A.; Sibilia, M.; Brostjan, C.; Fodinger, D.; Citi, S.; et al. Evidence That Cingulin Regulates Endothelial Barrier Function In Vitro and *In Vivo*. *Arter. Throm. Vas.* **2016**, *36*, 647–654. [CrossRef] [PubMed]

131. Chrifi, I.; Hermkens, D.; Brandt, M.M.; van Dijk, C.G.M.; Burgisser, P.E.; Haasdijk, R.; Pei, J.Y.; van de Kamp, E.H.M.; Zhu, C.B.; Blonden, L.; et al. Cgnl1, an endothelial junction complex protein, regulates GTPase mediated angiogenesis. *Cardiovasc. Res.* **2017**, *113*, 1776–1788. [CrossRef] [PubMed]
132. McKenzie, J.A.; Ridley, A.J. Roles of Rho/ROCK and MLCK in TNF-alpha-induced changes in endothelial morphology and permeability. *J. Cell. Physiol.* **2007**, *213*, 221–228. [CrossRef]
133. Persidsky, Y.; Heilman, D.; Haorah, J.; Zelivyanskaya, M.; Persidsky, R.; Weber, G.A.; Shimokawa, H.; Kaibuchi, K.; Ikezu, T. Rho-mediated regulation of tight junctions during monocyte migration across the blood-brain barrier in HIV-1 encephalitis (HIVE). *Blood* **2006**, *107*, 4770–4780. [CrossRef] [PubMed]
134. Wojciak-Stothard, B.; Potempa, S.; Eichholtz, T.; Ridley, A.J. Rho and Rac but not Cdc42 regulate endothelial cell permeability. *J. Cell Sci.* **2001**, *114*, 1343–1355. [PubMed]
135. Terry, S.; Nie, M.; Matter, K.; Balda, M.S. Rho signaling and tight junction functions. *Physiology* **2010**, *25*, 16–26. [CrossRef]
136. Yamamoto, M.; Ramirez, S.H.; Sato, S.; Kiyota, T.; Cerny, R.L.; Kaibuchi, K.; Persidsky, Y.; Ikezu, T. Phosphorylation of claudin-5 and occludin by rho kinase in brain endothelial cells. *Am. J. Pathol.* **2008**, *172*, 521–533. [CrossRef]
137. Izawa, Y.; Gu, Y.H.; Osada, T.; Kanazawa, M.; Hawkins, B.T.; Koziol, J.A.; Papayannopoulou, T.; Spatz, M.; Del Zoppo, G.J. beta1-integrin-matrix interactions modulate cerebral microvessel endothelial cell tight junction expression and permeability. *J. Cereb. Blood Flow Metab.* **2018**, *38*, 641–658. [CrossRef]
138. Faurobert, E.; Rome, C.; Lisowska, J.; Manet-Dupe, S.; Boulday, G.; Malbouyres, M.; Balland, M.; Bouin, A.P.; Keramidas, M.; Bouvard, D.; et al. CCM1-ICAP-1 complex controls beta 1 integrin-dependent endothelial contractility and fibronectin remodeling. *J. Cell Biol.* **2013**, *202*, 545–561. [CrossRef]
139. Millon-Fremillon, A.; Brunner, M.; Abed, N.; Collomb, E.; Ribba, A.S.; Block, M.R.; Albiges-Rizo, C.; Bouvard, D. Calcium and calmodulin-dependent serine/threonine protein kinase type II (CaMKII)-mediated intramolecular opening of integrin cytoplasmic domain-associated protein-1 (ICAP-1alpha) negatively regulates beta1 integrins. *J. Biol. Chem.* **2013**, *288*, 20248–20260. [CrossRef]
140. Stroeken, P.J.M.; Alvarez, B.; Van Rheenen, J.; Wijnands, Y.M.; Geerts, D.; Jalink, K.; Roos, E. Integrin cytoplasmic domain-associated protein-1 (ICAP-1) interacts with the ROCK-I kinase at the plasma membrane. *J. Cell. Physiol.* **2006**, *208*, 620–628. [CrossRef]
141. Wittchen, E.S.; Worthylake, R.A.; Kelly, P.; Casey, P.J.; Quilliam, L.A.; Burridge, K. Rap1 GTPase inhibits leukocyte transmigration by promoting endothelial barrier function. *J. Biol. Chem.* **2005**, *280*, 11675–11682. [CrossRef] [PubMed]
142. Ramos, C.J.; Lin, C.; Liu, X.; Antonetti, D.A. The EPAC-Rap1 pathway prevents and reverses cytokine-induced retinal vascular permeability. *J. Biol. Chem.* **2018**, *293*, 717–730. [CrossRef] [PubMed]
143. Wilson, C.W.; Parker, L.H.; Hall, C.J.; Smyczek, T.; Mak, J.; Crow, A.; Posthuma, G.; De Maziere, A.; Sagolla, M.; Chalouni, C.; et al. Rasip1 regulates vertebrate vascular endothelial junction stability through Epac1-Rap1 signaling. *Blood* **2013**, *122*, 3678–3690. [CrossRef] [PubMed]
144. Post, A.; Pannekoek, W.J.; Ross, S.H.; Verlaan, I.; Brouwer, P.M.; Bos, J.L. Rasip1 mediates Rap1 regulation of Rho in endothelial barrier function through ArhGAP29. *Proc. Natl. Acad. Sci. USA* **2013**, *110*, 11427–11432. [CrossRef] [PubMed]
145. Xu, K.; Sacharidou, A.; Fu, S.; Chong, D.C.; Skaug, B.; Chen, Z.J.; Davis, G.E.; Cleaver, O. Blood vessel tubulogenesis requires Rasip1 regulation of GTPase signaling. *Dev. Cell* **2011**, *20*, 526–539. [CrossRef]
146. de Kreuk, B.J.; Gingras, A.R.; Knight, J.D.R.; Liu, J.J.; Gingras, A.C.; Ginsberg, M.H. Heart of glass anchors Rasip1 at endothelial cell-cell junctions to support vascular integrity. *eLife* **2016**, *5*, e11394. [CrossRef]
147. Severson, E.A.; Parkos, C.A. Structural determinants of Junctional Adhesion Molecule A (JAM-A) function and mechanisms of intracellular signaling. *Curr. Opin. Cell Biol.* **2009**, *21*, 701–707. [CrossRef]
148. Giannotta, M.; Benedetti, S.; Tedesco, F.S.; Corada, M.; Trani, M.; D'Antuono, R.; Millet, Q.; Orsenigo, F.; Galvez, B.G.; Cossu, G.; et al. Targeting endothelial junctional adhesion molecule-A/EPAC/Rap-1 axis as a novel strategy to increase stem cell engraftment in dystrophic muscles. *EMBO Mol. Med.* **2014**, *6*, 239–258. [CrossRef]
149. Schneider, H.; Errede, M.; Ulrich, N.H.; Virgintino, D.; Frei, K.; Bertalanffy, H. Impairment of tight junctions and glucose transport in endothelial cells of human cerebral cavernous malformations. *J. Neuropathol. Exp. Neurol.* **2011**, *70*, 417–429. [CrossRef]

150. Jakimovski, D.; Schneider, H.; Frei, K.; Kennes, L.N.; Bertalanffy, H. Bleeding propensity of cavernous malformations: Impact of tight junction alterations on the occurrence of overt hematoma. *J. Neurosurg.* **2014**, *121*, 613–620. [CrossRef]
151. Lopez-Ramirez, M.A.; Fonseca, G.; Zeineddine, H.A.; Girard, R.; Moore, T.; Pham, A.; Cao, Y.; Shenkar, R.; de Kreuk, B.J.; Lagarrigue, F.; et al. Thrombospondin1 (TSP1) replacement prevents cerebral cavernous malformations. *J. Exp. Med.* **2017**, *214*, 3331–3346. [CrossRef] [PubMed]
152. Stamatovic, S.M.; Sladojevic, N.; Keep, R.F.; Andjelkovic, A.V. PDCD10 (CCM3) regulates brain endothelial barrier integrity in cerebral cavernous malformation type 3: Role of CCM3-ERK1/2-cortactin cross-talk. *Acta Neuropathol.* **2015**, *130*, 731–750. [CrossRef] [PubMed]
153. Johnson, A.M.; Roach, J.P.; Hu, A.; Stamatovic, S.M.; Zochowski, M.R.; Keep, R.F.; Andjelkovic, A.V. Connexin 43 gap junctions contribute to brain endothelial barrier hyperpermeability in familial cerebral cavernous malformations type III by modulating tight junction structure. *FASEB J.* **2018**, *32*, 2615–2629. [CrossRef] [PubMed]
154. Glading, A.J.; Ginsberg, M.H. Rap1 and its effector KRIT1/CCM1 regulate beta-catenin signaling. *Dis. Models Mech.* **2010**, *3*, 73–83. [CrossRef]
155. Wang, Y.; Li, Y.; Zou, J.; Polster, S.P.; Lightle, R.; Moore, T.; Dimaano, M.; He, T.C.; Weber, C.R.; Awad, I.A.; et al. The cerebral cavernous malformation disease causing gene KRIT1 participates in intestinal epithelial barrier maintenance and regulation. *FASEB J.* **2019**, *33*, 2132–2143. [CrossRef]
156. McCarthy, K.M.; Francis, S.A.; McCormack, J.M.; Lai, J.; Rogers, R.A.; Skare, I.B.; Lynch, R.D.; Schneeberger, E.E. Inducible expression of claudin-1-myc but not occludin-VSV-G results in aberrant tight junction strand formation in MDCK cells. *J. Cell Sci.* **2000**, *113*, 3387–3398.
157. Tang, A.T.; Sullivan, K.R.; Hong, C.C.; Goddard, L.M.; Mahadevan, A.; Ren, A.; Pardo, H.; Peiper, A.; Griffin, E.; Tanes, C.; et al. Distinct cellular roles for PDCD10 define a gut-brain axis in cerebral cavernous malformation. *Sci. Transl. Med.* **2019**, *11*. [CrossRef]

13

Potential for Tight Junction Protein–Directed Drug Development using Claudin Binders and Angubindin-1

Yosuke Hashimoto [1], Keisuke Tachibana [1], Susanne M. Krug [2], Jun Kunisawa [1,3,4,5,6], Michael Fromm [2] and Masuo Kondoh [1,*]

1 Graduate School of Pharmaceutical Sciences, Osaka University, Osaka 565-0871, Japan
2 Institute of Clinical Physiology, Charité–Universitätsmedizin Berlin, 12203 Berlin, Germany
3 Laboratory of Vaccine Materials, Center for Vaccine and Adjuvant Research and Laboratory of Gut Environmental System, National Institutes of Biomedical Innovation, Health and Nutrition (NIBIOHN), Osaka 567-0085, Japan
4 International Research and Development Center for Mucosal Vaccines, The Institute of Medical Sciences, The University of Tokyo, Tokyo 108-8639, Japan
5 Department of Microbiology and Immunology, Kobe University Graduate School of Medicine, Kobe 650-0017, Japan
6 Graduate School of Medicine and Graduate School of Dentistry, Osaka University, Osaka 565-0871, Japan
* Correspondence: masuo@phs.osaka-u.ac.jp

Abstract: The tight junction (TJ) is an intercellular sealing component found in epithelial and endothelial tissues that regulates the passage of solutes across the paracellular space. Research examining the biology of TJs has revealed that they are complex biochemical structures constructed from a range of proteins including claudins, occludin, tricellulin, angulins and junctional adhesion molecules. The transient disruption of the barrier function of TJs to open the paracellular space is one means of enhancing mucosal and transdermal drug absorption and to deliver drugs across the blood–brain barrier. However, the disruption of TJs can also open the paracellular space to harmful xenobiotics and pathogens. To address this issue, the strategies targeting TJ proteins have been developed to loosen TJs in a size- or tissue-dependent manner rather than to disrupt them. As several TJ proteins are overexpressed in malignant tumors and in the inflamed intestinal tract, and are present in cells and epithelia conjoined with the mucosa-associated lymphoid immune tissue, these TJ-protein-targeted strategies may also provide platforms for the development of novel therapies and vaccines. Here, this paper reviews two TJ-protein-targeted technologies, claudin binders and an angulin binder, and their applications in drug development.

Keywords: tight junction; claudin; angulin; drug development; angubindin-1; *Clostridium perfringens* enterotoxin; *Clostridium perfringens* iota-toxin; antibody

---

1. Introduction

The boundaries between the inside of the body and the outside environment in the airway and gastrointestinal tract, and between the systemic circulation and tissues in the brain, eye, testis, and placenta, are separated by epithelial and endothelial cell sheets, respectively. The paracellular spaces between the adjacent cells in these sheets are sealed by a structural and functional component called the tight junction (TJ) [1]. TJs control the diffusion of ions, solutes, and water across the paracellular space to maintain homeostasis, and the loss of TJ integrity appears to be associated with the development of intestinal diseases [2,3], atopic dermatitis [4], and psychiatric disorders [5,6]. TJs also prevent mucosal and epidermal absorption of drugs and the delivery of drugs from the systemic circulation to the brain, eye, testis, and placenta.

A freeze-fracture replica electron microscopy analysis has shown that TJs consist of a meshwork of proteins called TJ strands [7]. In epithelial cells, these TJ strands are located at the apical side of the lateral membrane. TJs include various membrane proteins—including claudins, TJ-associated MARVEL proteins (occludin, tricellulin, and marvelD3), junctional adhesion molecules, and angulins—and these membrane proteins are anchored to intracellular scaffold proteins, e.g., of the zonula occludens protein family [8,9]. The physiological characteristics of TJs are determined by the specific combinations and mixing ratios of these TJ proteins [10–12]. TJ strands are dynamic structures that are repeatedly breaking and annealing, which transiently loosens the TJ seal and allows the stepwise diffusion of solutes across the meshwork and through the paracellular space [13].

There are two types of TJs: Bicellular, where two cells meet, and tricellular, where three cells meet. Bicellular TJ strands extend horizontally along the apical membrane but extend vertically when they reach a tricellular contact. Tricellular TJs seal the tubular structure created at tricellular contacts by the three vertically extending TJ strands and the three adjoining cell membranes [14].

A modulation of the structure of TJs to loosen the paracellular space can be used to increase mucosal and epidermal drug absorption, as well as drug delivery to the brain. Currently, sodium caprate and mannitol are used clinically to enhance paracellular drug absorption and drug delivery to the brain, respectively [15,16]. However, sodium caprate causes mucosal damage and lacks tissue-specificity [15,17]. The mannitol widened the interendothelial TJs to a radius of approximately 20 nm, followed by deliver chemicals, peptides, antibodies, and viral vectors to the brain [18]. Research into understanding the biochemical structure of TJs and the physiological roles of the various TJ proteins has provided insights that have been applied to the development of TJ protein–targeted drugs. Here, the efficacy and safety of claudin and angulin binders for the development of TJ-directed drugs is reviewed.

2. Claudins and Angulins

2.1. Claudins

Claudins were identified in 1998 as components of TJ strands that are crucial for the sealing of the intercellular space [19]. Currently, the mammalian claudin family comprises 27 proteins [20]. Since 2014, the crystal structures of these claudins have been gradually elucidated [21–24]. Claudins are tetra-transmembrane proteins containing two loops: The first contains four β-strands and an α-helix (extracellular helix), and the second contains a β-strand and the cell-surface-exposed transmembrane 3 domain (Figure 1a). Almost all claudins have the zonula occludens-1 binding motif at their C-terminal end. Claudins have *cis*-interactions within a cell membrane as well as *trans*-interactions with claudins in adjacent cell membranes. Both the *cis*- and *trans*-interactions are required for the building of TJ strands. The crystal structures of claudins have revealed that the *cis*-interaction between the extracellular helix and the transmembrane 3 domain leads to the formation of claudin strands within the cell membrane [21–24]. Claudins have two flexible loops, defined as variable regions: The first is between the β1- and β2-strands in the first extracellular domain, and the second is between the transmembrane 3 domain and the β5-strand in the second extracellular domain, with these flexible loops being involved in the formation of *trans*-interactions between claudin strands [25,26]. These flexible regions also appear to determine the characteristics of claudin-based TJs, because their homology is low among claudin members [21].

2.2. Angulins

The tricellular TJ is a specialized structure at the point of contact among three cells. Tricellulin, a member of the tight junction-associated MARVEL protein (TAMP) family, is an essential component of tricellular TJs [27]. Tricellulin does not form homophilic *trans*-interactions nor does it localize at tricellular contacts by itself [10,28]. Instead, it has to be recruited to tricellular contacts by angulins. Angulins are type I trans-membrane proteins with an extracellular immunoglobulin-like domain and a cytosolic tail (Figure 1b) [29]. There are three angulins: Angulin-1 (also known as lipolysis-stimulated lipoprotein receptor, LSR); angulin-2 (also known as immunoglobulin-like domain containing receptor 1, ILDR-1); and angulin-3 (also known as immunoglobulin-like domain containing receptor 2, ILDR-2). The interaction between phosphorylated Ser288 in the C-terminal tail of angulin-1 and the C-terminal cytosolic tail of tricellulin is responsible for the recruitment of tricellulin to the tricellular TJ [28,30]. Angulin-1 and -2 have a much greater ability to recruit tricellulin than angulin-3 [29]. The extracellular domain of angulin-2 may form a trimeric structure at the tricellular contact. However, the underlying mechanism remains unclear yet [31].

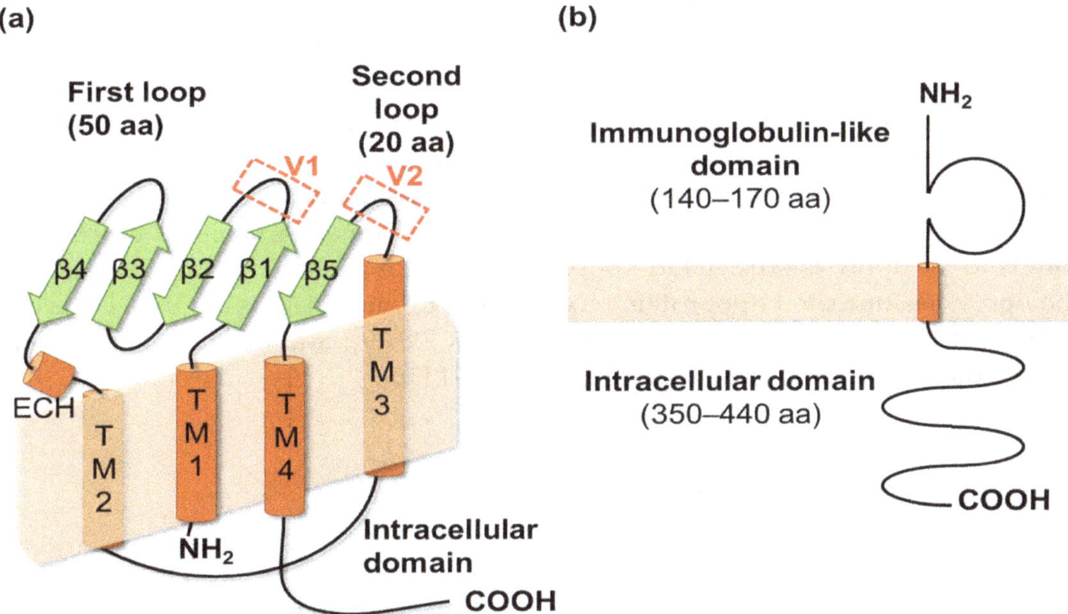

Figure 1. The representative structures of a claudin and an angulin. (**a**) Claudins are 20–27 kDa proteins containing four transmembrane (TM) domains and two extracellular loops. Extracellular loop 1 and 2 contains approximately 50 and 25 amino acids, respectively. TM domain 3 is much longer than the other three TM domains. Claudins also contain an extracellular helix (ECH) and two variable regions (V1 and V2). (**b**) Angulins are 60–70 kDa type I TM proteins containing an extracellular immunoglobulin-like domain and an intracellular tail. There is currently no structural information available for angulins. The secondary structural elements are shown as cylinders (α-helices) and arrows (β-strands). aa, amino acid.

3. Claudin and Angulin Binders

Currently, most binders that target TJ proteins are either fragments of bacterial toxins (first-generation binders) or monoclonal antibodies (mAbs; second-generation binders) [32–34].

3.1. First-Generation Binders: Fragments of Bacterial Toxins

Clostridium perfringens enterotoxin (CPE) has two domains: The N-terminal cytotoxic domain, which is involved in oligomerization and pore formation, and the C-terminal receptor binding domain (C-CPE) [35] (Figure 2a). The CPE receptor (CPE-R) was identified, and CPE-R has significant similarity to the rat androgen withdrawal apoptosis protein (RVP1) in 1997 [36]. Two years after the identification of CPE-R and RVP1, claudin-3 and -4 have been identified to be RVP1 and CPE-R, respectively [37]. C-CPE binds to claudin-3 and -4 [38]. However, C-CPE also binds to claudin-6, -7, -8, -9, -14, and -19 [22,39,40]. The affinity of C-CPE to claudin-4 is approximately 0.5 nM [41]. The treatment of MDCK cells with C-CPE decreases TJ integrity [38]. The research into the generation of claudin binders using C-CPE as a template for site-directed mutagenesis has produced several C-CPE mutants: One broad-spectrum binder to claudin-1 to -5 and four relatively specific binders for claudin-3, -4, and -5 (Table 1).

Table 1. List of claudin-binding mutants of C-terminal receptor binding domain of *Clostridium perfringens* enterotoxin.

Binder Type	Mutated Regions	Ref.
Negative-binding mutant	Y306A/L315A	[42]
Broad-spectrum binder (m19)	S304A/S305P/S307R/N309H/S313H	[43]
Enhancing specificity to claudin-3	L223A/D225A/R227A	[44]
Enhancing specificity to claudin-4	L254A/S256A/I258A/D284A	[44]
Enhancing specificity to claudin-5	Y306W/S313H	[45]
Improved specificity to claudin-5	N218Q/Y306W/S313H	[46]

Clostridium perfringens iota-toxin is a binary toxin composed of a cytotoxic domain (Ia) and a receptor-binding domain (Ib). Ib domain can be further classified into an Ia-binding domain (domain 1), oligomerization domain (domain 2), pore-formation domain (domain 3), and receptor-binding domain (domain 4). Domains 1 to 3 are involved in cytotoxicity [47]. Domain 4, which comprises amino acids 421–664, binds to the lipolysis-stimulated lipoprotein receptor without inducing cytotoxicity [48,49]. In 2013, the lipolysis-stimulated lipoprotein receptor was re-identified as angulin-1, the determining factor for the localization of tricellulin at tricellular TJs [28,29], and domain 4, the first tricellular TJ–specific modulator reported, was named angubindin-1 [50].

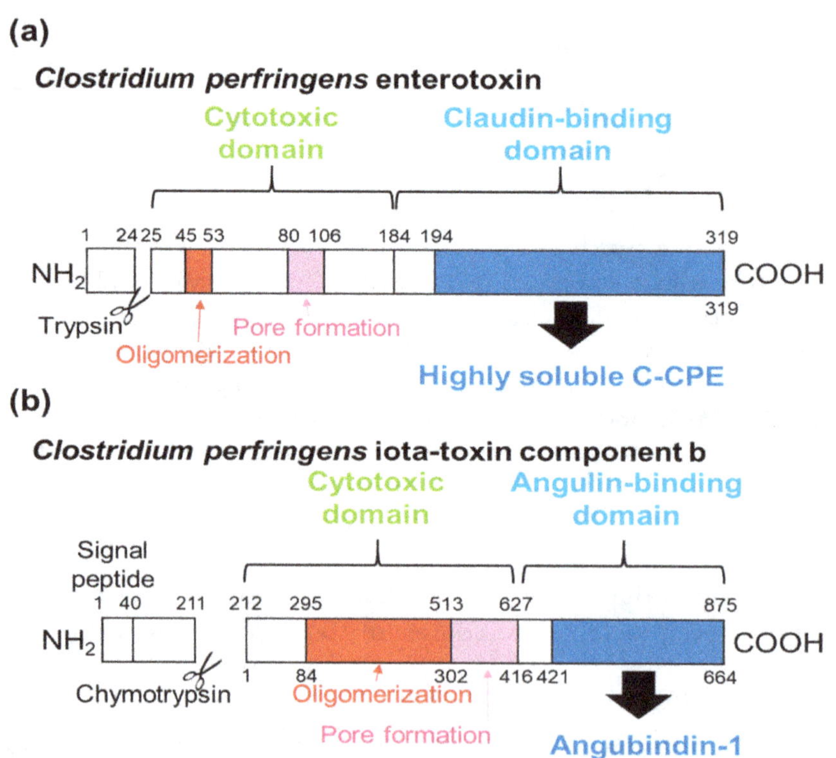

Figure 2. Schematic diagram showing the domains of first-generation claudin binders made from bacterial toxins. (**a**) *Clostridium perfringens* enterotoxin and (**b**) *C. perfringens* iota-toxin. Regions involved in oligomerization (red) and pore formation (pink) are indicated. Regions used as first-generation claudin binders are indicated in blue. The accession numbers for *C. perfringens* enterotoxin and iota-toxin are AOD41705 and CAA51960, respectively.

3.2. *Second-Generation Binders: Monoclonal Antibodies*

The TJ components are promising targets for the development of drugs to treat cancers and inflammatory bowel disease, for preventing infection by hepatitis C virus, and for the development of regenerative medicines. Antibodies are promising therapeutics for TJ-directed drug development because they bind to target proteins with high affinity and high specificity [51,52]. Thus, mAbs against the extracellular domains of the TJ components have been generated and their pharmaceutical activities are being investigated (Table 2).

Table 2. List of current monoclonal antibodies against tight junction components.

Indicated Application or Disease	Target	Monoclonal Antibody Name	Ref. or ClinicalTrials.gov Identifier
Modulation of epidermal barrier	Claudin-1	7A5	[53]
Modulation of blood–brain barrier	Claudin-5	R9, R2, 2B12	[54,55]
Inflammatory bowel disease	Claudin-2	1A2	[56]
Hepatitis C virus infection	Claudin-1 Occludin	OM-7D3-B3, 3A2 1-3, 67-2	[57,58] [59,60]
Gastric cancer (phase III study)	Claudin-18.2	IMAB362	NCT03504397
Pancreatic cancer (phase II study)	Claudin-18.2	IMAB362	NCT03816163
Germ cell tumor (phase II study)	Claudin-6	IMAB027	NCT03760081
Cancers (phase I or pre-clinical study)	Claudin-1 Claudin-2 Claudin-3 Claudin-4 Angulin-1 Angulin-3	3A2, 6F6 1A2 KMK3953, IgGH6 KM3934, 5D12 #1-25 BAY1905254	[61,62] [63] [64,65] [66,67] [68] NCT03666273
Regenerative medicine	Claudin-6	clone 342927	[69]

4. Drug Delivery Using Claudin and Angulin Binders

The authors have recently completed a series of studies examining TJ binders that has provided new insights for TJ-directed drug development. Here, the proofs-of-concept for TJ-directed drug development using first- and second-generation claudin and angulin binders are introduced.

4.1. *Mucosal Absorption*

Drug absorption across epithelia is either transcellular or paracellular [70]. One strategy for paracellular drug absorption is to loosen the TJs between adjacent epithelial cells. C-CPE increased jejunal absorption of dextran (4 kDa) 400-fold compared with sodium caprate, an absorption enhancer in current clinical use [17]. C-CPE also enhanced jejunal, nasal, and pulmonary absorption of a biologically active peptide [41]. The treatment of cells with angubindin-1 enhances the permeability of tricellular TJs to solutes up to 10 kDa (Figure 3). Angubindin-1 also enhanced jejunal absorption of dextran (4 kDa) [50]. These results demonstrate that modulation of bicellular and tricellular TJs could be useful strategy for the development of non-invasive drug-delivery systems.

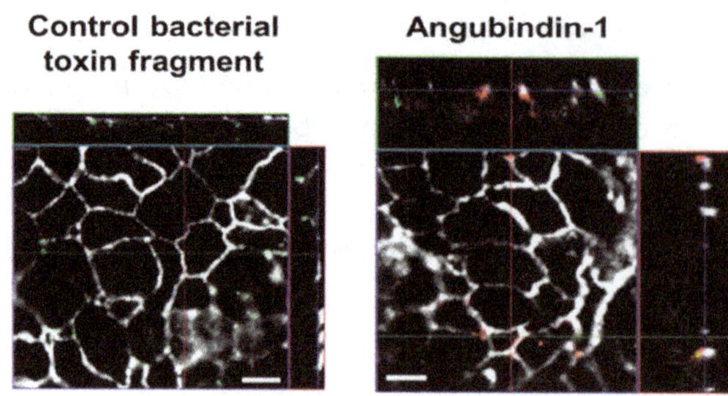

Figure 3. Visualization of the permeation of a macromolecule through tricellular TJs. HT-29/B6 cells were treated with control bacterial toxin fragment or angubindin-1 for 48 h. The cells were incubated with avidin and then with biotin-labeled tetramethylrhodamine-dextran 10 kDa (red) on the apical side of the insert for 1 h. The cells were then fixed and subjected to immunofluorescence analysis. The green signal represents tricellulin and the gray signal represents zonula occludens-1. Bars = 5 μm. The figure is reproduced from reference [50] with slight modifications and permission from the copyright holder.

4.2. Epidermal Absorption

The epidermis covers the outer body, preventing the passage of solutes and the absorption of drugs. However, epidermal administration is a potentially useful route of administration because it is noninvasive, can easily be stopped, and avoids first-pass metabolism [71]. The epidermal barrier comprises the stratum corneum and TJs in the stratum granulosum [72]. The analyses using knockout mice have revealed that claudin-1 is critical for TJ integrity in the stratum granulosum [73]. The treatment with an anti-claudin-1 mAb (7A5) weakened TJ integrity and enhanced the permeation of dextran (4 kDa) in an in vitro human epidermal model [53]. Thus, claudin-1-mediated modulation of the permeability of TJs in the stratum granulosum is a promising means of increasing epidermal drug absorption.

4.3. Cancer Targeting

Claudins are aberrantly expressed in many malignant tumors [74]. For example, in epithelium-derived tumors, TJ strands are disorganized and TJ proteins are distributed throughout the cell surface [75]. Claudin-3 and -4 are the most frequently overexpressed claudins in malignant tumors of prostate [76], breast [77], pancreatic [78], and ovarian cancers [79]. Claudin-1 and -2 are frequently overexpressed in colorectal cancers [80,81] and are associated with the promotion of tumor proliferation and invasiveness via the activation of intracellular signaling cascades [82–84].

The various claudin-targeting molecules, including toxins, toxin fragments, and antibodies, have been generated for claudin-targeted cancer therapy [32,33,85]. For example, CPE has been used as an anti-cancer agent against pancreatic cancer overexpressing claudin-4 [86]. One issue with claudin-targeted therapies is that claudins are expressed not only in malignant tissues, but also in non-malignant tissues. However, most claudins in non-malignant tissues are localized within TJ complexes, whereas their localization is often dysregulated from TJ complexes to the cell surface in malignant tissues [87,88]. This study found that C-CPE fused with protein synthesis inhibitory factor (C-CPE-PSIF) may recognize claudins with aberrant localization, resulting in less binding and therefore, less toxicity to the normal cells [89]. Caco-2 is a human colon carcinoma cell line. Caco-2 cells form a polarized cell monolayer with well-developed TJs when confluent, and they are frequently used as a model of polarized normal epithelial cells. The claudin-4 protein level in the confluent culture (normal epithelial-like cells) was higher than in the subconfluent culture (carcinoma cells), and C-CPE-PSIF was cytotoxic to preconfluent (immature TJs) but not to postconfluent Caco-2 cells (mature TJs) (Figure 4) [89,90]. Similarly, anti-claudin-4 mAbs systemically administered to mice preferentially accumulated in tumor tissue rather than in normal tissue [67,91]. Together, these results suggest that claudins are potential targets for the development of cancer-targeted therapies.

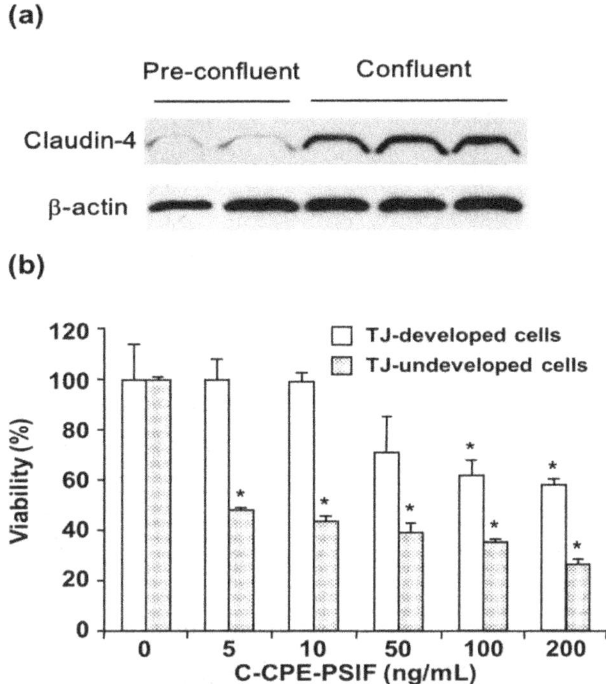

Figure 4. The effect of tight junction (TJ) maturity on the cytotoxicity of C-CPE fused with protein synthesis inhibitory factor (C-CPE-PSIF). Caco-2 cells were cultured to confluence or to preconfluence for 3 days to obtain cells with mature or immature TJs, respectively. (**a**) The cell lysates were subjected to western blotting. (**b**) The cells were treated with the indicated concentrations of C-CPE-PSIF for 48 h, and then cell viability was measured by WST-8 assay. The data are representative of at least three independent experiments. The data are shown as the mean ± S.D. ($n = 3$). * $p < 0.05$. The data are reproduced from reference [89] with slight modifications and permission from the copyright holder.

4.4. Targeting Tissues Involved in Immunological Processes

Mucosal vaccination may be a useful immunization strategy because it is non-invasive and it activates both the mucosal and systemic immune responses. Epithelial cells associated with the mucosa-associated lymphoid tissues (MALT) include Peyer's patches and nasopharynx and play pivotal roles in preventing the invasion of pathological microorganisms into the body by inducing the secretion of IgA [92]. MALT comprises various immune cells, including T cells, B cells, and dendritic cells, and is covered by follicle-associated epithelium. M cells are specialized epithelial cells in the follicle-associated epithelium that transport luminal antigens to immune cells in MALT by transcytosis [93]. In general, when antigen alone is orally or nasally administered, it fails to reach the MALT and so immune responses are not induced. Thus, the efficient delivery of antigen to MALT may provide effective mucosal vaccines.

Follicle-associated epithelium contains claudin-4-expressing cells, some of which are highly capable of capturing luminal antigen [94,95]. Claudin-4 is also expressed on the luminal surface of M cells [96]. Thus, claudin-4 targeting may be a promising strategy for delivering antigens to MALT. A nasally administered ovalbumin fused with C-CPE induced mucosal IgA production, systemic IgG production, and antigen-specific immune responses for preventing tumor growth (Figure 5) [97]. Of note, a simple mixture of C-CPE and antigen did not induce IgA production, indicating that the vaccination efficacy may be depending on the binding affinity of the C-CPE to claudins [97]. Nasal immunization with chimeric C-CPE-antigen did not induce mucosal injury [98]. The augmentation of the antigenicity of the first-generation binder C-CPE has been used to develop an adjuvant-free bivalent food poisoning vaccine [99]

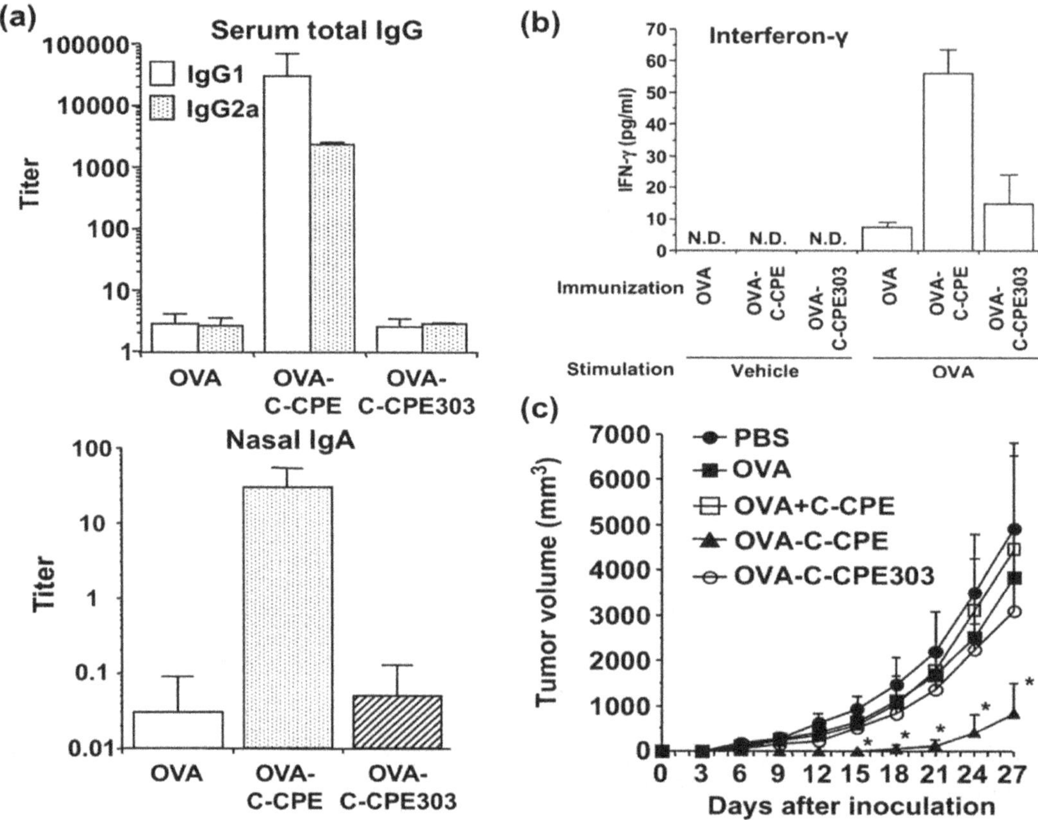

Figure 5. The effect of claudin-4-targeted mucosal immunization on systemic and mucosal immunity. (a and b) Mice were nasally immunized with ovalbumin (OVA), OVA-C-CPE, or OVA-C-CPE303 (which lacks the minimal claudin-binding region) (5 μg OVA in each formulation) once a week for 3 weeks. Seven days after the last immunization, serum and splenocytes were harvested. (**a**) The levels of serum IgG1 and IgG2a (upper panel) and nasal IgA (lower panel) were measured. (**b**) The splenocytes were stimulated with vehicle or OVA (1 mg/mL) for 24 h, and interferon-γ in the supernatant was measured. (**c**) Mice were nasally immunized with vehicle, OVA, a mixture of OVA and C-CPE, OVA-C-CPE, or OVA-C-CPE303 (5 μg OVA in each vaccine) once a week for 3 weeks. Seven days after the final immunization, the mice were injected subcutaneously with 1×10^6 OVA-expressing EL4 (H-2b) cells. The tumor volumes were measured over time. The data are shown as the mean ± S.D. ($n = 4$). * $p < 0.05$. The data are reproduced from reference [97] with slight modifications and permission from the copyright holder.

4.5. Targeting Inflamed Tissues

Ulcerative colitis is a chronic, relapsing inflammatory bowel disease characterized by severe diarrhea and mucosal inflammation in the colon. The disruption of the colonic mucosal barrier leads to the activation of immune responses against bacterial and food fragments in the colon, followed by the development of ulcerative colitis [2,3]. Although claudin-2 is rarely expressed in normal colonic epithelial cells, its expression in the colon is upregulated in ulcerative colitis patients [80]. Inflammatory cytokines, including tumor necrosis factor-α (TNF-α), decrease the epithelial barrier integrity and upregulate the expression of claudin-2 [100]. Claudin-2 decreases the integrity of TJs by facilitating the formation of discontinuous TJ strands [101]. This suggests that the inhibition of claudin-2 may restore the disrupted mucosal barrier. Indeed, an anti-claudin-2 mAb (1A2) ameliorated TNF-α-induced reduction of TJ integrity in Caco-2 cells. Moreover, the co-treatment of the cells with anti-claudin-2 mAb and an anti-TNF-α mAb showed an additive effect on the restoration of the barrier [56].

4.6. Drug Deliavery to the Brain

Unlike peripheral capillaries, those in the brain lack fenestrations and have well-developed TJs that form the blood–brain barrier (BBB). More than 98% of small-molecular-weight drugs cannot pass the BBB [102]. Claudin-5 and angulin-1 are abundantly expressed by brain endothelial cells in mice [103]. Claudin-5- or angulin-1-knockout mice have a size-selectively loosened BBB [104,105]. These data suggest that claudin-5 and angulin-1 are candidate targets for opening the BBB. Indeed, a C-CPE mutant that can bind to claudin-5, angubindin-1, and anti-claudin-5 mAb was able to reduce the transepithelial/transendothelial electrical resistance (TER) in an in vitro model of the BBB [54,106]. Furthermore, in the mice, an angubindin-1-, but not claudin-5-binding C-CPE mutant increased the permeability of the BBB to allow passage of a 16-mer gapmer antisense oligonucleotide (5.3 kDa) [106]. No obvious adverse effects were observed in the mice in these experiments.

5. Safety of Claudin- and Angulin-Targeted Therapies

A series of proof-of-concept studies examining claudin and angulin targeting has provided insights into enhancing drug absorption, treating cancer and inflammatory diseases, improving vaccines, and obtaining drug delivery to the brain. No apparent adverse effects were observed in these studies. However, claudins and angulins play roles in the formation of the intercellular seal between and among epithelial cells and endothelial cells in many tissues. Therefore, ensuring the safety of claudin- and angulin-targeted drugs is critical for future drug development.

The knockout and knockdown analyses of the genes encoding claudins and angulins have shown that there are risks associated with claudin- and angulin-targeted therapeutics (Table 3). For instance, the inhibitors of claudin-1 and -5 may induce atopic dermatitis and schizophrenia-like symptoms via the inhibition of the epidermal barrier and the BBB, respectively [4,6]. Claudin-2, -4, and angulin-2-targeted drugs may induce renal impairment with respect to the reabsorption of ions and water [107–109]. A deletion in exon 1 of claudin-1 results in neonatal ichthyosis and sclerosing cholangitis syndrome in humans [110]. A deletion of 1.5 to 3.0 Mb of human chromosome 22q11.2 that includes the *claudin-5* gene is associated with the development of schizophrenia [111]. A single nucleotide polymorphism in *claudin-5* is also associated with the development of schizophrenia [112,113].

Table 3. Phenotypes of representative claudin- or angulin-knockout or -knockdown mice.

	Phenotype of Knockout (KO) or Knockdown (KD) Mice	Ref.
Claudin-1	Atopic dermatitis (KD)	[4]
Claudin-2	Impaired renal Na^+, Cl^-, and water reabsorption (KO)	[108]
Claudin-3	Increased hepatocyte permeability to phosphate ion (KO)	[114]
Claudin-4	Impaired renal Ca^{2+} and Cl^- reabsorption (KO)	[107]
Claudin-5	Schizophrenia-like symptoms (KD)	[6]
Angulin-2	Impaired renal water reabsorption and colonic water absorption (KO)	[109]

The expression profiles of claudins and angulins differ among tissues (Table 4). The specific claudin ratio is critical for the functions of TJs [115]. Thus, the toxicity of claudin-directed drugs should be carefully investigated, especially if the target claudins are also expressed in non-target tissues. Claudins and angulins in TJs are embedded in the lateral cell membranes and extend into the intercellular space. The TJ cavity is estimated to be 0.5 nm under physiological conditions [116,117]. Large binders, such as antibodies, which are unable to access proteins embedded in TJs, are a promising modality for treating cancers and for improving the effectiveness of vaccines because in these conditions the target claudins are exposed on the cell surfaces [34,87,88,96]. TJ components are potent targets for the development of many novel therapies, but targets and drug modalities must be optimized to afford an acceptable risk–benefit balance.

Table 4. Claudin and angulin expression in representative tissues in a mouse or rat.

	Claudin					Angulin			Ref.
	1	2	3	4	5	1	2	3	
Epidermal cells (stratum granulosum)	+	-	-	+	-	+	-	-	[29,73]
Nasal epithelial cells	+	-	+	+	+	+	+	-	[29,118]
Lung (alveoli)	+	-	+	+	+	+	-	-	[29,119]
Small intestine (jejunum)	+	+	+	-	+	+	-	-	[29,109,120]
Colon (surface)	+	-	+	+	+	-	+	-	[29,109,120]
Liver	+	+	+	-	-	+	-	-	[29,121,122]
Kidney (glomerulus)	+	+	-	-	-	-	+	-	[29,123]
Kidney (proximal tube)	-	+	-	-	-	+	-	-	[29,123]
Kidney (thin ascending limb of the loop of Henle)	-	-	+	+	-	-	+	-	[29,123]
Kidney (collecting duct)	-	-	+	+	-	-	+	-	[29,109,123]
Brain endothelial cells	-	-	-	-	+	+	-	+	[103,105]
Brain ependymal cells	+	+	+	-	-	-	-	+	[29,124]
Lung endothelial cells	-	-	-	-	+	-	-	-	[125]

+, expressed; -, not detected.

6. Conclusions

TJ binders are classified as first-generation binders (toxins and their fragments), and second-generation binders (antibodies) [32,33,85]. The augmentation of the antigenicity of the first-generation binder, C-CPE, has been used to develop an adjuvant-free vaccine [97,99]. Second-generation binders are being used to develop cancer therapies (Table 2). For example, an anti-claudin-18.2 mAb is undergoing clinical study for the use in the treatment of gastric (phase III study) and pancreatic cancer (phase II study) [NCT03504397; NCT03816163].

The other application of TJ binders is to enhance the mucosal and epidermal absorption of drugs and to deliver drugs to the brain by modulating the permeability of TJs. The currently available TJ binders are toxin fragments and antibodies. However, the generation of novel TJ binders, such as peptides and chemicals, is now needed because of the potential antigenicity of toxins and the costs associated with antibody preparation. A high-throughput screening system for claudin-4 binders based on the time resolved fluorescence resonance energy transfer in a chemical library was developed [126]. In the future, the generations of peptide- and chemical-type of binders are expected to accelerate.

Author Contributions: Wrote or contributed to the writing of the manuscript: Y.H., K.T., S.M.K., J.K., M.F., and M.K.

Acknowledgments: We thank all members of Kondoh Lab and Kunisawa Lab for their useful comments and discussions.

Abbreviations

BBB	Blood–brain barrier
CPE	Clostridium perfringens enterotoxin
CPE-R	CPE receptor
C-CPE	C-terminal domain of CPE
C-CPE-PSIF	C-CPE fused with protein synthesis inhibitory factor
ECH	Extracellular helix
KD	Knockdown
KO	Knockout
mAb	Monoclonal antibody
MALT	Mucosa-associated lymphoid tissues
RVP1	Rat androgen withdrawal apoptosis protein
TER	Transepithelial/transendothelial electrical resistance
TJ	Tight junction
TM	Transmembrane
TNF-α	Tumor necrosis factor-α

References

1. Farquhar, M.G.; Palade, G.E. Junctional complexes in various epithelia. *J. Cell Biol.* **1963**, *17*, 375–412. [CrossRef] [PubMed]
2. de Souza, H.S.P.; Fiocchi, C. Immunopathogenesis of IBD: Current state of the art. *Nat. Rev. Gastro. Hepat.* **2016**, *13*, 13–27. [CrossRef] [PubMed]
3. Madsen, K.L.; Malfair, D.; Gray, D.; Doyle, J.S.; Jewell, L.D.; Fedorak, R.N. Interleukin-10 gene-deficient mice develop a primary intestinal permeability defect in response to enteric microflora. *Inflamm. Bowel Dis.* **1999**, *5*, 262–270. [CrossRef] [PubMed]
4. Tokumasu, R.; Yamaga, K.; Yamazaki, Y.; Murota, H.; Suzuki, K.; Tamura, A.; Bando, K.; Furuta, Y.; Katayama, I.; Tsukita, S. Dose-dependent role of claudin-1 in vivo in orchestrating features of atopic dermatitis. *Proc. Natl. Acad. Sci. USA* **2016**, *113*, E4061–E4068. [CrossRef] [PubMed]
5. Menard, C.; Pfau, M.L.; Hodes, G.E.; Kana, V.; Wang, V.X.; Bouchard, S.; Takahashi, A.; Flanigan, M.E.; Aleyasin, H.; LeClair, K.B.; et al. Social stress induces neurovascular pathology promoting depression. *Nat. Neurosci.* **2017**, *20*, 1752–1760. [CrossRef] [PubMed]
6. Greene, C.; Kealy, J.; Humphries, M.M.; Gong, Y.; Hou, J.; Hudson, N.; Cassidy, L.M.; Martiniano, R.; Shashi, V.; Hooper, S.R.; et al. Dose-dependent expression of claudin-5 is a modifying factor in schizophrenia. *Mol. Psychiatry* **2018**, *23*, 2156–2166. [CrossRef] [PubMed]
7. Staehelin, L.A.; Mukherjee, T.M.; Williams, A.W. Freeze-etch appearance of tight junctions in epithelium of small and large intestine mice. *Protoplasma* **1969**, *67*, 165–184. [CrossRef] [PubMed]
8. Van Itallie, C.M.; Tietgens, A.J.; Anderson, J.M. Visualizing the dynamic coupling of claudin strands to the actin cytoskeleton through ZO-1. *Mol. Biol. Cell* **2017**, *28*, 524–534. [CrossRef]
9. Umeda, K.; Ikenouchi, J.; Katahira-Tayama, S.; Furuse, K.; Sasaki, H.; Nakayama, M.; Matsui, T.; Tsukita, S.; Furuse, M.; Tsukita, S. ZO-1 and ZO-2 independently determine where claudins are polymerized in tight-junction strand formation. *Cell* **2006**, *126*, 741–754. [CrossRef]
10. Cording, J.; Berg, J.; Kading, N.; Bellmann, C.; Tscheik, C.; Westphal, J.K.; Milatz, S.; Günzel, D.; Wolburg, H.; Piontek, J.; et al. In tight junctions, claudins regulate the interactions between occludin, tricellulin and marvelD3, which, inversely, modulate claudin oligomerization. *J. Cell Sci.* **2013**, *126*, 554–564. [CrossRef]
11. Yamazaki, Y.; Tokumasu, R.; Kimura, H.; Tsukita, S. Role of claudin species-specific dynamics in reconstitution and remodeling of the zonula occludens. *Mol. Biol. Cell* **2011**, *22*, 1495–1504. [CrossRef] [PubMed]
12. Rossa, J.; Ploeger, C.; Vorreiter, F.; Saleh, T.; Protze, J.; Günzel, D.; Wolburg, H.; Krause, G.; Piontek, J. Claudin-3 and claudin-5 protein folding and assembly into the tight junction are controlled by non-conserved residues in the transmembrane 3 (TM3) and extracellular loop 2 (ECL2) segments. *J. Biol. Chem.* **2014**, *289*, 7641–7653. [CrossRef] [PubMed]
13. Tervonen, A.; Ihalainen, T.O.; Nymark, S.; Hyttinen, J. Structural dynamics of tight junctions modulate the properties of the epithelial barrier. *PLoS ONE* **2019**, *14*, e0214876. [CrossRef] [PubMed]
14. Staehelin, L.A. Further observations on fine-structure of freeze-cleaved tight junctions. *J. Cell Sci.* **1973**, *13*, 763–786. [PubMed]
15. Lindmark, T.; Söderholm, J.D.; Olaison, G.; Alvan, G.; Ocklind, G.; Artursson, P. Mechanism of absorption enhancement in humans after rectal administration of ampicillin in suppositories containing sodium caprate. *Pharm. Res.* **1997**, *14*, 930–935. [CrossRef] [PubMed]
16. Doolittle, N.D.; Miner, M.E.; Hall, W.A.; Siegal, T.; Hanson, E.J.; Osztie, E.; McAllister, L.D.; Bubalo, J.S.; Kraemer, D.F.; Fortin, D.; et al. Safety and efficacy of a multicenter study using intraarterial chemotherapy in conjunction with osmotic opening of the blood-brain barrier for the treatment of patients with malignant brain tumors. *Cancer* **2000**, *88*, 637–647. [CrossRef]
17. Kondoh, M.; Masuyama, A.; Takahashi, A.; Asano, N.; Mizuguchi, H.; Koizumi, N.; Fujii, M.; Hayakawa, T.; Horiguchi, Y.; Watanbe, Y. A novel strategy for the enhancement of drug absorption using a claudin modulator. *Mol. Pharmacol.* **2005**, *67*, 749–756. [CrossRef]
18. Rapoport, S. Osmotic opening of the blood-brain barrier: Principles, mechanism, and therapeutic applications. *Cell. Mol. Neurobiol.* **2000**, *20*, 217–230. [CrossRef]
19. Furuse, M.; Fujita, K.; Hiiragi, T.; Fujimoto, K.; Tsukita, S. Claudin-1 and -2: Novel integral membrane proteins localizing at tight junctions with no sequence similarity to occludin. *J. Cell Biol.* **1998**, *141*, 1539–1550. [CrossRef]

20. Mineta, K.; Yamamoto, Y.; Yamazaki, Y.; Tanaka, H.; Tada, Y.; Saito, K.; Tamura, A.; Igarashi, M.; Endo, T.; Takeuchi, K.; et al. Predicted expansion of the claudin multigene family. *FEBS Lett.* **2011**, *585*, 606–612. [CrossRef]
21. Suzuki, H.; Nishizawa, T.; Tani, K.; Yamazaki, Y.; Tamura, A.; Ishitani, R.; Dohmae, N.; Tsukita, S.; Nureki, O.; Fujiyoshi, Y. Crystal structure of a claudin provides insight into the architecture of tight junctions. *Science* **2014**, *344*, 304–307. [CrossRef] [PubMed]
22. Saitoh, Y.; Suzuki, H.; Tani, K.; Nishikawa, K.; Irie, K.; Ogura, Y.; Tamura, A.; Tsukita, S.; Fujiyoshi, Y. Structural insight into tight junction disassembly by *Clostridium perfringens* enterotoxin. *Science* **2015**, *347*, 775–778. [CrossRef] [PubMed]
23. Shinoda, T.; Shinya, N.; Ito, K.; Ohsawa, N.; Terada, T.; Hirata, K.; Kawano, Y.; Yamamoto, M.; Kimura-Someya, T.; Yokoyama, S.; et al. Structural basis for disruption of claudin assembly in tight junctions by an enterotoxin. *Sci. Rep.* **2016**, *6*, 33632. [CrossRef] [PubMed]
24. Nakamura, S.; Irie, K.; Tanaka, H.; Nishikawa, K.; Suzuki, H.; Saitoh, Y.; Tamura, A.; Tsukita, S.; Fujiyoshi, Y. Morphologic determinant of tight junctions revealed by claudin-3 structures. *Nat. Commun.* **2019**, *10*, 816. [CrossRef] [PubMed]
25. Piontek, J.; Winkler, L.; Wolburg, H.; Müller, S.L.; Zuleger, N.; Piehl, C.; Wiesner, B.; Krause, G.; Blasig, I.E. Formation of tight junction: Determinants of homophilic interaction between classic claudins. *FASEB J.* **2008**, *22*, 146–158. [CrossRef] [PubMed]
26. Suzuki, H.; Tani, K.; Tamura, A.; Tsukita, S.; Fujiyoshi, Y. Model for the architecture of claudin-based paracellular ion channels through tight junctions. *J. Mol. Biol.* **2015**, *427*, 291–297. [CrossRef] [PubMed]
27. Ikenouchi, J.; Furuse, M.; Furuse, K.; Sasaki, H.; Tsukita, S.; Tsukita, S. Tricellulin constitutes a novel barrier at tricellular contacts of epithelial cells. *J. Cell Biol.* **2005**, *171*, 939–945. [CrossRef] [PubMed]
28. Masuda, S.; Oda, Y.; Sasaki, H.; Ikenouchi, J.; Higashi, T.; Akashi, M.; Nishi, E.; Furuse, M. LSR defines cell corners for tricellular tight junction formation in epithelial cells. *J. Cell Sci.* **2011**, *124*, 548–555. [CrossRef]
29. Higashi, T.; Tokuda, S.; Kitajiri, S.; Masuda, S.; Nakamura, H.; Oda, Y.; Furuse, M. Analysis of the 'angulin' proteins LSR, ILDR1 and ILDR2-tricellulin recruitment, epithelial barrier function and implication in deafness pathogenesis. *J. Cell Sci.* **2013**, *126*, 966–977. [CrossRef]
30. Nakatsu, D.; Kano, F.; Taguchi, Y.; Sugawara, T.; Nishizono, T.; Nishikawa, K.; Oda, Y.; Furuse, M.; Murata, M. JNK1/2-dependent phosphorylation of angulin-1/LSR is required for the exclusive localization of angulin-1/LSR and tricellulin at tricellular contacts in EpH4 epithelial sheet. *Genes Cells* **2014**, *19*, 565–581. [CrossRef]
31. Kim, N.K.D.; Higashi, T.; Lee, K.Y.; Kim, A.R.; Kitajiri, S.; Kim, M.Y.; Chang, M.Y.; Kim, V.; Oh, S.H.; Kim, D.; et al. Downsloping high-frequency hearing loss due to inner ear tricellular tight junction disruption by a novel ILDR1 mutation in the Ig-like domain. *PLoS ONE* **2015**, *10*, e0116931. [CrossRef] [PubMed]
32. Hashimoto, Y.; Yagi, K.; Kondoh, M. Current progress in a second-generation claudin binder, anti-claudin antibody, for clinical applications. *Drug Discov. Today* **2016**, *21*, 1711–1718. [CrossRef] [PubMed]
33. Hashimoto, Y.; Yagi, K.; Kondoh, M. Roles of the first-generation claudin binder, *Clostridium perfringens* enterotoxin, in the diagnosis and claudin-targeted treatment of epithelium-derived cancers. *Pflugers Arch.* **2017**, *469*, 45–53. [CrossRef] [PubMed]
34. Hashimoto, Y.; Okada, Y.; Shirakura, K.; Tachibana, K.; Sawada, M.; Yagi, K.; Doi, T.; Kondoh, M. Anti-claudin antibodies as a concept for development of claudin-directed drugs. *J. Pharmacol. Exp. Ther.* **2019**, *368*, 179–186. [CrossRef] [PubMed]
35. Kitadokoro, K.; Nishimura, K.; Kamitani, S.; Fukui-Miyazaki, A.; Toshima, H.; Abe, H.; Kamata, Y.; Sugita-Konishi, Y.; Yamamoto, S.; Karatani, H.; et al. Crystal structure of *Clostridium perfringens* enterotoxin displays features of beta-pore-forming toxins. *J. Biol. Chem.* **2011**, *286*, 19549–19555. [CrossRef] [PubMed]
36. Katahira, J.; Inoue, N.; Horiguchi, Y.; Matsuda, M.; Sugimoto, N. Molecular cloning and functional characterization of the receptor for Clostridium perfringensenterotoxin. *J. Cell Biol.* **1997**, *136*, 1239–1247. [CrossRef] [PubMed]
37. Morita, K.; Furuse, M.; Fujimoto, K.; Tsukita, S. Claudin multigene family encoding four-transmembrane domain protein components of tight junction strands. *Proc. Natl. Acad. Sci. USA* **1999**, *96*, 511–516. [CrossRef]
38. Sonoda, N.; Furuse, M.; Sasaki, H.; Yonemura, S.; Katahira, J.; Horiguchi, Y.; Tsukita, S. *Clostridium perfringens* enterotoxin fragment removes specific claudins from tight junction strands: Evidence for direct involvement of claudins in tight junction barrier. *J. Cell Biol.* **1999**, *147*, 195–204. [CrossRef]

39. Winkler, L.; Gehring, C.; Wenzel, A.; Müller, S.L.; Piehl, C.; Krause, G.; Blasig, I.E.; Piontek, J. Molecular determinants of the interaction between *Clostridium perfringens* enterotoxin fragments and claudin-3. *J. Biol. Chem.* **2009**, *284*, 18863–18872. [CrossRef]
40. Fujita, K.; Katahira, J.; Horiguchi, Y.; Sonoda, N.; Furuse, M.; Tsukita, S. *Clostridium perfringens* enterotoxin binds to the second extracellular loop of claudin-3, a tight junction integral membrane protein. *FEBS Lett.* **2000**, *476*, 258–261. [CrossRef]
41. Uchida, H.; Kondoh, M.; Hanada, T.; Takahashi, A.; Hamakubo, T.; Yagi, K. A claudin-4 modulator enhances the mucosal absorption of a biologically active peptide. *Biochem. Pharmacol.* **2010**, *79*, 1437–1444. [CrossRef] [PubMed]
42. Takahashi, A.; Komiya, E.; Kakutani, H.; Yoshida, T.; Fujii, M.; Horiguchi, Y.; Mizuquchi, H.; Tsutsumi, Y.; Tsunoda, S.I.; Koizumi, N.; et al. Domain mapping of a claudin-4 modulator, the C-terminal region of C-terminal fragment of *Clostridium perfringens* enterotoxin, by site-directed mutagenesis. *Biochem. Pharmacol.* **2008**, *75*, 1639–1648. [CrossRef] [PubMed]
43. Takahashi, A.; Saito, Y.; Kondoh, M.; Matsushita, K.; Krug, S.M.; Suzuki, H.; Tsujino, H.; Li, X.R.; Aoyama, H.; Matsuhisa, K.; et al. Creation and biochemical analysis of a broad-specific claudin binder. *Biomaterials* **2012**, *33*, 3464–3474. [CrossRef] [PubMed]
44. Veshnyakova, A.; Piontek, J.; Protze, J.; Waziri, N.; Heise, I.; Krause, G. Mechanism of *Clostridium perfringens* enterotoxin interaction with claudin-3/-4 protein suggests structural modifications of the toxin to target specific claudins. *J. Biol. Chem.* **2012**, *287*, 1698–1708. [CrossRef] [PubMed]
45. Protze, J.; Eichner, M.; Piontek, A.; Dinter, S.; Rossa, J.; Blecharz, K.G.Z.; Vajkoczy, P.; Piontek, J.; Krause, G. Directed structural modification of *Clostridium perfringens* enterotoxin to enhance binding to claudin-5. *Cell. Mol. Life Sci.* **2015**, *72*, 1417–1432. [CrossRef] [PubMed]
46. Neuhaus, W.; Piontek, A.; Protze, J.; Eichner, M.; Mahringer, A.; Subileau, E.A.; Lee, I.F.M.; Schulzke, J.D.; Krause, G.; Piontek, J. Reversible opening of the blood-brain barrier by claudin-5-binding variants of *Clostridium perfringens* enterotoxin's claudin-binding domain. *Biomaterials* **2018**, *161*, 129–143. [CrossRef]
47. Nagahama, M.; Umezaki, M.; Oda, M.; Kobayashi, K.; Tone, S.; Suda, T.; Ishidoh, K.; Sakurai, J. *Clostridium perfringens* iota-toxin b induces rapid cell necrosis. *Infect. Immun.* **2011**, *79*, 4353–4360. [CrossRef]
48. Nagahama, M.; Yamaguchi, A.; Hagiyama, T.; Ohkubo, N.; Kobayashi, K.; Sakurai, J. Binding and internalization of *Clostridium perfringens* iota-toxin in lipid rafts. *Infect. Immun.* **2004**, *72*, 3267–3275. [CrossRef]
49. Papatheodorou, P.; Carette, J.E.; Bell, G.W.; Schwan, C.; Guttenberg, G.; Brummelkamp, T.R.; Aktories, K. Lipolysis-stimulated lipoprotein receptor (LSR) is the host receptor for the binary toxin *Clostridium difficile* transferase (CDT). *Proc. Natl. Acad. Sci. USA* **2011**, *108*, 16422–16427. [CrossRef]
50. Krug, S.M.; Hayaishi, T.; Iguchi, D.; Watari, A.; Takahashi, A.; Fromm, M.; Nagahama, M.; Takeda, H.; Okada, Y.; Sawasaki, T.; et al. Angubindin-1, a novel paracellular absorption enhancer acting at the tricellular tight junction. *J. Control. Release* **2017**, *260*, 1–11. [CrossRef]
51. Souriau, C.; Hudson, P.J. Recombinant antibodies for cancer diagnosis and therapy. *Expert Opin. Biol. Ther.* **2003**, *3*, 305–318. [CrossRef] [PubMed]
52. Espiritu, M.J.; Collier, A.C.; Bingham, J.P. A 21st-century approach to age-old problems: The ascension of biologics in clinical therapeutics. *Drug Discov. Today* **2014**, *19*, 1109–1113. [CrossRef] [PubMed]
53. Nakajima, M.; Nagase, S.; Iida, M.; Takeda, S.; Yamashita, M.; Watari, A.; Shirasago, Y.; Fukasawa, M.; Takeda, H.; Sawasaki, T.; et al. Claudin-1 binder enhances epidermal permeability in a human keratinocyte model. *J. Pharmacol. Exp. Ther.* **2015**, *354*, 440–447. [CrossRef] [PubMed]
54. Hashimoto, Y.; Shirakura, K.; Okada, Y.; Takeda, H.; Endo, K.; Tamura, M.; Watari, A.; Sadamura, Y.; Sawasaki, T.; Doi, T.; et al. Claudin-5-binders enhance permeation of solutes across the blood-brain barrier in a mammalian model. *J. Pharmacol. Exp. Ther.* **2017**, *363*, 275–283. [CrossRef] [PubMed]
55. Hashimoto, Y.; Zhou, W.; Hamauchi, K.; Shirakura, K.; Doi, T.; Yagi, K.; Sawasaki, T.; Okada, Y.; Kondoh, M.; Takeda, H. Engineered membrane protein antigens successfully induce antibodies against extracellular regions of claudin-5. *Sci. Rep.* **2018**, *8*, 8383. [CrossRef]
56. Takigawa, M.; Iida, M.; Nagase, S.; Suzuki, H.; Watari, A.; Tada, M.; Okada, Y.; Doi, T.; Fukasawa, M.; Yagi, K.; et al. Creation of a claudin-2 binder and its tight junction-modulating activity in a human intestinal models. *J. Pharmacol. Exp. Ther.* **2017**, *363*, 444–451. [CrossRef] [PubMed]

57. Fofana, I.; Krieger, S.E.; Grunert, F.; Glauben, S.; Xiao, F.; Fafi-Kremer, S.; Soulier, E.; Royer, C.; Thumann, C.; Mee, C.J.; et al. Monoclonal anti-claudin 1 antibodies prevent hepatitis C virus infection of primary human hepatocytes. *Gastroenterology* **2010**, *139*, 953–964. [CrossRef]
58. Fukasawa, M.; Nagase, S.; Shirasago, Y.; Iida, M.; Yamashita, M.; Endo, K.; Yagi, K.; Suzuki, T.; Wakita, T.; Hanada, K.; et al. Monoclonal antibodies against extracellular domains of claudin-1 block hepatitis C virus infection in a mouse model. *J. Virol.* **2015**, *89*, 4866–4879. [CrossRef]
59. Shimizu, Y.; Shirasago, Y.; Kondoh, M.; Suzuki, T.; Wakita, T.; Hanada, K.; Yagi, K.; Fukasawa, M. Monoclonal antibodies against occludin completely prevented hepatitis C virus infection in a mouse model. *J. Virol.* **2018**, *92*. [CrossRef]
60. Okai, K.; Ichikawa-Tomikawa, N.; Saito, A.C.; Watabe, T.; Sugimoto, K.; Fujita, D.; Ono, C.; Fukuhara, T.; Matsuura, Y.; Ohira, H.; et al. A novel occludin-targeting monoclonal antibody prevents hepatitis C virus infection in vitro. *Oncotarget* **2018**, *9*, 16588–16598. [CrossRef]
61. Hashimoto, Y.; Tada, M.; Iida, M.; Nagase, S.; Hata, T.; Watari, A.; Okada, Y.; Doi, T.; Fukasawa, M.; Yagi, K.; et al. Generation and characterization of a human-mouse chimeric antibody against the extracellular domain of claudin-1 for cancer therapy using a mouse model. *Biochem. Biophys. Res. Commun.* **2016**, *477*, 91–95. [CrossRef] [PubMed]
62. Cherradi, S.; Ayrolles-Torro, A.; Vezzo-Vie, N.; Gueguinou, N.; Denis, V.; Combes, E.; Boissiere, F.; Busson, M.; Canterel-Thouennon, L.; Mollevi, C.; et al. Antibody targeting of claudin-1 as a potential colorectal cancer therapy. *J. Exp. Clin. Cancer Res.* **2017**, *36*, 89. [CrossRef] [PubMed]
63. Hashimoto, Y.; Hata, T.; Tada, M.; Iida, M.; Watari, A.; Okada, Y.; Doi, T.; Kuniyasu, H.; Yagi, K.; Kondoh, M. Safety evaluation of a human chimeric monoclonal antibody that recognizes the extracellular loop domain of claudin-2. *Eur. J. Pharm. Sci.* **2018**, *117*, 161–167. [CrossRef] [PubMed]
64. Ando, H.; Suzuki, M.; Kato-Nakano, M.; Kawamoto, S.; Misaka, H.; Kimoto, N.; Furuya, A.; Nakamura, K. Generation of specific monoclonal antibodies against the extracellular loops of human claudin-3 by immunizing mice with target-expressing cells. *Biosci. Biotechnol. Biochem.* **2015**, *79*, 1272–1279. [CrossRef] [PubMed]
65. Romani, C.; Cocco, E.; Bignotti, E.; Moratto, D.; Bugatti, A.; Todeschini, P.; Bandiera, E.; Tassi, R.; Zanotti, L.; Pecorelli, S.; et al. Evaluation of a novel human IgG1 anti-claudin3 antibody that specifically recognizes its aberrantly localized antigen in ovarian cancer cells and that is suitable for selective drug delivery. *Oncotarget* **2015**, *6*, 34617–34628. [CrossRef]
66. Suzuki, M.; Kato-Nakano, M.; Kawamoto, S.; Furuya, A.; Abe, Y.; Misaka, H.; Kimoto, N.; Nakamura, K.; Ohta, S.; Ando, H. Therapeutic antitumor efficacy of monoclonal antibody against Claudin-4 for pancreatic and ovarian cancers. *Cancer Sci.* **2009**, *100*, 1623–1630. [CrossRef]
67. Hashimoto, Y.; Kawahigashi, Y.; Hata, T.; Li, X.; Watari, A.; Tada, M.; Ishii-Watabe, A.; Okada, Y.; Doi, T.; Fukasawa, M.; et al. Efficacy and safety evaluation of claudin-4-targeted antitumor therapy using a human and mouse cross-reactive monoclonal antibody. *Pharmacol. Res. Perspect.* **2016**, *4*, e00266. [CrossRef]
68. Hiramatsu, K.; Serada, S.; Enomoto, T.; Takahashi, Y.; Nakagawa, S.; Nojima, S.; Morimoto, A.; Matsuzaki, S.; Yokoyama, T.; Takahashi, T.; et al. LSR antibody therapy inhibits ovarian epithelial tumor growth by inhibiting lipid uptake. *Cancer Res.* **2018**, *78*, 516–527. [CrossRef]
69. Ben-David, U.; Nudel, N.; Benvenisty, N. Immunologic and chemical targeting of the tight-junction protein Claudin-6 eliminates tumorigenic human pluripotent stem cells. *Nat. Commun.* **2013**, *4*, 1992. [CrossRef]
70. Powell, D.W. Barrier Function of Epithelia. *Am. J. Physiol.* **1981**, *241*, G275–G288. [CrossRef]
71. Tran, T.N. Cutaneous drug delivery: An update. *J. Investig. Dermatol. Symp. Proc.* **2013**, *16*, S67–S69. [CrossRef] [PubMed]
72. Tsukita, S.; Furuse, M. Claudin-based barrier in simple and stratified cellular sheets. *Curr. Opin. Cell Biol.* **2002**, *14*, 531–536. [CrossRef]
73. Furuse, M.; Hata, M.; Furuse, K.; Yoshida, Y.; Haratake, A.; Sugitani, Y.; Noda, T.; Kubo, A.; Tsukita, S. Claudin-based tight junctions are crucial for the mammalian epidermal barrier: A lesson from claudin-1-deficient mice. *J. Cell Biol.* **2002**, *156*, 1099–1111. [CrossRef] [PubMed]
74. Osanai, M.; Takasawa, A.; Murata, M.; Sawada, N. Claudins in cancer: Bench to bedside. *Pflugers Arch.* **2017**, *469*, 55–67. [CrossRef] [PubMed]
75. Oliveira, S.S.; Morgado-Diaz, J.A. Claudins: Multifunctional players in epithelial tight junctions and their role in cancer. *Cell. Mol. Life Sci.* **2007**, *64*, 17–28. [CrossRef] [PubMed]

76. Long, H.Y.; Crean, C.D.; Lee, W.H.; Cummings, O.W.; Gabig, T.G. Expression of *Clostridium perfringens* enterotoxin receptors claudin-3 and claudin-4 in prostate cancer epithelium. *Cancer Res.* **2001**, *61*, 7878–7881. [PubMed]
77. Kominsky, S.L.; Vali, M.; Korz, D.; Gabig, T.G.; Weitzman, S.A.; Argani, P.; Sukumar, S. *Clostridium perfringens* enterotoxin elicits rapid and specific cytolysis of breast carcinoma cells mediated through tight junction proteins claudin 3 and 4. *Am. J. Pathol.* **2004**, *164*, 1627–1633. [CrossRef]
78. Michl, P.; Barth, C.; Buchholz, M.; Lerch, M.M.; Rolke, M.; Holzmann, K.H.; Menke, A.; Fensterer, H.; Giehl, K.; Lohr, M.; et al. Claudin-4 expression decreases invasiveness and metastatic potential of pancreatic cancer. *Cancer Res.* **2003**, *63*, 6265–6271.
79. Zhu, Y.H.; Brannstrom, M.; Janson, P.O.; Sundfeldt, K. Differences in expression patterns of the tight junction proteins, claudin 1, 3, 4 and 5, in human ovarian surface epithelium as compared to epithelia in inclusion cysts and epithelial ovarian tumours. *Int. J. Cancer* **2006**, *118*, 1884–1891. [CrossRef]
80. Weber, C.R.; Nalle, S.C.; Tretiakova, M.; Rubin, D.T.; Turner, J.R. Claudin-1 and claudin-2 expression is elevated in inflammatory bowel disease and may contribute to early neoplastic transformation. *Lab. Investig.* **2008**, *88*, 1110–1120. [CrossRef]
81. Kinugasa, T.; Huo, Q.; Higash, D.; Shibaguchi, H.; Kuroki, M.; Tanaka, T.; Futami, K.; Yamashita, Y.; Hachimine, K.; Maekawa, S.; et al. Selective up-regulation of claudin-1 and claudin-2 in colorectal cancer. *Anticancer Res.* **2007**, *27*, 3729–3734. [CrossRef]
82. Dhawan, P.; Singh, A.B.; Deane, N.G.; No, Y.; Shiou, S.R.; Schmidt, C.; Neff, J.; Washington, M.K.; Beauchamp, R.D. Claudin-1 regulates cellular transformation and metastatic behavior in colon cancer. *J. Clin. Investig.* **2005**, *115*, 1765–1776. [CrossRef] [PubMed]
83. Dhawan, P.; Ahmad, R.; Chaturvedi, R.; Smith, J.J.; Midha, R.; Mittal, M.K.; Krishnan, M.; Chen, X.; Eschrich, S.; Yeatman, T.J.; et al. Claudin-2 expression increases tumorigenicity of colon cancer cells: Role of epidermal growth factor receptor activation. *Oncogene* **2011**, *30*, 3234–3247. [CrossRef] [PubMed]
84. Tabaries, S.; Dong, Z.; Annis, M.G.; Omeroglu, A.; Pepin, F.; Ouellet, V.; Russo, C.; Hassanain, M.; Metrakos, P.; Diaz, Z.; et al. Claudin-2 is selectively enriched in and promotes the formation of breast cancer liver metastases through engagement of integrin complexes. *Oncogene* **2011**, *30*, 1318–1328. [CrossRef] [PubMed]
85. Hashimoto, Y.; Fukasawa, M.; Kuniyasu, H.; Yagi, K.; Kondoh, M. Claudin-targeted drug development using anti-claudin monoclonal antibodies to treat hepatitis and cancer. *Ann. N. Y. Acad. Sci.* **2017**, *1397*, 5–16. [CrossRef] [PubMed]
86. Michl, P.; Buchholz, M.; Rolke, M.; Kunsch, S.; Lohr, M.; McClane, B.; Tsukita, S.; Leder, G.; Adler, G.; Gress, T.M. Claudin-4: A new target for pancreatic cancer treatment using *Clostridium perfringens* enterotoxin. *Gastroenterology* **2001**, *121*, 678–684. [CrossRef] [PubMed]
87. Morin, P.J. Claudin proteins in human cancer: Promising new targets for diagnosis and therapy. *Cancer Res.* **2005**, *65*, 9603–9606. [CrossRef] [PubMed]
88. Kominsky, S.L. Claudins: Emerging targets for cancer therapy. *Expert Rev. Mol. Med.* **2006**, *8*, 1–11. [CrossRef]
89. Saeki, R.; Kondoh, M.; Kakutani, H.; Tsunoda, S.; Mochizuki, Y.; Hamakubo, T.; Tsutsumi, Y.; Horiguchi, Y.; Yagi, K. A novel tumor-targeted therapy using a claudin-4-targeting molecule. *Mol. Pharmacol.* **2009**, *76*, 918–926. [CrossRef]
90. Meunier, V.; Bourrie, M.; Berger, Y.; Fabre, G. The human intestinal epithelial cell line Caco-2; pharmacological and pharmacokinetic applications. *Cell Biol. Toxicol.* **1995**, *11*, 187–194. [CrossRef]
91. Torres, J.B.; Knight, J.C.; Mosley, M.J.; Kersemans, V.; Koustoulidou, S.; Allen, D.; Kinchesh, P.; Smart, S.; Cornelissen, B. Imaging of claudin-4 in pancreatic ductal adenocarcinoma using a radiolabelled anti-claudin-4 monoclonal antibody. *Mol. Imaging Biol.* **2018**, *20*, 292–299. [CrossRef] [PubMed]
92. Kunisawa, J.; Fukuyama, S.; Kiyono, H. Mucosa-associated lymphoid tissues in the aerodigestive tract: Their shared and divergent traits and their importance to the orchestration of the mucosal immune system. *Curr. Mol. Med.* **2005**, *5*, 557–572. [CrossRef] [PubMed]
93. Kraehenbuhl, J.P.; Neutra, M.R. Epithelial M cells: Differentiation and function. *Annu. Rev. Cell Dev. Biol.* **2000**, *16*, 301–332. [CrossRef] [PubMed]
94. Nagatake, T.; Fujita, H.; Minato, N.; Hamazaki, Y. Enteroendocrine cells are specifically marked by cell surface expression of claudin-4 in mouse small intestine. *PLoS ONE* **2014**, *9*, e90638. [CrossRef] [PubMed]
95. Tamagawa, H.; Takahashi, I.; Furuse, M.; Yoshitake-Kitano, Y.; Tsukita, S.; Ito, T.; Matsuda, H.; Kiyono, H. Characteristics of claudin expression in follicle-associated epithelium of Peyer's patches: Preferential localization of claudin-4 at the apex of the dome region. *Lab. Investig.* **2003**, *83*, 1045–1053. [CrossRef]

96. Ye, T.; Yue, Y.; Fan, X.M.; Dong, C.S.; Xu, W.; Xiong, S.D. M cell-targeting strategy facilitates mucosal immune response and enhances protection against CVB3-induced viral myocarditis elicited by chitosan-DNA vaccine. *Vaccine* **2014**, *32*, 4457–4465. [CrossRef]
97. Kakutani, H.; Kondoh, M.; Fukasaka, M.; Suzuki, H.; Hamakubo, T.; Yagi, K. Mucosal vaccination using claudin-4-targeting. *Biomaterials* **2010**, *31*, 5463–5471. [CrossRef]
98. Suzuki, H.; Kakutani, H.; Kondoh, M.; Watari, A.; Yagi, K. The safety of a mucosal vaccine using the C-terminal fragment of *Clostridium perfringens* enterotoxin. *Pharmazie* **2010**, *65*, 766–769. [CrossRef]
99. Suzuki, H.; Hosomi, K.; Nasu, A.; Kondoh, M.; Kunisawa, J. Development of adjuvant-free bivalent food poisoning vaccine by augmenting the antigenicity of *Clostridium perfringens* enterotoxin. *Front. Immunol.* **2018**, *9*, 2320. [CrossRef]
100. Mankertz, J.; Amasheh, M.; Krug, S.M.; Fromm, A.; Amasheh, S.; Hillenbrand, B.; Tavalali, S.; Fromm, M.; Schulzke, J.D. TNF alpha up-regulates claudin-2 expression in epithelial HT-29/B6 cells via phosphatidylinositol-3-kinase signaling. *Cell Tissue Res.* **2009**, *336*, 67–77. [CrossRef]
101. Furuse, M.; Furuse, K.; Sasaki, H.; Tsukita, S. Conversion of zonulae occludentes from tight to leaky strand type by introducing claudin-2 into Madin-Darby canine kidney I cells. *J. Cell Biol.* **2001**, *153*, 263–272. [CrossRef]
102. Pardridge, W.M. The blood-brain barrier: Bottleneck in brain drug development. *NeuroRx* **2005**, *2*, 3–14. [CrossRef]
103. Vanlandewijck, M.; He, L.; Mae, M.A.; Andrae, J.; Ando, K.; Del Gaudio, F.; Nahar, K.; Lebouvier, T.; Lavina, B.; Gouveia, L.; et al. A molecular atlas of cell types and zonation in the brain vasculature. *Nature* **2018**, *554*, 475–480. [CrossRef]
104. Nitta, T.; Hata, M.; Gotoh, S.; Seo, Y.; Sasaki, H.; Hashimoto, N.; Furuse, M.; Tsukita, S. Size-selective loosening of the blood-brain barrier in claudin-5-deficient mice. *J. Cell Biol.* **2003**, *161*, 653–660. [CrossRef]
105. Sohet, F.; Lin, C.; Munji, R.N.; Lee, S.Y.; Ruderisch, N.; Soung, A.; Arnold, T.D.; Derugin, N.; Vexler, Z.S.; Yen, F.T.; et al. LSR/angulin-1 is a tricellular tight junction protein involved in blood-brain barrier formation. *J. Cell Biol.* **2015**, *208*, 703–711. [CrossRef]
106. Zeniya, S.; Kuwahara, H.; Daizo, K.; Watari, A.; Kondoh, M.; Yoshida-Tanaka, K.; Kaburagi, H.; Asada, K.; Nagata, T.; Nagahama, M.; et al. Angubindin-1 opens the blood-brain barrier in vivo for delivery of antisense oligonucleotide to the central nervous system. *J. Control. Release* **2018**, *283*, 126–134. [CrossRef]
107. Fujita, H.; Hamazaki, Y.; Noda, Y.; Oshima, M.; Minato, N. Claudin-4 deficiency results in urothelial hyperplasia and lethal hydronephrosis. *PLoS ONE* **2012**, *7*, e52272. [CrossRef]
108. Muto, S.; Hata, M.; Taniguchi, J.; Tsuruoka, S.; Moriwaki, K.; Saitou, M.; Furuse, K.; Sasaki, H.; Fujimura, A.; Imai, M.; et al. Claudin-2-deficient mice are defective in the leaky and cation-selective paracellular permeability properties of renal proximal tubules. *Proc. Natl. Acad. Sci. USA* **2010**, *107*, 8011–8016. [CrossRef]
109. Gong, Y.F.; Himmerkus, N.; Sunq, A.; Milatz, S.; Merkel, C.; Bleich, M.; Hou, J.H. ILDR1 is important for paracellular water transport and urine concentration mechanism. *Proc. Natl. Acad. Sci. USA* **2017**, *114*, 5271–5276. [CrossRef]
110. Hadj-Rabia, S.; Baala, L.; Vabres, P.; Hamel-Teillac, D.; Jacquemin, E.; Fabre, M.; Lyonnet, S.; De Prost, Y.; Munnich, A.; Hadchouel, M.; et al. Claudin-1 gene mutations in neonatal sclerosing cholangitis associated with lchthyosis: A tight junction disease. *Gastroenterology* **2004**, *127*, 1386–1390. [CrossRef]
111. Arinami, T. Analyses of the associations between the genes of 22q11 deletion syndrome and schizophrenia. *J. Hum. Genet.* **2006**, *51*, 1037–1045. [CrossRef]
112. Wei, J.; Hemmings, G.P. A study of the combined effect of the CLDN5 locus and the genes for the phospholipid metabolism pathway in schizophrenia. *Prostaglandins Leukot. Essent. Fatty Acids* **2005**, *73*, 441–445. [CrossRef]
113. Sun, Z.Y.; Wei, J.; Xie, L.; Shen, Y.; Liu, S.Z.; Ju, G.Z.; Shi, J.P.; Yu, Y.Q.; Zhang, X.; Xu, Q.; et al. The CLDN5 locus may be involved in the vulnerability to schizophrenia. *Eur. Psychiatry* **2004**, *19*, 354–357. [CrossRef]
114. Tanaka, H.; Imasato, M.; Yamazaki, Y.; Matsumoto, K.; Kunimoto, K.; Delpierre, J.; Meyer, K.; Zerial, M.; Kitamura, N.; Watanabe, M.; et al. Claudin-3 regulates bile canalicular paracellular barrier and cholesterol gallstone cores formation in mice. *J. Hepatol.* **2018**, *69*, 1308–1316. [CrossRef]
115. Furuse, M.; Sasaki, H.; Tsukita, S. Manner of interaction of heterogeneous claudin species within and between tight junction strands. *J. Cell Biol.* **1999**, *147*, 891–903. [CrossRef]

116. Knipp, G.T.; Ho, N.F.H.; Barsuhn, C.L.; Borchardt, R.T. Paracellular diffusion in Caco-2 cell monolayers: Effect of perturbation on the transport of hydrophilic compounds that vary in charge and size. *J. Pharm. Sci.* **1997**, *86*, 1105–1110. [CrossRef]
117. Watson, C.J.; Rowland, M.; Warhurst, G. Functional modeling of tight junctions in intestinal cell monolayers using polyethylene glycol oligomers. *Am. J. Physiol. Cell Physiol.* **2001**, *281*, C388–C397. [CrossRef]
118. Steinke, A.; Meier-Stiegen, S.; Drenckhahn, D.; Asan, E. Molecular composition of tight and adherens junctions in the rat olfactory epithelium and fila. *Histochem. Cell Biol.* **2008**, *130*, 339–361. [CrossRef]
119. Schlingmann, B.; Overgaard, C.E.; Molina, S.A.; Lynn, K.S.; Mitchell, L.A.; Dorsainvil White, S.; Mattheyses, A.L.; Guidot, D.M.; Capaldo, C.T.; Koval, M. Regulation of claudin/zonula occludens-1 complexes by hetero-claudin interactions. *Nat. Commun.* **2016**, *7*, 12276. [CrossRef]
120. Markov, A.G.; Veshnyakova, A.; Fromm, M.; Amasheh, M.; Amasheh, S. Segmental expression of claudin proteins correlates with tight junction barrier properties in rat intestine. *J. Comp. Physiol. B* **2010**, *180*, 591–598. [CrossRef]
121. Inai, T.; Sengoku, A.; Guan, X.; Hirose, E.; Iida, H.; Shibata, Y. Heterogeneity in expression and subcellular localization of tight junction proteins, claudin-10 and -15, examined by RT-PCR and immunofluorescence microscopy. *Arch. Histol. Cytol.* **2005**, *68*, 349–360. [CrossRef]
122. Rahner, C.; Mitic, L.L.; Anderson, J.M. Heterogeneity in expression and subcellular localization of claudins 2, 3, 4, and 5 in the rat liver, pancreas, and gut. *Gastroenterology* **2001**, *120*, 411–422. [CrossRef]
123. Kiuchi-Saishin, Y.; Gotoh, S.; Furuse, M.; Takasuga, A.; Tano, Y.; Tsukita, S. Differential expression patterns of claudins, tight junction membrane proteins, in mouse nephron segments. *J. Am. Soc. Nephrol.* **2002**, *13*, 875–886.
124. Steinemann, A.; Galm, I.; Chip, S.; Nitsch, C.; Maly, L.P. Claudin-1, -2 and -3 are selectively expressed in the epithelia of the choroid plexus of the mouse from early development and into adulthood while claudin-5 is restricted to endothelial cells. *Front. Neuroanat.* **2016**, *10*, 16. [CrossRef]
125. He, L.Q.; Vanlandewijck, M.; Mae, M.A.; Andrae, J.; Ando, K.; Del Gaudio, F.; Nahar, K.; Lebouvier, T.; Lavina, B.; Gouveia, L.; et al. Single-cell RNA sequencing of mouse brain and lung vascular and vessel-associated cell types. *Sci. Data* **2018**, *5*, 180160. [CrossRef]
126. Watari, A.; Kodaka, M.; Matsuhisa, K.; Sakamoto, Y.; Hisaie, K.; Kawashita, N.; Takagi, T.; Yamagishi, Y.; Suzuki, H.; Tsujino, H.; et al. Identification of claudin-4 binder that attenuates tight junction barrier function by TR-FRET-based screening assay. *Sci. Rep.* **2017**, *7*, 14514. [CrossRef]

14

Role of Claudin Proteins in Regulating Cancer Stem Cells and Chemoresistance-Potential Implication in Disease Prognosis and Therapy

Saiprasad Gowrikumar [1], Amar B. Singh [1,2,3] and Punita Dhawan [1,2,3,*]

1 Department of Biochemistry and Molecular Biology, University of Nebraska Medical Center, Omaha, NE 68198-5870, USA; sai.gowrikumar@unmc.edu (S.G.); amar.singh@unmc.edu (A.B.S.)
2 VA Nebraska-Western Iowa Health Care System, Omaha, NE 68105, USA
3 Department of Biochemistry and Molcular Biology, Fred and Pamela Buffet Cancer Center, University of Nebraska Medical Center, Omaha, NE 68105, USA
* Correspondence: punita.dhawan@unmc.edu.

Abstract: Claudins are cell–cell adhesion proteins, which are expressed in tight junctions (TJs), the most common apical cell-cell adhesion. Claudin proteins help to regulate defense and barrier functions, as well as differentiation and polarity in epithelial and endothelial cells. A series of studies have now reported dysregulation of claudin proteins in cancers. However, the precise mechanisms are still not well understood. Nonetheless, studies have clearly demonstrated a causal role of multiple claudins in the regulation of epithelial to mesenchymal transition (EMT), a key feature in the acquisition of a cancer stem cell phenotype in cancer cells. In addition, claudin proteins are known to modulate therapy resistance in cancer cells, a feature associated with cancer stem cells. In this review, we have focused primarily on highlighting the causal link between claudins, cancer stem cells, and therapy resistance. We have also contemplated the significance of claudins as novel targets in improving the efficacy of cancer therapy. Overall, this review provides a much-needed understanding of the emerging role of claudin proteins in cancer malignancy and therapeutic management.

Keywords: claudins; cancer; stem cell; chemoresistance

1. Introduction

1.1. Tight Junctions

Tight junctions (TJs) are the sites where tissues interface directly with the external environment or internal compartments that are contiguous with the external environment and are lined by mucosal surfaces, where epithelial cells act insulation for the internal organ. These structures not only provide a protective layer but also act as a selective barrier between the body and the gut lumen that restricts free exchange across the paracellular space [1,2]. There are three main transport mechanisms across the epithelial layers, which include the trans-cellular pathway (passive diffusion), carrier dependent pathway (carrier or receptors), and the paracellular pathway (passage through spaces between cells). Among these transport mechanisms, the apical junctional complex, a crucial factor for the paracellular pathway, is composed of three junctions from apical to basal are known as the tight junction (zonula occludens), adherens junction (zonula adherens), and desmosome (macula adherens) [3]. The TJs are intercellular junctions, which act as permeability barriers in epithelial cells [1]. The tight junction proteins are diverse and include occludins (the first one to be found), claudins, tricellulin, cingulin, and junctional adhesion molecules (JAM). These proteins interact within themselves and with the cellular cytoskeleton to form a complex architecture [4–8]. Among these TJ proteins, claudins are key proteins, acting as both pores and barriers, aiding the paracellular pathway between epithelial cells [9,10].

1.2. *Claudins*

The functionalities of claudins are as follows: (1) Fence function, responsible for maintaining polarity by differentiating apical and basolateral cell domains; (2) Signaling molecule, involved in cell growth, survival, proliferation, and differentiation; (3) Barrier function, this gate function separates compartments with fluids to avoid intermixing [11]. Claudins were identified as a major integral membrane protein by Tsukita and his colleagues in 1998, before which the only known tight junction protein was occludin [12,13]. Studies conducted to overexpress claudins in fibroblasts, which do not have tight junctions, were able to reconstitute tight junction-like networks of strands, which shows the importance of claudins in tight junction assembly [14]. Several claudin isoforms have been identified in mammals. These have high sequence homology in the first and fourth transmembrane domains and extracellular loops. Further, the homologous classic claudins include claudins 1–10, 14, 15, 17, and 19, and non-classical claudins comprised of claudins 11–13, 16, 18, and 20–27, which are less homologous [15].

The structure of claudins is comprised of four transmembrane domains, the intracellular N and C termini, and the two extracellular loops (ECLs). The claudin structure encompasses N-termini (7 amino acids), C-termini (25–55 amino acids), and loops containing 25–55 aminoacids. The ECLs are involved in barrier and pore formations. There are two ECLs, ECL1 consists of ~50 amino acids with two conserved cysteines involved in the barrier function. Negative and positive charges in ECL1 contribute to pore formation. The schematic representation of the structure of claudins and its classification is depicted (Figure 1). The ECL2 is responsible for homo and heterotypic interactions and was recently shown to be involved in host cell binding and cytotoxicity for the *Clostridium perfringens* enterotoxin. The ECL2 usually has ~25 amino acids, but fewer in claudin-11 and more in claudin-18 [16]. Claudins interact with other TJ-associated proteins through carboxy-terminal tails, which contain a PDZ-domain binding motif [17].

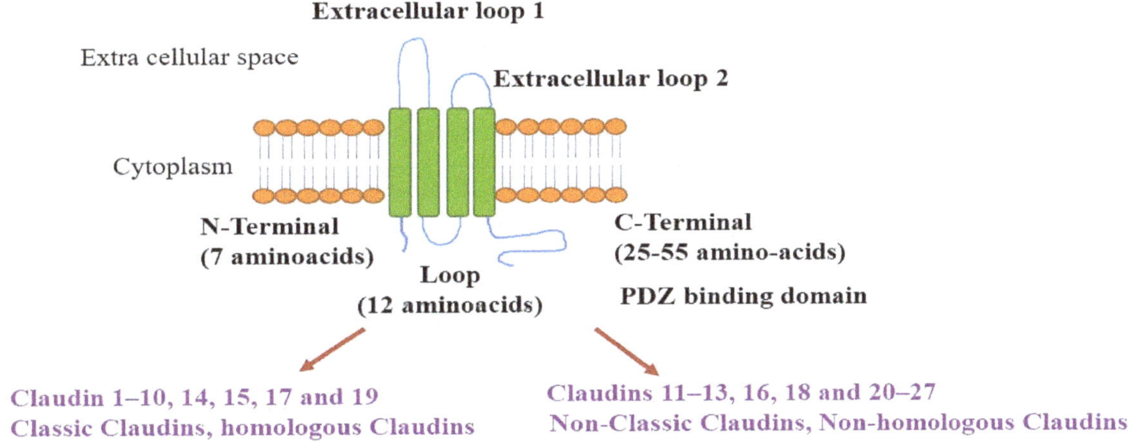

Figure 1. Structural organization of claudin proteins (monomer), and its classification based on homologous sequences between them. Colour code: Green- transmembrane domains; Orange: Bilipid layer, Blue–Extracellular loops/N and C termini.

2. Claudins as Oncogenic Signal Transducer

The expression of claudins varies among different tissue types [18]. As an important structure in regulating paracellular permeability, claudin overexpression influences trans-epithelial resistance (TER) and ion permeability [19–22]. Aberrant expressions of claudins have been reported in various cancers. Some of the claudins known to be frequently dysregulated in cancers are claudin-1, -3, -4, and -7 [23]. A large body of evidence highlights claudins as pro and anti-tumorigenic factors [24–31]. The potential of claudins to act as proto-oncogene or tumor promotor in various cancers are summarized in Table 1. In addition, several recent studies have also demonstrated the importance of claudins as tumor suppressors [24–31]. A recent study by Chang et al. in 2019 provided evidence for intestinal hyperplasia and adenomas in claudin-7 knockdown mice [32]. Consistent with this, claudin-7 was downregulated in colon cancer patient samples as compared to normal tissue [33]. These effects of claudin-7 were achieved by inhibiting phosphorylation and nuclear localization of Akt. Conversely, claudin-7 association with Epithelial cell adhesion molecule (EPCAM) supports proliferation, upregulation of anti-apoptotic proteins, and drug resistance [33]. Claudin-18 knockout mice spontaneously developed lung adenocarcinomas, and its mRNA expression was decreased in lung adenocarcinomas. Claudin-18 inhibits Akt signaling through modulation of yes-associated protein/Taz (Yap/Taz) and insulin-like growth factor (IGF-1R) signaling in lung cancer [34]. Further, the depletion of claudin-3 induced tumor burden by enhancing β-catenin activity through (IL)-6/STAT3 signaling in colon cancer [35]. Yet another study by Che et al. in 2018 [36] identified claudin-3 as a suppressor of lung squamous cell carcinoma cells, in which overexpression of claudin-3 inhibited invasion, migration, and EMT of lung squamous cell carcinoma. Similarly, claudin-4 accelerates cell migration and invasion in ovarian tumor cell lines, in support of this, peptide-mediated silencing of claudin-4 in ovarian cancer cells exhibited lower tumor burden [37]. Claudin-6 was shown to be a tumor suppressor through genetic manipulation studies in cervical carcinoma cells wherein loss of claudin-6 exacerbated cell proliferation and tumor growth [38,39]. An array of articles from Dhawan et al., have proved a significant role of claudin-1 as a tumor promoter in colon cancer [40,41]. In one of their reports, increased claudin-1 expression was causally associated with metastasis [40]. In contrast to claudin-1, claudin-7 has an inverse role on EMT, wherein it causes mesenchymal to epithelial transformation (MET) in Rab25 dependent manner to combat colon cancer [42]. Similarly, claudin-2 is upregulated in colon cancer and is involved in cancer progression. Claudin-2 suppression in colon cancer cells has led to decreased cell proliferation through the modulation of EGF signaling [43]. Opposite colon cancer, claudin-1 is frequently down-regulated in invasive human breast cancer. Recently, mutations of claudin-1 have been reported in breast cancer, which has led to claudin-1 transcript variants shorter than classical claudin-1 transcript [44]. Taken together, it appears that the deregulated claudin composition in any given epithelial cells sheet may modify the signaling and associated changes in protein partnering to modulate oncogenesis.

Table 1. Claudins as tumour promotor/suppressor.

Claudins Subtype	Cancer Type	Proto-Oncogene	Reference
Claudin-6	Gastric cancer	Tumour promotor	[25]
Claudin-1	Colon cancer	Tumour promotor	[40,45]
Claudin-3	Ovarian cancer	Tumour promotor	[22]
Claudin-4	Ovarian cancer	Tumour promotor	[22]
Claudin-6	Breast cancer, Gastric cancer	Tumour promotor	[26,27]
Claudin-7	Colon cancer	Tumour promotor	[46]
Claudin-2	Lung cancer	Tumour promotor	[28]
Claudin-1	Gastric cancer	Tumour suppressor	[47]
Claudin-1	Lung cancer	Tumour suppressor	[29]
Claudin-3	Ovarian cancer	Tumour suppressor	[31]
Claudin-4	Ovarian cancer	Tumour suppressor	[31]
Claudin-7	Lung cancer	Tumour suppressor	[30]
Claudin-11	Gastric cancer	Tumour suppressor	[48]
Claudin-2	Osteosarcoma	Tumour suppressor	[49]

To glimpse how claudins can achieve its pro or anti-tumorigenic effect, understanding the regulation of claudins in normal and cancer cells is essential. Recently it has been demonstrated that claudins are not a static and rigid seal of the paracellular space; rather, they are dynamically capable of responding to various biochemical and mechanical stimuli through reshaping and remodeling [50,51]. Epigenetic regulation of claudins has recently gained significant importance. The claudin-3 promotor is known to possess low DNA methylation and high histone H3 acetylation for its expression in ovarian cancer cells [52]. DNA hypomethylation of the claudin-4 promotor is an important factor for its high expressions in gastric cancer [53]. Downregulation of claudin 1 via DNA promoter methylation is reported in estrogen receptor-positive breast cancer [54]. Claudins are also regulated at the transcriptional level by different transcription factors. A study has reported novel post-transcriptional regulation of claudin-1 in colon cancer cells [55], the authors documented the role of histone deacetylase (HDAC)-dependent histone acetylation as a key post-transcriptional regulation over claudin-1 expression, as found through HDAC inhibitor studies. Studies demonstrate the interaction of Slug and Snail (transcriptional factors) with the E-box element in the claudin-1 promoter causes inhibition of claudin-1. Snail is known to act as a transcription factor causing repression of E-cadherin (E-CAD) and has a potential role in promoting tumorigenesis. Slug is also a pivotal transcription factor involved in cell migration during embryogenesis and in tumor cell invasion and migration [56]. Yet another transcriptional factor known to be associated with claudin-1 is Runt-related transcription factor 3 (RUNX3), which is a gastric tumor suppressor [47]. Caudal homeobox proteins (Cdx1 & Cdx2) and GATA binding protein 4, GATA4) are known activators of claudin-1 promoters in colon cancer [57]. Sp1 is a transcriptional factor known to regulate claudin-3 and claudin-4 promoter activity in ovarian cancer [52,58]. Apart from these transcriptional regulations, claudins are also known to be regulated by post-translational modifications involved in their protein localization, interaction with other proteins, and overall turnover [59,60]. The post-translational modification of claudins includes palmitoylation, O-glycosylation, and phosphorylation [61,62]. Phosphorylation is one of the key regulatory modifications for the regulation of intracellular localization and degradation of claudins.

Claudins are phosphorylated by many different enzymes like protein-kinase A/C, protein phosphatase 2A and mitogen-activated protein kinase (MAPK) [63,64]. The localization or dissociation of claudins to TJs is regulated by phosphorylation. For phosphorylated claudin-1, -5, and -16 are localized in the TJs while in contrast, phosphorylated claudin-3 and -4 dissociate from TJs [64–66]. Furthermore, the rho family of small dimeric G proteins mediated phosphorylation of claudin-5 at T207 was recently reported [67]. The phosphorylation of claudin-1 at different serine sites (192, 205, 206, and T191) regulates its assembly at tight junctions [68]. The cAMP-dependent protein kinase (PKA) is known to phosphorylate of claudin-3 at amino acid 192 at the C terminus. Claudin-4 is phosphorylated by atypical PKC (aPKC) at serine195 [65]. Another important posttranslational modification playing a key role in claudin regulation is palmitoylation. Emerging articles have demonstrated the importance of palmitoylation in claudin localization into tight junctions. In claudin-5 self-assembly, palmitoylation restricts specific protein-protein conformations, as reported by Rajagopal et al. [61]. Claudin-7 interacts directly with EpCAM along the basal membrane. Palmitoylation regulates the ability of claudin-7 to interact with integrins, recruiting EpCAM, and concomitantly associate with the actin cytoskeleton [69].

3. Claudins and Stem Cells

Stem cells are crucial for the development and homeostasis of many different tissues. Stem cells are also involved in cell replacement therapies in the case of cell damage or degeneration [70]. Pluripotency of stem cells is defined as self-renewing and differentiating potential into all three germ layers. Human pluripotent stem cells are very promising in regenerative medicine. The stem cell further differentiates into a wide variety of cells under the influence of diverse signaling molecules, growth factors, and transcription factors [71]. Recent research is focused on understanding the signals, which maintain pluripotency or differentiation potential. Various intrinsic and extrinsic factors are involved in stem cell maintenance, self-renewal, and differentiation [71]. On the other hand, stem cells are also an important factor for many tumors. Dysregulated pluripotent stem cells in tumors are more aggressive and have the potential to reform the whole tumor [72]. Thus, it becomes important to selectively remove undifferentiated human pluripotent stem cells (hPSCs) from differentiated cultures. For achieving this, selective pluripotent-specific cell surface markers are needed, which can separate undifferentiating from the differentiated cells. While screening for a highly specific marker protein specific for the undifferentiated hPSCs, Uri Ben-David et al. [73] found claudin-6 to be highly specific for undifferentiated hPSCs. The expression of claudin-6 was 90-fold higher in undifferentiated hPSCs than in differentiated cells. The proof for the involvement of claudins in epithelial differentiation from embryonic stem cells was first reported by Sugimoto et al. [74], where the potential of claudin-6 to trigger epithelial morphogenesis in mouse stem cells was reported. Also, claudin-6 regulated other tight-junction and microvillus molecules claudin-7, occludin, Zonula occludens (ZO-1α+), and ezrin/radixin/moesin-binding phosphoprotein50, which strongly proved the role of claudins in epithelial differentiation [74]. This was further supported by other studies, which also showed the expression of claudin-6 is an early marker in embryonic stem cells [75,76]. Differentiation of Human Embryonic Stem Cells to Hepatocyte-Like Cells resulted in a decrease in stem cell markers Oct3/4 and Nanog as expected. Along with stem cell markers, claudin-1 declined eventually, whereas claudin-4 increased and was highest at the end stage of differentiation [77].

A growing body of evidence focuses on cancer stem cells in cancer biology. The drawbacks of cancer treatment failures and drug resistance are proved to originate from cancer stem cells, which are a small subpopulation in tumors. Recently the factors regulating cancer stem cells have gained significant importance and opened new avenues for targeted therapies and thus decrease the chance of recurrence of the disease [78]. Cancer stem cells (CSCs) represent a small group of cells in typically heterogeneous tumors, which possess tumor-initiating and self-renewal properties, giving rise to non-tumorigenic progeny. CSCs are enriched after chemotherapy and lead to therapy failure and thus recurrence of cancer. The role of CSCs in tumor relapse, metastasis, and therapeutic refractoriness is well described [79]. The role of claudins in cancer stem cell (CSC) biology is gaining much attention. The WNT pathway is well known to provide the key signals for achieving this particular phenotype. It is also established that the *Wnt* signal transduction pathway is important in normal and malignant stem cells [80]. Recent articles have highlighted the link between claudin and the *Wnt*/β-catenin signaling pathway and the role of CSC in this cross-talk. Claudin-1 and claudin-2 transcription is regulated by WNT*t* signaling, and they are known to regulate the β-Catenin- T-cell factor/lymphoid enhancer-binding factor (TCF/LEF) signaling pathway to regulate CSC [81,82]. In contrast, other claudins negatively regulate WNTsignaling cascades, such as loss of claudin-3 inducing WNT/β-catenin activation, thus aiding in the promotion of colon cancer [35]. Darido et al. provided evidence for Tcf-4 and Sox-9 regulating the expression of claudin-7 [46]. In addition, studies by Prat et al. discovered a new claudin-low molecular subtype of breast cancer [83]. The key characteristics of this subtype are low expression of tight junction and junction adherens proteins (claudin-3, -4 and -7, and E-cadherin), and enriched in stem cell and EMT features. Patients having high-grade invasive ductal breast carcinoma in this subgroup had a poor prognosis, absence of luminal differentiation markers, enhanced EMT markers, expression of immune response genes, and most closely resembled mammary epithelial stem cells. This suggested that low claudin cells might emerge from more immature stem or progenitor cells and comprise cancer stem cells. Thus, identification of the low claudin subtype in breast cancer has shown the potential of claudins in regulating stem cells. In addition, claudin-3 is known to play an oncogenic role in non-small cell lung cancer (NSCLC). One of the major contributing factors for the role of claudin-3 is regulation of cancer stemness and chemoresistance in non-squamous NSCLC. The depletion of claudin-3 was able to combat the formation of spheres and tumor formation as well as increased sensitivity to cisplatin [84]. Further, claudin-3 inhibition by small-molecule inhibitors including withaferin A, estradiol and fulvestrant, suppressed cancer stemness and combated chemoresistance, giving strong evidence for the role of claudin-3 in inducing stemness. Another claudin playing an important role in stem cell regulation is claudin-18 in lung cancer [85], which has been reported to have a role in the aberrant proliferation of alveolar epithelial type II (AT2) cells, resulting in lung enlargement and parenchymal expansion by restrictions on stem/progenitor cell proliferation. Recently, claudin-2 was shown to be restricted in the stem/progenitor cell compartment of intestinal crypts. It enriches aldehyde dehydrogenase $ALDH^{High}$ cancer stem-like cells in heterogeneous colorectal cancer cell populations through the regulation of Yes-associated protein (YAP) activity and miR-222-3p expression [86]. Overall, these studies give an overview of the potential role of claudins in stem cell biology. The role of claudins in the regulation of stem cells is summarized in Table 2. The claudin mediated enrichment of stem cells provides a new axis-of-evil for a preferential therapeutic target, which has potential clinical consequences.

Table 2. Claudins and stemness.

Claudin Subtype	Stem Cell Related Functions	References
Claudin-6	Early marker in embryonic stem cell.High expression in undifferentiated human pluripotent stem cells (hPSCs). Trigger epithelial morphogenesis in mouse stem cells.	[73,74]
Claudin-1 and 2	Known to regulate the β-Catenin-TCF/LEF signaling pathway to regulate CSC.	[81]
Claudin low subtype in breast cancer	Enriched in stem cells and more EMT.	[83]
Claudin-3	Regulation on cancer stemness and chemoresistance in non-small cell lung cancer (NSCLC).	[84]
Claudin-18	Triggers lung enlargement and parenchymal expansion by restrictions on stem/progenitor cell proliferation.	[85]
Claudin-2	Enrich $ALDH^{High}$ cancer stem-like cells in heterogeneous colorectal cancer cell populations.	[86]

4. Claudins in Chemoresistance

Most cancer patients initially respond to chemotherapy. Eventually, cancer relapses due to chemoresistance resulting in treatment failure causing death. The mechanisms of chemoresistance in cancers are still largely unknown [87]. Since the role of claudins in the regulation of cancer stem cells is well documented, their correlation with drug resistance and distant metastasis is inevitable and obvious [49,88,89]. In brief, claudin-3 and -6 are correlated with lymph node metastasis in squamous cell lung carcinomas [90,91]. Claudin-4 is highly expressed in primary and metastatic prostate cancer [92] and gastric cancer [93,94]. Claudin-1 and -7 have proved to have an inverse role in colon cancer, wherein claudin-1 elevates the metastasis of colon cancer cells. On the other hand, suppression of claudin-7 leads to liver metastasis [40,42]. Epithelial to mesenchymal transition (EMT) is a piece of vital machinery responsible for invasiveness and initiation of metastasis and chemoresistance of cancer cells. Claudins are known inducers of EMT in cancers. Claudin-1 is known to induce EMT in colon, liver, nasopharyngeal carcinoma, and breast cancers [40,95,96]. At the same time, claudin-7 is reported to be involved in establishing MET in colon cancer [36,42]. Claudin-3 suppresses EMT in lung cancer cells [36]. Overall, the potential role of claudins in EMT, Metastasis and CSC enrichment provides the rationale for exploring them as a key factor in establishing drug resistance. Claudins as chemo-resistance modulators is an emerging field of research. In a recent article, the potential of claudin-6 in enhancing chemoresistance to Adriamycin in triple-negative breast cancer (TNBC) was documented [97]. This effect of claudin-6 was mediated through its regulation over the AF-6/extracellular signal–regulated kinases (AF-6/ERK signaling pathway and up-regulation of cancer stem cells. Claudin-3 is also identified as a molecule to combat cisplatin chemoresistance in non-squamous lung carcinoma [84]. Here, claudin-3 overexpressing lung cancer cells were insensitive to cisplatin treatment compared to control cells. Adding to this, knockdown of claudin-3 or claudin-4 in ovarian cancer cells induced resistance to cisplatin by the regulation of Cu transporter CTR1 [98]. Another study by a Japanese group of researchers reported a high expression of claudin-4 in the ovarian cancer tissues of platinum-resistant patients [99]. In lung cancer, claudin-1 is a key deciding factor for metastasis and a responsible factor for drug resistance towards cisplatin through the up-regulation of Unc-51 Like Autophagy Activating Kinase 1 (ULK1) phosphorylation [100,101]. It is also known to enhance drug resistance in liver cancer cells by modulating autophagy to achieve drug resistance. The role of claudin-7 in drug resistance [102] has also been reported, wherein decreased drug resistance, increased apoptosis and diminished anti-apoptotic PI3K/Akt pathway was achieved by knocking down claudin-7, proving the potentiality in chemo-resistance [103]. It is well known that EpCAM associates with claudin-7 and is known to be involved in cancer metastasis. Florian et al. [69] have provided evidence for the increased migratory potential of pancreatic cancer cells upon EPCAM and claudin-7 association influencing cell-cell adhesion. Interestingly, the EPCAM and claudin-7 association seems to enhance drug resistance against cisplatin through enhancing MAPK and c-Jun N-terminal kinases (JNK) pathways. Altogether, these studies indicate the important role of claudins contributing to drug resistance in cancer cells. The pictorial representation of the role of claudins as a stem cell regulator and its impact in chemoresistance is shown in Figure 2.

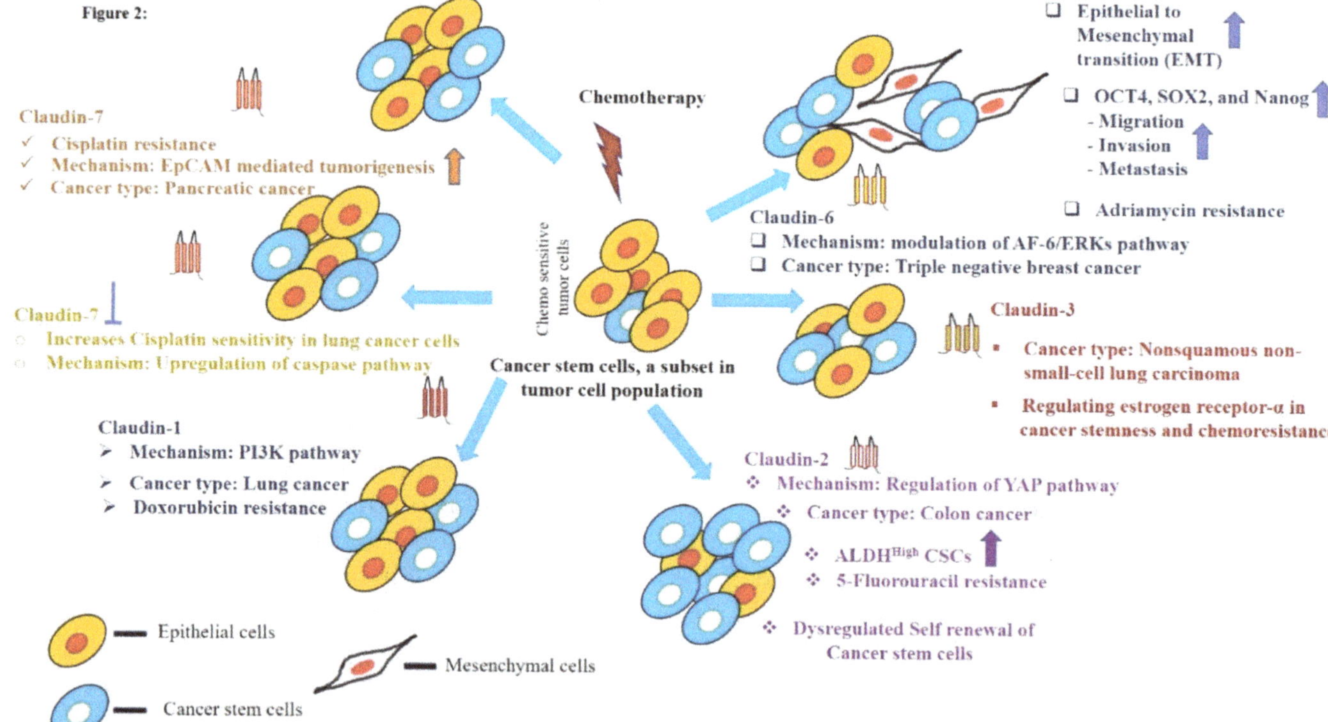

Figure 2. The central role of claudins in the regulation of epithelial to mesenchymal transition (EMT), cancer stem cells, and chemoresistance in various cancers. ⊥- inhibition of claudin-7. The arrows indicate the upregulation and higher enrichment of the mentioned signaling molecules, colour is respective of each claudin.

5. Claudins in Prognosis

Emerging data defining mechanisms through which claudins augment cancer metastasis provides the rationale for exploring claudins as prognostic factors and therapeutic targets in cancer. The importance of claudins is established using cancer cell models, mouse models, and human patient samples. Target molecules for cancer surveillance in high-risk populations are desperately warranted. As a vital emerging modulator in molecular or cellular pathways related to cancers, claudins could be targeted or used as biomarkers for prognosis, diagnosis, and treatments. A number of recent studies have projected a role for claudins as key prognosis factors in cancers. In one of the study Lechpammer et al. [104] demonstrated the potential of claudins as a diagnostic and prognostic factor in renal cell carcinoma. Claudin-1, -3, -4, -7, and -8 were studied in human renal cell carcinomas and oncocytomas. The data from their research showed an inverse correlation between claudin-3 and -4 expression with overall survival in clear cell renal cell carcinomas, and these claudins could be considered for prognosis in renal cell carcinomas. Claudin-7 and 8 can be implied as useful markers in the identification of renal cell carcinomas from oncocytomas [105].

Claudin-6 was reported as a prognosis factor in NSCLC patients. In this report, the patients with low claudin-6 had a lower survival rate than the patients with high claudin-6. [91] reported low claudin-6 as an independent indicator of prognosis in NSCLC patients. In this study, they documented low claudin-6 in 61 of 123 NSCLC tissue samples, and patients with low claudin-6 expression correlated with lower survival rates than those with high claudin-6 expression. The influence of claudin-3, claudin-7, and claudin-18 in gastric cancer patients were also studied [106]. Claudin-3 and claudin-7 were expressed in 25.4% and 29.9% of the gastric cancer tissues, respectively. However, 51.5% of gastric cancer tissues exhibited reduced expression of claudin-18. Claudin-7 expression correlated with shorter overall survival in gastric cancer patients, while the overall survival was increased in patients with claudin-18 expression. Recently, claudin-3 and -7 are also considered as novel prognostic factors in triple-negative breast cancer (TNBC) through its aberrant immunohistochemical expressions [107]. Claudin-3 cytoplasmic expression is an indicator of poor survival in triple-negative breast cancer. In addition, epigenetic modifications of claudins are reported to be a promising prognosis marker of various cancers. Zhenzhen et al. [106] recently demonstrated that the methylation of claudin-3 is a prognostic factor in gastric adenocarcinoma.

Further, the serum levels of claudin-7 among patients with colorectal cancer (CRC) was significantly reduced and correlated with high tumor stage and high carcinoembryonic antigen levels [108]. Claudin-7 was found to be downregulated in CRC, as reported by Bhat et al. [42], and associated with diminished EMT and tumor progression. These studies give a strong rationale to consider claudin-7 as a biomarker for predicting the development, proliferation, and prognosis of CRC. A claudin-low molecular subtype of breast cancer has been described with a concomitant upregulation of several EMT markers and an enrichment in stem cell features [109]. In an interesting article by Danzinger et al., the importance of claudin-3 in triple-negative breast cancer (TNBC) was documented. It was reported that claudin-3 expression was correlated with a Breast cancer type 1 (BRCA1) mutation [107]. This could help in guiding the decision for BRCA testing for triple-negative breast cancer (TNBC). Also, the expression of claudin-11 has been suggested as a biomarker for advanced-stage cutaneous squamous carcinoma, and reflects the distinct stages of tumor development and differentiation [110]. The clinical significance of claudin-11 was addressed in Laryngeal Squamous Cell Carcinoma (LSCC) by Nissinen et al. [110]. In this study, elevated promoter methylation of claudin-11 in tumor tissues was observed. Patients with lymph node metastasis with an advanced clinical stage showed more methylation in the claudin-11 promoter, which associated with poor overall survival of LSCC patients. In TNBCs, claudin-1, -3, -4, and -7 higher expression rates are more frequent than in other subtypes [111]. Claudin-4 high/claudin-1 low, claudin-4 high/claudin-7 low, and claudin-4 high/claudin-1 low/claudin-7 low types were also significantly correlated with lymph node metastasis, and showed worse survival. Apart from this, a recent article from Upadhaya et al. documented the therapeutic potential of claudin-1 in oral epithelial dysplasia and oral squamous cell carcinoma [112]. Overall the differential expression pattern of claudins may reflect the distinct stages of tumor development and differentiation and have been implied as prognostic factors for early determination of the tumor state.

6. Claudins as Therapeutic Agents

So far, over 100 monoclonal antibody (mAb) products are in clinical trials [113]. In an oncology setting, these monoclonal antibodies can mediate antibody-dependent cellular cytotoxicity (ADCC) and complement-dependent cytotoxicity (CDC) against cancer cells [114]. There is a long-lasting history of antibody-mediated targeting of claudin-1 against hepatitis C virus (HCV) infections, and wherein many researchers have provided proof for the importance of claudins in HCV infections as viral entry point [115,116]. A study by Fofana et al. [117] designed monoclonal antibodies against claudin-1 to combat HCV entry. It was promising to see the antibodies raised against claudin-1 was able to block HCV entry. A recent study by Colpitts et al. has documented the humanization of a claudin-1-specific monoclonal antibody and was investigated in a large panel of primary human hepatocytes, and was found to be very promising for clinical HCV prevention and cure [118]. These studies hold significance because these antibodies could prevent HCV infection after liver transplantation, and virus spread in chronically infected patients. These antibodies are now being tested in cancer models. Claudins, as a potential target in antibody-based therapies for carcinomas, was investigated by Offner et al. [119]. In this study, the antibodies were raised against the extracellular domains of claudin-1, -3, and -4. Recently Romani et al. engineered a fully human anti-claudin-3 IgG1 antibody (IgGH6) [120], which is specific to claudin-3 and no cross-reactivity with other claudins was observed. Recent work by Cherradi et al. [121] investigated the importance of claudin-1 in different colorectal cancer (CRC) molecular subtypes. There is a differential expression pattern of claudin-1 based on the subtype. A murine monoclonal antibody against the extracellular part of human claudin-1 (6 F6 mAb) was designed and generated, which was specifically able to pick claudin-1 positive CRC cell lines, and no other cross-reactivity was observed. Furthermore, 6 F6 mAb was able to combat colony formation, xenograft growth and metastasis of claudin-1 positive CRC cells suggesting its utility as a therapeutic. Fujiwara et al. recently targeted claudin-4 in CRC using an anti-claudin-4 extracellular domain antibody [122]. The efficacy of the anti-claudin-4 antibody is promising and observed to enhance the anti-tumorigenic potential of 5-fluorouracil (FU) and anti-EGFR antibodies. These works demonstrate the proof of concept for exploiting claudins as targets for monoclonal antibodies in therapies.

Some of the monoclonal antibodies against claudins, including anti-claudin-18.2 (IMAB362-claudin-18.2) and the anti-claudin-6 (IMAB027-claudin-6), have also found their way into clinical trials [123]. Claudin-18.2 is expressed on the outer cell membrane of gastric cancer cells and binds to monoclonal antibodies. The IMAB362 was proven to be clinically safe as the patients were devoid of any side effects. Also, IMAB027 is in an ongoing clinical trial for recurrent advanced ovarian cancer (NCT02054351), and patients have not demonstrated any adverse effects [123]. Clinical trials for claudiximab (claudin-18 targeting) in advanced gastroesophageal cancer patients are also underway [124]. Recently, claudiximab was reported to be a first-in-class chimeric monoclonal antibody for the treatment of gastric cancer targeting claudin-18, which is an important factor in gastric cancer metastasis. This is just the beginning of an exciting journey and more research is warranted to revolutionize claudins targeted monoclonal antibodies in cancer therapy.

Another avenue to exploit Claudins as a therapeutics is their ability to behave as receptors for microbes. *Clostridium perfringens* enterotoxin (CPE) has the potential to bind with claudin receptors. CPE binds to the C-terminal CPE domain at both the first and second extracellular loops (ECL-1 and ECL-2) of claudins [125]. The affinity of CPE to claudins causes a pore leading to calcium influx responsible for host cell death. The claudin–CPE interaction is gaining significance in receptor decoy therapeutics for potential applications in gastrointestinal disease, cancer therapy/diagnoses, and drug delivery [125]. Claudin-3 and claudin-4 have been widely demonstrated to function as CPE receptors [126,127]. The binding ability of CPE to claudins, especially claudin-3 and claudin-4, has raised a great opportunity to target cancers with dysregulated claudin-3 and -4 cancers, especially breast, ovarian, and pancreatic cancers. The binding of CPE to claudin-3 and -4 was documented to induce dose-dependent cytolysis in breast cancer cells expressing claudin-3 and -4 [128]. Recent studies have exploited the CPE mediated targeting of claudin-3 and 4 cancers to target therapy-resistant ovarian cancer, pancreatic, and breast cancer xenografts possessing increased expressions of claudin-3 and -4. In one of the studies, the possibility of CPE binding claudin-3 as a visualization tool for identifying of micrometastatic chemotherapy-resistant ovarian cancer has been demonstrated [129]. The applicability of CPE, claudin-3, and -4 interactions is exploited in gene therapy against colon cancer. Recombinant (recCPE) and optimized CPE expressing vector (optCPE) were demonstrated to have a cytotoxic effect in claudin overexpressing colon cancer cells [130,131]. Further, the recent identification of the crystal structure of claudin-9 revealed that human claudin-9 has high-affinity for the CPE receptor and treatment with CPE caused cell death in human claudin-9 expressing cells [132]. In continuation of these studies, an interesting approach of nanoparticle-based targeting of cancer cells was documented by researchers, wherein the C-terminus of the CPE was conjugated to gold nanoparticles (AuNPs). This combination binds to claudin expressing tumor cells and kills the cells using gold nanoparticle-mediated laser perforation (GNOME-LP) technique [133,134]. Thus, the clinical relevance and functional importance of claudins in diverse cancers make them potential therapeutic targets.

7. Claudins as a Visualization Tool

The use of monoclonal antibodies against claudins have also been utilized in imaging modalities. Recently, claudin-4 was studied as an imaging tool for x-ray computed tomography (CT) in the prognosis of pancreatic ductal adenocarcinoma (PDAC) [135]. Claudin-4 is a known biomarker in PDAC detection. In this study, researchers reported a novel radiolabeled anti-claudin-4 monoclonal antibody in detecting PDAC using single-photon emission computed tomography (SPECT) imaging. The results showed promising uptake of anti-claudin-4 monoclonal antibody by PDAC tumors and were helpful in early detection and characterization of PDAC malignancy. Also, the researcher later targeted the extracellular domain of claudin-4 (4D3) with monoclonal antibodies (4D3) in combating bladder and lung cancer [136].

Colonoscopic aided screening and polyp and tumor removal have led to the reduced incidence and mortality of colorectal cancer (CRC). However, the lack of specificity is a major pitfall in these approaches and makes them less effective. It is especially difficult to detect the regions of flat dysplasia or serrated polyps, which also possess malignant potential. Thus, a targeted approach for advanced endoscopic techniques is a cornerstone requirement. A promising approach was recently demonstrated for the real-time endoscopic imaging of colonic adenomas [137]. In this study, the researchers exploited claudin-1 as a potential target in endoscopic imaging of colonic adenomas. As claudin-1 is highly expressed in the early development of CRC, endoscopic imaging might be useful for detecting either polypoid or flat precancerous lesions that are difficult to visualize [138]. Peptide (peptide sequence—RTSPSSR), specific to claudin-1, was developed against the extracellular loop of claudin-1. This peptide showed greater intensity for human adenomas, hyperplastic polyps and sessile serrated adenomas thus proposing the possibility of using claudin-1 peptide aided endoscopic imaging for the future clinical translation to detect precancerous lesions. Recently another study by our group demonstrated the significance of claudin-1 as a useful target for near-infrared antibody-based imaging for visualization of colorectal tumors [138]. When animals injected with colon cancer cells subcutaneously were imaged using claudin-1 antibody conjugated LI-COR IR800DyeCW through a LI-COR Pearl Trilogy Fluorescence Imaging System, the system was able to target tumors specifically. These studies pave the way for using claudins as a tool for fluorescence-guided surgery, which will help in more specific targeting of the tumors in a stage-specific manner. A comprehensive representation encompassing the role of claudins and the monoclonal antibodies against claudins as therapeutic and detection tools is given in Figure 3, and the role of claudins as a therapeutic, prognostic, and detection agents is tabulated in Table 3.

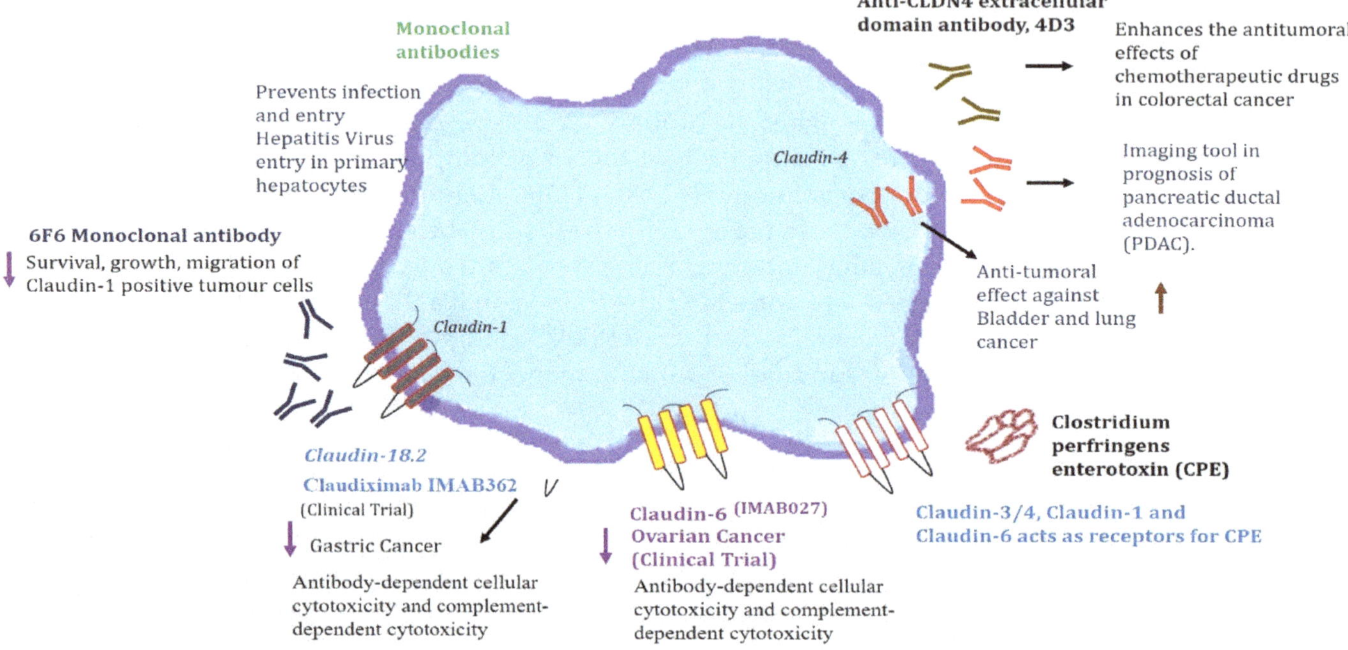

Figure 3. Claudins as an employable platform for prognostic, diagnostic, and therapeutic targets. The upward arrow indicated upregulation and downwards arrow indicated downregulation of the mentioned signaling events.

Table 3. Claudins as prognostic, therapeutic and detection agents.

Claudins Subtype	Disease Type	Therapeutic Agent	Clinical Application	Reference
Claudin-1	Hepatitis C virus infection	Residues within the first extracellular loop. Humanization of a claudin-1-specific monoclonal antibody.	Hepatitis C virus co-receptor. Clinical prevention and cure of Hepatitis C virus(HCV) infection.	[139] [118]
Claudin-6	Ovarian cancer	*Clostridium perfringens* enterotoxin (CPE) cytotoxicity.	CPE-mediated cytotoxicity in Ovarian cancer.	[127]
Claudin-3	Ovarian cancer uterine carcinomas	Human anti-claudin-3 IgG1 antibody.	Candidate for antibody-drug conjugate therapeutic applications.	[120,140]
Claudin-1	Colon cancer	Human claudin-1 (6F6 mAb).	Suppressed survival, growth, and migration of claudin-1 positive cells. Suppressed tumor growth and liver metastasis formation.	[121]
Claudin-4	Colorectal cancer	Anti-claudin-4 extracellular domain antibody.	Enhancer of anti-tumoral effects of chemotherapeutic agents.	[122]
Claudin-4	Pancreatic Cancer (PDAC)	Indium-111 tagged anti-claudin-4 monoclonal antibody.	X-ray computed tomography sided detection of PDAC.	[135]
Claudin-18.2	Gastric and gastroesophageal junction cancer	Chimeric monoclonal antibody that binds to claudin-18.2 (NCT03504397)	Cell death through antibody-dependent cellular cytotoxicity and complement-dependent cytotoxicity.	[123]
Claudin-4	Pancreatic cancer	Claudin-4 binder C-CPE 194	Enhances Tazeffects of anticancer agents via a MAPK pathway.	[141]
Claudin-3 and 4	Prostate cancer	Claudin-3 and claudin-4 targeted *Clostridium perfringens* protoxin	Selectively cytotoxic to PSA-producing prostate cancer cells.	[126]
Claudin-1	Colon cancer	Peptide RTSPSSR, specific to claudin-1 against the extracellular loop of claudin-1.	Specific to human adenomas, hyperplastic polyps, and sessile serrated adenomas.	[137]
Claudin-1	Colon cancer	Claudin-1 antibody conjugated with LI-COR IR800DyeCW	Near-infrared antibody-based imaging for visualization of colorectal tumors.	[138]
Claudin-9	Hepatitis C virus infection	Residues N38 and V45 in the first extracellular loop (EL1) of claudin-9 are responsible for HCV entry. Also found in PBMS (peripheral blood mononuclear cell) contributing to extrahepatic HCV infection.	It can be implicated in the development of drugs to block HCV entry into the liver and peripheral blood mononuclear cell (PBMS).	[142]

Table 3. *Cont.*

Claudins Subtype	Disease Type	Therapeutic Agent	Clinical Application	Reference
Claudin-11	Gastric Cancer	Hyper-methylation of claudin-11 promotor region leads to significant downregulation in gastric cancer.	Identification of the associated signaling cascades might lead to novel approaches in diagnosis and therapy for gastric cancer.	[48]
Claudin-7	Non-small cell lung cancer (NSCLC)	Reduced expression—Poor outcome Claudin-7 low NSCLC—Poor survival. Claudin-7 high NSCLC—High Survival.	Biomarker and a potential therapeutic target in patients with NSCLC.	[143]
Claudin-7	Epithelial Ovarian cancer	Claudin-7 transcripts were significantly enhanced in epithelial ovarian carcinoma patients. Silencing claudin-7 displayed enhanced sensitivity to Cisplatin treatment.	Independent prognostic factor and a key protein in regulating response to platinum-based chemotherapy in the treatment of epithelial ovarian cancer (EOC).	[144]
Claudin-2	Irritable bowel disease (IBD)	Anti-claudin-2 mAb 1A2	Prevent *cis*- and *trans*-interactions of claudin-2, attenuating the formation of leaky tight junction (TJ) seals.	[145]

8. Future Perspectives

The quest for prognostic, diagnostic, and therapeutic markers for many cancers is of high importance. More reliable and earlier detection markers have implications for diagnostic and therapeutic targeting. As the role of claudins in the regulation and enrichment of cancer stem cells and chemo-resistance becomes obvious, targeting claudins for diminishing cancer stem cells, which are cancer-propagating subsets of malignant cells, would be very useful. The potential of the claudin–cancer stem cell axis provides great potential for combating invasive, metastatic, and drug resistance phenotypes of various cancers. Future studies focusing on the role of claudins in cancer stem cells will be warranted to specifically target these populations to curb down residual tumor cells left after standard therapies.

Claudins are gaining their importance as detection and therapeutic agents. Future engineering of more monoclonal antibodies against claudins will have potential applications in targeted therapy, and claudin assisted endoscopy, imaging of various tumors. Also, the antibody-based detections will provide ample opportunity for the early diagnosis of any inflammatory diseases before they reach cancer status. The ongoing clinical trials for monoclonal antibodies against claudins might lead to claudin directed immunotherapies. Recently, small molecules inhibitors have been gaining more attention in cancer biology, as they aid in targeted therapy. No known small molecule inhibitors are currently being researched for claudins. Thus, in the future, screening for more potent inhibitors against claudins is warranted. Overall, to strengthen the therapeutic window of claudins, a more translational view of claudins by researchers is warranted.

Author Contributions: Conceptualization, S.G., P.D., and A.B.S.; methodology, S.G. and P.D.; resources, S.G. and P.D.; writing—original draft preparation, S.G.; writing—review and editing, S.G., A.B.S., and P.D.; supervision, P.D.; project administration, P.D. and A.B.S.; funding acquisition, P.D. and A.B.S. All authors have read and agreed to the published version of the manuscript.

References

1. Farquhar, M.G.; Palade, G.E. Junctional complexes in various epithelia. *J. Cell Biol.* **1963**, *17*, 375–412. [CrossRef]
2. Schneeberger, E.E.; Lynch, R.D. The tight junction: a multifunctional complex. *Am. J. Physiol.-Cell Physiol.* **2004**, *286*, C1213–C1228. [CrossRef]
3. Niessen, C.M. Tight junctions/adherens junctions: basic structure and function. *J. Investig. Dermatol.* **2007**, *127*, 2525–2532. [CrossRef]
4. Gonzalez-Mariscal, L.; Betanzos, A.; Nava, P.; Jaramillo, B.E. Tight junction proteins. *Prog. Biophys. Mol. Biol.* **2003**, *81*, 1–44. [CrossRef]
5. Stevenson, B.R.; Siliciano, J.D.; Mooseker, M.S.; Goodenough, D.A. Identification of ZO-1: a high molecular weight polypeptide associated with the tight junction (zonula occludens) in a variety of epithelia. *J. Cell Biol.* **1986**, *103*, 755–766. [CrossRef] [PubMed]
6. Citi, S.; Sabanay, H.; Jakes, R.; Geiger, B.; Kendrick-Jones, J. Cingulin, a new peripheral component of tight junctions. *Nature* **1988**, *333*, 272–276. [CrossRef] [PubMed]
7. Furuse, M.; Itoh, M.; Hirase, T.; Nagafuchi, A.; Yonemura, S.; Tsukita, S.; Tsukita, S. Direct association of occludin with ZO-1 and its possible involvement in the localization of occludin at tight junctions. *J. Cell Biol.* **1994**, *127*, 1617–1626. [CrossRef] [PubMed]
8. Nunes, F.D.; Lopez, L.N.; Lin, H.W.; Davies, C.; Azevedo, R.B.; Gow, A.; Kachar, B. Distinct subdomain organization and molecular composition of a tight junction with adherens junction features. *J. Cell Sci.* **2006**, *119*, 4819–4827. [CrossRef] [PubMed]
9. Gunzel, D.; Yu, A.S. Claudins and the modulation of tight junction permeability. *Physiol. Rev.* **2013**, *93*, 525–569. [CrossRef]
10. Angelow, S.; Yu, A.S. Claudins and paracellular transport: an update. *Curr. Opin. Nephrol. Hypertens.* **2007**, *16*, 459–464. [CrossRef]
11. Krause, G.; Protze, J.; Piontek, J. Assembly and function of claudins: Structure-function relationships based on homology models and crystal structures. *Semin. Cell Dev. Biol.* **2015**, *42*, 3–12. [CrossRef] [PubMed]

12. Tsukita, S.; Furuse, M. Occludin and claudins in tight-junction strands: leading or supporting players? *Trends Cell Biol.* **1999**, *9*, 268–273. [CrossRef]
13. Tsukita, S.; Furuse, M. Overcoming barriers in the study of tight junction functions: from occludin to claudin. *Genes Cells* **1998**, *3*, 569–573. [CrossRef] [PubMed]
14. Furuse, M.; Sasaki, H.; Fujimoto, K.; Tsukita, S. A single gene product, claudin-1 or -2, reconstitutes tight junction strands and recruits occludin in fibroblasts. *J. Cell Biol.* **1998**, *143*, 391–401. [CrossRef] [PubMed]
15. Lal-Nag, M.; Morin, P.J. The claudins. *Genome Biol.* **2009**, *10*, 235. [CrossRef]
16. Angelow, S.; Ahlstrom, R.; Yu, A.S. Biology of claudins. *Am. J. Physiol. Renal Physiol.* **2008**, *295*, F867–F876. [CrossRef]
17. Gunzel, D.; Fromm, M. Claudins and other tight junction proteins. *Compr. Physiol.* **2012**, *2*, 1819–1852. [CrossRef]
18. Soini, Y. Expression of claudins 1, 2, 3, 4, 5 and 7 in various types of tumours. *Histopathology* **2005**, *46*, 551–560. [CrossRef]
19. Schlingmann, B.; Molina, S.A.; Koval, M. Claudins: Gatekeepers of lung epithelial function. *Semin. Cell Dev. Biol.* **2015**, *42*, 47–57. [CrossRef]
20. Amasheh, S.; Meiri, N.; Gitter, A.H.; Schoneberg, T.; Mankertz, J.; Schulzke, J.D.; Fromm, M. Claudin-2 expression induces cation-selective channels in tight junctions of epithelial cells. *J. Cell Sci.* **2002**, *115*, 4969–4976. [CrossRef]
21. Wang, Y.; Mumm, J.B.; Herbst, R.; Kolbeck, R.; Wang, Y. IL-22 Increases Permeability of Intestinal Epithelial Tight Junctions by Enhancing Claudin-2 Expression. *J. Immunol.* **2017**, *199*, 3316–3325. [CrossRef] [PubMed]
22. Agarwal, R.; D'Souza, T.; Morin, P.J. Claudin-3 and claudin-4 expression in ovarian epithelial cells enhances invasion and is associated with increased matrix metalloproteinase-2 activity. *Cancer Res.* **2005**, *65*, 7378–7385. [CrossRef] [PubMed]
23. Morin, P.J. Claudin proteins in human cancer: promising new targets for diagnosis and therapy. *Cancer Res.* **2005**, *65*, 9603–9606. [CrossRef] [PubMed]
24. Kage, H.; Flodby, P.; Zhou, B.; Borok, Z. Dichotomous roles of claudins as tumor promoters or suppressors: Lessons from knockout mice. *Cell. Mol. Life Sci.* **2019**, *76*, 4663–4672. [CrossRef] [PubMed]
25. Kohmoto, T.; Masuda, K.; Shoda, K.; Takahashi, R.; Ujiro, S.; Tange, S.; Ichikawa, D.; Otsuji, E.; Imoto, I. Claudin-6 is a single prognostic marker and functions as a tumor-promoting gene in a subgroup of intestinal type gastric cancer. *Gastric Cancer* **2019**, 1–15. [CrossRef]
26. Yafang, L.; Qiong, W.; Yue, R.; Xiaoming, X.; Lina, Y.; Mingzi, Z.; Ting, Z.; Yulin, L.; Chengshi, Q. Role of Estrogen Receptor-alpha in the Regulation of Claudin-6 Expression in Breast Cancer Cells. *J. Breast Cancer* **2011**, *14*, 20–27. [CrossRef]
27. Rendon-Huerta, E.; Teresa, F.; Teresa, G.M.; Xochitl, G.S.; Georgina, A.F.; Veronica, Z.Z.; Montano, L.F. Distribution and expression pattern of claudins 6, 7, and 9 in diffuse- and intestinal-type gastric adenocarcinomas. *J. Gastrointest Cancer* **2010**, *41*, 52–59. [CrossRef]
28. Ikari, A.; Watanabe, R.; Sato, T.; Taga, S.; Shimobaba, S.; Yamaguchi, M.; Yamazaki, Y.; Endo, S.; Matsunaga, T.; Sugatani, J. Nuclear distribution of claudin-2 increases cell proliferation in human lung adenocarcinoma cells. *Biochim. Biophys. Acta* **2014**, *1843*, 2079–2088. [CrossRef]
29. Chao, Y.C.; Pan, S.H.; Yang, S.C.; Yu, S.L.; Che, T.F.; Lin, C.W.; Tsai, M.S.; Chang, G.C.; Wu, C.H.; Wu, Y.Y.; et al. Claudin-1 is a metastasis suppressor and correlates with clinical outcome in lung adenocarcinoma. *Am. J. Respir. Crit. Care Med.* **2009**, *179*, 123–133. [CrossRef]
30. Lu, Z.; Ding, L.; Hong, H.; Hoggard, J.; Lu, Q.; Chen, Y.H. Claudin-7 inhibits human lung cancer cell migration and invasion through ERK/MAPK signaling pathway. *Exp. Cell Res.* **2011**, *317*, 1935–1946. [CrossRef]
31. Shang, X.; Lin, X.; Alvarez, E.; Manorek, G.; Howell, S.B. Tight junction proteins claudin-3 and claudin-4 control tumor growth and metastases. *Neoplasia* **2012**, *14*, 974–985. [CrossRef] [PubMed]
32. Xu, C.; Wang, K.; Ding, Y.H.; Li, W.J.; Ding, L. Claudin-7 gene knockout causes destruction of intestinal structure and animal death in mice. *World J. Gastroenterol.* **2019**, *25*, 584–599. [CrossRef] [PubMed]
33. Nubel, T.; Preobraschenski, J.; Tuncay, H.; Weiss, T.; Kuhn, S.; Ladwein, M.; Langbein, L.; Zoller, M. Claudin-7 regulates EpCAM-mediated functions in tumor progression. *Mol. Cancer Res.* **2009**, *7*, 285–299. [CrossRef] [PubMed]

34. Shimobaba, S.; Taga, S.; Akizuki, R.; Hichino, A.; Endo, S.; Matsunaga, T.; Watanabe, R.; Yamaguchi, M.; Yamazaki, Y.; Sugatani, J.; et al. Claudin-18 inhibits cell proliferation and motility mediated by inhibition of phosphorylation of PDK1 and Akt in human lung adenocarcinoma A549 cells. *Biochim. Biophys. Acta* **2016**, *1863*, 1170–1178. [CrossRef]
35. Ahmad, R.; Kumar, B.; Chen, Z.; Chen, X.; Muller, D.; Lele, S.M.; Washington, M.K.; Batra, S.K.; Dhawan, P.; Singh, A.B. Loss of claudin-3 expression induces IL6/gp130/Stat3 signaling to promote colon cancer malignancy by hyperactivating Wnt/beta-catenin signaling. *Oncogene* **2017**, *36*, 6592–6604. [CrossRef]
36. Che, J.; Yue, D.; Zhang, B.; Zhang, H.; Huo, Y.; Gao, L.; Zhen, H.; Yang, Y.; Cao, B. Claudin-3 Inhibits Lung Squamous Cell Carcinoma Cell Epithelial-mesenchymal Transition and Invasion via Suppression of the Wnt/beta-catenin Signaling Pathway. *Int. J. Med. Sci.* **2018**, *15*, 339–351. [CrossRef]
37. Hicks, D.A.; Galimanis, C.E.; Webb, P.G.; Spillman, M.A.; Behbakht, K.; Neville, M.C.; Baumgartner, H.K. Claudin-4 activity in ovarian tumor cell apoptosis resistance and migration. *BMC Cancer* **2016**, *16*, 788. [CrossRef]
38. Zhang, X.; Ruan, Y.; Li, Y.; Lin, D.; Quan, C. Tight junction protein claudin-6 inhibits growth and induces the apoptosis of cervical carcinoma cells in vitro and in vivo. *Med. Oncol.* **2015**, *32*, 148. [CrossRef]
39. Zhang, X.; Ruan, Y.; Li, Y.; Lin, D.; Liu, Z.; Quan, C. Expression of apoptosis signal-regulating kinase 1 is associated with tight junction protein claudin-6 in cervical carcinoma. *Int. J. Clin. Exp. Pathol.* **2015**, *8*, 5535–5541.
40. Dhawan, P.; Singh, A.B.; Deane, N.G.; No, Y.; Shiou, S.R.; Schmidt, C.; Neff, J.; Washington, M.K.; Beauchamp, R.D. Claudin-1 regulates cellular transformation and metastatic behavior in colon cancer. *J. Clin. Investig.* **2005**, *115*, 1765–1776. [CrossRef]
41. Singh, A.B.; Sharma, A.; Smith, J.J.; Krishnan, M.; Chen, X.; Eschrich, S.; Washington, M.K.; Yeatman, T.J.; Beauchamp, R.D.; Dhawan, P. Claudin-1 up-regulates the repressor ZEB-1 to inhibit E-cadherin expression in colon cancer cells. *Gastroenterology* **2011**, *141*, 2140–2153. [CrossRef] [PubMed]
42. Bhat, A.A.; Pope, J.L.; Smith, J.J.; Ahmad, R.; Chen, X.; Washington, M.K.; Beauchamp, R.D.; Singh, A.B.; Dhawan, P. Claudin-7 expression induces mesenchymal to epithelial transformation (MET) to inhibit colon tumorigenesis. *Oncogene* **2015**, *34*, 4570–4580. [CrossRef] [PubMed]
43. Dhawan, P.; Ahmad, R.; Chaturvedi, R.; Smith, J.J.; Midha, R.; Mittal, M.K.; Krishnan, M.; Chen, X.; Eschrich, S.; Yeatman, T.J.; et al. Claudin-2 expression increases tumorigenicity of colon cancer cells: role of epidermal growth factor receptor activation. *Oncogene* **2011**, *30*, 3234–3247. [CrossRef] [PubMed]
44. Blanchard, A.A.; Zelinski, T.; Xie, J.; Cooper, S.; Penner, C.; Leygue, E.; Myal, Y. Identification of Claudin 1 Transcript Variants in Human Invasive Breast Cancer. *PLoS ONE* **2016**, *11*, e0163387. [CrossRef]
45. Oku, N.; Sasabe, E.; Ueta, E.; Yamamoto, T.; Osaki, T. Tight junction protein claudin-1 enhances the invasive activity of oral squamous cell carcinoma cells by promoting cleavage of laminin-5 gamma2 chain via matrix metalloproteinase (MMP)-2 and membrane-type MMP-1. *Cancer Res.* **2006**, *66*, 5251–5257. [CrossRef]
46. Darido, C.; Buchert, M.; Pannequin, J.; Bastide, P.; Zalzali, H.; Mantamadiotis, T.; Bourgaux, J.F.; Garambois, V.; Jay, P.; Blache, P.; et al. Defective claudin-7 regulation by Tcf-4 and Sox-9 disrupts the polarity and increases the tumorigenicity of colorectal cancer cells. *Cancer Res.* **2008**, *68*, 4258–4268. [CrossRef]
47. Chang, T.L.; Ito, K.; Ko, T.K.; Liu, Q.; Salto-Tellez, M.; Yeoh, K.G.; Fukamachi, H.; Ito, Y. Claudin-1 has tumor suppressive activity and is a direct target of RUNX3 in gastric epithelial cells. *Gastroenterology* **2010**, *138*, 255–265 e251-253. [CrossRef]
48. Agarwal, R.; Mori, Y.; Cheng, Y.; Jin, Z.; Olaru, A.V.; Hamilton, J.P.; David, S.; Selaru, F.M.; Yang, J.; Abraham, J.M.; et al. Silencing of claudin-11 is associated with increased invasiveness of gastric cancer cells. *PLoS One* **2009**, *4*, e8002. [CrossRef]
49. Zhang, X.; Wang, H.; Li, Q.; Li, T. CLDN2 inhibits the metastasis of osteosarcoma cells via down-regulating the afadin/ERK signaling pathway. *Cancer Cell Int.* **2018**, *18*, 160. [CrossRef]
50. Yamazaki, Y.; Tokumasu, R.; Kimura, H.; Tsukita, S. Role of claudin species-specific dynamics in reconstitution and remodeling of the zonula occludens. *Mol. Biol. Cell* **2011**, *22*, 1495–1504. [CrossRef]
51. Matsuda, M.; Kubo, A.; Furuse, M.; Tsukita, S. A peculiar internalization of claudins, tight junction-specific adhesion molecules, during the intercellular movement of epithelial cells. *J. Cell Sci.* **2004**, *117*, 1247–1257. [CrossRef] [PubMed]
52. Honda, H.; Pazin, M.J.; D'Souza, T.; Ji, H.; Morin, P.J. Regulation of the CLDN3 gene in ovarian cancer cells. *Cancer Biol. Ther.* **2007**, *6*, 1733–1742. [CrossRef] [PubMed]

53. Kwon, M.J.; Kim, S.H.; Jeong, H.M.; Jung, H.S.; Kim, S.S.; Lee, J.E.; Gye, M.C.; Erkin, O.C.; Koh, S.S.; Choi, Y.L.; et al. Claudin-4 overexpression is associated with epigenetic derepression in gastric carcinoma. *Lab. Investig.* **2011**, *91*, 1652–1667. [CrossRef] [PubMed]
54. Di Cello, F.; Cope, L.; Li, H.; Jeschke, J.; Wang, W.; Baylin, S.B.; Zahnow, C.A. Methylation of the claudin 1 promoter is associated with loss of expression in estrogen receptor positive breast cancer. *PLoS ONE* **2013**, *8*, e68630. [CrossRef] [PubMed]
55. Krishnan, M.; Singh, A.B.; Smith, J.J.; Sharma, A.; Chen, X.; Eschrich, S.; Yeatman, T.J.; Beauchamp, R.D.; Dhawan, P. HDAC inhibitors regulate claudin-1 expression in colon cancer cells through modulation of mRNA stability. *Oncogene* **2010**, *29*, 305–312. [CrossRef] [PubMed]
56. Martinez-Estrada, O.M.; Culleres, A.; Soriano, F.X.; Peinado, H.; Bolos, V.; Martinez, F.O.; Reina, M.; Cano, A.; Fabre, M.; Vilaro, S. The transcription factors Slug and Snail act as repressors of Claudin-1 expression in epithelial cells. *Biochem. J.* **2006**, *394*, 449–457. [CrossRef]
57. Bhat, A.A.; Sharma, A.; Pope, J.; Krishnan, M.; Washington, M.K.; Singh, A.B.; Dhawan, P. Caudal homeobox protein Cdx-2 cooperates with Wnt pathway to regulate claudin-1 expression in colon cancer cells. *PLoS ONE* **2012**, *7*, e37174. [CrossRef]
58. Honda, H.; Pazin, M.J.; Ji, H.; Wernyj, R.P.; Morin, P.J. Crucial roles of Sp1 and epigenetic modifications in the regulation of the CLDN4 promoter in ovarian cancer cells. *J. Biol. Chem.* **2006**, *281*, 21433–21444. [CrossRef]
59. Shigetomi, K.; Ikenouchi, J. Regulation of the epithelial barrier by post-translational modifications of tight junction membrane proteins. *J. Biochem.* **2018**, *163*, 265–272. [CrossRef]
60. Van Itallie, C.M.; Anderson, J.M. Claudin interactions in and out of the tight junction. *Tissue Barriers* **2013**, *1*, e25247. [CrossRef]
61. Rajagopal, N.; Irudayanathan, F.J.; Nangia, S. Palmitoylation of Claudin-5 Proteins Influences Their Lipid Domain Affinity and Tight Junction Assembly at the Blood-Brain Barrier Interface. *J. Phys. Chem. B* **2019**, *123*, 983–993. [CrossRef] [PubMed]
62. Butt, A.M.; Khan, I.B.; Hussain, M.; Idress, M.; Lu, J.; Tong, Y. Role of post translational modifications and novel crosstalk between phosphorylation and O-beta-GlcNAc modifications in human claudin-1, -3 and -4. *Mol. Biol. Rep.* **2012**, *39*, 1359–1369. [CrossRef] [PubMed]
63. French, A.D.; Fiori, J.L.; Camilli, T.C.; Leotlela, P.D.; O'Connell, M.P.; Frank, B.P.; Subaran, S.; Indig, F.E.; Taub, D.D.; Weeraratna, A.T. PKC and PKA phosphorylation affect the subcellular localization of claudin-1 in melanoma cells. *Int. J. Med. Sci.* **2009**, *6*, 93–101. [CrossRef] [PubMed]
64. D'Souza, T.; Indig, F.E.; Morin, P.J. Phosphorylation of claudin-4 by PKCepsilon regulates tight junction barrier function in ovarian cancer cells. *Exp. Cell Res.* **2007**, *313*, 3364–3375. [CrossRef]
65. D'Souza, T.; Agarwal, R.; Morin, P.J. Phosphorylation of claudin-3 at threonine 192 by cAMP-dependent protein kinase regulates tight junction barrier function in ovarian cancer cells. *J. Biol. Chem.* **2005**, *280*, 26233–26240. [CrossRef]
66. Akizuki, R.; Shimobaba, S.; Matsunaga, T.; Endo, S.; Ikari, A. Claudin-5, -7, and -18 suppress proliferation mediated by inhibition of phosphorylation of Akt in human lung squamous cell carcinoma. *Biochim. Biophys. Acta Mol. Cell Res.* **2017**, *1864*, 293–302. [CrossRef]
67. Yamamoto, M.; Ramirez, S.H.; Sato, S.; Kiyota, T.; Cerny, R.L.; Kaibuchi, K.; Persidsky, Y.; Ikezu, T. Phosphorylation of claudin-5 and occludin by rho kinase in brain endothelial cells. *Am. J. Pathol.* **2008**, *172*, 521–533. [CrossRef]
68. Ahmad, W.; Shabbiri, K.; Ijaz, B.; Asad, S.; Sarwar, M.T.; Gull, S.; Kausar, H.; Fouzia, K.; Shahid, I.; Hassan, S. Claudin-1 required for HCV virus entry has high potential for phosphorylation and O-glycosylation. *Virol. J.* **2011**, *8*, 229. [CrossRef]
69. Heiler, S.; Mu, W.; Zoller, M.; Thuma, F. The importance of claudin-7 palmitoylation on membrane subdomain localization and metastasis-promoting activities. *Cell Commun. Signal.* **2015**, *13*, 29. [CrossRef]
70. Biteau, B.; Hochmuth, C.E.; Jasper, H. Maintaining tissue homeostasis: dynamic control of somatic stem cell activity. *Cell Stem Cell* **2011**, *9*, 402–411. [CrossRef]
71. Romito, A.; Cobellis, G. Pluripotent Stem Cells: Current Understanding and Future Directions. *Stem Cells Int.* **2016**, *2016*, 9451492. [CrossRef] [PubMed]
72. Duinsbergen, D.; Salvatori, D.; Eriksson, M.; Mikkers, H. Tumors originating from induced pluripotent stem cells and methods for their prevention. *Ann. N.Y. Acad. Sci.* **2009**, *1176*, 197–204. [CrossRef] [PubMed]

73. Ben-David, U.; Nudel, N.; Benvenisty, N. Immunologic and chemical targeting of the tight-junction protein Claudin-6 eliminates tumorigenic human pluripotent stem cells. *Nat. Commun.* **2013**, *4*, 1992. [CrossRef] [PubMed]
74. Sugimoto, K.; Ichikawa-Tomikawa, N.; Satohisa, S.; Akashi, Y.; Kanai, R.; Saito, T.; Sawada, N.; Chiba, H. The tight-junction protein claudin-6 induces epithelial differentiation from mouse F9 and embryonic stem cells. *PLoS ONE* **2013**, *8*, e75106. [CrossRef] [PubMed]
75. Wang, L.; Xue, Y.; Shen, Y.; Li, W.; Cheng, Y.; Yan, X.; Shi, W.; Wang, J.; Gong, Z.; Yang, G.; et al. Claudin 6: a novel surface marker for characterizing mouse pluripotent stem cells. *Cell Res.* **2012**, *22*, 1082–1085. [CrossRef] [PubMed]
76. Turksen, K.; Troy, T.C. Claudin-6: a novel tight junction molecule is developmentally regulated in mouse embryonic epithelium. *Dev. Dyn.* **2001**, *222*, 292–300. [CrossRef]
77. Erdelyi-Belle, B.; Torok, G.; Apati, A.; Sarkadi, B.; Schaff, Z.; Kiss, A.; Homolya, L. Expression of Tight Junction Components in Hepatocyte-Like Cells Differentiated from Human Embryonic Stem Cells. *Pathol. Oncol. Res.* **2015**, *21*, 1059–1070. [CrossRef]
78. Abdullah, L.N.; Chow, E.K. Mechanisms of chemoresistance in cancer stem cells. *Clin. Transl. Med.* **2013**, *2*, 3. [CrossRef]
79. Phi, L.T.H.; Sari, I.N.; Yang, Y.G.; Lee, S.H.; Jun, N.; Kim, K.S.; Lee, Y.K.; Kwon, H.Y. Cancer Stem Cells (CSCs) in Drug Resistance and their Therapeutic Implications in Cancer Treatment. *Stem Cells Int.* **2018**, *2018*, 5416923. [CrossRef]
80. Nusse, R. Wnt signaling and stem cell control. *Cell Res.* **2008**, *18*, 523–527. [CrossRef]
81. Miwa, N.; Furuse, M.; Tsukita, S.; Niikawa, N.; Nakamura, Y.; Furukawa, Y. Involvement of claudin-1 in the beta-catenin/Tcf signaling pathway and its frequent upregulation in human colorectal cancers. *Oncol. Res.* **2001**, *12*, 469–476. [CrossRef] [PubMed]
82. Gowrikumar, S.; Ahmad, R.; Uppada, S.B.; Washington, M.K.; Shi, C.; Singh, A.B.; Dhawan, P. Upregulated claudin-1 expression promotes colitis-associated cancer by promoting beta-catenin phosphorylation and activation in Notch/p-AKT-dependent manner. *Oncogene* **2019**, *38*, 5321–5337. [CrossRef] [PubMed]
83. Prat, A.; Parker, J.S.; Karginova, O.; Fan, C.; Livasy, C.; Herschkowitz, J.I.; He, X.; Perou, C.M. Phenotypic and molecular characterization of the claudin-low intrinsic subtype of breast cancer. *Breast Cancer Res.* **2010**, *12*, R68. [CrossRef] [PubMed]
84. Ma, L.; Yin, W.; Ma, H.; Elshoura, I.; Wang, L. Targeting claudin-3 suppresses stem cell-like phenotype in nonsquamous non-small-cell lung carcinoma. *Lung Cancer Manag.* **2019**, *8*, LMT04. [CrossRef] [PubMed]
85. Zhou, B.; Flodby, P.; Luo, J.; Castillo, D.R.; Liu, Y.; Yu, F.X.; McConnell, A.; Varghese, B.; Li, G.; Chimge, N.O.; et al. Claudin-18-mediated YAP activity regulates lung stem and progenitor cell homeostasis and tumorigenesis. *J. Clin. Investig.* **2018**, *128*, 970–984. [CrossRef]
86. Paquet-Fifield, S.; Koh, S.L.; Cheng, L.; Beyit, L.M.; Shembrey, C.; Molck, C.; Behrenbruch, C.; Papin, M.; Gironella, M.; Guelfi, S.; et al. Tight Junction Protein Claudin-2 Promotes Self-Renewal of Human Colorectal Cancer Stem-like Cells. *Cancer Res.* **2018**, *78*, 2925–2938. [CrossRef]
87. Zheng, H.C. The molecular mechanisms of chemoresistance in cancers. *Oncotarget* **2017**, *8*, 59950–59964. [CrossRef]
88. Wang, K.; Li, T.; Xu, C.; Ding, Y.; Li, W.; Ding, L. Claudin-7 downregulation induces metastasis and invasion in colorectal cancer via the promotion of epithelial-mesenchymal transition. *Biochem. Biophys. Res. Commun.* **2019**, *508*, 797–804. [CrossRef]
89. Tabaries, S.; Siegel, P.M. The role of claudins in cancer metastasis. *Oncogene* **2017**, *36*, 1176–1190. [CrossRef]
90. Zhang, L.; Wang, Y.; Zhang, B.; Zhang, H.; Zhou, M.; Wei, M.; Dong, Q.; Xu, Y.; Wang, Z.; Gao, L.; et al. Claudin-3 expression increases the malignant potential of lung adenocarcinoma cells: role of epidermal growth factor receptor activation. *Oncotarget* **2017**, *8*, 23033–23047. [CrossRef]
91. Wang, Q.; Zhang, Y.; Zhang, T.; Han, Z.G.; Shan, L. Low claudin-6 expression correlates with poor prognosis in patients with non-small cell lung cancer. *Onco Targets Ther.* **2015**, *8*, 1971–1977. [CrossRef] [PubMed]
92. Landers, K.A.; Samaratunga, H.; Teng, L.; Buck, M.; Burger, M.J.; Scells, B.; Lavin, M.F.; Gardiner, R.A. Identification of claudin-4 as a marker highly overexpressed in both primary and metastatic prostate cancer. *Br. J. Cancer* **2008**, *99*, 491–501. [CrossRef] [PubMed]
93. Hwang, T.L.; Changchien, T.T.; Wang, C.C.; Wu, C.M. Claudin-4 expression in gastric cancer cells enhances the invasion and is associated with the increased level of matrix metalloproteinase-2 and -9 expression. *Oncol. Lett.* **2014**, *8*, 1367–1371. [CrossRef] [PubMed]

94. Ohtani, S.; Terashima, M.; Satoh, J.; Soeta, N.; Saze, Z.; Kashimura, S.; Ohsuka, F.; Hoshino, Y.; Kogure, M.; Gotoh, M. Expression of tight-junction-associated proteins in human gastric cancer: downregulation of claudin-4 correlates with tumor aggressiveness and survival. *Gastric Cancer* **2009**, *12*, 43–51. [CrossRef]
95. Stebbing, J.; Filipovic, A.; Giamas, G. Claudin-1 as a promoter of EMT in hepatocellular carcinoma. *Oncogene* **2013**, *32*, 4871–4872. [CrossRef]
96. Zhou, B.; Moodie, A.; Blanchard, A.A.; Leygue, E.; Myal, Y. Claudin 1 in Breast Cancer: New Insights. *J. Clin. Med.* **2015**, *4*, 1960–1976. [CrossRef]
97. Yang, M.; Li, Y.; Ruan, Y.; Lu, Y.; Lin, D.; Xie, Y.; Dong, B.; Dang, Q.; Quan, C. CLDN6 enhances chemoresistance to ADM via AF-6/ERKs pathway in TNBC cell line MDAMB231. *Mol. Cell. Biochem.* **2018**, *443*, 169–180. [CrossRef]
98. Shang, X.; Lin, X.; Manorek, G.; Howell, S.B. Claudin-3 and claudin-4 regulate sensitivity to cisplatin by controlling expression of the copper and cisplatin influx transporter CTR1. *Mol. Pharmacol.* **2013**, *83*, 85–94. [CrossRef]
99. Yoshida, H.; Sumi, T.; Zhi, X.; Yasui, T.; Honda, K.; Ishiko, O. Claudin-4: a potential therapeutic target in chemotherapy-resistant ovarian cancer. *Anticancer Res.* **2011**, *31*, 1271–1277.
100. Zhao, Z.; Li, J.; Jiang, Y.; Xu, W.; Li, X.; Jing, W. CLDN1 Increases Drug Resistance of Non-Small Cell Lung Cancer by Activating Autophagy via Up-Regulation of ULK1 Phosphorylation. *Med. Sci. Monit.* **2017**, *23*, 2906–2916. [CrossRef]
101. Akizuki, R.; Maruhashi, R.; Eguchi, H.; Kitabatake, K.; Tsukimoto, M.; Furuta, T.; Matsunaga, T.; Endo, S.; Ikari, A. Decrease in paracellular permeability and chemosensitivity to doxorubicin by claudin-1 in spheroid culture models of human lung adenocarcinoma A549 cells. *Biochim. Biophys. Acta Mol. Cell Res.* **2018**, *1865*, 769–780. [CrossRef] [PubMed]
102. Philip, R.; Heiler, S.; Mu, W.; Buchler, M.W.; Zoller, M.; Thuma, F. Claudin-7 promotes the epithelial-mesenchymal transition in human colorectal cancer. *Oncotarget* **2015**, *6*, 2046–2063. [CrossRef] [PubMed]
103. Hoggard, J.; Fan, J.; Lu, Z.; Lu, Q.; Sutton, L.; Chen, Y.H. Claudin-7 increases chemosensitivity to cisplatin through the upregulation of caspase pathway in human NCI-H522 lung cancer cells. *Cancer Sci.* **2013**, *104*, 611–618. [CrossRef] [PubMed]
104. Lechpammer, M.; Resnick, M.B.; Sabo, E.; Yakirevich, E.; Greaves, W.O.; Sciandra, K.T.; Tavares, R.; Noble, L.C.; DeLellis, R.A.; Wang, L.J. The diagnostic and prognostic utility of claudin expression in renal cell neoplasms. *Mod. Pathol.* **2008**, *21*, 1320–1329. [CrossRef]
105. Osunkoya, A.O.; Cohen, C.; Lawson, D.; Picken, M.M.; Amin, M.B.; Young, A.N. Claudin-7 and claudin-8: immunohistochemical markers for the differential diagnosis of chromophobe renal cell carcinoma and renal oncocytoma. *Hum. Pathol.* **2009**, *40*, 206–210. [CrossRef]
106. Yang, L.; Sun, X.; Meng, X. Differences in the expression profiles of claudin proteins in human gastric carcinoma compared with nonneoplastic mucosa. *Mol. Med. Rep.* **2018**, *18*, 1271–1278. [CrossRef]
107. Danzinger, S.; Tan, Y.Y.; Rudas, M.; Kastner, M.T.; Weingartshofer, S.; Muhr, D.; Singer, C.F.; kConFab, I. Differential Claudin 3 and EGFR Expression Predicts BRCA1 Mutation in Triple-Negative Breast Cancer. *Cancer Investig.* **2018**, *36*, 378–388. [CrossRef]
108. Karabulut, M.; Alis, H.; Bas, K.; Karabulut, S.; Afsar, C.U.; Oguz, H.; Gunaldi, M.; Akarsu, C.; Kones, O.; Aykan, N.F. Clinical significance of serum claudin-1 and claudin-7 levels in patients with colorectal cancer. *Mol. Clin. Oncol.* **2015**, *3*, 1255–1267. [CrossRef]
109. Sabatier, R.; Finetti, P.; Guille, A.; Adelaide, J.; Chaffanet, M.; Viens, P.; Birnbaum, D.; Bertucci, F. Claudin-low breast cancers: clinical, pathological, molecular and prognostic characterization. *Mol. Cancer* **2014**, *13*, 228. [CrossRef]
110. Nissinen, L.; Siljamaki, E.; Riihila, P.; Piipponen, M.; Farshchian, M.; Kivisaari, A.; Kallajoki, M.; Raiko, L.; Peltonen, J.; Peltonen, S.; et al. Expression of claudin-11 by tumor cells in cutaneous squamous cell carcinoma is dependent on the activity of p38delta. *Exp. Dermatol.* **2017**, *26*, 771–777. [CrossRef]
111. Dias, K.; Dvorkin-Gheva, A.; Hallett, R.M.; Wu, Y.; Hassell, J.; Pond, G.R.; Levine, M.; Whelan, T.; Bane, A.L. Claudin-Low Breast Cancer; Clinical & Pathological Characteristics. *PLoS ONE* **2017**, *12*, e0168669. [CrossRef]
112. Upadhaya, P.; Barhoi, D.; Giri, A.; Bhattacharjee, A.; Giri, S. Joint detection of claudin-1 and junctional adhesion molecule-A as a therapeutic target in oral epithelial dysplasia and oral squamous cell carcinoma. *J. Cell. Biochem.* **2019**, *120*, 18117–18127. [CrossRef] [PubMed]
113. Chau, C.H.; Steeg, P.S.; Figg, W.D. Antibody-drug conjugates for cancer. *Lancet* **2019**, *394*, 793–804. [CrossRef]

114. Shuptrine, C.W.; Surana, R.; Weiner, L.M. Monoclonal antibodies for the treatment of cancer. *Semin. Cancer Biol.* **2012**, *22*, 3–13. [CrossRef]
115. Mailly, L.; Xiao, F.; Lupberger, J.; Wilson, G.K.; Aubert, P.; Duong, F.H.T.; Calabrese, D.; Leboeuf, C.; Fofana, I.; Thumann, C.; et al. Clearance of persistent hepatitis C virus infection in humanized mice using a claudin-1-targeting monoclonal antibody. *Nat. Biotechnol.* **2015**, *33*, 549–554. [CrossRef]
116. Fukasawa, M.; Nagase, S.; Shirasago, Y.; Iida, M.; Yamashita, M.; Endo, K.; Yagi, K.; Suzuki, T.; Wakita, T.; Hanada, K.; et al. Monoclonal antibodies against extracellular domains of claudin-1 block hepatitis C virus infection in a mouse model. *J. Virol.* **2015**, *89*, 4866–4879. [CrossRef]
117. Fofana, I.; Krieger, S.E.; Grunert, F.; Glauben, S.; Xiao, F.; Fafi-Kremer, S.; Soulier, E.; Royer, C.; Thumann, C.; Mee, C.J.; et al. Monoclonal anti-claudin 1 antibodies prevent hepatitis C virus infection of primary human hepatocytes. *Gastroenterology* **2010**, *139*, 953–964, 964. e4. [CrossRef]
118. Colpitts, C.C.; Tawar, R.G.; Mailly, L.; Thumann, C.; Heydmann, L.; Durand, S.C.; Xiao, F.; Robinet, E.; Pessaux, P.; Zeisel, M.B.; et al. Humanisation of a claudin-1-specific monoclonal antibody for clinical prevention and cure of HCV infection without escape. *Gut* **2018**, *67*, 736–745. [CrossRef]
119. Offner, S.; Hekele, A.; Teichmann, U.; Weinberger, S.; Gross, S.; Kufer, P.; Itin, C.; Baeuerle, P.A.; Kohleisen, B. Epithelial tight junction proteins as potential antibody targets for pancarcinoma therapy. *Cancer Immunol. Immunother.* **2005**, *54*, 431–445. [CrossRef]
120. Romani, C.; Cocco, E.; Bignotti, E.; Moratto, D.; Bugatti, A.; Todeschini, P.; Bandiera, E.; Tassi, R.; Zanotti, L.; Pecorelli, S.; et al. Evaluation of a novel human IgG1 anti-claudin3 antibody that specifically recognizes its aberrantly localized antigen in ovarian cancer cells and that is suitable for selective drug delivery. *Oncotarget* **2015**, *6*, 34617–34628. [CrossRef]
121. Cherradi, S.; Ayrolles-Torro, A.; Vezzo-Vie, N.; Gueguinou, N.; Denis, V.; Combes, E.; Boissiere, F.; Busson, M.; Canterel-Thouennon, L.; Mollevi, C.; et al. Antibody targeting of claudin-1 as a potential colorectal cancer therapy. *J. Exp. Clin. Cancer Res.* **2017**, *36*, 89. [CrossRef] [PubMed]
122. Fujiwara-Tani, R.; Sasaki, T.; Luo, Y.; Goto, K.; Kawahara, I.; Nishiguchi, Y.; Kishi, S.; Mori, S.; Ohmori, H.; Kondoh, M.; et al. Anti-claudin-4 extracellular domain antibody enhances the antitumoral effects of chemotherapeutic and antibody drugs in colorectal cancer. *Oncotarget* **2018**, *9*, 37367–37378. [CrossRef] [PubMed]
123. Sahin, U.; Schuler, M.; Richly, H.; Bauer, S.; Krilova, A.; Dechow, T.; Jerling, M.; Utsch, M.; Rohde, C.; Dhaene, K.; et al. A phase I dose-escalation study of IMAB362 (Zolbetuximab) in patients with advanced gastric and gastro-oesophageal junction cancer. *Eur. J. Cancer* **2018**, *100*, 17–26. [CrossRef] [PubMed]
124. Singh, P.; Toom, S.; Huang, Y. Anti-claudin 18.2 antibody as new targeted therapy for advanced gastric cancer. *J. Hematol. Oncol.* **2017**, *10*, 105. [CrossRef]
125. Shrestha, A.; Uzal, F.A.; McClane, B.A. The interaction of Clostridium perfringens enterotoxin with receptor claudins. *Anaerobe* **2016**, *41*, 18–26. [CrossRef]
126. Romanov, V.; Whyard, T.C.; Waltzer, W.C.; Gabig, T.G. A claudin 3 and claudin 4-targeted Clostridium perfringens protoxin is selectively cytotoxic to PSA-producing prostate cancer cells. *Cancer Lett.* **2014**, *351*, 260–264. [CrossRef]
127. Lal-Nag, M.; Battis, M.; Santin, A.D.; Morin, P.J. Claudin-6: a novel receptor for CPE-mediated cytotoxicity in ovarian cancer. *Oncogenesis* **2012**, *1*, e33. [CrossRef]
128. Kominsky, S.L.; Vali, M.; Korz, D.; Gabig, T.G.; Weitzman, S.A.; Argani, P.; Sukumar, S. Clostridium perfringens enterotoxin elicits rapid and specific cytolysis of breast carcinoma cells mediated through tight junction proteins claudin 3 and 4. *Am. J. Pathol.* **2004**, *164*, 1627–1633. [CrossRef]
129. Cocco, E.; Shapiro, E.M.; Gasparrini, S.; Lopez, S.; Schwab, C.L.; Bellone, S.; Bortolomai, I.; Sumi, N.J.; Bonazzoli, E.; Nicoletti, R.; et al. Clostridium perfringens enterotoxin C-terminal domain labeled to fluorescent dyes for in vivo visualization of micrometastatic chemotherapy-resistant ovarian cancer. *Int. J. Cancer* **2015**, *137*, 2618–2629. [CrossRef]
130. Pahle, J.; Menzel, L.; Niesler, N.; Kobelt, D.; Aumann, J.; Rivera, M.; Walther, W. Rapid eradication of colon carcinoma by Clostridium perfringens Enterotoxin suicidal gene therapy. *BMC Cancer* **2017**, *17*, 129. [CrossRef]
131. Pahle, J.; Aumann, J.; Kobelt, D.; Walther, W. Oncoleaking: Use of the Pore-Forming Clostridium perfringens Enterotoxin (CPE) for Suicide Gene Therapy. *Methods Mol. Biol.* **2015**, *1317*, 69–85. [CrossRef] [PubMed]
132. Vecchio, A.J.; Stroud, R.M. Claudin-9 structures reveal mechanism for toxin-induced gut barrier breakdown. *Proc. Natl. Acad. Sci. USA* **2019**, *116*, 17817–17824. [CrossRef] [PubMed]

133. Becker, A.; Leskau, M.; Schlingmann-Molina, B.L.; Hohmeier, S.C.; Alnajjar, S.; Murua Escobar, H.; Ngezahayo, A. Functionalization of gold-nanoparticles by the Clostridium perfringens enterotoxin C-terminus for tumor cell ablation using the gold nanoparticle-mediated laser perforation technique. *Sci. Rep.* **2018**, *8*, 14963. [CrossRef] [PubMed]
134. Becker, A.; Lehrich, T.; Kalies, S.; Heisterkamp, A.; Ngezahayo, A. Parameters for Optoperforation-Induced Killing of Cancer Cells Using Gold Nanoparticles Functionalized With the C-terminal Fragment of Clostridium Perfringens Enterotoxin. *Int. J. Mol. Sci.* **2019**, *20*. [CrossRef] [PubMed]
135. Torres, J.B.; Knight, J.C.; Mosley, M.J.; Kersemans, V.; Koustoulidou, S.; Allen, D.; Kinchesh, P.; Smart, S.; Cornelissen, B. Imaging of Claudin-4 in Pancreatic Ductal Adenocarcinoma Using a Radiolabelled Anti-Claudin-4 Monoclonal Antibody. *Mol. Imaging Biol.* **2018**, *20*, 292–299. [CrossRef] [PubMed]
136. Kuwada, M.; Chihara, Y.; Luo, Y.; Li, X.; Nishiguchi, Y.; Fujiwara, R.; Sasaki, T.; Fujii, K.; Ohmori, H.; Fujimoto, K.; et al. Pro-chemotherapeutic effects of antibody against extracellular domain of claudin-4 in bladder cancer. *Cancer Lett.* **2015**, *369*, 212–221. [CrossRef] [PubMed]
137. Rabinsky, E.F.; Joshi, B.P.; Pant, A.; Zhou, J.; Duan, X.; Smith, A.; Kuick, R.; Fan, S.; Nusrat, A.; Owens, S.R.; et al. Overexpressed Claudin-1 Can Be Visualized Endoscopically in Colonic Adenomas In Vivo. *Cell. Mol. Gastroenterol. Hepatol.* **2016**, *2*, 222–237. [CrossRef]
138. Hollandsworth, H.M.; Lwin, T.M.; Amirfakhri, S.; Filemoni, F.; Batra, S.K.; Hoffman, R.M.; Dhawan, P.; Bouvet, M. Anti-Claudin-1 Conjugated to a Near-Infrared Fluorophore Targets Colon Cancer in PDOX Mouse Models. *J. Surg. Res.* **2019**, *242*, 145–150. [CrossRef]
139. Evans, M.J.; von Hahn, T.; Tscherne, D.M.; Syder, A.J.; Panis, M.; Wolk, B.; Hatziioannou, T.; McKeating, J.A.; Bieniasz, P.D.; Rice, C.M. Claudin-1 is a hepatitis C virus co-receptor required for a late step in entry. *Nature* **2007**, *446*, 801–805. [CrossRef]
140. Romani, C.; Comper, F.; Bandiera, E.; Ravaggi, A.; Bignotti, E.; Tassi, R.A.; Pecorelli, S.; Santin, A.D. Development and characterization of a human single-chain antibody fragment against claudin-3: A novel therapeutic target in ovarian and uterine carcinomas. *Am. J. Obstet. Gynecol.* **2009**, *201*, 70 e71–e79. [CrossRef]
141. Kono, T.; Kondoh, M.; Kyuno, D.; Ito, T.; Kimura, Y.; Imamura, M.; Kohno, T.; Konno, T.; Furuhata, T.; Sawada, N.; et al. Claudin-4 binder C-CPE 194 enhances effects of anticancer agents on pancreatic cancer cell lines via a MAPK pathway. *Pharmacol. Res. Perspect.* **2015**, *3*, e00196. [CrossRef] [PubMed]
142. Zheng, A.; Yuan, F.; Li, Y.; Zhu, F.; Hou, P.; Li, J.; Song, X.; Ding, M.; Deng, H. Claudin-6 and claudin-9 function as additional coreceptors for hepatitis C virus. *J. Virol.* **2007**, *81*, 12465–12471. [CrossRef] [PubMed]
143. Yamamoto, T.; Oshima, T.; Yoshihara, K.; Yamanaka, S.; Nishii, T.; Arai, H.; Inui, K.; Kaneko, T.; Nozawa, A.; Woo, T.; et al. Reduced expression of claudin-7 is associated with poor outcome in non-small cell lung cancer. *Oncol. Lett.* **2010**, *1*, 501–505. [CrossRef] [PubMed]
144. Kim, C.J.; Lee, J.W.; Choi, J.J.; Choi, H.Y.; Park, Y.A.; Jeon, H.K.; Sung, C.O.; Song, S.Y.; Lee, Y.Y.; Choi, C.H.; et al. High claudin-7 expression is associated with a poor response to platinum-based chemotherapy in epithelial ovarian carcinoma. *Eur. J. Cancer* **2011**, *47*, 918–925. [CrossRef]
145. Takigawa, M.; Iida, M.; Nagase, S.; Suzuki, H.; Watari, A.; Tada, M.; Okada, Y.; Doi, T.; Fukasawa, M.; Yagi, K.; et al. Creation of a Claudin-2 Binder and Its Tight Junction-Modulating Activity in a Human Intestinal Model. *J. Pharmacol. Exp. Ther.* **2017**, *363*, 444–451. [CrossRef]

Permissions

All chapters in this book were first published by MDPI; hereby published with permission under the Creative Commons Attribution License or equivalent. Every chapter published in this book has been scrutinized by our experts. Their significance has been extensively debated. The topics covered herein carry significant findings which will fuel the growth of the discipline. They may even be implemented as practical applications or may be referred to as a beginning point for another development.

The contributors of this book come from diverse backgrounds, making this book a truly international effort. This book will bring forth new frontiers with its revolutionizing research information and detailed analysis of the nascent developments around the world.

We would like to thank all the contributing authors for lending their expertise to make the book truly unique. They have played a crucial role in the development of this book. Without their invaluable contributions this book wouldn't have been possible. They have made vital efforts to compile up to date information on the varied aspects of this subject to make this book a valuable addition to the collection of many professionals and students.

This book was conceptualized with the vision of imparting up-to-date information and advanced data in this field. To ensure the same, a matchless editorial board was set up. Every individual on the board went through rigorous rounds of assessment to prove their worth. After which they invested a large part of their time researching and compiling the most relevant data for our readers.

The editorial board has been involved in producing this book since its inception. They have spent rigorous hours researching and exploring the diverse topics which have resulted in the successful publishing of this book. They have passed on their knowledge of decades through this book. To expedite this challenging task, the publisher supported the team at every step. A small team of assistant editors was also appointed to further simplify the editing procedure and attain best results for the readers.

Apart from the editorial board, the designing team has also invested a significant amount of their time in understanding the subject and creating the most relevant covers. They scrutinized every image to scout for the most suitable representation of the subject and create an appropriate cover for the book.

The publishing team has been an ardent support to the editorial, designing and production team. Their endless efforts to recruit the best for this project, has resulted in the accomplishment of this book. They are a veteran in the field of academics and their pool of knowledge is as vast as their experience in printing. Their expertise and guidance has proved useful at every step. Their uncompromising quality standards have made this book an exceptional effort. Their encouragement from time to time has been an inspiration for everyone.

The publisher and the editorial board hope that this book will prove to be a valuable piece of knowledge for researchers, students, practitioners and scholars across the globe.

List of Contributors

Enrique Gamero-Estevez
Department of Human Genetics, McGill University, Montreal, QC H4A 3J1, Canada
Research Institute of the McGill University Health Centre, Glen Site, Montreal, QC H4A 3J1, Canada

Sero Andonian
Division of Urology, McGill University, Montreal, QC H4A 3J1, Canada

Bertrand Jean-Claude
Research Institute of the McGill University Health Centre, Glen Site, Montreal, QC H4A 3J1, Canada
Department of Medicine, McGill University, Montreal, QC H4A 3J1, Canada

Indra Gupta and Aimee K. Ryan
Department of Human Genetics, McGill University, Montreal, QC H4A 3J1, Canada
Research Institute of the McGill University Health Centre, Glen Site, Montreal, QC H4A 3J1, Canada
Departments of Pediatrics, McGill University, Montreal, QC H4A 3J1, Canada

Susanne Milatz
Institute of Physiology, Kiel University, Christian-Albrechts-Platz 4, 24118 Kiel, Germany

Udo Heinemann
Macromolecular Structure and Interaction Laboratory, Max Delbrück Center for Molecular Medicine, 13125 Berlin, Germany

Anja Schuetz
Protein Production & Characterization Platform, Max Delbrück Center for Molecular Medicine, 13125 Berlin, Germany

Natascha Roehlen, Armando Andres Roca Suarez, Houssein El Saghire, Catherine Schuster and Joachim Lupberger
Institut de Recherche sur les Maladies Virales et Hépatiques, Inserm UMR1110, F-67000 Strasbourg, France
Université de Strasbourg, F-67000 Strasbourg, France

Antonio Saviano and Thomas F. Baumert
Institut de Recherche sur les Maladies Virales et Hépatiques, Inserm UMR1110, F-67000 Strasbourg, France
Université de Strasbourg, F-67000 Strasbourg, France
Pôle Hepato-digestif, Institut Hopitalo-universitaire, Hôpitaux Universitaires de Strasbourg, F-67000 Strasbourg, France

Mónica Díaz-Coránguez, Xuwen Liu and David A. Antonetti
Department of Ophthalmology and Visual Sciences, University of Michigan, Kellogg Eye Center, Ann Arbor, MI 48105, USA

Rodney Tatum and John Hoggard
Department of Anatomy and Cell Biology, Brody School of Medicine, East Carolina University, Greenville, NC 27834, USA

Junming Fan
Department of Anatomy and Cell Biology, Brody School of Medicine, East Carolina University, Greenville, NC 27834, USA
Institute of Hypoxia Medicine, School of Basic Medical Sciences,Wenzhou Medical University, Wenzhou 325035, China

Yan-Hua Chen
Department of Anatomy and Cell Biology, Brody School of Medicine, East Carolina University, Greenville, NC 27834, USA
East Carolina Diabetes and Obesity Institute, East Carolina University, Greenville, NC 27834, USA

Wolfgang Stremmel
Institute of Pharmacy and Molecular Biotechnology, University of Heidelberg, D-69120 Heidelberg, Germany

Simone Staffer
University Clinics of Heidelberg, D-69120 Heidelberg, Germany

Ralf Weiskirchen
Institute of Molecular Pathobiochemistry, Experimental Gene Therapy and Clinical Chemistry, RWTH University Hospital Aachen, D-52074 Aachen, Germany

Christian K. Tipsmark and Laura V. Ellis
Department of Biological Sciences, University of Arkansas, SCEN 601, Fayetteville, AR 72701, USA

Steffen S. Madsen
Department of Biological Sciences, University of Arkansas, SCEN 601, Fayetteville, AR 72701, USA
Department of Biology, University of Southern Denmark, Campusvej 55, 5230 Odense M, Denmark

Andreas M. Nielsen
Department of Biology, University of Southern Denmark, Campusvej 55, 5230 Odense M, Denmark

Maryline C. Bossus
Department of Biological Sciences, University of Arkansas, SCEN 601, Fayetteville, AR 72701, USA
Department of Math and Sciences, Lyon College, 2300 Highland Rd, Batesville, AR 72501, USA

Christina Baun and Thomas L. Andersen
Department of Nuclear Medicine, Odense University Hospital, Sdr. Boulevard 29, 5000 Odense C, Denmark

Jes Dreier and Jonathan R. Brewer
Department of Biochemistry and Molecular Biology, University of Southern Denmark, Campusvej 55, 5230 Odense M, Denmark

Susanne Janke, Sonnhild Mittag, Juliane Reiche and Otmar Huber
Department of Biochemistry II, Jena University Hospital, Friedrich Schiller University Jena, 07743 Jena, Germany

John Mackay Søfteland and Mihai Oltean
The Transplant Institute, Sahlgrenska University Hospital, 413 45 Gothenburg, Sweden
Laboratory for Transplantation and Regenerative Medicine, Institute of Clinical Sciences, Sahlgrenska Academy at the University of Gothenburg, Sahlgrenska Science Park Medicinaregatan 8, 413 90 Gothenburg, Sweden

Arvind Manikantan Padma and Mats Hellström
Laboratory for Transplantation and Regenerative Medicine, Institute of Clinical Sciences, Sahlgrenska Academy at the University of Gothenburg, Sahlgrenska Science Park Medicinaregatan 8, 413 90 Gothenburg, Sweden

Anna Casselbrant
Department of Gastrosurgical Research and Education, Institute of Clinical Sciences, Sahlgrenska Academy at the University of Gothenburg, Sahlgrenska University Hospital, 41345 Gothenburg, Sweden

Ali-Reza Biglarnia and Johan Linders
Department of Transplantation, Skåne University Hospital, 205 02 Malmö, Sweden

Antonio Pesce
Department of Medical and Surgical Sciences and Advanced Technologies, University of Catania, Via Santa Sofia 86, 95123 Catania, Italy

Lucian Petru Jiga
Department for Plastic, Aesthetic, Reconstructive and Hand Surgery, Evangelisches Krankenhaus Oldenburg, Medical Campus University of Oldenburg, Steinweg 13-17, 26122 Oldenburg, Germany

Bogdan Hoinoiu and Mihai Ionac
Pius Branzeu Center for Laparoscopic Surgery and Microsurgery, University of Medicine and Pharmacy, P-ta. E. Murgu 2, 300041 Timisoara, Romania

Shu Wei
Graduate Program in Public Health and Preventive Medicine, Wuhan University of Science and Technology, Wuhan 430081, China
Department of Pathology, The University of Chicago, Chicago, IL 60615, USA

Ye Li and Christopher R.Weber
Department of Pathology, The University of Chicago, Chicago, IL 60615, USA

Sean P. Polster and Issam A. Awad
Section of Neurosurgery, Department of Surgery, The University of Chicago, Chicago, IL 60615, USA

Le Shen
Department of Pathology, The University of Chicago, Chicago, IL 60615, USA
Section of Neurosurgery, Department of Surgery, The University of Chicago, Chicago, IL 60615, USA

Yosuke Hashimoto, Keisuke Tachibana and Masuo Kondoh
Graduate School of Pharmaceutical Sciences, Osaka University, Osaka 565-0871, Japan

Susanne M. Krug and Michael Fromm
Institute of Clinical Physiology, Charité-Universitätsmedizin Berlin, 12203 Berlin, Germany

Jun Kunisawa
Graduate School of Pharmaceutical Sciences, Osaka University, Osaka 565-0871, Japan
Laboratory of Vaccine Materials, Center for Vaccine and Adjuvant Research and Laboratory of Gut Environmental System, National Institutes of Biomedical Innovation, Health and Nutrition (NIBIOHN), Osaka 567-0085, Japan
International Research and Development Center for Mucosal Vaccines, The Institute of Medical Sciences, The University of Tokyo, Tokyo 108-8639, Japan
Department of Microbiology and Immunology, Kobe University Graduate School of Medicine, Kobe 650-0017, Japan
Graduate School of Medicine and Graduate School of Dentistry, Osaka University, Osaka 565-0871, Japan

Saiprasad Gowrikumar
Department of Biochemistry and Molecular Biology, University of Nebraska Medical Center, Omaha, NE 68198-5870, USA

Amar B. Singh and Punita Dhawan
Department of Biochemistry and Molecular Biology, University of Nebraska Medical Center, Omaha, NE 68198-5870, USA
VA Nebraska-Western Iowa Health Care System, Omaha, NE 68105, USA
Department of Biochemistry and Molcular Biology, Fred and Pamela Buffet Cancer Center, University of Nebraska Medical Center, Omaha, NE 68105, USA

Index

www.ingramcontent.com/pod-product-compliance
Lightning Source LLC
Chambersburg PA
CBHW080503200326
41458CB00012B/4071
9781641167635